Formaldehyde tank cars

FORMALDEHYDE

By

J. FREDERIC WALKER

Chemical Research Division, Electrochemicals Department
E. I. du Pont de Nemours & Company, Inc.
Niagara Falls, N. Y.

SECOND EDITION

American Chemical Society
Monograph Series

BOOK DIVISION
REINHOLD PUBLISHING CORPORATION

Also publishers of Chemical Engineering Catalog, Materials & Methods—the Magazine
of Materials Engineering, Advertising Management for ACS.

330 West Forty-second Street   New York 36, USA.

1953

Printed in the U.S.A. by
WAVERLY PRESS, Inc., Baltimore, Md.

GENERAL INTRODUCTION

American Chemical Society's Series of Chemical Monographs

By arrangement with the Interallied Conference of Pure and Applied Chemistry, which met in London and Brussels in July, 1919, the American Chemical Society was to undertake the production and publication of Scientific and Technologic Monographs on chemical subjects. At the same time it was agreed that the National Research Council, in cooperation with the American Chemical Society and the American Physical Society, should undertake the production and publication of Critical Tables of Chemical and Physical Constants. The American Chemical Society and the National Research Council mutually agreed to care for these two fields of chemical progress. The American Chemical Society named as Trustees, to make the necessary arrangements of the publication of the Monographs, Charles L. Parsons, secretary of the Society, Washington, D. C.; the late John E. Teeple, then treasurer of the Society, New York; and the late Professor Gellert Alleman of Swarthmore College. The Trustees arranged for the publication of the ACS Series of (a) Scientific and (b) Technological Monographs by the Chemical Catalog Company, Inc. (Reinhold Publishing Corporation, successor) of New York.

The Council of the American Chemical Society, acting through its Committee on National Policy, appointed editors (the present list of whom appears at the close of this sketch) to select authors of competent authority in their respective fields and to consider critically the manuscripts submitted.

The first Monograph of the Series appeared in 1921. After twenty-three years of experience certain modifications of general policy were indicated. In the beginning there still remained from the preceding five decades a distinct though arbitrary differentiation between so-called "pure science" publications and technologic or applied science literature. By 1944 this differentiation was fast becoming nebulous. Research in private enterprise had grown apace and not a little of it was pursued on the frontiers of knowledge. Furthermore, most workers in the sciences were coming to see the artificiality of the separation. The methods of both groups of workers are the same. They employ the same instrumentalities, and frankly recognize that their objectives are common, namely, the search for new knowledge for the service of man. The officers of the Society therefore combined the two editorial Boards in a single Board of twelve representative members.

Also in the beginning of the Series, it seemed expedient to construe rather broadly the definition of a Monograph. Needs of workers had to be recognized. Consequently among the first hundred Monographs appeared works in the form of treatises covering in some instances rather broad areas. Because such necessary works do not now want for publishers, it is considered advisable to hew more strictly to the line of the Monograph character, which means more complete and critical treatment of relatively restricted areas, and, where a broader field needs coverage, to subdivide it into logical subareas. The prodigious expansion of new knowledge makes such a change desirable.

These Monographs are intended to serve two principal purposes: first, to make available to chemists a thorough treatment of a selected area in form usable by persons working in more or less unrelated fields to the end that they may correlate their own work with a larger area of physical science discipline; second, to stimulate further research in the specific field treated. To implement this purpose the authors of Monographs are expected to give extended references to the literature. Where the literature is of such volume that a complete bibliography is impracticable, the authors are expected to append a list of references critically selected on the basis of their relative importance and significance.

Preface to the Second Edition

Formaldehyde production in the United States alone has approximately doubled since the first edition of "Formaldehyde" was published in 1944. Improved methods of manufacture, new uses and important contributions to fundamental chemistry have accompanied this industrial expansion. Formaldehyde is not only an important low-cost chemical intermediate, it is also a synthetic tool like hydrogenation, oxidation, etc. A knowledge of formaldehyde chemistry is almost essential to the chemical investigator who wishes to construct new molecules to meet technological requirements.

As with the previous edition of this Monograph, our objective is to give a systematic summary of the salient facts and theories of formaldehyde chemistry. It is not exhaustive. A detailed enumeration of literature references would require many volumes and its utility would be limited by its very size.

In presenting the second edition of this Monograph, the writer wishes to express his gratitude to the individuals and organizations who have given valuable assistance. Special thanks are due to T. R. Aalto of the Heyden Chemical Corporation, R. W. Kixmiller and W. H. Trotter of the Celanese Corporation of America and J. C. Walker of the Cities Service Oil Company. Indebtedness is acknowledged to the many readers of the first edition who supplied valuable criticisms, corrections and reprints of their own contributions to the knowledge of formaldehyde chemistry.

The author again expresses his obligations to his friends and associates in the Du Pont Company. In this regard, individual recognition is due to A. W. Anderson, P. R. Austin, G. T. Barnhill, G. F. Biehn, B. C. Bren, R. E. Brooks, R. E. Burk, A. F. Chadwick, A. G. Cole, R. L. Craven, F. W. Gänder, W. B. Gilbert, W. E. Grigsby, C. H. Hamblet, R. O. Humphrey, O. Johnson, R. D. Kralovec, M. A. Kubico, H. M. Kvalnes, J. H. McCormick, F. C. McGrew, J. Mitchell, Jr., S. D. Nogare, W. A. Payne, E. A. Smith, D. M. Smith, W. A. Stanton, C. W. Tucker, and F. A. Wolff.

J. Frederic Walker

Niagara Falls, N. Y.
March 30, 1953

Preface to the First Edition

The growing importance of formaldehyde as a commercial chemical, its many unique characteristics, and its varied applications have created a definite need for a systematic and critical account of formaldehyde chemistry. This Monograph is presented in the hope of fulfilling this need. Previously no single book dealing with this subject has been available in the English language. In 1908 a book on formaldehyde by Orlov was published in Russia and appeared in a German translation by C. Kietabl in 1909 ("Formaldehyd", by J. E. Orlov, translated by C. Kietabl, Verlag von Johann Ambrosius Barth, Leipzig, 1909). More recently, Arthur Menzel edited and revised a treatise on formaldehyde by Vanino and Seitter which was apparently first published in the early years of the twentieth century ("Der Formaldehyd", by A. Menzel, Hartleben's Verlag, Wien und Leipzig, 1927). Both of these books are now inadequate because of the rapid scientific advances of the last fifteen years.

In presenting this Monograph, the writer wishes to express his gratitude to a host of friends without whose help and cooperation its preparation would have been impossible. Special thanks are due to his associates in the Electrochemicals Department of the du Pont Company for their help and inspiration in the initiation, planning, and prosecution of the work. Thanks are also due to Dr. G. C. Bailey and Dr. B. S. Lacy for their assistance in preparing the section dealing with formaldehyde manufacture. The author is also grateful to Dr. Lacy for his aid in clarifying the problems involved in the analysis and interpretation of physical, chemical and thermodynamic data. It is with pleasure that he acknowledges the help and assistance of Mrs. Ruth Goodrick and Miss Janet Searles for their loyal aid in editing the manuscript and collaborating in the search for source material.

Special thanks for comment, advice, and supplementary information are gratefully acknowledged to J. F. T. Berliner, T. L. Davis, A. D. Gilbert, W. E. Gordon, J. W. Hill, E. F. Izard, F. L. Koethen, E. R. Laughlin, H. A. Lubs, A. M. Neal, R. N. Pease, J. R. Sabina, G. L. Schwartz, V. B. Sease, and J. R. Weber.

J. FREDERIC WALKER

Niagara Falls, N. Y.
April 4, 1944

INTRODUCTION

Formaldehyde chemistry is complicated by the fact that, although formaldehyde is well known in the form of its aqueous solutions and solid polymers, it is seldom encountered in pure monomeric form. Substances with which the chemist must deal divide mainly in two groups: (I) compounds with the type formula, $(CH_2O)_n$ and (II) addition compounds or solvates of which the most important are the simple hydrate, methylene glycol $[CH_2(OH)_2]$ and the hydrated linear polymers or polyoxymethylene glycols whose type formula is $HO(CH_2O)_n \cdot H$. Group I includes formaldehyde monomer itself, the relatively little known cyclic polymers, trioxane $[(CH_2O)_3]$ and tetraoxymethylene $[(CH_2O)_4]$, and the anhydrous polyoxymethylene formed when monomeric formaldehyde polymerizes. Group II includes (A) methylene glycol and the low molecular weight polyoxymethylene glycols present in formaldehyde solutions, (B) the mixture of solid polyoxymethylene glycols known commercially as paraformaldehyde, and (C) the so-called alpha-polyoxymethylenes, $HO(CH_2O)_n \cdot H$, in which n is greater than 100. The anhydrous high molecular weight polymer may be regarded as the end product of this series. Polymeric compounds which are functional derivatives of polyoxymethylene glycols are also occasionally encountered. All of these hydrates, polymers, and polymer derivatives are reversible and react as formaldehyde, differing only in the readiness with which they split up or hydrolyze to yield formaldehyde or its simple reactive solvates.

In discussing formaldehyde chemistry, we shall first give attention to the methods by which it is normally produced (Chapter 1). Following this, we shall deal with the physical and thermodynamic properties of the various formaldehyde substances: the simple monomer (Chapter 2), formaldehyde solutions (Chapters 3–6), and polymers (Chapter 7). Chapters 8–16 are devoted to the chemical properties of formaldehyde and its reactions with various types of inorganic and organic chemicals. Chapters 17 and 18 deal with formaldehyde detection and analysis.

Hexamethylenetetramine (Chapter 19), formed by the reaction of formaldehyde and ammonia, must receive special consideration in any discussion of formaldehyde substances. Although this compound, $(CH_2)_6N_4$, is a distinct chemical entity, it plays an important industrial role as a special form of formaldehyde. Chemically it is an ammono analogue of the cyclic formaldehyde polymers in which trivalent nitrogen has replaced bivalent oxygen.

The industrial applications of commercial formaldehyde substances, viz., formaldehyde solution, formaldehyde polymers and hexamethylenetetramine are discussed in Chapters 20 and 21. In this connection, it must be pointed out that this Monograph makes no attempt to replace the many excellent works of reference which deal with formaldehyde resins. Although the resin industry is the principal formaldehyde consumer and accordingly may not be neglected, it is our purpose to emphasize the applications of formaldehyde which have received less attention in previous publications.

J. Frederic Walker

Niagara Falls, N. Y.
January, 1953

CONTENTS

CHAPTER 1

FORMALDEHYDE PRODUCTION

Since its discovery in the latter half of the nineteenth century, formaldehyde has become an industrial product of outstanding commercial value. According to U. S. Tariff Commission figures, production in the United States alone reached a total of over 600 million pounds, calculated as 37 per cent commercial solution, in 1948. The total manufacturing capacity of United States plants was then estimated as 750 to 900 million pounds per year[62, 82]. More recent estimates (Feb., 1951) indicate a potential of 1270 million pounds[17a].

Commercially formaldehyde is manufactured and marketed chiefly in the form of an aqueous solution containing 37 per cent by weight of dissolved formaldehyde (CH_2O). This solution commonly contains sufficient methanol (8 to 15 per cent) to prevent precipitation of polymer under ordinary conditions of transportation and storage. Unstabilized solutions containing less than 1 per cent methanol and up to 50 per cent CH_2O are also handled commercially, but only where conditions permit transportation and sale in bulk, since they must be kept warm to prevent polymerization. Thirty per cent formaldehyde containing less than 1 per cent methanol has a limited application. This solution does not require the presence of methanol or other solution stabilizers. Economics of shipping and storage favor growing use of the higher strength solutions for large-scale industrial use.

Formaldehyde is also marketed in the form of its polymeric hydrate, paraformaldehyde, $HO \cdot (CH_2O)_n \cdot H$ or $(CH_2O)_n \cdot H_2O$, sometimes erroneously designated as "trioxymethylene". This polymer behaves chemically as a solid, substantially anhydrous form of formaldehyde. Although the usual commercial product contains 95 to 96 per cent CH_2O, polymers ranging from 90 to 98 per cent CH_2O are also marketed. Paraformaldehyde is manufactured from formaldehyde solution and is consequently the more expensive form (pages 119–129). Trioxane, $(CH_2O)_3$, the true cyclic trimer of formaldehyde (pages 146–153) is also manufactured on a limited scale.

Hexamethylenetetramine, $(CH_2)_6N_4$, which is formed by the reaction of formaldehyde and ammonia (Cf. Chapter 19) reacts as formaldehyde in many instances and therefore may be regarded as a special form of formaldehyde from the standpoint of use.

Growth of Formaldehyde Industry

Formaldehyde manufacture was begun in the United States on a limited scale in 1901. At this time, it was used chiefly as a disinfectant and embalming agent. However, the development of the synthetic resin industry initiated by the discoveries of Dr. Leo H. Baeckeland in 1909 soon resulted in an increased commercial demand on the infant industry. The immense growth of resins based on formaldehyde has since exercised a continually growing influence on formaldehyde manufacture. When World War I

TABLE 1. FORMALDEHYDE: DOMESTIC PRODUCTION AND PRICE[3, 62, 63, 73]

Unit: 1000 lbs of 37% (by wt.)

Year	Production[a]	Price[b]
1914	8,426	$0.085
1924	26,155	0.097
1929	51,786	0.091
1930	40,763	0.072
1933	52,236	0.060
1939	134,479	0.058
1940	180,885	0.058
1941	309,912	0.055
1942	347,463	0.055
1943	522,920	0.055
1944	522,440	0.054
1945	509,602	0.032
1946	460,048	0.032
1947	615,853	0.036
1948	623,837	0.037
1949	549,744	0.037
1950	835,142	0.037
1951	968,605	Ca. 0.04

[a] 1914–46: Tariff Commission, Annual Reports; 1947–48: Facts for Industry, Series 6-2-50, 6-2-55; 1949–51: U. S. Tariff Commission.

[b] 1914: "contract price," approximate only; 1924–44: bbl.; 1945–51: tanks, works.

commenced in 1914, the requirements for all chemicals increased, and formaldehyde was no exception. World War II pushed growing production facilities to the limit causing a severe shortage which eventually resulted in a considerable postwar expansion. During World War II, the United States War Department built two plants for the manufacture of formaldehyde and hexamethylenetetramine which was nitrated to produce the explosive, cyclonite, also known as RDX. These were the Cherokee Ordnance Works at Danville, Pennsylvania, and the Morgantown Ordnance Works at Morgantown, West Virginia. The plants were constructed and operated by Heyden and du Pont respectively[50].

Peace-time uses of formaldehyde are still dominated by the plastics industry which consumed over 130 million pounds of commercial 37 per cent solution in 1940 and still employs approximately 75 per cent of the total volume manufactured[62]. Available United States production figures from 1941 to 1951 are shown in Table 1 coupled with price figures which show, as would be expected, an almost continual decrease with increasing production. Authentic data on foreign production are not available.

The drop in production for the year 1949 marked the end of wartime shortages and a reduced consumption by synthetic resin producers in the first half of the year. However, increasing resin and plastics operations resulted in a definite increase in the latter months of 1949 and 1950[63].

The number of producers and plants manufacturing formaldehyde has almost trebled in the last ten years. In 1940, there were 5 producers and 7 plants in operation; in September 1951, this number had increased to the 14 producers and 20 plants listed below[3]:

Producers

Allied Chemical and Dye Corporation (Semet Solvay Division)	South Point, Ohio
Bakelite Corporation, Carbide and Carbon Chemical Corporation	Bound Brook, N. J.
The Borden Company	Bainbridge, N. Y.
	Demopolis, Ala.
	Springfield, Ore.
Celanese Corporation	Bishop, Tex.
Cities Service Corporation	Tallant, Okla.
Commercial Solvents	Agnew, Calif.
E. I. du Pont de Nemours & Co., Inc.	Belle, W. Va.
	Perth Amboy, N. J.
	Toledo, Ohio
Durez Plastics and Chemicals, Inc.	North Tonawanda, N. Y.
Heyden Chemical Corporation	Fords, N. J.
	Garfield, N. J.
Kay-Fries Chemical Company	West Haverstraw, N. Y.
Monsanto Chemical Company	Indian Orchard, Mass.
Rohm & Haas Company	Bristol, Pa.
	Bridesburg, Pa.
Spencer Chemical Company	Calumet City, Ill.
Watson-Park Company	Ballard Vale, Mass.

The large petrochemical plant of the McCarthy Chemical Co., at Winnie, Texas, is omitted from the above list since it was shut down in the early months of 1950[2].

Early History of Formaldehyde

With the publication of Liebig's research on acetaldehyde in 1835[40], the groundwork for the comprehension of the chemical nature of aldehydes was clearly set forth. In the years that followed, other aliphatic aldehydes were discovered and were easily recognized as belonging to this group of

Figure 1. Alexander Mikhailovich Butlerov, (1828–1886) deserves to be remembered as the discoverer of formaldehyde. In 1859, he published an accurate description of formaldehyde solution, formaldehyde gas and formaldehyde polymer with an account of chemical reactions including the formation of hexamethylenetetramine on reaction with ammonia. Although he did not characterize his methylene oxide polymer as a formaldehyde polymer, he was definitely aware of the fact that it behaved like the unknown "formyl aldehyde". The picture shown above was published in a book entitled, "A. M. Butlerov, 1828–1928" (Leningrad, 1929). This book contains six chapters by different authors dealing with his life and work.

Published by H. M. Leicester, *J. Chem. Ed.*, **17**, 203–9 (1940). Print used was given to author by Dr. Tenney L. Davis.

chemical compounds. Propionaldehyde, butyraldehyde, and isovaleraldehyde, among others, were discovered before 1860. Formaldehyde, however, remained unknown. The ease with which methanol passes on oxidation from formaldehyde to formic acid and thence to carbon dioxide and water made the isolation of formaldehyde difficult.

Formaldehyde was first prepared by Butlerov[16, 80] in 1859 as the product of an attempted synthesis of methylene glycol [$CH_2(OH)_2$]. The preparation was carried out by hydrolyzing methylene acetate previously obtained by the reaction of methylene iodide with silver acetate. Butlerov noticed the characteristic odor of the formaldehyde solution thus produced, but was unable to isolate the unstable glycol which decomposes to give formal-

Figure 2. August Wilhelm von Hofmann (1818–1892), who first prepared formaldehyde from methanol, is shown above in a portrait painted by Angeli in February, 1890. Hofmann's preparation clearly established both the structure and identity of formaldehyde as the first member of the aldehyde group.

Print reproduced from 1902 Sonderheft, Berichte der Deutschen Chemischen Geselschaft.

dehyde and water. Butlerov also prepared a solid polymer of formaldehyde by reacting methylene iodide and silver oxalate. He showed that this compound was a polymer of oxymethylene, (CH_2O), but failed to realize that it depolymerized on vaporization. He also obtained the new polymer by the reaction of methylene iodide and silver oxide, which gave additional evidence of its structure. He showed that it formed a crystalline product with ammonia (hexamethylenetetramine) and even stated that its reactions were such as one might expect from the unknown "formyl aldehyde".

In 1868, A. W. Hofmann[29] prepared formaldehyde by passing a mixture of methanol vapors and air over a heated platinum spiral, and definitely identified it. This procedure was the direct forebear of modern methods of formaldehyde manufacture. To Hofmann, the teacher, it seemed bad pedagogy that the first member of the aldehyde family should remain unknown, and he accordingly supplied the missing information.

Present Methods of Manufacture

Today, formaldehyde is produced principally from methanol but increasing quantities (about 20 per cent in 1948) are derived from the direct oxidation of hydrocarbon gases. The development of the petrochemical industry in the United States is also observed in the mounting production of synthetic methanol from natural gas. Edgar[23] points out that in 1946, 71 per cent of the carbon monoxide employed for methanol manufacture was derived from coal or coke, whereas in 1948, 77 per cent was made from natural gas.

The methanol process gives essentially pure formaldehyde containing some methanol, with traces of formic acid as a primary product, whereas hydrocarbon oxidation processes give a mixture of aldehydes, alcohols, ketones, fatty acids and other products which must be refined for sale.

Although a number of other methods for producing formaldehyde, including the hydrogenation of carbon oxides, the pyrolytic decomposition of formates, etc., have been described in the chemical literature or disclosed in patents, they do not appear to have achieved commercial importance. Of these latter procedures, the reduction of carbon monoxide alone merits special attention.

Reduction of Carbon Oxides. Because of its potential simplicity and the low cost of the raw materials involved, the preparation of formaldehyde by the direct reduction of carbon oxides with hydrogen has been the subject of considerable study, and numerous patents covering various techniques are recorded in the chemical literature.

According to Ipatieff and Monroe[32], the production of methanol from carbon dioxide and hydrogen under pressure probably involves formation of formaldehyde as an intermediate whose conversion to methanol may then proceed along two paths. Part of the methanol is formed by simple hydrogenation and part by the Cannizzaro reaction of formaldehyde with itself. With copper-alumina catalysts, the reactions proceed at temperatures of 282 to 487°C and pressures of 117 to 410 atmospheres.

The equilibrium constant for the reduction of carbon monoxide was determined by Newton and Dodge[51] for the reaction,

$$CO + H_2 \rightleftharpoons CH_2O.$$

According to their findings as based on experiments at 250°C:

$$Kp = \frac{P_{CH_2O}}{P_{CO} \cdot P_{H_2}} = \log_{10}^{-1}\left[\frac{374}{T} - 5.431\right]$$

Yield calculations based on this formula as cited by F. Fischer[26] are indicated in Table 2 which was prepared by H. Koch. As indicated by these data, impractically high pressures would be required to obtain useful yields. In addition, an extremely active and selective catalyst would be needed to obtain equilibrium at a reasonable rate and at the same time avoid hydrogenation of the formaldehyde produced. In the present state of our knowledge this reduction reaction would appear to be hopelessly unfavorable as a means of formaldehyde synthesis.

Production of Formaldehyde from Methanol. Processes for the production of formaldehyde from methanol were the first developed and still remain

TABLE 2. FORMALDEHYDE YIELDS CALCULATED FROM EQUILIBRIUM
VALUES FOR THE REDUCTION OF CARBON MONOXIDE[26]

°C	Absolute	Mol Per Cent CH₂O			
		1 atm	100 atm	1,000 atm	10,000 atm
27	300	0.002	0.16	1.6	13
127	400	0.001	0.06	0.6	5.5
227	500	0.0005	0.01	0.3	3

the major source of commercial formaldehyde. In general, they are carried out by passing a mixture of methanol and air over a heated stationary catalyst at approximately atmospheric pressure and scrubbing the off-gases with water to obtain aqueous formaldehyde. Procedures now in use include two general methods. The first or classical procedure makes use of a silver or copper catalyst and employs a rich mixture of methanol with air. Off-gases from this process contain 18 to 20 per cent hydrogen and less than 1 per cent oxygen, as well as minor amounts of carbon oxides and methane. The second method, which makes use of an oxide catalyst, e.g., iron-molybdenum oxide, employs a lean methanol-air mixture and produces a formaldehyde solution that is substantially free of unreacted methanol[31, 50]. The off-gases from this process contain unreacted oxygen and no appreciable concentration of hydrogen.

Reaction Mechanism. The formation of formaldehyde from methanol and air was first believed to be a gas phase oxidation process, as indicated in the following equation:

$$CH_3OH + \tfrac{1}{2}O_2 \rightarrow CH_2O + H_2O + 38 \text{ kcal.}$$

This is still believed to be the mechanism followed when an oxide catalyst is employed. However, mechanism studies of the reaction and products obtained by the metal catalyzed process demonstrated that under these conditions it is either a dehydrogenation, followed by oxidation of hydrogen to the extent that oxygen is present in the gaseous mixture as shown below[39, 55, 60],

$$CH_3OH \rightleftharpoons CH_2O + H_2 - 20 \text{ kcal}$$

$$H_2 + \tfrac{1}{2}O_2 \rightarrow H_2O + 58 \text{ kcal}$$

or a combination of the dehydrogenation and oxidation reactions. Oxidation reactions supply heat to make the process self-sustaining and are also believed to keep the catalyst active and displace the dehydrogenation equilibrium to the right.

A recent German study[55] gives evidence that the silver catalyzed process may depend exclusively on dehydrogenation, the combustion of hydrogen off-setting the endothermic nature of this reaction. Results of this work indicate that the temperature at the silver contact is a more important factor in determining the formaldehyde yield than the exact methanol-air ratio. This is further demonstrated by the fact that if the reaction temperature is raised by preheating the feed gases or by insulation, the yield is increased by substantially the same amount as would be obtained by effecting a similar rise of temperature by the use of additional oxygen. However, an irreducible minimum of oxygen is required since heating must not exceed temperatures at which there is substantial decomposition of the methanol or formaldehyde. Yield calculations based on dehydrogenation alone show a systematic variation with temperature, whereas calculations based on oxidation and dehydrogenation show unpredictable variations.

Undesirable reactions, which must be avoided by proper control of temperature and other factors if high yields are to be obtained, include: (a) pyrolytic decomposition of formaldehyde and (b) further oxidation of formaldehyde to formic acid, carbon oxides, and water.

(a) $CH_2O \rightarrow CO + H_2$

(b) $CH_2O + \tfrac{1}{2}O_2 \rightarrow HCOOH \text{ or } CO + H_2O$

   $CH_2O + O_2 \rightarrow CO_2 + H_2O$

Decomposition of formaldehyde by heat proceeds at a slow but measurable rate at 300°C, increasing rapidly at temperatures above 400°C (page 159).

Formation of formaldehyde by the dehydrogenation of methanol was studied in 1910 by Sabatier and Mailhe[60], who showed it to be a reversible reaction. More recently, Newton and Dodge[51] quantitatively determined the equilibrium relations for this reaction and found that

$$Kp = \frac{P_{CH_2O} P_{H_2}}{P_{CH_3OH}} = \log_{10}^{-1}\left[-\frac{4600}{T} + 6.470 \right]$$

From this equation it is possible to calculate that, for equilibrium conditions at atmospheric pressure, the dehydrogenation of methanol vapors to formaldehyde and hydrogen would be about 50 per cent at 400°C, 90 per cent at 500°C, and 99 per cent at 700°C. These yields are not readily obtained by dehydrogenation alone, since in the absence of air this reaction is relatively slow compared with the rate of decomposition of formaldehyde to carbon monoxide and hydrogen.

Water vapor has a favorable and apparently specific effect on the metal catalyzed formaldehyde process. Its action was first noticed by Trillat[70] in 1903 and was later reported by Thomas[65] in his research on the silver catalyzed process. The major effects noted are that it lowers the reaction temperature and increases the net yield, apparently by reduction of the undesirable reactions noted in the preceding paragraph. Uhl and Cooper[71] report results of this type on adding increasing concentrations of water vapor to an approximately 40:60 mixture of methanol and air.

Development of the Methanol Process. In the original method by which Hofmann[29] first produced formaldehyde, a mixture of air and methanol vapors obtained by drawing air through a reservoir of liquid methanol was passed into a flask containing a hot platinum spiral. Once initiated, the reaction was self-sustaining, as manifested by the glowing spiral. The product was dissolved in water contained in a series of gas-washing bottles. Because of difficulties with explosions in carrying out this procedure, Volhard[74] devised a scheme employing a spirit lamp filled with methanol and provided with a wick contiguous to the platinum spiral. Koblukov[35] subsequently found that the platinum could be replaced by platinized asbestos placed in a heated tube.

By heating methanol in a reservoir over a water bath, Tollens[67] was able to regulate the air-methanol vapor ratio, which was found to have a direct effect on the formaldehyde yield. This and minor refinements[68] were later applied to Loew's[41] procedure which replaced the platinum catalyst by copper gauze. Loew drew a rapid current of dry air through cold methanol (at about 18°C) and passed the resultant mixture through a 30-cm hard-glass tube containing a cylinder of coarse copper gauze 5 cm long. Brass gauze surrounding that portion of the tube which contained the copper catalyst was gently heated. Passage of the mixed reaction gases over the catalyst caused it to glow with a brightness which depended upon the rate of flow. The reaction products emerging from the glass tube were conducted through a large empty vessel and two flasks half-filled with water. Continuous operation of the process yielded solutions containing as much as 15 to 20 per cent formaldehyde. When the methanol was heated by Tollens[68] to 45–50°C and the reaction gases passed over the copper catalyst (surrounded by asbestos diaphragms to prevent explosions), a 30 per cent conversion to formaldehyde was obtained.

With the pioneering study on the preparation of formaldehyde completed by the demonstration in 1886 that it could be produced in continuous manner[41, 68], production on a commercial scale became feasible. Industrial developments are difficult to trace in detail, since practical advances are usually kept secret. It therefore becomes necessary to rely on patent literature for such information, which is fragmentary at best.

In 1889, both France and Germany granted August Trillat[69] the first patents to cover a process for the manufacture of formaldehyde. The procedure, which represented little improvement over the methods of Loew[41] and Tollens[63], consisted in discharging methanol vapors in the form of a spray into the open end of an externally heated copper tube packed with coke or broken tile, both the coke and the copper serving as catalysts. Although rights under these patents were sold to Meister, Lucius and Brüning at Höchst a.M., the process does not appear to have been employed commercially.

According to Bugge[15], the firm of Mercklin and Lösekann, founded in 1888 at Seelze, near Hanover, Germany, started the commercial manufacture of formaldehyde in 1889. Shipments of 5 to 20 kg were made to various factories and to university laboratories. Also prepared for the chemical market were such formaldehyde derivatives as paraformaldehyde, hexamethylenetetramine, and anhydroformaldehyde aniline.

At approximately this period, investigations were being conducted on the use of formaldehyde as a disinfectant. Meister, Lucius, and Brüning were customers of Mercklin and Lösekann, although the former company also carried out work of their own on formaldehyde manufacture. Formaldehyde purchases of the Badische Aniline and Soda Fabrik amounted to 1000 marks per month as early as 1891. The aldehyde was used chiefly for hardening gelatin films and for syntheses in the medicinal and dyestuff fields. With the patenting in 1901[24] of a method for synthesizing phenylglycine (an intermediate for indigo) from aniline, formaldehyde, and an alkali or alkaline-earth cyanide, there resulted a marked increase in the demand for formaldehyde.

The apparatus finally developed by Mercklin and Lösekann is shown in Figure 3[15]. It will be noted that the lower two-thirds of the reaction vessel is provided with vertical copper rods extending down into the methanol; between the upper extremities of the rods is packed the contact mass, i.e.,

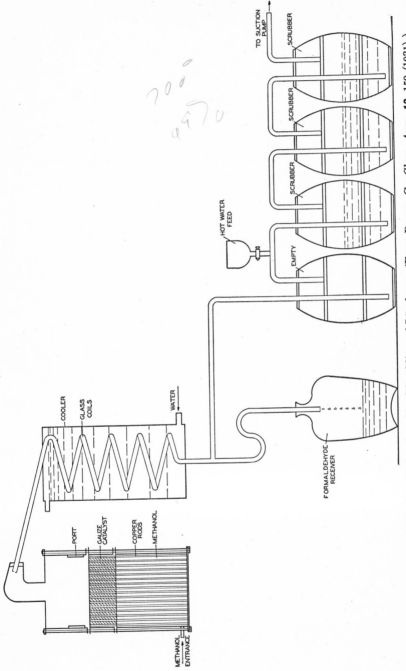

Figure 3. Formaldehyde unit developed by Mercklin and Lösekann. (From Bugge, G., *Chem. App.*, **18**, 159 (1931).)

spirals of copper gauze. The catalyst was heated from outside ports which could be closed after the unit was in operation. Air drawn into the apparatus carried methanol vapors over the catalyst, and the reaction products passed to the condensing system which comprised a glass condenser and several wooden scrubbers. It was possible to produce 30 per cent formaldehyde solution at a rate of 50 kg per 24 hours in this piece of equipment. Mercklin and Lösekann eventually had twelve of these units in operation.

Other companies started formaldehyde production about 1895, or perhaps even earlier. The better known manufacturers were: Verein für Chemische Industrie and Hugo Blank at Hoherlehme, near Berlin, in Germany; Firma Dr. Sieber at Attisholz, and Konstanzer Holzverkohlungs-Industrie A. G., in Switzerland. Mercklin and Lösekann went out of business in 1901. A few years later the firm of F. H. Meyer in Hanover-Hainholz, Germany, developed an improved formaldehyde unit[72] based on the work reported by Orlov[53, 54]. In time, the Holzverkohlungs-Industrie, A. G., manufacturer of methanol, also became one of the big producers of formaldehyde. In addition to the above concerns, large dyestuff companies which later amalgamated as the I. G. Farbenindustrie manufactured formaldehyde for their own consumption.

Industrial research on the manufacture of formaldehyde was initiated in 1898 by the firm of Hugo Blank under the direction of H. Finkenbeiner and O. Blank. Investigations extending over a period of several years covered the preparation of catalysts, temperatures of operation, time of reaction, and absorption of product. The technical formaldehyde manufacturing unit developed in the early nineteen hundreds as a result of this work was later duplicated in France, Belgium, and the United States. With this installation, air was drawn into a copper vaporizer containing methanol held at a fixed temperature. The vapors formed were passed through a preheater, and then through a tube containing a heated spiral of copper gauze. The contact tube was heated to start the reaction. During the reaction it was sufficiently air-cooled to prevent fusion of the contact mass. Scrubbing equipment comprised a bank of glass gas-washing bottles in conjunction with a coil condenser. This unit produced approximately 230 pounds per day of 40 per cent formaldehyde.

A comprehensive study of the oxidation of methanol was made by various investigators during the early period of commercial development. Trillat[69] published several articles dealing with the state of division of the catalyst, its temperature and composition. Improvements in operation and equipment were also reported by Klar and Schulze[37], Brochet[13], and Orlov[53, 54]. In 1910, O. Blank[9] patented the use of a silver catalyst in Germany. A year later Le Blanc and Plaschke[39] reported that formaldehyde yields obtained with silver catalysts were higher than those with copper. Thomas[65] reported laboratory-scale results on the preparation of formaldehyde by oxidizing methanol in the presence of copper, silver, and gold catalysts. For the silver catalyst these results approach those reported by Homer in 1941[31]. In 1913, the silver catalyst was introduced in United States operations with the patent of Kuznezow[38].

Commercial production of formaldehyde in the United States started around 1901. At that time, Buffalo was a center for the refining and marketing of products from the destructive distillation of wood. Many small stills in New York and Pennsylvania shipped their first distillate to Buffalo for refining. This business was started in 1880 by E. B. Stevens, who was associated with George N. Pierce, later founder of the Pierce-Arrow Motor Corporation. The plant organized by Pierce and Stevens was first named the Buffalo Alcholine Works, later the Manhattan Spirits Co., then the Wood Products Distilling Co. (It was sold in 1907 to the U. S. Industrial Alcohol Co.) In 1901 the company had purchased apparatus for the manufacture of formaldehyde from the Van Heyden Company in Germany; the equipment was operated until 1903, when this part of the plant and business was sold to the Heyden Chemical Co. and transferred to Garfield, N. J. About this same time, Harrison Brothers in Philadelphia bought a German formaldehyde process and operated it in connection with their paint works. Both the Heyden Company and the Perth Amboy Chemical Works (a subsidiary of The Roessler and Hasslacher Chemical Co., now the Electrochemicals Department of E. I. du Pont de Nemours & Co., Inc.) began manufacture of formaldehyde in 1904.

Recent Improvements in the Production of Formaldehyde from Methanol. With the development of larger-scale manufacturing equipment, improvements were made in the method for vaporizing alcohol and in the scrubbing systems. Apparatus of aluminum, stoneware, and other more durable materials replaced the glassware previously employed. Improvements also became necessary in the control of heat from the exothermic reaction. Gauze of the type employed by Blank[9] was found to disintegrate or fuse together so firmly that no more vapors could be drawn through it. Since this was particularly troublesome with high air-methanol ratios, low ratios were employed in order to keep the catalyst active over an extensive period, the excess methanol being subsequently distilled from the formaldehyde. A British plant employing this method of procedure is described by H. W. Homer[31]. As illustrated in Figure 4[31], a compressor forces air through a methanol vaporizer held at constant temperature. This gives an accurately controlled gas mixture, which is then preheated and passed through the catalyst contact. The catalyst is heated from an outside source to start the process. The gas mixture leaving the catalyst passes into a system involving scrubbers, multi-tubular coolers, and a distilling column as shown in the diagram. Formaldehyde containing some methanol is delivered to storage tanks and nearly pure methanol is returned to the vaporizer.

In recent years, considerable progress has been made in the methods employed for the manufacture of formaldehyde from methanol. This has

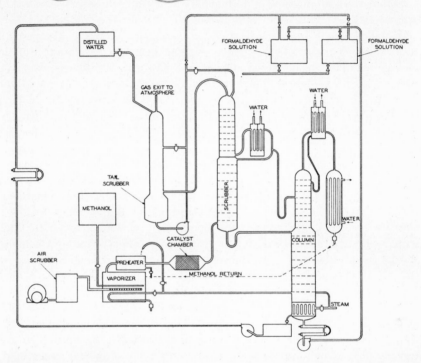

Figure 4. British unit for conversion of methanol to formaldehyde. (From Homer H. W., *J. Chem. Soc.*, Transactions and Communications, 1941, p. 217 (4T).)

been brought about by the use of efficient catalysts, improved methods of control, and engineering economies implicit in large-scale manufacture. The efficiency of the process depends in large part on maintaining the optimum ratio of methanol to air. Thomas[65] demonstrated that with the silver catalyst, gaseous mixtures containing from 0.5 to 0.3 gram oxygen per gram of methanol give manufacturing yields ranging from 73 to 60 per cent and net yields from 83 to 92 per cent of theoretical. A methanol-air mixture of the type cited contains 30 to 41 per cent of methanol by volume. With increasing concentrations of methanol the manufacturing yield gradually falls off, but the net yield increases. Homer[31] states that net yields of around 90 per cent are obtained in commercial practice. Catalyst efficiency is readily determined by analysis of the exit gases as indicated by the figures in Table 3[31]. From these figures the net yield of formaldehyde can be calculated in the manner illustrated. It will be apparent that the oxygen of the incoming air is almost completely consumed and that a high hydrogen content must be maintained to secure good yields. This means that the dehydrogenation reaction should be encouraged as much as is consistent with practical conversion.

Modern formaldehyde plants using the classical metal catalyst process

TABLE 3. ANALYSES OF FORMALDEHYDE PROCESS EXIT GASES
(Data from H. W. Homer[31])

	Composition in Volume Per cent	
	Efficient Catalyst	Old or Poisoned Catalyst
CO_2.............................	4.8	5.5
CO.............................	0.2	0.6
CH_4.............................	0.3	0.4
O_2.............................	0.3	0.3
H_2.............................	20.2	17.5
N_2 (by difference)...........	74.2	75.7

The per cent of theoretical yield can be calculated from the analysis shown for the efficient catalyst catalyst as follows:

$$\% \text{ Yield} = 100 - 100 \left[\frac{\%CO_2 + \%CO + \%CH_4}{0.528(\%N_2) + \%H_2 + 2(\%CH_4) - \%CO - 2(\%CO_2) - 2(\%O_2)} \right]$$

$$\% \text{ Yield} = 100 - 100 \left[\frac{4.8 + 0.2 + 0.3}{0.528 \times 74.2 + 20.2 + 2 \times 0.3 - 0.2 - 2 \times 4.8 - 2 \times 0.3} \right]$$

$$\text{Yield} = 89.3\% \text{ of theoretical}$$

are similar in essence to the British process previously described but differ in methods of reaction control, product isolation, etc. An American plant designed and constructed by the Chemical Construction Corp. for Durez Plastics and Chemicals, Inc., at North Tonawanda, New York, is described in considerable detail by T. R. Olive[52]. This plant, which has a rated capacity of 91,000 pounds of 37 per cent formaldehyde per day, commenced operations in the spring of 1948. Methanol from storage tanks is fed to a steam-heated, horizontal, cylindrical vaporizer designed to produce methanol vapor at 85°C and 18 psi. This vaporizer is automatically controlled to maintain continuous delivery at the required rate through a steam heated superheater. Air is drawn into the process through a wire mesh filter and a packed scrubbing column where it is washed with caustic soda to remove carbon dioxide and sulfurous compounds since the latter are stated to be injurious to catalyst activity. The air is then heated to about

70°C and mixed in controlled ratio with methanol vapor. The mixture which contains approximately one volume of air for each volume of methanol vapor is filtered and fed to the reactors which contain a bed of prepared silver catalyst. These reactors are jacketed vessels which operate in the range 450 to 650°C, the preferred temperature being around 635°. The construction allows control of reaction temperatures by means of bayonet-type cooling tubes, which can be fed with either water or steam. A flame arrestor at the entrance of each reactor prevents flash backs. Initiation of the reaction is carried out manually with an electric igniter. Gases from the reactor burners are rapidly quenched in primary scrubbing towers packed with Raschig rings in which liquor is constantly recirculated and cooled with spill or trombone-type coolers. The solution introduced at the top of the primary absorbers contains about 28 to 30 per cent formaldehyde and 20 to 22 per cent methanol in water. Gases leaving the primary absorbers are blown through a single secondary scrubber which is fed with distilled water. The methanol-containing formaldehyde solution or F-M is removed from the primary absorbers to a storage tank and from here it eventually passes to a vacuum distilling column which yields an over-head product of approximately 99 per cent methanol and a bottom product of formaldehyde solution containing less than 1 per cent methanol. The formaldehyde from this column goes to the product storage tank which is steam heated to prevent polymer precipitation. A related process of this type employed by the Spencer Chemical Co. has recently been described by Hader, Wallace and McKinney[26a].

Under some conditions of processing, a relatively low-methanol product is obtained directly and distillation of an F-M fluid intermediate is unnecessary. A wartime German process of this type yielding a 30 per cent formaldehyde is described by W. Wolstenholme *et al.* in a British Intelligence Objectives Sub-Committee report[83]. Air, saturated with a mixture of methanol and water vapor in a 60:40 ratio, is passed over a granular silver catalyst. The reaction temperature is in the range 640 to 660°C. The product gases are condensed and scrubbed to yield a 30 per cent by weight formaldehyde solution containing 0.5 to 2.0 per cent methanol and not more than 0.05 per cent formic acid.

The Springfield, Massachusetts, plant constructed by the Monsanto Chemical Co. as described by Pope[58] gives a low-methanol solution of somewhat higher than the usual strength that can be diluted to give a 37 per cent formaldehyde containing about 2 per cent methanol or adjusted with methanol to a 37 per cent stabilized solution containing about 7 per cent methanol. A flowsheet for this process is shown in Figure 5. As indicated, methanol is drawn from storage to a process supply tank which holds a supply sufficient for one shift's operation. Methanol from the supply

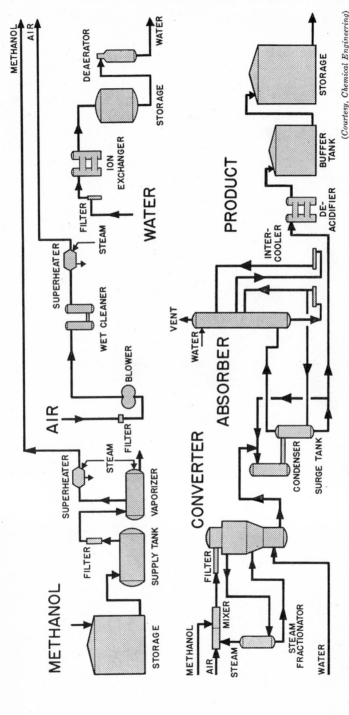

Figure 5. Formaldehyde process flow sheet.[58] This process employs methanol as a raw material but does not require distillation equipment to remove methanol from the final product.

(Courtesy, Chemical Engineering)

tank is passed through a porous stone filter to a vaporizer heated with steam at approximately 15 psi. The vapor is superheated to 115°C and passed to the mixer where it is combined with moist air which has been drawn through a maze filter and forced at 11 to 12 psi through a water scrubber and superheater which warms it to around 107°C. Steam also enters the mixer and becomes a part of the feed gases which enter the reactor (Figure 6) through a porous metal filter. The process employs approximately four tons of air per hour, around 2 tons of methanol and about 1⅓ tons steam.

(Courtesy, Chemical Engineering)
Figure 6. Reactor for conversion of methanol to formaldehyde.

The gases pass through the catalyst bed and are then quenched in a shell-and-tube heat exchanger. Reactor gases cooled by a spray of condensed product enter the condenser at 120 to 170°C. The surge tank separates condensate, which goes to one of three buffer tanks, from gas which is fed to the bubble plate absorber. The absorber is equipped with coolers and fed with water at about 30°C. The product goes to the buffer tank at a rate of about 5 tons per hour, passing through an ion-exchanger to remove formic acid. The buffer tanks are agitated and are used to adjust the solution to the desired formaldehyde and methanol concentrations. These tanks, as well as the final storage tanks, are steam heated to maintain the requisite storage temperature. All the water employed in the process is de-ionized and the steam used for the reaction is generated by the heat evolved by the reaction itself. Automatic recorder and temperature control instrumentation is used throughout the process. Materials of construction include copper and copper alloys for handling methanol, stainless steel, Type 347, for reactor, condenser and surge tank, and mild steel lined with plastic for storage tanks.

Numerous modifications and specific improvements in the classical formaldehyde process are covered by patents. Examples include a special

(Courtesy, Heyden Chemical Corporation, Garfield, New Jersey)

Figure 7. Formaldehyde absorber columns and control instrument panels in conversion from synthetic methanol.

reactor designed by Schideler and Richardson[61] and Meath's[44] two-phase process in which the gases from a silver catalyst reactor are cooled, blended with additional air and then passed through a second reactor.

The oxide catalyst process for formaldehyde manufacture differs from the classical procedure in that it employs a metal oxide, or mixtures containing metal oxides, as a catalyst, operates with a feed gas containing methanol with a large excess of air, and forms directly a low-methanol product[31, 47]. Catalysts patented for processes of this type include vanadium oxide[5], mixtures involving salts or oxides of vanadium with other metal oxides[19, 33], iron-molybdenum oxide mixtures[46], tungsten-molybdenum oxide[4], etc.

According to Homer[31], the reaction obtained with these catalysts is purely one of oxidation.

$$CH_3OH + \tfrac{1}{2}O_2 \rightarrow CH_2O + H_2O$$

Homer also states that in this type of process a large excess of air is em-

(Courtesy, E. I. du Pont de Nemours & Co., Inc.)
Figure 8. Methanol process formaldehyde plant at Toledo, Ohio.

ployed and that little if any methyl alcohol remains to be removed from the formaldehyde solution obtained. Inspection of the specific patents cited shows that the methanol-air mixtures favored for use with these catalysts contain from 5 to 10 per cent methanol by volume. While manufacturing yields are said to be higher than those obtained with silver catalysts in some instances, the cost of pumping the larger amount of air and the need for greater condensing capacity is said to offset the gain in yield.

Special modifications of the oxide catalyst process which have been patented include a procedure developed by Brondyke and Monier[14] for recycling spent gases and a special multistage converter in which increments of methanol are progressively added to the convertor gases following a procedure developed by Field[25].

Payne[57] claims a combination catalyst process in which a mixture of methanol vapors and air is passed over a silver catalyst, after which additional oxygen-containing gas is added and the unconverted methanol is

(*Courtesy, E. I. du Pont de Nemours & Co., Inc.*)
Figure 9. Formaldehyde tank truck.

oxidized in the presence of hydrogen produced in the first step without appreciable oxidation of this hydrogen to water.

In considering all types of methanol conversion processes, attention must be given to the flammability of mixtures of methanol vapor and air in order to avoid explosions. Data on these mixtures indicate that at temperatures of 60°C or less, flammability commences at methanol concentrations of 6 to 8 per cent by volume and ceases at concentrations variously reported from 25 to 37 per cent. In a review of the literature on this subject, Coward and Jones[18] point out that the range of flammability increases with increasing temperature. According to one investigator, on the lean side the limiting methanol vapor concentration is reported to fall from 7.5 per cent at 50°C to 5.9 at 250°C, whereas on the rich side it rises from 24.9 per cent at 100°C to 36.8 per cent at 200°C.

The Production of Formaldehyde from Hydrocarbon Gases. Hydrocarbon oxidation processes involve a controlled reaction of a hydrocarbon gas with air or oxygen, followed by abrupt cooling and condensation of the products usually by scrubbing with water to give a crude solution which must then be refined to isolate formaldehyde from the other products. The latter may include methanol, methylal, acetaldehyde, *n*-propanol, isopropanol, propionaldehyde, acetone, butyl alcohols, 2 and 3 carbon glycols, tetrahydrofuran, various organic acids, etc. Formaldehyde is not the major product of these petrochemical processes, the ratio of the various products being dependent on the hydrocarbon raw materials employed and the reaction techniques.

Oxidation of Methane. Although methane is the simplest hydrocarbon gas and would be expected to give formaldehyde uncontaminated with more complex organic products, its partial oxidation is extremely difficult. Blair and Wheeler[8] demonstrated that it is not oxidized at an appreciable rate below 600°C whereas, as previously pointed out, formaldehyde begins to decompose at temperatures considerably below this figure. The reactions which take place are illustrated below:

(a) $CH_4 + \frac{1}{2}O_2 \rightarrow CH_3OH$

(b) $CH_3OH + \frac{1}{2}O_2 \rightarrow CH_2O + H_2O$

(c) $CH_2O \rightarrow CO + H_2$

(d) $CH_2O + \frac{1}{2}O_2 \rightarrow CO + H_2O$

(e) $CH_2O + O_2 \rightarrow CO_2 + H_2O$

The indications are that the main difficulty lies not so much in preventing the thermal decomposition (c) as in controlling the oxidation reaction (d). In general, production of formaldehyde greatly exceeds that of methanol at ordinary pressures. Formation of methanol instead of formaldehyde is favored by the use of high pressures and low ratios of air to methane.

Reviewing the research on the methane oxidation process, Mayor[42, 43] points out that, in view of the secondary reactions, exposures of gases to reaction temperatures must be brief; also, only a small portion of methane is converted to formaldehyde per pass.

Of particular interest in connection with methane oxidation is the use of gaseous catalysts. Bibb and Lucas[7] have demonstrated that a low concentration of nitric acid vapor obtained by passing the gases to be oxidized through concentrated nitric acid at room temperature acts as an effective reaction catalyst. Gross conversions of methane to formaldehyde equivalent to approximately 4 per cent are reported by exposing mixtures of air and methane plus nitric acid vapor to a temperature of 700 to 750°C for one-half second. Medvedev[45] reports that small quantities of hydrogen chloride facilitate the conversion of methane-air mixtures in the presence of solid catalysts such as phosphates of tin, iron and aluminum. According to Mayor[43], the best results have been obtained by the use of a solid catalyst in conjunction with an oxide of nitrogen.

Experiments on the oxidation of methane with oxygen reported by Patry and Monceaux[56] indicate somewhat better results than those obtained with air, particularly when nitric oxides are employed as catalysts. A lower ignition temperature is obtained and yields of approximately 16 per cent formaldehyde are reported at 570°C using about 16 volumes of oxygen per 100 volumes of methane.

Commercial formaldehyde processes based on the oxidation of substantially pure methane have not been developed in the United States. However, a process of this type, developed by Gütehoffnungshuette, A.-G., has been operated in Germany. This process as described in a United States FIAT report by Holm and Reichl[30] involves the gas phase oxidation of a natural gas containing approximately 98 per cent methane with air in the presence of 0.08 volume per cent nitric acid based on the methane-air feed. Methane and air in the volume ratio of 3.7 to 1.0 are added separately to recycled process off-gases in such proportions that one volume of total feed is mixed with 9 volumes of off-gas. The resultant mixture is heated to 400°C, mixed with nitric acid catalyst, and fed into a ceramic-lined steel tube reactor in which the temperature ranges from 400 to 600°C. On leaving the reactor the gases are chilled to 200°C by heat exchange with the composite feed and passed through a water scrubber to remove formaldehyde. A portion of the off-gas is then withdrawn for use as fuel to heat the reactor, and replaced with an equal volume of fresh feed for recycling. Net yields of 35 per cent formaldehyde are claimed for the reaction process but actual conversions allowing for use of methane in firing the reactor are reported as about 10 per cent of the theoretical. Around 3 per cent of the methane, which is converted to products other than formaldehyde, goes to methanol, the remainder to carbon oxides and a trace of formic acid.

Holm and Reichl[30] also describe another German methane oxidation process in the laboratory stage which was claimed to give a 90 per cent yield of formaldehyde. This procedure employed oxygen containing approximately 1 per cent ozone and a supported barium peroxide catalyst activated with approximately 0.5 per cent of silver oxide. The preferred methane raw material, a coke-oven gas containing about 70 per cent methane, was mixed with about 50 per cent its volume of the ozonized oxygen and a reaction temperature of about 110°C was claimed. The crude methane contained

3 per cent ethylene plus nitrogen (12 per cent), carbon monoxide (11 per cent) and hydrogen (4 per cent). Conversions of methane to formaldehyde were stated to approach 25 per cent with this raw material, whereas pure methane gave a yield of only 5 per cent.

Oxidation of Higher Hydrocarbon Gases. Industrial processes for the production of formaldehyde from hydrocarbons in the United States are based principally on the direct oxidation of gaseous hydrocarbons containing more than one carbon atom. The oxidation of ethane, propane and butane can be controlled to yield formaldehyde under conditions similar to those employed with methane. In general, the higher hydrocarbon gases can be oxidized at a much lower temperature than methane, or even ethane, and fair rates of reaction can be obtained at temperatures at which aldehydes, ketones and alcohols may be isolated without prohibitive losses through decomposition.

Research on the production of formaldehyde from methane, propane, and the hydrocarbon mixtures encountered in natural gas has been described by Bibb.[6] Studies involving ethane, propane, and higher paraffins are also reported by Wiezevich and Frolich[81], who used iron, nickel, aluminum, and other metals as catalysts and employed pressures up to 135 atmospheres.

According to Meyer[47], the Hanlon-Buchanon Corp., which became a part of the Warren Petroleum Corp. in 1946, became interested in the utilization of propane and butane as chemical raw materials in the late 1920's and installed a direct oxidation pilot plant at Iowa State University. A semi-commercial plant was built in Louisiana on the basis of this work but was abandoned in 1933 because of market conditions and the fact that processes were not developed for isolating commercial grade chemicals from the crude oxidation mixture.

The Cities Service Oil Company has been a pioneer in the industrial development of the direct oxidation of the paraffin hydrocarbon gases[1, 75—78] to yield substantial amounts of commercially acceptable grades of methanol, formaldehyde, and acetaldehyde in conjunction with smaller amounts of acetone, dimethyl acetal and higher alcohols. In 1925, J. C. Walker of Cities Service and his associates found that the corrosion of pipelines caused by natural gas containing free oxygen could be minimized by passing the gas over a contact catalyst, a procedure which led to the formation of the chemicals just mentioned.[76] Typical liquid products were found to contain 34 to 36 per cent methanol, 20 to 23 per cent formaldehyde and 5 to 6 per cent acetaldehyde, together with varying amounts of higher alcohols, ketones, aldehydes and water. According to patent specifications, mixed catalysts comprising aluminum phosphate and metal oxides may be employed for the oxidation reaction and pressures in the neighborhood of 7 to 20 atmospheres with temperatures of 430 to 480°C are stated to be satisfactory.[78] The first commercial plant to employ an oxidation process of this type was constructed at Tallant, Oklahoma, in 1926[1] (Figure 10). This has since been expanded and three additional oxidation plants have been built in

(Courtesy, Chem. & Met. Eng.)
Figure 10. Converter for production of formaldehyde from natural gas.

conjunction with a modern chemicals refinery to produce commercial-specification chemicals from the crude oxidation products.

Following the inauguration of an extensive research program on the manufacture of petroleum products in 1932, the Celanese Corporation erected an oxidation pilot plant in 1941, followed by the construction of a full-scale plant in Bishop, Texas, in 1943[34]. Production was initiated with the manufacture of acetaldehyde, acetic acid, formaldehyde and methanol. Other products including methylal, acetone, n-propanol, isopropanol, butyl alcohols, glycols such as propylene glycol, etc., have since been added.

According to Hightower[27], propane and butane, the basic hydrocarbon raw materials, are brought to the plant as liquids and stored in pressure tanks. Oxidation equipment is so arranged that either of these gases or a mixture of the two can be processed as desired. A mixture of compressed hydrocarbon gas and air is partially oxidized using excess hydrocarbon to facilitate control of the exothermic reaction. The gases from this reaction are then cooled and scrubbed with water, after which the liquid product

(Courtesy, Celanese Corporation of America)

Figure 11. Aerial view of Celanese Corporation of America Plant at Bishop, Texas. This plant produces organic chemicals including formaldehyde by the controlled oxidation of butane and propane.

goes to a separator from which dilute formaldehyde and a mixture of alcohols, ketones, higher aldehydes, etc., emerge as distinct fractions. The former is concentrated and purified to give a commercial grade solution and the latter is refined into its various isolable constituents. The process is illustrated by the flow sheet in Figure 14[27]. The refining procedures include fractionation, azeotropic distillation, extractive distillation and liquid-liquid extraction[21a, 27].

Materials of construction include steel and stainless steel where contact with aqueous condensates is involved. Outdoor refinery structure is used with the metal equipment protected by aluminum paint.

A typical example from a Celanese patent by Bludworth[10] states that one volume (one part by weight) of gaseous butane preheated to 300°C is mixed with 110 volumes (34 parts by weight) of steam and 10 volumes (5 parts by weight) of air at about 400°C, and 300 to 400 pounds pressure, the reaction being quenched with water after approximately one second. Formaldehyde and organic acids are removed together by scrubbing the reaction products with water. Pure formaldehyde is obtained by refining this solution. A hundred pounds of butane are reported to yield 15.2 pounds of formaldehyde, 19.6 pounds of acetaldehyde, 7 pounds of acetone, 19 pounds of methanol, 1 pound of propanol, 0.5 pound of butanol and 11.4

(*Courtesy, Celanese Corporation of America*)

Figure 12. Facilities to recover and purify organic chemicals produced from propane and butane. Formaldehyde production unit is in the center foreground.

pounds of organic acids. Oxidation of a 5 to 7 carbon gasoline is described in a later patent by the same inventor[11]. A patent by Kotzebue[36], issued in 1950, claims a more efficient procedure for oxidizing propane and butane in which the products undergo an initial reaction with oxygen in the presence of an inert diluent at 600 to 900°F followed by a final oxidation with pure oxygen at 900 to 1250°F.

Numerous patents covering a variety of processes for the oxidation of hydrocarbon gases have been assigned to other companies, e.g., Monsanto[64, 21], Godfrey L. Cabot, Inc.[66, 59], and Stanolind Oil and Gas Co.[43].

The McCarthy Chemical Co. which operated for a short period, closing down in 1950 as mentioned previously, employed a pressure oxidation of natural gas from which liquefiables were removed. Oxygen of 90 to 95 per cent purity was employed and the primary products were methanol, acetaldehyde and some formaldehyde. Additional formaldehyde was then obtained from methanol by the conventional process using a copper cat-alyst[28].

Figure 13. Distillation towers for refining petrochemicals. Formaldehyde concen-tration and recovery unit is in the immediate foreground.

Purification of Formaldehyde

The use of ion-exchange resins for the removal of formic acid and metallic impurities from formaldehyde has proved practical in some instances. However, it is most commonly employed solely for the removal of formic acid and, as has already been pointed out, may be built into the manu-facturing process as a final purification step. Zowader[84] states that most manufacturers can produce solution that will meet specifications without employing a special step for the removal of acids by careful processing and the use of pure methanol as a raw material. Nevertheless, anion ex-change resins fill a definite need for producing a product with extremely

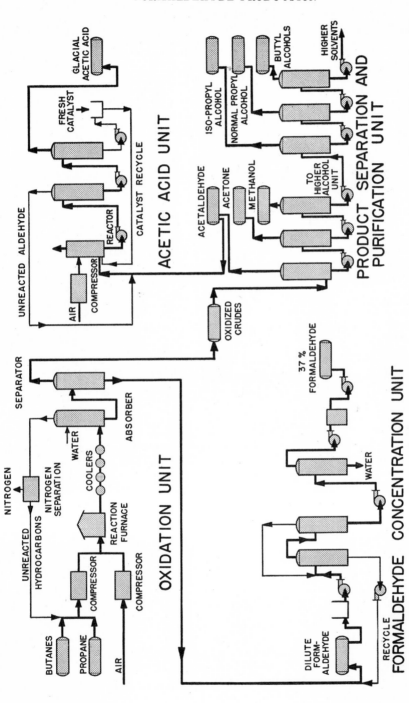

(Courtesy of Chemical Engineering)

Figure 14. Flow sheet of hydrocarbon oxidation process for the manufacture of formaldehyde and other petrochemicals from propane and butane.[27]

low acidity or obtaining specification material from an off-grade formaldehyde. In the methanol process, the product runs high in formic acid if the methanol raw material is impure. As Myers[49] indicates, crude formaldehyde may sometimes contain as much as 0.05 to 0.08 per cent formic acid and although it may be removed by active carbon, this technique is expensive and inefficient. Anion exchange resins having a capacity up to and above 2 lb formic acid per cu ft are reported to be available for formaldehyde purification. Development of a commercial size unit for treating formaldehyde with both cation and anion exchange resins is described by Cristy and Lembcke[20]. Zowader[84] describes a procedure involving the use of anion exchange resin alone. These investigators point out that care must be taken to see that the pH of the treated solution is not raised too high since this catalyzes precipitation of polymer (page 76). Use of sodium carbonate and sodium bicarbonate is recommended for regeneration of the spent resin. Cristy and Lembcke recommend washing the treated resin until an effluent of pH 8.0 is obtained and report that when caustic soda is used as a regenerant, an undesirably elevated pH is obtained. Care must be taken to avoid resins which cause coloration of the formaldehyde or which swell on use. A slight coloration can be removed with active carbon. A maximum temperature of 35°C is recommended for the influent formaldehyde by one investigator. Cristy and Lembcke[20] report a 96 per cent recovery of formaldehyde of satisfactory strength and flow rates of 0.5 gal/sq ft/min. The same investigators state that loss of resin by attrition is small and an original resin bed showed satisfactory performance after operating for three years.

Removal of formic acid from formaldehyde with a phenol-formaldehyde resin containing polyalkylene diamine groups is described in a patent by Cheetham and Myers[17]. A British patent[12] discloses the use of a formaldehyde resin derived from guanidine, biguanide or guanyl urea.

Shipping

Formaldehyde is shipped in tank cars, tank wagons, drums, barrels, carboys and bottles. In recent years, it has been shipped by tanker from Texas manufacturing plants to distribution centers on the Eastern Seaboard (Figure 15). Containers are preferably stainless steel (18 per cent chromium—8 per cent nickel), ordinary steel lined with rubber or resins, aluminum or glass. Tank cars are usually insulated to prevent undue cooling with consequent precipitation of polymer in transit. Heating may also be employed for this purpose. These precautions are of special importance when low-methanol (i.e., unstabilized or "uninhibited" solution is being handled). Although wooden barrels are sometimes employed, they tend to discolor the solution. Detailed information on materials of construction for formaldehyde containers will be found under Formaldehyde Storage, Chapter 3, pages 79–80.

(Courtesy, Celanese Corporation of America)

Figure 15. Formaldehyde tanker. Loading formaldehyde on specially constructed tanker at Corpus Christi for transfer to an Eastern Terminal at New Haven, Connecticut. This vessel carries four million pounds of formaldehyde per trip in heated stainless steel tanks.

References

1. Anon, *Chem. & Met. Eng.*, **49**, 154–9 (Sept., 1942).
2. Anon, *Chem. Eng.*, **57**, Chementator, 67 (Feb., 1950), 80, 84 (April, 1950).
3. Anon., *Chem. Eng.*, **58**, 369 (1951).
4. Arnold, H. R., (to E. I. du Pont de Nemours & Co., Inc.), U. S. Patents 2,320,253 (1943) and 2,439,880 (1948).
5. Bailey, G. C., and Craver, A. E., (to The Barrett Co.), U. S. Patent 1,383,059 (1921).
6. Bibb, C. H., *Ind. Eng. Chem.*, **24**, 10 (1932). Cf. U. S. Patents 1,392,886 (1922); Re. 15,789 (1924); 1,547,725 (1925).
7. Bibb, C. H., and Lucas, H. J., *Ind. Eng. Chem.*, **21**, 633 (1929).
8. Blair, E. W., and Wheeler, T. S., *J. Soc. Chem. Ind.*, **41**, 303–310T, (1922).
9. Blank, O., German Patent 228,697 (1910).
10. Bludworth, J. E., (to Celanese Corp. of America), U. S. Patent 2,128,908 (1938).
11. Bludworth, J. E., (to Celanese Corp. of America), U. S. Patent 2,396,710 (1945).
12. British Patent, (to American Cyanamid Co.), 573,051 (1945).
13. Brochet, A., *Compt. rend.*, **119**, 122 (1894); *Ibid.*, **121**, 133 (1895).

14. Brondyke, W. F., and Monier, J. A., (to E. I. du Pont de Nemours & Co., Inc)., U. S. Patent 2,436,287 (1948).
15. Bugge, G., *Chem. App.*, **18**, 157–60 (1931).
16. Butlerov (Butlerow), A., *Ann.*, **111**, 242–52 (1859).
17. Cheetham, H. C., and Myers, R. J., (to the Resinous Products and Chem. Co.), U. S. Patent 2,341,907 (1944).
17a. *Chemical Industries Week*, **68**, 12 (Feb., 1951).
18. Coward, H. F., and Jones, G. W., "Limits of Inflammability of Gases and Vapors," Bull. 279, U. S. Bur. Mines, pp. 81–2, U. S. Government Printing Office, Washington, D. C. (1939).
19. Craver, A. E., (to Weiss and Downs, Inc.), U. S. Patent 1,851,754 (1932).
20. Cristy, G. A., and Lembcke, R. C., *Chem. Eng. Progress*, **44**, 417 (1948).
21. Derby, W. R., (to Monsanto Chemical Co.), U. S. Patent 2,376,668 (1945).
21a. Dice, H. K., (to Celanese Corp. of America), U. S. Patent 2,570,215 (1951).
22. Du Pont, *Chem. Inds.* **51**, 487 (1942).
23. Edgar, J. L., *Chemistry & Industry*, p. 472, July 1, 1950.
24. Farbwerke vorm. Meister, Lucius, and Brüning, British Patent 22,733 (1901); German Patent 135,332 (1902).
25. Field, E., (to E. I. du Pont de Nemours & Co., Inc.), U. S. Patent 2,504,402 (1950).
26. Fischer, F., *Oel u. Kohle*, **39**, 521–2 (1943).
26a. Hader, R. N., Wallace, R. D., and McKinney, R. W., *Ind. Eng. Chem.*, **44**, 1508–18 (1952).
27. Hightower, J. V., *Chem. Eng.*, **55**, 105–7, 136–9 (July, 1948).
28. Hightower, J. V., *Chem. Eng.*, **56**, 92–4, 132–5 (Jan., 1949).
29. Hofmann, A. W., *Ann.*, **145**, 357 (1868); *Ber.*, **2**, 152–160 (1869).
30. Holm, M. M., and Reichl, E. H., FIAT Report, No. 1085, Office of Military Government for Germany (U. S.), March 31, 1947.
31. Homer, H. W., *J. Soc. Chem. Ind.*, **60**, 213–8T (1941).
32. Ipatieff, V. N., and Monroe, G. S., *J. Am. Chem. Soc.*, **67**, 2168–71 (1945).
33. Jaeger, A. O., (to The Selden Company), U. S. Patent 1,709,853 (1929).
34. Kirkpatrick, S. D., *Chem. Eng.*, **56**, 91–102 (Nov., 1949).
35. Koblukov, J., *J. Russ. Phys.-Chem. Soc.*, **14**, 194 (1882); *Ber.*, **15**, 1448 (1882).
36. Kotzebue, M. H., (to Celanese Corp. of America), U. S. Patent 2,495,332 (1950).
37. Klar, M., and Schulze, C., German Patent 106,495 (1898).
38. Kuznezow, M. J., (to Perth Amboy Chemical Works), U. S. Patent 1,067,665 (1913).
39. LeBlanc, M., and Plaschke, E., *Z. Elektrochem.*, **17**, 45–57 (1911).
40. Liebig, J., *Ann.*, **14**, 133–67 (1835).
41. Loew, O., *J. prakt. Chem.* (2), **33**, 321–51 (1886).
42. Mayor, Y., *Rev. Chim. Ind.* (*Paris*), **46**, 34–40, 70–6, 110–6, 136–40 (1937).
43. Mayor, Y., *L'Ind. Chim.*, **26**, 291–2 (1939).
44. Meath, W. B., (to Allied Chem. & Dye Corp.), U. S. Patent 2,462,413 (1949).
45. Medvedev, S. S., *Trans. Karpov Chem. Inst.*, **1925**, No. 4, 117–25; *C. A.*, **20**, 2273 (1926).
46. Meharg, V. E., and Adkins, H., (to Bakelite Corp.), U. S. Patent 1,913,405 (1933).
47. Meyer, R. E., *Chem. Eng. News*, **28**, 1910 (1950).
48. Michael, V. F., and Phinney, J. A., (to Stanolind Oil and Gas Co.), U. S. Patent 2,482,284 (1949).
49. Myers, F. J., *Ind. Eng. Chem.*, **35**, 861 (1943).
50. Neidig, C. P., *Chem. Ind.*, **61**, 214–7 (1947).

51. Newton, R. H., and Dodge, B. F., *J. Am. Chem. Soc.*, **55**, 4747 (1933).
52. Olive, T. R., *Chem. Eng.*, **56**, 130–3, 146–9 (Feb., 1949).
53. Orloff, J. E.,* "Formaldehyd," pp. 185–242, J. A. Barth, Leipzig (1909).
54. Orlov, E. I.,* *J. Russ. Phys.-Chem. Soc.*, **39**, 855–68, 1023–44, 1414–39 (1907); **40**, 796–9 (1908).
55. P. B. Report 73452 (Sator), March 5, 1938.
56. Patry, M., and Monceaux, P., *Compt. rend.*, **223**, 329–31 (1946).
57. Payne, W. A., (to E. I. du Pont de Nemours & Co., Inc.), U. S. Patent 2,519,788 (1950).
58. Pope, L. B., *Chem. Eng.*, **57**, 102–5 (Jan. 1950).
59. Rossman, R. P. (to Godfrey L. Cabot, Inc.), U. S. Patent 2,467,993 (1949).
60. Sabatier, P., and Mailhe, A., *Ann. chim. phys.* (8), **20**, 344 (1910).
61. Schideler, P., and Richardson, R. S., (to Chem. Construction Corp), U. S. Patent 2,406,908 (1946).
62. Skeen, J. R., *Chem. Eng. News*, **26**, 3344–5 (1948).
63. Stenerson, H., *Chem. Eng. News*, **28**, 812 (1950).
64. Thomas, C. A., (to Monsanto Chem. Co.), U. S. Patent 2,365,851 (1944).
65. Thomas, M. D., *J. Am. Chem. Soc.*, **42**, 867 (1920).
66. Sherwood, T. K., (to Godfrey L. Cabot, Inc.), U. S. Patent 2,412,014 (1946).
67. Tollens, B., *Ber.*, **15**, 1629 (1882).
68. Tollens, B., *Ber.*, **19**, 2133–5 (1886).
69. Trillat, A., French Patent 199,919 (1889); German Patent 55,176 (1889).
70. Trillat, A., *Bull. soc. chim.*, *III*, **27**, 797–803 (1902); **29**, 35–47 (1903).
71. Uhl, H. B., and Cooper, I. H., (to Heyden Chem. Corp.), U. S. Patent 2,465,498 (1949).
72. Ullman, F., "Enzyklopädie der technischen Chemie," 2nd Ed., pp. 416–7, Urban & Schwarzenberg, Berlin-Vienna (1930).
73. United States Dept. Commerce, Bureau of Foreign and Domestic Commerce, "Statistical Abstract of the United States," p. 827, Washington, United States Government Printing Office (1929).
74. Volhard, J., *Ann.*, **176**, 128–135 (1875).
75. Walker, J. C., (to Empire Oil & Refining Co.), U. S. Patents 2,007,115–6 (1935).
76. *Ibid.*, U. S. Patent 2,042,134 (1936).
77. Walker, J. C., (to Cities Service Oil Company), U. S. Patent 2,153,526 (1939).
78. *Ibid.*, U. S. Patent 2,186,688 (1940).
79. Walker, J. C., and Malakoff, H. L., *Oil Gas J.*, **45**, 59 (Dec. 21, 1946).
80. Walker, J. F., *J. Chem. Ed.*, **10**, 546–51 (1933).
81. Wiezevich, P. J., and Frolich, P. K., *Ind. Eng. Chem.*, **26**, 267 (1934).
82. Williams, A. A., *Chem. Eng.*, **55**, 112–3 (1948).
83. Wolstenholme, W., Fray, A., and Nickels, K., P. B. 79,300 (BIOS Final Report No. 1331, Item No. 22), Office of Technical Services, Dept. of Commerce, Washington, D. C.
84. Zowader, H., *Chem. Eng. Progress*, **45**, 279–80 (1949).

* References 53 and 54 are by the same author. The discrepancy in spelling is due to the difference in English and German transliteration from Russian.

CHAPTER 2

MONOMERIC FORMALDEHYDE

Although not commercially available in this form, monomeric formalde-hyde is important both from a theoretical and a practical standpoint. It serves as a basis for the determination of fundamental physical constants and is involved wherever formaldehyde is employed in the gaseous state. A knowledge of the pure monomer is a necessary preface to the understand-ing of formaldehyde solutions and polymers.

Pure, dry formaldehyde is a colorless gas which condenses on chilling to give a liquid that boils at −19°C and freezes to a crystalline solid at −118°C[42]. Both liquid and gas polymerize readily at ordinary and low temperatures and can be kept in the pure monomeric state only for a limited time. Because of these facts formaldehyde is sold and transported only in solution or in the polymerized state. In its aqueous solutions, formaldehyde is almost completely hydrated. At ordinary temperatures these hydrates have a relatively high degree of stability, although from a chemical standpoint they are extremely reactive. When required, mono-meric formaldehyde is best prepared from the commercial solution or poly-mer at the point of use and employed directly for the purpose at hand.

Formaldehyde Gas

Monomeric formaldehyde gas is characterized by a pungent odor and is extremely irritating to the mucous membranes of the eyes, nose, and throat even when present in concentrations as low as 20 parts per million[11]. In this connection, it should be noted that the polymeric vapors of trioxane, $(CH_2O)_3$, the little-known trimer of formaldehyde, are not irritating but possess a pleasant, chloroform-like odor.

Pure, dry formaldehyde gas shows no visible polymerization at tempera-tures of 80 to 100°C and obeys the ideal gas laws without pronounced deviation[50]. However, its stability is dependent on purity, even a trace of water provoking rapid polymerization. At ordinary temperatures the dry gas polymerizes slowly, building up a white film of polyoxymethylene on the walls of the containing vessel. Kinetic studies indicate that this transformation takes the form of a surface reaction. The studies of Trautz and Ufer[50], and Spence[40] indicated a unimolecular reaction at high pres-sures and a polymolecular one at pressures below 200 mm. A more recent investigation by Sauterey[37a] implies that the reaction is dependent upon

34

the nature of the surface on which the polymerization takes place and that a reaction order cannot be defined. It is not accelerated by ultra-violet light[40, 50]. In the presence of water vapor and other polar impurities formaldehyde gas is stable only at pressures of 2 to 3 mm or concentrations of about 0.4 per cent at ordinary temperatures. Sauterey[37a] concludes that the effect of water vapor on polymerization is due to the same mechanism as that controlling polymer formation in aqueous solutions.

The marked promoting effect of small quantities of formic and acetic acid on the polymerization of monomeric formaldehyde vapor at 18 and 100°C has been studied in detail by Carruthers and Norrish[5]. These investigators found that formic acid disappeared from the gas phase during polymerization at a rate nearly proportional to the rate of formaldehyde removal and assumed that polyoxymethylene chains originated from the hydroxyl group of this acid. The high efficiency of formic acid as a polymerization promoter is explained by addition to the polyoxymethylene chains with the formation of two hydroxyl groups and consequent branching. This is indicated in the following equations in which n equals an unknown whole number:

$$HCOOH + CH_2O \rightarrow HCOOCH_2OH$$

$$HCOOCH_2OH + nCH_2O \rightarrow HCOO(CH_2O)_nCH_2OH$$

$$HCOO(CH_2O)_nCH_2OH + HCOOH \rightarrow HCOO(CH_2O)_nCH_2OCH(OH)_2$$

$$HCOO(CH_2O)_nCH_2OCH(OH)_2 + 2nCH_2O \rightarrow$$

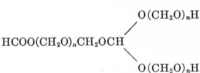

According to Carruthers and Norrish, the polymerization is bimolecular with reference to formaldehyde. The process is heterogeneous and occurs at a cold surface.

Formaldehyde gas that is 90 to almost 100 per cent pure must be kept at temperatures in the neighborhood of 100 to 150°C or above if polymerization is to be avoided. As previously pointed out, chemical decomposition of formaldehyde gas is not appreciable at temperatures below 400°C[1].

Formaldehyde gas is flammable, having a heat of combustion of 4.47 kcal per gram[53]. It forms explosive mixtures with air or oxygen. According to studies carried out in our laboratories, the flammable limits of mixtures of formaldehyde gas and air range from compositions containing approximately 0.39 to 13 volumes of air per volume of formaldehyde gas. This means that formaldehyde-air mixtures containing from 7 to 72 per cent formaldehyde should be regarded as potentially explosive. The initial combustion temperature of formaldehyde-air mixtures is reported to be approximately 300°C, as measured by passing mixtures of formaldehyde and air through a heated tube[4, 41].

Although monomeric formaldehyde is obtained whenever formaldehyde solution or its linear polymers are subjected to vaporization, gas having a high formaldehyde concentration is best obtained by the action of heat on a paraformaldehyde containing at least 95 per cent available formaldehyde or the more highly polymerized polyoxymethylenes, which contain less than 1 per cent combined water. Studies in our laboratories have demonstrated that a smooth and rapid generation of gas is readily obtained by use of white mineral oil (b.p. approx. 350°C or 662°F) as a heat transfer agent. For this purpose an oil slurry containing 10 to 50 per cent of paraformaldehyde is heated in the range 115 to 140°C. Continuous generation of monomeric formaldehyde by metering a suspension of polymer in an inert carrier (e.g., di-(2-ethylhexyl)phthalate) into a heated generator has been described in a patent by Miller[24]. Gases obtained by these procedures contain small proportions of water and polymerize rapidly at temperatures below 100°C. Dry gas is best prepared from pure anhydrous liquid formaldehyde. Chemical drying agents cannot be used for removing moisture from formaldehyde gas, because they function as polymerization catalysts.

Formaldehyde gas is readily soluble in water and polar solvents such as alcohols, but is only slightly soluble in such liquids as acetone, benzene, chloroform, and ether[51].

Heat Capacity of Formaldehyde Gas. In the absence of any experimental data, we have calculated heat capacity figures for formaldehyde by the method described by Dobratz[9] based on spectroscopic measurements[20]. This calculation leads to the following expressions for heat capacity at constant pressure in cal/gram mol formaldehyde gas as a function of its temperature in degrees centigrade and partial pressure in atmospheres.

$$(C_p)_{p=0} = 9.48 + 0.00914t - 0.00000064t^2$$

Values at pressures other than zero can be calculated by making a reasonable assumption for the critical values for formaldehyde, *e.g.*, $t_k = 141°C$ and $P_k = 65$ atm. With these we obtain:

$$(C_p)_{p=p} = 9.48 + 0.00914t - 0.00000064t^2 + 0.08p \left(\frac{413}{273 + t} \right)^3$$

Values calculated from these expressions for moderate pressures and temperatures are shown in Table 4. Values for C_p are probably reasonably accurate up to temperatures of 400°C. Since errors, even in values calculated for zero pressure, may be in tenths rather than hundredths of a unit, it is obvious that we may ignore the effect of pressure, up to at least one atmosphere, and assume that $(C_p)_{p=1} = (C_p)_{p=0}$ for temperatures of 100 to 400°C. For low partial pressures, such as 0.1 atm, we may assume that $(C_p)_{p=0} = (C_p)_{p=p}$ at all temperatures from 0 to 400°C.

Heat of Formation and Free Energy of Monomeric Gaseous Formaldehyde. Within the limits of experimental error, the most accurate value for the heat of formation of monomeric gaseous formaldehyde determinable from existent data is 28 kcal per mol at 18°C.

$$C_{Graphite(Rossini)} + H_2 + \tfrac{1}{2}O_2 \to CH_2O\,(g) + 28 \text{ kcal}$$

The heat of combustion of the monomeric gas was determined by von Wartenberg and Lerner-Steinberg[53] in 1925 as 134.1 kcal.

$$CH_2O\ (g) + O_2 \to H_2O\ (liq.) + CO_2 + 134.1 \text{ kcal}$$

The heat of formation as calculated from this figure, using Rossini's recent

TABLE 4. CALCULATED HEAT CAPACITIES (C_p) FOR FORMALDEHYDE GAS

Temperature t(°C)	Cp (g cal/g mol)		
	$p = 0$	$p = 0.1$ atm	$p = 1.0$ atm
0	9.48	9.51	9.75
100	10.39	10.39	10.49
200	11.28	11.28	11.33
300	12.16	12.16	12.19
400	13.03	13.03	13.05

values[35] for the heats of formation of carbon dioxide (94.03 kcal) and water (68.31 kcal) at 18°C, is 28.3 kcal per mol, or 28.5 kcal as based on the beta-graphite standard. Calculation of the heat of formation from the I.C.T. value[16] for the free energy of carbon monoxide, based on equilibrium measurements and the Newton and Dodge formula[26] for the variation of the carbon monoxide-hydrogen-formaldehyde equilibrium with temperature, gives a value of 27.7 kcal at 18°C (corrected to the newer Rossini standards mentioned above) which serves as an independent check on the von Wartenberg value. This confirmation is important because von Wartenberg, Muchlinski, and Riedler[52] had previously published an extremely erratic value (4.0 kcal) for the heat of formation of formaldehyde.

Values for the heat of formation of formaldehyde at various temperatures can be readily calculated by means of equilibrium data. However, since variation with temperature is less than 1 kcal in the range 18 to 700°C, and since the accuracy of available data does not permit us to express the heat of formation figure in fractions of a kcal, variations from the 28 kcal value may be disregarded as negligible within these limits.

In 1897 Delépine[6] reported that the heat of formation of formaldehyde was 25.4 kcal. This figure, determined indirectly from the heat of combustion of hexamethylenetetramine and the heat of reaction of formaldehyde and ammonia, is naturally less accurate than the later measurements. Recalculated on the basis of Rossini's values, this figure becomes 24.7 kcals.

The free energy of formaldehyde, based on the most accurate data now available, should be taken as -27 kcal per gram mol at 25°C. When this figure is calculated from the I.C.T. value for the free energy of carbon monoxide and the Newton and Dodge equilibrium measurements for the reaction $CO + H_2 \rightleftarrows CH_2O$, at two temperatures, using their expression, $\log_{10} K = (374/T) - 5.431$[26], the value obtained is -26.6 kcal per gram mol[20]. This figure has been corrected with respect to Rossini's recent heat figure for carbon dioxide gas (see above). H. W. Thompson[47] has calculated the variation of the free energy of formaldehyde with temperature with the aid of spectroscopic data, *viz.*, the vibration frequencies reported by Nielsen and Ebers[29]. The fundamental free-energy value employed in this work is an average of values calculated from available data. Free-energy figures given by Thompson[47] for various temperatures from 25 to 1227°C (based on the I.C.T. beta-graphite standard) are as follows:

Temperature		ΔF
(°C)	(°Abs)	(kcal per g mol)
25.16	298.16	-27.17
27	300	-27.16
127	400	-26.6
227	500	-25.2
427	700	-24.4
727	1000	-21.0
1227	1500	-17.3

On the basis of Rossini's recent heat figure for carbon dioxide, the value for ΔF at 25°C would be -27.0 kcal. Free-energy values calculated with the aid of earlier spectroscopic data were reported by Stevenson and Beach[45], whose corresponding value for ΔF at 25°C is -27.1.

Ionization Potentials for Formaldehyde Gas. These potentials were measured by Sugden and Price[46] using the electron impact method. As indicated by breaks in the accelerating potential positive ion curve, they are shown below together with comparative data for acetaldehyde and acetone.

Compound	Ionization Potentials in Volts			
CH_2O	10.8 ± 0.1	11.8 ± 0.2	13.1 ± 0.2	
CH_3CHO	10.4 ± 0.1	11.3 ± 0.1	12.3 ± 0.1	13.5 ± 0.2
CH_3COCH_3	10.2 ± 0.1	11.3 ± 0.1	12.2 ± 0.1	13.6 ± 0.2

The effect of the alkyl constituents in acetaldehyde and acetone is apparently related to induction or charge transfer effects in the molecule. An ionization potential of 11.3 ± 0.5 volts for formaldehyde was reported by Jewitt[17] in 1934.

Absorption Spectra and Chemical Structure of Formaldehyde Vapor. The ultraviolet absorption spectrum of formaldehyde vapor has been studied by Purvis and McClellan[32], Schou and Henri[38], Herzberg[13], Dieke and Kistiakowsky[8], and Price[33]. This spectrum consists of 35 to 40 bands between 3700 and 2500 Ångström units with a maximum absorption at 2935. According to Henri, the bands are classified in eleven groups, the first seven of which contain rotational bands, whereas the others which are less definite are of the predissociated type. Studies by Henri and Schou have shown that the absorption spectrum of a solution of liquid formaldehyde in hexane at −70°C is essentially the same as the spectrum of the monomeric gas, with the exception of a displacement toward the red of 5 Ångström units.

Price[33] has studied the absorption spectrum in the far ultraviolet between 2300 and 1000 Ångströms. From these data, he has calculated the strength of the carbon-oxygen bond in the formaldehyde molecule as equivalent to 164 kcal per mol.

Calculations of the interatomic distances in the Y-shaped formaldehyde molecule, based on measurements of the ultraviolet absorption bands, are shown below, as reported by Dieke and Kistiakowsky, and Henri and Schou. All values are given in Ångström units.

	Values reported by Dieke and Kistiakowsky	Values reported by Henri and Schou
Carbon to oxygen	1.185	1.09
Carbon to hydrogen	1.15	1.3
Hydrogen to hydrogen	1.88	1.38

The carbon-to-oxygen distance has also been calculated by Luis Bru[3] on the basis of electron diffraction data. His value for this figure is 1.15 ± 0.05. According to Torkington[49], resonance in the O^-—CH_2^+ bond accounts for an increase in the H—C—H angle amounting to 3°26′ more than the normal 120° figure. A molecular orbital structure for formaldehyde has been evolved by Mulliken[25] and application of the Orbital Valency Force Field to the formaldehyde molecule is reviewed by Linnett[23].

Dieke and Kistiakowsky photographed the absorption spectrum under high dispersion so that the rotational structure could be carefully analyzed. From these photographs, they calculated the molecular constants for the normal state of the molecule and six vibrational states on the excited electronic levels. Further information on the rotation spectrum based on measurements in the microwave region are reported by Lawrance and co-workers[21, 22] and by Bragg and Sharbaugh[2].

The major infrared absorption bands for formaldehyde as charted by Randall, Fowler, Fuson and Dangl[34] are as follows:

μ	Wave No. Cm^{-1}	Atom Grouping Involved
3.54	2825	CH_2
3.60	2780	CH_2
5.74	1744	CO
6.65	1503	CH_2
7.81	1280	CH_2
8.57	1167	CH_2

The 3.54 μ and 5.74 μ bands are the strongest. These data are based principally on the work of Nielsen and co-workers[27, 28, 30, 14] who have made an exhaustive study in this field supplemented by a comparative study of deutero-formaldehyde, CD_2O. Other studies in the infrared have been reported by Salant and West[36], Titeica[48], Nordsieck[31] and Singer[39]. Salant and West report a number of weak bands at around 1 to 2 μ and Singer reports no appreciable absorption in the region between 10 to 20 μ.

The Raman spectrum of aldehydes[19] shows carbonyl frequencies in the neighborhood of $\Delta \bar{\nu}$ 1720 and additional characteristic frequencies at $\Delta \bar{\nu}$ 510 and 1390 in the simpler aldehydes. In the case of gaseous formaldehyde[15], there is apparently a shift corresponding to $\Delta \bar{\nu}$ 1768.

Properties of Liquid Formaldehyde

As previously stated, liquid formaldehyde polymerizes rapidly even at temperatures of $-80°C$ unless extremely pure. However, even the purest samples show signs of polymerization after about four hours. When the liquid is heated in a sealed tube, polymerization attains almost explosive violence. Satisfactory agents for inhibiting polymerization have not yet been discovered, although Spence[40] states that quinol increases stability to a slight extent.

According to Kekulé[18], the density of the pure liquid is 0.9151 at $-80°C$ and 0.8153 at $-20°C$. The mean coefficient of expansion calculated from these values is 0.00283, a figure similar to the expansion coefficients of liquid sulfur dioxide and liquid ammonia.

The partial pressure of the pure liquid has been recently measured over a wide range of temperatures by Spence and Wild (Table 5)[42]. According to these investigators, the partial pressure of liquid formaldehyde may be accurately calculated with the following equation:

$$\log_{10} P_{atm} = -1429/T + 1.75 \log T - 0.0063T + 3.0177$$

The heat of vaporization of liquid formaldehyde at its boiling point (−19.2°C), as calculated from the partial pressure, is 5.570 kcal per gram mol. Trouton's constant, $\Delta H/T$, is 21.9 entropy units. The critical temperature of the liquid has not been determined because of its rapid polymerization on being warmed.

At low temperatures liquid formaldehyde is miscible in all proportions with a wide variety of nonpolar organic solvents, such as toluene, ether, chloroform, and ethyl acetate. According to Sapgir[37], solutions of liquid formaldehyde in acetaldehyde obey the laws set down for ideal solutions. Although the solutions obtained with the above mentioned solvents are somewhat more stable in most cases than the pure aldehyde, they also precipitate polymer on storage. Liquid formaldehyde is only slightly

TABLE 5. PARTIAL PRESSURE OF LIQUID FORMALDEHYDE[*]

Temperature (°C)	Pressure (mm Hg)	Temperature (°C)	Pressure mm Hg
−109.4	0.95	−65.3	58.95
−104.4	1.85	−64.6	61.65
−98.3	3.60	−63.7	65.20
−95.2	4.85	−55.8	111.0
−89.1	8.68	−54.0	124.7
−85.6	12.25	−49.3	163.1
−78.9	21.02	−40.6	266.6
−78.3	22.11	−39.1	290.6
−71.9	35.40	−34.3	368.9
−68.5	46.43	−22.3	664.3

* Data from R. Spence and W. Wild[42].

miscible with petroleum ether or p-cymene[51]. Polar solvents, such as alcohols, amines, or acids, either act as polymerization catalysts or react to form methylol or methylene derivatives.

In some respects, liquid formaldehyde is comparatively inert chemically. It does not appear to react with elemental sodium, sodium hydroxide, potassium carbonate, or phosphorus pentoxide at or below its boiling point. Many reagents appear to function chiefly as polymerication catalysts. It dissolves iodine to give a yellow-orange solution which polymerizes rapidly. The liquid does not react with ice and may be distilled from it with only slight polymerization taking place. However, water in ether solution reacts readily with formation of the hydrated polymer, paraformaldehyde[51].

Preparation of Liquid Formaldehyde

Anhydrous liquid formaldehyde is most readily prepared by vaporizing alkali-precipitated alpha-polyoxymethylene (page 131), condensing the vapors thus obtained and redistilling the crude liquid condensate. This procedure may be carried out with the apparatus shown in Figure 16a, which is made up principally of "Pyrex" test tubes equipped with side-

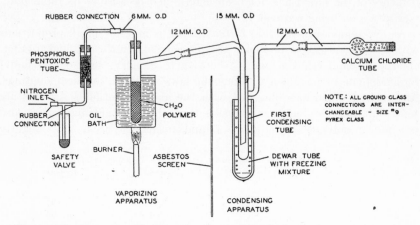

Figure 16a. Apparatus for preparing liquid formaldehyde. Preparation of crude liquid. (From Walker, J. F., *J. Am. Chem. Soc.*, **55**, 2825 (1935).

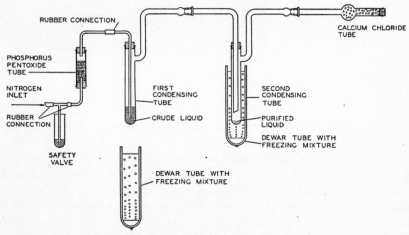

Figure 16b. Apparatus for preparing liquid formaldehyde. Purification step. (From Walker, J. F., *J. Am. Chem. Soc.*, **55**, 2825 (1935).

arms and connected by means of interchangeable ground-glass joints[51]. In preparing the crude liquid (upper figure), the vaporizing tube is charged to about one-half its height with the alkali-precipitated polymer. The system, which must be dried thoroughly before use, is then swept out with dry nitrogen. When this has been done, the oil bath is heated to 150°C and the condensing tube is chilled by means of a bath of solid carbon dioxide and methanol. The polymer is then vaporized in a slow stream of nitrogen which is maintained throughout the run. The temperature of the oil bath is raised gradually until practically all the polymer has vaporized. Formation of polymer on the walls of the tubes leading from the vaporizer to the condensing tubes should be kept to a minimum by heating the exposed portion of these tubes. This is best accomplished by winding the tubes with resistance wire and heating with electricity. The crude liquid, which is opalescent with precipitated polymer, is then purified by distillation in a slow current of nitrogen (Fig. 16b). Dry formaldehyde gas polymerizes very slowly, and heating of the tubes is not necessary in this step. Addition of a small amount of phosphorus pentoxide to the crude liquid before distillation gives a product of extremely high purity. Approximately 10 to 12 cc of pure liquid formaldehyde may be obtained from 20 g of polymer.

Spence and Wild[43] recommended a somewhat more complicated process for preparing liquid formaldehyde, involving evaporation of polymer at reduced pressure and condensation of the product as a solid in a receiver cooled with liquid air. Impurities are removed by a separator inserted between vaporizer and receiver. This separator is a 10-mm tube folded into three U-tubes in a compact block 20-cm high. The lower half of the separator is immersed in a solid carbon dioxide-ether bath, and the upper half is heated electrically so that it can be maintained at 100 to 120°C.

Alkali-precipitated polyoxymethylene is superior to paraformaldehyde for the preparation of liquid formaldehyde because it contains only about 0.1 per cent combined water, whereas paraformaldehyde contains as much as 2 to 4 per cent water even after desiccation. In the liquid formaldehyde process the combined water is set free when the polymer is depolymerized by heat, and forms nonvolatile hydrated polymers during the condensation process. These polymers, which have the type formula $HO \cdot (CH_2O)_n \cdot H$, are left behind when the crude liquid is distilled or in the case of the Spence and Wild procedure, are held back by the separator. Since one mol of water removes many mols of formaldehyde, the use of a polymer containing little or no water is essential if even fair yields of liquid are to be obtained. Spence and Wild employed desiccated paraformaldehyde as a source of formaldehyde in their procedure, and it is possible that this may explain the necessity for a more complicated procedure. Acid-precipitated alpha-polyoxymethylene (page 131) cannot be used for liquid formaldehyde preparation because traces of acid present in the vapor produced from this polymer accelerate polymerization to such an extent that little or no liquid is obtained.

The preparation of liquid formaldehyde has also been described by

Kekulé[18], Delépine[7], Harries[12], Ghosh and Mali[10], Trautz and Ufer[50] and Staudinger.[44]

References

1. Bone, W. A., and Smith, H. L., *J. Chem. Soc.*, **87**, 910 (1905).
2. Bragg, J. K., and Sharbaugh, A. H., *Phys. Rev.*, **75**, 1774–5 (1949).
3. Bru, L., *Anales soc. espan. fis. y quim.*, **30**, 483–5 (1932).
4. Callendar, H. L., *Engineering*, **123**, 147–8 (1927).
5. Carruthers, J. E., and Norrish, R. G. W., *Trans. Faraday Soc.*, **32**, 195–208 (1936).
6. Delépine, M., *Compt. rend.*, **124**, 816 (1897).
7. Delépine, M., *Ann. chim.* (7), **15**, 530 (1898).
8. Dieke, G. H., and Kistiakowsky, G. B., *Proc. Natl. Acad. Sci. U. S.*, **18**, 367–72 (1932); *Phys. Rev.*, **45**, 4–28 (1934).
9. Dobratz, C. J., *Ind. Eng. Chem.*, **33**, 759 (1941).
10. Ghosh, J., and Mali, S., *Quart. J. Ind. Chem. Soc.*, **1**, 41 (1924).
11. Haggard, H. W., *J. Ind. Hyg.*, **5**, 390 (1923).
12. Harries, C., *Ber.*, **34**, 635 (1901).
13. Herzberg, G., *Trans. Faraday Soc.*, **27**, 378–84 (1931).
14. Herzberg, G., "Infrared and Raman Spectra of Polyatomic Molecules," p. 300, New York, D. Van Nostrand Co., Inc. (1945).
15. Hibben, J. H., "The Raman Effect and its Chemical Applications," p. 187, New York, Reinhold Publishing Corp. (1939).
16. "International Critical Tables," Vol. 7, p. 243, New York, McGraw-Hill Book Co. (1930).
17. Jewitt, T. N., *Phys. Rev.*, **46**, 616 (1934).
18. Kekulé, A., *Ber.*, **25**, (3), 2435 (1892).
19. Kohlrausch, K. W. F., and Köppl, F., *Z. physik. Chem.*, **24B**, 370 (1934).
20. Lacy, B. S., (Electrochemicals Dept., E. I. du Pont de Nemours & Co., Inc.), unpublished communication.
21. Lawrance, R. B., and co-workers, *Bull. Am. Phys. Soc.*, **24**, 29 (1949).
22. Lawrance, R. B., *Bull. Am. Phys. Soc.*, **25**, 43 (1950).
23. Linnett, J. W., and co-workers, *Trans. Faraday Soc.*, **45**, 832–44 (1949).
24. Miller, A. R., Jr., (to B. F. Goodrich Co.), U. S. Patent 2,460,592 (1949).
25. Mulliken, R. S., *J. Chem. Phys.*, **3**, 514 (1935).
26. Newton, R. H., and Dodge, B. F., *J. Am. Chem. Soc.*, **55**, 4747 (1937).
27. Nielsen, H. H., *Phys. Rev.*, **46**, 117–21 (1934).
28. Nielsen, H. H., and Ebers, E. S., *J. Chem. Phys.*, **5**, 822–7 (1937).
29. Nielsen, H. H., and Ebers, E. S., *J. Chem. Phys.*, **6**, 311–5 (1938).
30. Nielsen, H. H., and Patty, J. R., *Phys. Rev.*, **37**, 1708 (1931); **39**, 957–966 (1932).
31. Nordsieck, A., *Phys. Rev.*, **45**, 133–4 (1934).
32. Purvis, J. E., and McClelland, N. P., *J. Chem. Soc.*, **101**, 1810–13 (1912); Purvis, J. E., *J. Chem. Soc.*, **127**, 9–14 (1925).
33. Price, W. C., *Phys. Rev.*, **46**, 529 (1934); *J. Chem. Phys.*, **3**, 256–9 (1935).
34. Randall, H. M., Fowler, R. G., Fuson, N., and Dangl, J. R., "Infrared Determination of Organic Structures," pp. 49, 51, 53, 55 and 56, New York, D. Van Nostrand Co., Inc. (1949).
35. Rossini, F. D., *Chem. Rev.*, **27**, 7 (1940).
36. Salant, E. O., and West, W., *Phys. Rev.*, **33**, 640 (1929).
37. Sapgir, S., *Bull. soc. chim. Belg.*, **38**, 392–408 (1929).
37a. Sauterey, R., *Ann. chim.* (12), **7**, 24–72 (1952).

38. Schou, S. A., and Henri, V., *Compt. rend.*, **182**, 1612–4 (1926); **186**, 690–2, 1050–2 (1928); *Nature*, **118**, 225 (1926); *Z. Physik*, **49**, 774–826 (1928). Henri, V., *J. chim. phys.*, **25**, 665–721 (1928); **26**, 1–43 (1929).
39. Singer, W. E., *Phys. Rev.*, **71**, 531–3 (1947).
40. Spence, R., *J. Chem. Soc.*, **1933**, 1193.
41. Spence, R., *J. Chem. Soc.*, **1936**, 649–57.
42. Spence, R., and Wild, W., *J. Chem. Soc.*, **1935**, 506–9.
43. Spence, R., and Wild, W., *J. Chem. Soc.*, **1935**, 338–340.
44. Staudinger, H., "Die Hochmolekularen Organischen Verbindungen," p. 280, Berlin, Julius Springer, 1932.
45. Stevenson, D. P., and Beach, J. Y., *J. Chem. Phys.*, **6**, 25, 341 (1938).
46. Sugden, T. M., and Price, W. C., *Trans. Faraday Soc.*, **44**, 116–123 (1948).
47. Thompson, H. W., *Trans. Faraday Soc.*, **37**, 251–6 (1941).
48. Titeica, R., *Compt. rend.*, **195**, 307–9 (1932); *Bul. Bilunar Soc. Fiz. Romania*, **57**, 31–8 (1933); *Ann. phys.* (*11*), **1**, 533–621 (1934).
49. Torkington, P., *Nature*, **163**, 446 (1949).
50. Trautz, M., and Ufer, E., *J. prakt. Chem.* (*2*), **113**, 105–36 (1926).
51. Walker, J. F., *J. Am. Chem. Soc.*, **55**, 2821–5 (1933).
52. Wartenberg, H. v., Muchlinsky, A., and Riedler, G., *Angew. Chem.*, **37**, 457–9 (1924).
53. Wartenberg, H. V., and Lerner-Steinberg, B., *Angew. Chem.*, **38**, 591–2 (1925).

CHAPTER 3

STATE OF DISSOLVED FORMALDEHYDE

Formaldehyde solutions possess many unusual characteristics. These properties, which at first glance seem almost anomalous, are based on simple chemical facts which, once understood, serve as a key to knowledge of their behavior.

In general, formaldehyde solutions may be classified as belonging to two distinct types: (a) true solutions in which the dissolved formaldehyde is present in monomeric form, and (b) solutions in which the solute is chemically combined with the solvent. Solutions of the former type are obtained with nonpolar solvents. The latter type, which includes the familiar water solutions, are obtained with polar solvents.

Solutions in Nonpolar Solvents

At ordinary temperatures, anhydrous formaldehyde gas is only moderately soluble in nonpolar solvents such as ethyl ether, chloroform, or toluene. Concentrated solutions may be readily obtained by mixing these solvents with liquid formaldehyde at low temperatures; but on warming polymerization and volatilization of the solute take place, leaving only small amounts of the dissolved gas at room temperature. When a solution of liquid formaldehyde in ethyl ether is brought to room temperature in a sealed tube, the formaldehyde polymerizes, giving finely divided polyoxymethylene in practically quantitative yield[34]. Although solutions of formaldehyde in nonpolar solvents are definitely more stable at low temperatures than pure liquid formaldehyde itself, the dissolved formaldehyde gradually polymerizes, even at $-80°C$. As previously pointed out (page 39), the absorption spectrum of a solution of formaldehyde in hexane is essentially identical with that of the anhydrous monomeric gas, indicating that a solution of this type may be regarded as a true solution of formaldehyde, CH_2O. Because of their lack of stability and low formaldehyde content at ordinary temperatures, these solutions are of little practical importance.

Solutions in Polar Solvents

In polar solvents, such as water or methyl alcohol, concentrated solutions of formaldehyde are readily obtained at room temperatures, and precipitation of polymer is evident only at comparatively high concentrations. Furthermore, although the dissolved aldehyde is completely available for chemical reaction, the solutions do not give up their quota of formaldehyde gas even on being warmed. The Solution of anhydrous formaldehyde in these solvents is a highly exothermic reaction, whereas little or no heat is evolved in the case of nonpolar solvents. Because of their commercial importance, aqueous formaldehyde solutions have been the subject of intense research, and the results of this work have given us a comparatively complete picture of their composition, which may well serve as a model for the understanding of solutions in other polar solvents.

Water Solutions of Formaldehyde

Anhydrous formaldehyde gas dissolves in water with the evolution of 15 kcal per gram mol—a heat value which is reported to be independent of the concentration of the solution formed up to approximately 30 per cent[12, 13, 34]. Formaldehyde solutions are also produced by heating formaldehyde polymers with water. Concentrated solutions containing up to and above 50 per cent by weight of formaldehyde are readily obtained. At ordinary temperatures solutions containing over 30 per cent by weight become cloudy on standing and precipitate polymer. Clear liquid formaldehyde-water systems containing up to around 95 per cent formaldehyde are known. However, the temperature necessary to maintain solution clarity and prevent separation of solid polymer increases from room temperature to around 120°C as the solution concentration is increased (page 99). Commericial formaldehyde, which contains 37 per cent formaldehyde by weight and remains clear at ordinary temperatures, contains 8 to 15 per cent methanol, which prevents polymer precipitation.

State of Formaldehyde in Aqueous Solution

Research on the state of formaldehyde in aqueous solution may be summarized as showing that the dissolved aldehyde is present largely as an equilibrium mixture of the monohydrate, methylene glycol, $CH_2(OH)_2$, and a series of low molecular weight polymeric hydrates, or polyoxymethylene glycols, having the type formula $HO \cdot (CH_2O)_n \cdot H$. The state of equilibrium is determined by the temperature and formaldehyde content of the solution. High temperatures and low formaldehyde concentrations favor the monohydrate; low temperatures and high concentrations favor the polymeric hydrates. Polymers having a degree of polymerization greater than the trimeric hydrate, $HO(CH_2O)_3H$, are probably only partially soluble at room temperature and precipitate from solution when formed to any considerable extent. A small quantity of unhydrated monomer, CH_2O, is present but the equilibrium concentration is apparently well under 0.1 per cent even in concentrated solutions.

Information on the state of formaldehyde in the gas phase in equilibrium with aqueous solutions will be reviewed in detail in connection with distillation phenomena (see Chapter 6). Studies in this field demonstrate that the major portion of the formaldehyde vapor consists of unhydrated monomer, probably in equilibrium with a small concentration of methylene glycol vapor.

Methylene Glycol

As previously indicated, formaldehyde monohydrate has the chemical structure of methylene glycol, $CH_2(OH)_2$, the primary member of the homologous series of glycols. Its formation by the reaction of formaldehyde with water is analogous to that of ethylene glycol by the reaction of ethylene oxide with water, as shown below:

$$H_2C{:}O \;+\; H_2O \;\rightarrow\; H_2C(OH)_2$$ 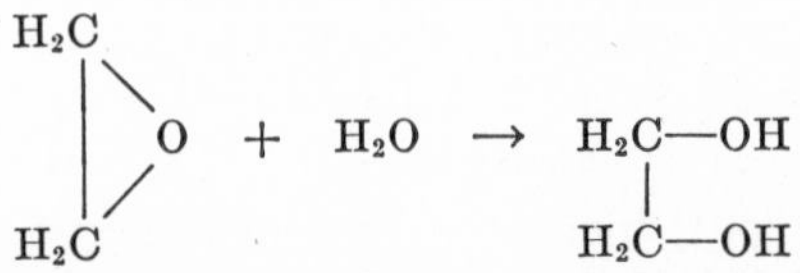
Formaldehyde *Water* *Methylene glycol*

$$
\begin{array}{c}
H_2C \\
\diagdown \\
\Big\rangle O \;+\; H_2O \;\rightarrow\; \; H_2C{-}OH \\
\diagup \qquad\qquad\qquad\quad | \\
H_2C \qquad\qquad\qquad H_2C{-}OH
\end{array}
$$
Ethylene oxide *Water* *Ethylene glycol*

Proof of this interpretation of formaldehyde solvation is found in the fact that, although Auerbach and Barschall[1] have shown by cryoscopic measurements that dissolved formaldehyde is apparently completely monomeric in solutions containing 2 per cent or less dissolved formaldehyde, ultraviolet absorption and Raman spectra of these solutions reveal that the carbonyl grouping (C:O) characteristic of the formaldehyde molecule is almost completely absent. Schou[26] has demonstrated that although a solution of formaldehyde in hexane shows the characteristic spectrum for formaldehyde, ($H_2C{:}O$), aqueous solutions give only a general absorption not differentiated into lines and bands, indicating almost complete absence of unhydrated formaldehyde. Comparison of ultraviolet absorption spectra for solutions of chloral, (CCl_3CHO), and its stable isolable hydrate, ($CCl_3CH(OH)_2$), showed that these spectra were analogous to those obtained with hexane and water solutions of formaldehyde respectively.

Raman spectra of water solutions of formaldehyde also show the substantial or complete absence of monomeric unhydrated formaldehyde. As Nielsen[24] points out, four investigators[18, 22, 23, 31] of the Raman spectra of aqueous formaldehyde, following different methods of procedure, drew the unanimous conclusion that this solution does not give rise to the fundamental frequencies of the simple formaldehyde molecule, CH_2O. According to Hibben[18], the principal lines of the Raman spectra for formaldehyde solution correspond both in intensity and frequency to those of ethylene glycol rather than to those of an aliphatic aldehyde.

Baccaredda[2] reports that velocity measurements of ultrasonic waves in aqueous formaldehyde also give definite indication that the formaldehyde is present in combined form. However, this investigator concludes that methylene glycol itself is associated with water since the ultrasonic velocities in dilute solutions are higher than would be expected for an ideal solution of this compound.

Chemical evidence that formaldehyde solution contains methylene glycol may be found in the fact that formaldehyde solutions are obtained when methylene acetate and methylal are hydrolyzed:

$$CH_3COOCH_2OOCCH_3 \; + \; H_2O \; \rightarrow \; CH_2(OH)_2 \; + \; 2CH_3COOH$$

Methylene acetate   *Water*   *Methylene glycol*   *Acetic acid*

$$CH_3OCH_2OCH_3 \; + \; H_2O \; \rightarrow \; CH_2(OH)_2 \; + \; 2CH_3OH$$

Methylal   *Water*   *Methylene glycol*   *Methanol*

By ether extraction of a 30 per cent formaldehyde solution, Staudinger[28] succeeded in isolating a viscous liquid containing 58 per cent combined formaldehyde. According to theory, pure methylene glycol should contain 62.5 per cent formaldehyde (CH_2O). Staudinger's oil was easily soluble in water and ethyl acetate, but showed only a limited solubility in ether and benzene. On standing, it polymerized quickly to a solid mixture of polymeric formaldehyde hydrates. All attempts to isolate methylene glycol in a pure state have been unsuccessful. It should be noted, however, that the properties of the above-mentioned oil are similar to those of ethylene glycol, which shows similar solubility relationships and is extremely difficult to crystallize.

Unhydrated Formaldehyde in Aqueous Solution

Although Raman spectra give no indication of unhydrated formaldehyde monomer in aqueous solutions, ultraviolet absorption spectra indicate its presence in small concentrations and this is further supported by polarographic investigations.

Bieber and Trümpler[8] have demonstrated the equilibrium between formaldehyde monomer and methylene glycol in dilute solutions and determined the approximate value of the equilibrium constant for the hydration reaction.

$$CH_2O \; + \; H_2O \; \rightleftharpoons \; CH_2(OH)_2$$

This was done by ultraviolet spectrography using a long, 70-cm absorption cell and a 0.87 molal solution of formaldehyde gas in water. Previous measurements by Schou[26], which served as a basis for his calculation that the maximum ratio of unhydrated to hydrated molecules was 1:1200, were made with a 6-cm cell and more concentrated solutions which contained an appreciable proportion of hydrated polymers.

Bieber and Trümpler's results showed that although the 0.87 molal solution gave no selective absorption at ordinary temperatures, the characteristic C=O absorption band became evident on heating and increased in depth with increasing temperature. The absorption maximum indicated a slightly shorter wave length than the 2940 Ångström value reported by Schou for solutions of formaldehyde in hexane. However, this is in agreement with similar measurements by Schou for acetaldehyde in hexane and water which indicate a comparable shift in wave length. By extrapolation from known extinction values of higher aldehydes, Bieber and Trümpler estimated a value of 13.5 for the molar extinction value of formaldehyde in water and used this value to calculate dissolved monomer concentrations using the equation:

$$\log \frac{I_0}{I} = \epsilon.c.d$$

In this equation I_0 indicates the intensity of the light source, I the intensity of the light leaving the solution, ϵ the extinction constant of formaldehyde, c the concentration of the absorbing substance in mols per liter and d the length of the absorption cell in centimeters.

Values for maximum absorption wave length, unhydrated formaldehyde monomer concentration and the equilibrium constant, $(CH_2O)(H_2O)/(CH_2(OH)_2)$ as reported by Bieber and Trümpler for several temperatures are shown in Table 6.

Using these approximate values for the equilibrium constant at various temperatures, Bieber and Trümpler's calculated value for the molar heat of hydration of formaldehyde monomer was 14.6 kcal which is in good agreement with the accepted value of 15 kcal. A calculation of the CH_2O—$CH_2(OH)_2$ equilibrium constant at 20°C gave the approximate value of 10^{-4}.

Absorption spectra studies carried out in the writer's laboratory by Chadwick[9] with 2 to 28 per cent formaldehyde using a Beckmann spectrophotometer and a 10-cm cell showed a specific absorption with a maximum in the 2800 to 2850 Ångström region at 30°C, which moved to the 2900 to 2950 region at 60°C. The solutions also showed considerable nonspecific absorption which was independent of temperature but increased with increasing formaldehyde concentration. After correcting for this factor in the specific absorption peaks, estimations for the concentration of unhydrated monomer using 13.5 as the molar extinction constant for formaldehyde gave the values in Table 7.

Polarographic studies by Bieber and Trümpler[3-8], and Vesely and Brdicka[32] indicate that unhydrated monomer is reduced polarographically, whereas methylene glycol is not. Accordingly, it is concluded that the polarogram for formaldehyde in dilute aqueous solution shows a limiting current which is chiefly dependent on the dehydration rate of methylene glycol. These findings will be reviewed in dealing with the kinetics of solution equilibrium (pages 55–60).

TABLE 6. FORMALDEHYDE MONOMER CONTENT AND METHYLENE GLYCOL-CH$_2$O EQUILIBRIUM CONSTANT FOR 0.87 MOLAL AQUEOUS FORMALDEHYDE[8]

Solution Temperature (°C)	Maximum Absorption Wave Length (Ångström Units)	CH$_2$O Content (Mols per Liter)	Equilibrium Constant K
54	2880	1.37×10^{-3}	1.58×10^{-3}
58	2885	1.86×10^{-3}	2.14×10^{-3}
64	2890	2.62×10^{-3}	3.01×10^{-3}

TABLE 7. PROXIMATE UNHYDRATED MONOMER CONTENT OF AQUEOUS FORMALDEHYDE AT VARYING CONCENTRATIONS AND TEMPERATURES[9]

Total Content Dissolved CH$_2$O (Wt. %)	Monomer Content in Wt. %	
	(30°C)	(60°C)
2	0.001	0.004
5	0.002	0.007
15	0.005	0.017
28.4	0.011	0.026

Polymeric Hydrates and Solution Equilibrium

Evidence of the presence of polymers in aqueous formaldehyde was first obtained by Tollens and Mayer[30] in 1888 and Eschweiler and Grossmann[16] in 1890, who observed that molecular weight values for formaldehyde considerably above 30 were obtained on measuring the freezing point of recently diluted formaldehyde solutions. The fact that these solutions gave the normal value of 30 after standing for some time showed that the polymers gradually dissociated after dilution. In 1905, Auerbach and Barschall[1] made an extensive study of this phenomenon and determined the average molecular weight of dissolved formaldehyde at room temperature (approx. 20°C) in solutions of various concentrations up to approximately 38 per cent. All measurements were made in dilute solutions by the cryoscopic method. The concentrated solutions studied were diluted with ice-water for freezing-point measurement. The success of this method was attributed to the fact that depolymerization on dilution was slow at ice-water temperatures. Only in the case of the most concentrated solutions was difficulty encountered. With such solutions a series of molecular weight measurements were made at various time intervals after dilution. By plotting these values against time and extrapolating to zero time, the molecular weight value for the instant of dilution was estimated. These data are shown in Table 8.

Auerbach found that these cryoscopic data were consistent with a chemical equilibrium between formaldehyde hydrate (methylene glycol) and a trimeric hydrate, (trioxymethylene glycol), as shown in the following equation:

$$3CH_2(OH)_2 \rightleftharpoons HOCH_2OCH_2OCH_2OH + 2H_2O$$

TABLE 8. APPARENT VARIATIONS IN THE MOLECULAR WEIGHT OF DISSOLVED FORMALDEHYDE*

Formaldehyde Concentration (%)	Mol. Wt. Determined by Freezing-point Measurements
2.4	30.0
5.9	32.5
11.2	36.2
14.7	38.5
21.1	43.0
22.6	44.6
24.0	44.9
28.2	48.5
33.8	53.3

* Data of Auerbach[1].

Mass action constants based on this equation showed almost no variational trend with increase of formaldehyde concentration up to 30 per cent, which was not the case with simple polymerization equations such as $2CH_2O \rightleftharpoons (CH_2O)_2$, $3CH_2O \rightleftharpoons (CH_2O)_3$, and $4CH_2O \rightleftharpoons (CH_2O)_4$. The latter equations, first studied by Auerbach, led him to the conclusion that the true state of affairs was something of a mean between the equations involving the trimer and tetramer. For this reason, he experimented with equations involving hydrates and arrived at the equation involving methylene and trioxymethylene glycols.

The equilibrium constant for the methylene glycol-trioxymethylene glycol-water equilibrium as determined by Auerbach is shown below:

$$K = 0.026 = \frac{[CH_2(OH)_2]^3}{[HO(CH_2O)_3H]\cdot(H_2O)^2}$$

Concentrations in the mass action expression are measured in terms of mols per liter.

On the basis of these findings, Auerbach concluded that polymeric hydrates other than the trimeric hydrate $HO \cdot (CH_2O)_3 \cdot H$ are not present in appreciable proportions in solutions containing up to 30 per cent formaldehyde. Deviations encountered in the mass action constants for more concentrated solutions were attributed to the presence of substantial concentrations of higher polymers. If Auerbach's conclusions are accepted, the fraction of dissolved formaldehyde in the methylene glycol form can be calculated from the apparent molecular weights. Values obtained in this way are shown in Table 9.

TABLE 9. RELATION OF FORMALDEHYDE PARTIAL PRESSURES AT 20°C TO APPARENT MOL FRACTION OF METHYLENE GLYCOL AS CALCULATED FROM AUERBACH'S DATA

Grams CH₂O per 100 g Solution	P_{CH_2O} (mm Hg*)	K Values for Henry's Law†	Fraction of CH₂O in form of Methylene Glycol‡	K' Modified Henry's Laws§
2.4	0.11	7.5	0.98	4.6
5.9	0.24	6.8	0.87	4.4
11.2	0.39	5.6	0.72	4.3
14.7	0.49	5.2	0.64	4.5
21.1	0.61	4.4	0.53	4.3
22.6	0.67	4.5	0.49	4.7
24.0	0.70	4.4	0.48	4.6
28.2	0.80	4.1	0.41	5.0
33.8	0.93	4.0	0.33	5.5

Average $K' = 4.5$

* Based on the data of Ledbury and Blair.

† Constant-K $K = \dfrac{P_{CH_2O}}{X_{CH_2O}}$.

‡ Calculated from Auerbach's data.

§ Constant-K' $K' = \dfrac{P_{CH_2O}}{X_{CH_2(OH)_2}}$.

Further confirmation of formaldehyde solution hypotheses can be found in the study of the partial pressure of formaldehyde over its aqueous solutions. Compound formation in solution is manifested by a pronounced negative deviation from Raoult's law. A deviation from Henry's law, which becomes apparent at a 4 per cent concentration and increases with increasing formaldehyde content, bears witness to the steadily growing proportion of polymeric hydrates. The writer[35] has found that this deviation, as shown by Ledbury and Blair's data for the partial pressure of formaldehyde solutions at 20°C (page 92), is in quantitative agreement with Auerbach's data in that the partial pressure of formaldehyde is almost directly proportional to the apparent mol fraction of methylene glycol as calculated from the cryoscopic measurements. As a result of this relationship, the partial pressure of formaldehyde may be expressed by the equation, $P_{CH_2O} = K'X_{CH_2(OH)_2}$, in which P_{CH_2O} equals the partial pressure of formaldehyde, K' is a modified Henry's law constant, and $X_{CH_2(OH)_2}$ represents the apparent mol fraction of methylene glycol. According to our findings, K' equals approximately 4.5 at 20°C when the partial pressure is expressed in millimeters of mercury. The data on which this determination is based will be found in Table 9.

Assuming that a similar relationship prevails at higher temperatures, approximate values for the percentage of dissolved formaldehyde in methylene glycol form can be estimated from partial-pressure values. Calculations of this sort indicate that at 45°C all the dissolved formaldehyde is in the form of methylene glycol at concentrations up to 4 per cent and that at 100°C, this range is extended to about 22 per cent. Rough estimates for the constant K' at 45 and 100°C are 25 and 340, respectively. Calculations of the molal heat of monomer hydration estimated from the effect of heat on the modified Henry's law constant roughly approximate the 15 kcal value for the heat of solution of anhydrous formaldehyde gas[35]. Although the partial pressure of formaldehyde is undoubtedly determined by the concentration of unhydrated, dissolved monomer, it is also a function of the methylene glycol content under equilibrium conditions. This is in agreement with Bieber and Trümpler's demonstration of the methylene glycol-monomer equilibrium.

Skrabal and Leutner[27] disagree with Auerbach's conclusions on the basis that reactions of the type which are believed to take place in formaldehyde solutions normally proceed in the step-wise fashion indicated below:

$$2CH_2(OH)_2 \rightleftharpoons HOCH_2OCH_2OH + H_2O$$

$$HOCH_2OCH_2OH + CH_2(OH)_2 \rightleftharpoons HOCH_2OCH_2OCH_2OH + H_2O$$

etc.

The absence of dioxymethylene glycol molecules would accordingly appear to be highly irregular. However, we may also regard Auerbach's conclusion as a simplifying assumption that is in fair statistical agreement. At concentrations up to about 30 per cent, the apparent trioxymethylene glycol concentration may approximate the combined effect of $(CH_2O)_2 \cdot H_2O$, $(CH_2O)_3 \cdot H_2O$ and small concentrations of higher polymeric hydrates.

Baccaredda's study of the velocity of ultrasonic waves in pure formaldehyde solutions containing up to 50 per cent CH_2O indicate that in the more concentrated solutions, the formaldehyde is largely in the form of polymeric hydrates[2]. Values for ultrasonic velocities at 20°C in solutions corresponding to the compositions $CH_2O \cdot H_2O$, $2CH_2O \cdot H_2O$ and $3CH_2O \cdot H_2O$ were approximated by extrapolation from data obtained: (1) from less concentrated solutions at the former temperature, and (2) from 50 per cent solutions at higher temperatures. These velocity values were then compared with hypothetical figures calculated for ultrasonic waves in the hydrates: $CH_2O \cdot H_2O$, $(CH_2O)_2H_2O$ and $(CH_2O)_3 \cdot H_2O$. As seen in Table 10, the extrapolated experimental values fall between those calculated for the dimeric and trimeric hydrates.

Sauterey[25, 25a] claims that the magnetic susceptibility values for aqueous formaldehyde at various concentrations and pH values indicate that the solution contains an equilibrium mixture of methylene glycol and poly-oxymethylene glycols, and that high pH values favor the momomeric methylene glycol.

Chemical derivatives of polyoxymethylene glycols involving two, three, and four formaldehyde units have been isolated from reactions involving formaldehyde polymers and serve as indirect evidence for the existence of the parent glycols. The diacetate $(CH_3OOCH_2OCH_2OOCCH_3)$ and the

TABLE 10. COMPARISON OF ULTRASONIC VELOCITY DATA FOR FORMALDEHYDE
SOLUTIONS WITH CALCULATED VALUES FOR SIMPLE CH_2O HYDRATES[2]

Composition of Media		Ultrasonic Velocities in Meters per Second	
Ratio CH_2O/H_2O	Per Cent CH_2O	Extrapolated Exp. Value	Calculated for Indicated Hydrate
1	62.5	1633	1534 $(CH_2O \cdot H_2O)$
2	77.0	1639	1603 $(CH_2O)_2.H_2O$
3	83.4	1642	1666 $(CH_2O)_3.H_2O$

dimethyl ether $(CH_3OCH_2OCH_2OCH_3)$ of dioxymethylene glycol are well known. The production of triformals on reaction of an excess of 30 per cent formaldehyde with amines (page 282) may be an indication of the existence of trioxymethylene glycol in these solutions. Higher members of the homologous polyoxymethylene diacetates and dimethyl ethers were isolated and characterized by Staudinger and Lüthy[29] (Chapter 7).

Kinetics of Changes in Solution Equilibrium

As previously pointed out, the state of equilibrium between mono- and polymeric formaldehyde hydrates is determined by the temperature and concentration of the formaldehyde solution. On dilution, the polymeric hydrates depolymerize and a new state of equilibrium is attained. This fact, primarily demonstrated by cryoscopic data, is also supported by thermochemical evidence. The work of Delépine[13] has shown that when a formaldehyde solution is diluted with water there is a gradual absorption of heat following the initial heat of dilution, which is instanteously liberated. The absorption of heat is caused by the gradual depolymerization of the polymeric hydrates.

Neither cyroscopic nor thermochemical measurements afford an accurate means for studying the depolymerization reaction. Fortunately, however, the progress of this reaction can be followed by measuring small changes in refractive index and density which follow dilution of formaldehyde solutions. The changes in refractive index can be accurately measured with a liquid interferometer[19]. The density changes can be measured with a dilatometer[27]. These methods give consistent results which are in substantial agreement with cryoscopic findings and afford an accurate basis for the study of the kinetics of the reactions involved.

According to Wadano and co-workers[33], who made an extended study of formaldehyde solution kinetics based on interferometer measurements, the rate constants for the depolymerization reaction on dilution to low concentrations (3 per cent or less) are those of a monomolecular reaction. Lesser degrees of dilution give rate figures which are influenced by reverse reactions of a higher order. The temperature coefficient of the reaction rate is calculated as 2.7 per 10°C. This means that the rate is almost tripled for a temperature rise of ten degrees. The hydrogen and hydroxyl ion concentrations of the solution have a considerable influence on this rate, an effect which is at a minimum in the pH range 2.6 to 4.3 and increases rapidly with higher or lower pH values. The hydroxyl ion was found to be 10^7 times stronger in its effect than the hydrogen ion. Mean rate measurements for solution having a pH of 4.22 were 0.0126 at 20°C and 0.0218 at 25.5°C. Neutral salts such as potassium chloride have no noticeable influence on the rate. However, methanol and similar substances which lower the ion product of water, lower the rate, as will be seen below:

% Added Methanol	Rate Constant for Dilution from 36.5 to 2.92%
0	0.0126
2	0.0106
4	0.0056

At low temperatures, the depolymerization which follows dilution of formaldehyde is a very slow reaction. A solution in the pH range 2.6 to 4.3 requires *more than 50 hours* to attain equilibrium after dilution from 36.5 to 3 per cent at 0°C.

Wadano points out that these kinetic studies cast considerable light on the mechanism of the depolymerization reaction, showing that it possesses a marked similarity to the hydrolysis of acids and amides, and to the mutarotation of sugar. These reactions also have a pH range of minimum rate and possess activation heats of approximately the same magnitude. The heat of activation for the depolymerization of dissolved polymeric formaldehyde hydrates was calculated by Wadano from rate measurements as 17.4 kcal per gram mol of formaldehyde (CH_2O). The activation heat for ester hydrolysis is 16.4 to 18.5 kcal, amide hydrolysis 16.4 to 26.0 kcal and mutarotation 17.3 to 19.3 kcal. The figures for glucoside splitting are much higher, indicating a fundamental difference. It is highly probable that formaldehyde hydrates are amphoteric in the pH range for minimum reaction rates, but form anions and cations under more acidic or alkaline conditions. Polymerization and depolymerization reactions probably involve these dissociated molecules and ions.

Wadano concludes that the following equilibria occur in dilute formaldehyde and are predominant in the pH ranges indicated:

pH Range	Equilibrium
Below 2.6	$H_2C(OH)_2 \rightleftharpoons H_2C^+\!-\!OH + OH^-$
2.6 to 4.5	$H_2C(OH)_2 \rightleftharpoons H_2C^+\!-\!O^- + OH^- + H^+$
Above 4.5	$H_2C(OH)_2 \rightleftharpoons H_2C(OH)O^- + H^+$

Equilibria of the following sort are postulated for the more concentrated solutions:

pH Range	Equilibrium
Below 2.6	$HOCH_2OCH_2OCH_2OH \rightleftharpoons HOCH_2OCH_2OH_2^+ + OH^-$
2.6 to 4.5	$HOCH_2OCH_2OCH_2OH \rightleftharpoons CH_2^+OCH_2OCH_2O^- + H^+ + OH^-$
Above 4.5	$HOCH_2OCH_2OCH_2OH \rightleftharpoons HOCH_2OCH_2OCH_2O^- + H^+$

With this in mind, the formation of the hydrated trimer in the acid range may be envisaged as a combination of the small concentrations of anion and amphoteric ions with the more abundant cations.

$$HOCH_2O^- + CH_2^+O^- + HOCH_2^+ \rightarrow HOCH_2OCH_2OCH_2OH$$

Formation of insoluble high molecular weight hydrates may take place in a similar manner. However, in this case polymer is precipitated until the equilibrium concentrations of higher polymers do not exceed their solubility saturation value at the prevailing temperature (Cf. Sauterey[25a]).

The kinetics of polymerization reactions in aqueous formaldehyde are probably similar to those involved in the depolymerization that accompanies dilution. Reaction rates are apparently slow and decrease rapidly with decrease in temperature. This is indicated by the fact that a 50 per cent solution that has been solidified or jelled by cooling can be readily clarified by warming, even after storage for several weeks in the neighborhood of 0°C, whereas a similar solution that has been kept at 35°C will not give a clear solution even after prolonged heating. Chilling precipitates the higher and less soluble polymers that are already present in the concentrated solution but depresses the rate of polymerization to the still higher polymers prescribed for equilibrium at the lower temperature. The latter polymers must depolymerize before dissolving at the original solution temperature and this reaction is slow. Furthermore, since heating solid formaldehyde polymers tends to increase the degree of polymerization, clarification is often practically impossible. The influence of temperature on polymerization rates is indicated by Yate's[36] data on the time required for clarification of 50, 60 and 70 per cent solutions after storage at various temperatures. Table 11 shows the time required for complete clarification of a 25-g sample of solution in a 23- by 145-mm test tube after immersion in a 90 to 92°C water bath following 15 days' storage at the indicated temperatures. Table 12 shows the maximum storage period after which clarification can be effected under the same conditions in not over 8 minutes. In all cases, the concentrated solutions were completely clear before subjection to low temperature storage. The minimum clarification

TABLE 11. CLARIFICATION TIME FOR CONCENTRATED FORMALDEHYDE AFTER 15 DAYS' STORAGE AT LOW TEMPERATURES

Concentration of Original CH₂O Solution (%)	Clarification Time in Minutes for 25-g Sample at 90–92°C After Storage at:		
	25–30°C	17–19°C	5–15°C
50	13	7	4
60	29	9	7
70	Over 9 hrs	41	14

TABLE 12. EFFECT OF LOW TEMPERATURE STORAGE TIME ON CLARIFICATION OF CONCENTRATED FORMALDEHYDE

Storage Temperature (°C)	Maximum Storage Time in Days After Which 25-g Sample Can Be Clarified in not Over 8 Minutes at 90–92°C	
	50% Solution	60% Solution
0–7	86	49
17–19	Approx. 15	Approx. 15
35	6	2.4

temperature for these solutions approximated the minimum temperature for storage without precipitation of polymer.

Table 12 gives a rough measure of the effect of temperature on the rate of polymerization. Since polymerization as well as depolymerization varies with pH and, in general, proceeds at a minimum rate in the 2.6 to 4.5 range, variations in the above data will be noted for minor pH variations in these limits; higher and lower pH values cause practically irreversible polymerization on cooling. A 60 per cent solution adjusted to a pH of 6.3 could not be clarified after 80 minutes at 90°C after 2 days' storage at 5 to 15°C using Yates' technique.

A detailed study of the polarographic behavior of aqueous formaldehyde coupled with supporting studies with the oscillograph and ultraviolet absorption spectrum by Bieber and Trümpler[3-6] gives excellent evidence that polarography can supply a measure of the rate of dehydration of methylene glycol to the monomeric aldehyde in these solutions. The half-wave potential of formaldehyde in aqueous solution is dependent upon pH, temperature and formaldehyde concentration but is not a function of the diffusion of methylene glycol. Probability favors the hypothesis that monomeric formaldehyde is the only solution constituent that can be reduced at the dropping mercury electrode. Spectrographic data have demonstrated that the equilibrium concentration of dissolved monomer is extremely small but once consumed at the electrode surface it can be generated from glycol at the controlling rate of the equilibrium reaction. Bieber and Trümpler conclude from the study of oscillographic potential-time curves that a short-lived reduction intermediate is probably formed reversibly and then goes over to the irreversible end product.

Studies of the influence of pH on the formaldehyde polarogram indicate that the dehydration rate is minimal in the 5 to 7 pH range and becomes increasingly rapid as the pH varies from these limits. The effect of temperature on the temperature coefficient of the reaction as estimated by the limiting current indicates that it decreases with increasing temperature and pH until the diffusion of the total dissolved aldehyde becomes the sole controlling factor. At high pH values, the limiting current is a linear function of the total dissolved formaldehyde. A similar relation prevails at lower pH values in well buffered solutions although the limiting current becomes progressively less as this value approaches 7. However, there is no proportionality between concentration and limiting current in unbuffered neutral solutions. This is explained by the fact that hydroxyl ions are formed in the cathodic reduction.

$$CH_2O + 2e^- + 2H_2O \rightarrow -CH_3OH + 2OH^-$$

Trümpler and Bieber point out that these ions accumulate at the electrode raising the local pH. This allows a further increase in current which pushes the pH value still higher so that the effect is essentially autocatalytic. The increase in limiting current as the pH drops below 5 may be explained as hydrogen ion catalysis of the dehydration reaction.

The influence of pH on the dehydration of methylene glycol in aqueous formaldehyde has been directly demonstrated by Trümpler and Bieber[3]. This was done by employing the dynamic procedure employed by Ledbury and Blair (see page 91) for measuring formaldehyde partial pressures. This involves measurement of the formaldehyde content of an inert gas after passage through a scrubber filled with formaldehyde solution. The rate at which formaldehyde is removed by the gas stream is proportional to the formaldehyde partial pressure under equilibrium conditions. However, as the rate of gas flow is increased, solutions of identical concentration but varying pH show different values for the rate of CH_2O removal and these are proportional to the variation of limiting current with pH as measured by the polarograph. This means that at high scrubbing rates, equilibrium cannot be attained and formaldehyde removal is determined by the rate of dehydration of methylene glycol. As would be expected, the variations of formaldehyde removal with pH become increasingly apparent as the rate of scrubbing is increased from 8 liters of nitrogen per hour to 25 liters of nitrogen per hour. This finding also explains the fact that Ledbury and Blair were unable to get satisfactory partial pressure values for formaldehyde solutions at low temperatures with scrubbing rates that proved satisfactory at ordinary temperatures. At these temperatures, the rate of dehydration was too slow to allow the attainment of equilibrium and the scrubbing rate had to be lowered until equilibrium was the controlling factor.

A study of the urea-formaldehyde reaction in aqueous solution by Crowe and Lynch[10] indicates that this reaction is controlled by the dehydration rate of methylene glycol and correlates with polarographic findings. However, in this case the formation of urea anions (NH_2CONH^-) is also a factor.

The hydration reaction rate undoubtedly plays an important part in the removal of formaldehyde by scrubbing from gaseous mixtures. The qualitative effect of pH and temperature on this reaction must of necessity be similar to the effect of these same variables on dehydration although absolute rate values undoubtedly differ. In this case, the rate of solution of monomer and the solubility of the monomer are additional factors.

Thermochemistry of Changes in Solution Equilibria

As previously stated, Delépine[13] reported that concentrated formaldehyde solutions evolve heat instantaneously on dilution, and then gradually absorb heat at a constantly decreasing rate until a final state of equilibrium is attained. The primary heat effect is undoubtedly a normal heat of dilution. However, although this heat quantity can be determined with a fair degree of accuracy, the figures obtained are undoubtedly lowered by the secondary endothermic reaction, probably depolymerization. The approximate quantities of heat immediately evolved on diluting 30 and 15 per cent formaldehyde solutions to 3 per cent, as measured by Delépine, are 0.45 and 0.33 kcal, respectively, per mol of formaldehyde involved.

Since the heat absorption following dilution proceeds slowly, accurate measurement of the net heat change for dilution and establishment of equilibrium is impossible. Accordingly, in the hope of obtaining a comparative measure of this net change for dilution of formaldehyde solutions of various concentrations, Delépine made use of the fact that an equilibrium is obtained instantaneously when formaldehyde is added to dilute caustic (page 178). Making use of this phenomenon, he found that so long as the final concentrations of caustic and formaldehyde in the diluted solutions were identical in each experiment, the quantities of heat evolved on diluting formaldehyde solutions containing one gram mol of formaldehyde with dilute caustic were equivalent, even though the concentrations of the solutions varied from 1.5 to 30 per cent. Heat evolution caused by changes in caustic concentration was shown to be inappreciable under the conditions of experiment, and it was also demonstrated that the Cannizzaro reaction did not take place to any measurable extent. As a result of these findings, Delépine concluded that the final value for the heat evolved when one gram of formaldehyde is dissolved in a formaldehyde solution is independent of concentration up to concentrations of 30 per cent.

Solutions of Formaldehyde in Alcohols

Formaldehyde is readily soluble in alcohols with evolution of heat to give solutions which do not readily give up formaldehyde gas. In these solutions, the dissolved aldehyde is apparently in the form of simple hemiacetals having the type formula, $ROCH_2OH$. In analogy with aqueous solutions, these solvates are probably in equilibrium with polyoxymethylene derivatives, such as $ROCH_2OCH_2OH$, $ROCH_2OCH_2OCH_2OH$, etc., in the more concentrated solutions. The evidence also indicates that the hemiacetals or alcoholates of formaldehyde are more stable than the corresponding hydrates. It is also possible that solution equilibrium is more favorable to the simple hemiacetals.

In 1898, Delépine[11] observed that a concentrated solution of formaldehyde gas in methanol boiled at 96°C, 26°C higher than the normal boiling point of the pure solvent. More recently, it has been noted[34] that when pure dry methanol is added to an approximately equal volume of liquid formaldehyde at −80°C, a clear mobile liquid may be first obtained without noticeable heat evolution. However, on standing a violent reaction (probably hemiacetal formation) takes place; heat is evolved and the mixture solidifies. On warming to room temperature, the solid melts and a clear solution is obtained. Methanol solutions containing well over 50 per cent dissolved formaldehyde do not precipitate polymer on standing at ordinary temperatures. Solutions of this type containing 40 and 55 per cent formaldehyde, respectively, and 10 per cent or less water have recently been offered for sale in commerical quantities. Similar solutions containing 40 per cent formaldehyde in propanol, *n*-butanol and isobutanol have also been advertised.

Approximate heats of solution for gaseous formaldehyde in methanol, *n*-propanol, and *n*-butanol at 23°C[34] are shown below:

Solvent	Δ H (Kcal per Mol CH_2O)
Methanol	15.0
n-Propanol	14.2
n-Butanol	14.9

The heat of solution of formaldehyde gas in water obtained in the same apparatus under similar conditions was 14.8 kcal per mol.

It will be noted from these data that approximately the same quantity of heat is liberated when formaldehyde is dissolved either in alcohols or in water. The heat of polymerization for formaldehyde, although somewhat lower than the 15 kcal value originally reported by Delépine[14], is also of the same order of magnitude. Since the carbonyl linkage in the formaldehyde molecule is saturated with formation of an —O—CH₂—O— grouping in all of these cases, the similarity may be explained by the hypothesis that this is the principal factor in determining the quantity of heat evolved.

The affinity of formaldehyde for methanol is also demonstrated by its utilization as a stabilizer to prevent the precipitation of polymer in concentrated aqueous formaldehyde. This probably indicates the formation of the hemiacetal with consequent lowering of the concentration of the less soluble polyoxymethylene glycols. Also pertinent is the fact that Bieber and Trümpler[7] have found that the formaldehyde wave in the polarogram of aqueous formaldehyde is lowered by addition of methanol.

Vapor density studies by Hall and Piret[17] show that on vaporization, the hemiacetals in alcoholic formaldehyde solutions are less completely dissociated than the methylene glycol from aqueous solutions especially at the lower temperatures. Under identical conditions, the degree of dissociation seems to rise to a maximum with propyl alcohol and then decreases with further increase in the molecular weight of the alcohol involved. Calculations by the same investigators based on the effect of temperature on the vapor equilibrium also indicate that the heat of reaction of normal alcohols containing up to six carbon atoms as well as isopropanol and isobutanol approximates 14.8 kcals per gram mol.

The fact that methanol in aqueous formaldehyde apparently raises the partial pressure of formaldehyde (see pp. 73–74) is probably due to the fact that the partial pressure of the hemiacetal is greater than that of the formaldehyde hydrates. The methods of determining formaldehyde in the gas phase for partial pressure determinations do not differentiate between free and solvated aldehyde.

The relative affinity of formaldehyde for water and alcohols is indicated quantitatively by the data of Johnson and Piret[20] on the partition coefficients of formaldehyde between water and water-insoluble alcohols as compared with similar data involving nonpolar solvents. Alcoholic extracts probably contain an equilibrium mixture of hemiacetal, methylene glycol and water, whereas solvents such as chloroform, ethers and esters probably extract principally methylene glycol. In this work 2.5 and 25 per cent formaldehyde was extracted at 25°C with an equal volume of solvent. Typical results are shown in Table 13. The partition coefficient, K, is the ratio of the concentration of formaldehyde in the organic phase to its concentration in the aqueous phase at equilibrium. A study of the effect of temperature on these coefficients indicated a gradual linear decrease with increasing temperatures between 0 and 50°C. Variations of pH between 3 and 7 showed no appreciable effect. Lower values led to loss of available formaldehyde as formals, whereas higher values resulted in formation of methanol and formates by the Cannizzaro reaction.

Formaldehyde is also soluble in many other polar solvents. Syrupy

TABLE 13. PARTITION COEFFICIENTS OF FORMALDEHYDE SOLUTION AT 25°C[20]

Solvent	Initial Aqueous Solution % CH_2O	K	Per cent Total CH_2O in Org. Phase	CH_2O Concn.		Density	
				Organic Phase	Aqueous Phase	Organic Phase	Aqueous Phase
n-Butyl alcohol	2.455	3.13	80.6	2.060	0.568	0.8552	0.9906
	24.88	1.88	83.5	16.61	8.06	0.9207	1.0120
sec-Butyl alcohol	2.455	1.21	64.1	1.534	1.150	0.8804	0.9745
	24.88	Single Phase		—	—	—	—
n-Amyl alcohol	2.533	3.15	78.5	2.234	0.598	0.8407	0.9985
	24.70	1.67	73.7	17.20	9.07	0.8983	1.0230
sec-Amyl alcohol	2.455	1.03	51.9	1.481	1.207	0.8346	0.9972
	24.88	0.874	55.3	14.12	13.81	0.8858	1.0363
tert.-Amyl alcohol	2.455	0.686	44.7	1.184	1.517	0.8635	0.9884
	24.88	0.851	80.7	13.68	15.16	0.9461	1.0010
n-Hexyl alcohol	2.455	2.84	76.8	2.101	0.621	0.8368	1.0019
	25.31	1.39	67.6	17.01	10.60	0.8900	1.0289
n-Octyl alcohol	2.533	2.26	71.4	2.069	0.765	0.8363	1.0020
	25.42	1.06	61.6	16.07	12.51	0.8743	1.0382
Benzyl alcohol	2.533	3.35	80.4	1.810	0.562	1.0436	1.0028
	25.37	1.78	78.3	15.06	8.86	1.0713	1.0266
Methyl ethyl ketone	2.485	0.192	8.97	0.396	1.781	0.8330	0.9645
	24.50	Single Phase		—	—	—	—
Chloroform	2.559	0.00785	0.768	0.0134	2.509	1.4780	1.0060
	25.13	0.0126	1.23	0.227	24.70	1.4796	1.0792
Diisopropyl ether	2.513	0.0547	5.20	0.181	2.387	0.7260	1.0056
	24.88	0.00856	0.817	0.305	24.18	0.7286	1.0717

liquids obtained by the reaction of formaldehyde with formamide and acetamide[21] may be regarded as solutions of formaldehyde in these solvents or as solutions of methylol amides. In the case of acetamide, the simple methylol derivative has been isolated as a crystalline solid melting at 50 to 52°C[15]. Further information concerning the more stable methylol derivatives will be found in the section devoted to formaldehyde reactions.

References

1. Auerbach, F., and Barschall, H., "Studien über Formaldehyd," Part I, Formaldehyd in wässriger Lösung, pp. 10–23, Berlin, Julius Springer, 1905 (Reprint from *Arb. kaiserl. Gesundh.*, **22**, (3)).

2. Baccaredda, M., *Gazz. chim. ital.*, **78**, 735–42 (1948).
3. Bieber, R., and Trümpler, G., *Helv. Chim. Acta*, **30**, 706–33 (1947).
4. *Ibid.*, pp. 971–90.
5. *Ibid.*, pp. 1109–13.
6. *Ibid.*, pp. 1286–94.
7. *Ibid.*, pp. 1534–42.
8. *Ibid.*, pp. 1860–5.
9. Chadwick, A. F. (Electrochemicals Dept., E. I. du Pont de Nemours & Co., Inc.), unpublished communication.
10. Crowe, G. A., Jr., and Lynch, C. C., *J. Am. Chem. Soc.*, **71**, 3731–3 (1949).
11. Delépine, M., *Ann. de chimie* (7), **15**, 554 (1898).
12. Delépine, M., *Compt. rend.*, **124**, 816 (1897).
13. *Ibid.*, **124**, 1454 (1897).
14. *Ibid.*, **124**, 1528 (1897).
15. Einhorn, A., and Ladisch, C., *Ann.*, **343**, 265 (1906).
16. Eschweiler, W., and Grossmann, G., *Ann.*, **258**, 103 (1890).
17. Hall, M. W., and Piret, E. L., *Ind. Eng. Chem.*, **41**, 1277–86 (1949).
18. Hibben, J. H., *J. Am. Chem. Soc.*, **53**, 2418–9 (1931).
19. Hirsch, P., and Kossuth, A. E., *Ferment-Forschung*, **6**, 302 (1922).
20. Johnson, H. G., and Piret, E. L., *Ind. Eng. Chem.*, **40**, 743–6 (1948).
21. Kalle & Co., German Patent 164,610 (1902).
22. Kohlrausch, K. F. W., and Koppl, F., *Z. physik. Chem.*, **24B**, 370 (1924).
23. Krishnamurti, P., *Indian J. Phys.*, **6**, 309 (1931).
24. Nielsen, H., and Ebers, E. S., *J. Chem. Phys.*, **5**, 823 (1937).
25. Sauterey, R., *Compt. rend.*, **229**, 884–6 (1949).
25a. Sauterey, R., *Ann. de chimie* (*12*), **7**, 1–74 (1952).
26. Schou, S. A., *J. chim. phys.*, **26**, 72–6 (1929).
27. Skrabal, A., and Leutner, R., *Oesterr. Chem. Ztg.*, **40**, 235–6 (1937).
28. Staudinger, H., "Die Hochmolekularen Organischen Verbindungen," p. 249, Berlin, Julius Springer (1932).
29. Staudinger, H., and Lüthy, M., *Helv. Chim. Acta*, **8**, 41 (1925).
30. Tollens, B., and Mayer, F., *Ber.*, **21**, 1566, 3503 (1888).
31. Trumpy, B., *Klg. Norske Videnskap Selskap Shrifter*, **9**, 1–20 (1935).
32. Vesely, K., and Brdicka, R., *Coll. Czech Chem. Communs.*, **12**, 313–32 (1947); *Chem. Abs.*, **43**, 25 (1949).
33. Wadano, M., Trogus, C., and Hess, K., *Ber.*, **67**, 174 (1934).
34. Walker, J. F., *J. Am. Chem. Soc.*, **55**, 2821, 2825 (1933).
35. Walker, J. F., *J. Phys. Chem.*, **35**, 1104–13 (1931).
36. Yates, E. S., (to E. I. du Pont de Nemours & Co., Inc.), U. S. Patent 2,440,732 (1948).

CHAPTER 4

COMMERCIAL FORMALDEHYDE SOLUTIONS

Formaldehyde is principally marketed in the form of aqueous solutions at concentrations ranging from 30 to 50 per cent by weight dissolved formaldehyde. Best known and most commonly encountered is the standard 37 per cent by weight solution containing sufficient methanol to prevent precipitation of solid polymer under ordinary conditions of shipping and storage. Formerly classified as U.S.P. grade[30], this solution has now been dropped from the "United States Pharmacopeia" and is listed in the "National Formulary"[20] as N.F. Formaldehyde Solution or Liquor Formaldehydi. In recent years, increasing quantities of formaldehyde have been sold as low methanol, unstabilized or "uninhibited" solution at concentrations 37 to 50 per cent by weight[12, 22]. These solutions contain approximately 1 per cent or less methanol and must be kept warm to prevent polymer precipitation. Solution of this type containing 30 per cent formaldehyde is stable at ordinary temperatures but is seldom encountered in the United States in commercial usage. As previously pointed out (page 61), solutions of formaldehyde in methanol, propanol and butanol were offered commercially in 1950.

In Europe, the standard 37 per cent by weight formaldehyde was first known under the trade names "Formalin" and "Formol"[23], and was generally described as 40 per cent formaldehyde, since it contained approximately 40 grams of formaldehyde per 100 cc. This volume per cent measure of formaldehyde concentration has now been generally abandoned both in Great Britain and the United States for the more accurate weight per cent measure. Volume per cent figures are unsatisfactory, as concentrations measured in this way vary with temperature and are also influenced by the presence of methanol or other solution stabilizers. Unless otherwise specified, all formaldehyde concentration figures in this book are reported in weight per cent.

Specifications and Purity

In general, commercial formaldehyde solutions are of high purity capable of meeting stringent product requirements. N.F. Formaldehyde Solution[20] is described as containing "not less than 37 per cent formaldehyde (CH$_2$O), with variable amounts of methanol to prevent polymerization." It is normally a clear, colorless solution possessing the characteristic pungent odor of formaldehyde. The following figures are representative of typical specifications for this grade of solution (see also pages 393–394).

Strength: 37.0 to 37.3 per cent CH$_2$O by weight

Methanol content: Normally varies from 6.0 to 15 per cent by weight according to trade requirements

Acidity: Usually about 0.02–0.04 per cent calculated as formic acid (0.1–0.2 cc N KOH per 20 cc)

Iron: Less then 1.0 ppm

Copper: Less than 1.0 ppm

Aluminum: About 3 ppm

Heavy metals: Trace

Organic and inorganic impurities other than those listed above are substantially absent. However, solutions stored for prolonged periods may develop measurable amounts of methylal. The N.F. requirement for acidity[20] is that 20 cc of solution shall not consume more than 1 cc of normal alkali. This would be equivalent to 0.2 per cent formic acid and is well above ordinary commercial specifications. However, some resin manufacturers demand a formaldehyde having an acid content of not more than 0.02 per cent. The total solids content of commercial formaldehyde is usually about 50 ppm or less.

Formaldehyde manufactured by hydrocarbon oxidation sometimes possesses a foreign odor reminiscent of cracked petroleum and may give a yellow or brown coloration when mixed with an equal volume of concentrated sulfuric acid.

Low methanol formaldehyde solutions containing 37 to 50 per cent formaldehyde meet specifications similar to those given for the N.F. product. The acidity of high grade 45 to 50 per cent solutions is usually around 0.05 per cent calculated as formic acid. All formaldehyde solutions are slightly acid and have a pH in the range 2.8 to 4.0.

Physical Properties

Density and Refractivity. The density and refractive index of commercial formaldehyde solutions vary with methanol content and temperature. Table 14, prepared from the data of Natta and Baccaredda[21], gives these figures for 37 per cent formaldehyde solution containing 0 to 20 per cent methanol at 18°C. The influence of minor variations in methanol content on the density of low methanol 37 per cent formaldehyde is shown in Figure 17. Specific gravity data for solutions ranging from 36.8 to 37.6 per cent formaldehyde and containing from 0 to 15 per cent methanol are shown in Table 15. Low methanol 30 per cent formaldehyde has a density of 1.0910 and a refractive index of 1.3676 at 18°C[21].

With the following formula it is possible to calculate the approximate density of formaldehyde solutions containing up to 15 per cent methanol at temperatures in the neighborhood of 18°C.

$$\text{Density} = 1.00 + F\,\frac{3}{1000} - M\,\frac{2}{1000}$$

where F = weight per cent HCHO and M = weight per cent CH_3OH.

TABLE 14. INFLUENCE OF METHANOL CONTENT ON DENSITY AND REFRACTIVE INDEX
OF 37 PER CENT FORMALDEHYDE[21]

Methanol (%)	Density (18°C)	Refractive Index (18°C)
0	1.1128	1.3759
5	1.1009	1.3766
10	1.0890	1.3772
15	1.0764	1.3776
20	1.0639	1.3778

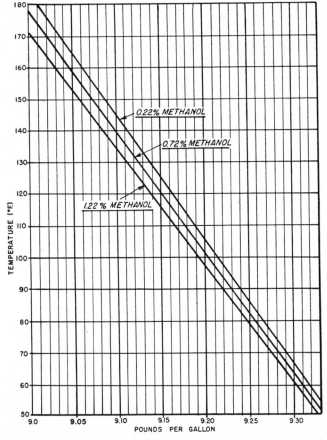

(*Courtesy E. I. du Pont de Nemours & Co.*)

Figure 17. Influence of temperature and methanol concentration on density of low methanol 37% formaldehyde. (Density referred to water at 60° F.)

TABLE 15.—FORMALDEHYDE–METHANOL–WATER SOLUTIONS
Specific Gravities in Air at 25°/25°C
Interpolation Values: 1% CH₃OH = 0.0025 and 1% HCHO = 0.0030

%/wt. HCHO

%/wt. CH₃OH	36.8	36.9	37.0	37.1	37.2	37.3	37.4	37.5	37.6
0	1.1121	1.1124	1.1127	1.1130	1.1133	1.1136	1.1139	1.1142	1.1145
.1	.1118	.1121	.1124	.1127	.1130	.1133	.1136	.1139	.1142
.2	.1115	.1118	.1121	.1124	.1127	.1130	.1133	.1136	.1139
.3	.1112	.1115	.1118	.1121	.1124	.1127	.1130	.1133	.1136
.4	.1110	.1113	.1116	.1119	.1122	.1125	.1128	.1131	.1134
.5	1.1108	1.1111	1.1114	1.1117	1.1120	1.1123	1.1126	1.1129	1.1132
.6	.1105	.1108	.1111	.1114	.1117	.1120	.1123	.1126	.1129
.7	.1102	.1105	.1108	.1111	.1114	.1117	.1120	.1123	.1126
.8	.1100	.1103	.1106	.1109	.1112	.1115	.1118	.1121	.1124
.9	.1098	.1101	.1104	.1107	.1110	.1113	.1116	.1119	.1122
1.0	1.1095	1.1098	1.1101	1.1104	1.1107	1.1110	1.1113	1.1116	1.1119
1.1	.1092	.1095	.1098	.1101	.1104	.1107	.1110	.1113	.1116
1.2	.1090	.1093	.1096	.1099	.1102	.1105	.1108	.1111	.1114
1.3	.1087	.1090	.1093	.1096	.1099	.1102	.1105	.1108	.1111
1.4	.1085	.1088	.1091	.1094	.1097	.1100	.1103	.1106	.1109
1.5	1.1082	1.1085	1.1088	1.1092	1.1095	1.1098	1.1101	1.1104	1.1107
1.6	.1080	.1083	.1086	.1089	.1092	.1095	.1098	.1101	.1104
1.7	.1077	.1080	.1083	.1086	.1089	.1092	.1095	.1098	.1101
1.8	.1075	.1078	.1081	.1084	.1087	.1090	.1093	.1096	.1099
1.9	.1072	.1075	.1078	.1081	.1084	.1087	.1090	.1093	.1096
2.0	1.1070	1.1073	1.1076	1.1079	1.1082	1.1085	1.1088	1.1091	1.1094
2.1	.1068	.1071	.1073	.1076	.1079	.1082	.1085	.1088	.1091
2.2	.1065	.1068	.1071	.1074	.1077	.1080	.1083	.1086	.1089
2.3	.1062	.1065	.1068	.1071	.1074	.1077	.1080	.1083	.1086
2.4	.1060	.1063	.1066	.1069	.1072	.1075	.1078	.1081	.1084
2.5	1.1058	1.1061	1.1064	1.1067	1.1070	1.1073	1.1076	1.1079	1.1082
2.6	.1055	.1058	.1061	.1064	.1067	.1070	.1073	.1076	.1079
2.7	.1052	.1055	.1058	.1061	.1064	.1067	.1070	.1073	.1076
2.8	.1050	.1053	.1056	.1059	.1062	.1065	.1068	.1071	.1074
2.9	.1048	.1051	.1054	.1057	.1060	.1063	.1066	.1069	.1072
3.0	1.1045	1.1048	1.1051	1.1054	1.1057	1.1060	1.1063	1.1066	1.1069
3.1	.1042	.1045	.1048	.1051	.1054	.1057	.1060	.1063	.1066
3.2	.1040	.1043	.1046	.1049	.1052	.1055	.1058	.1061	.1064
3.3	.1037	.1940	.1043	.1046	.1049	.1052	.1055	.1058	.1061
3.4	.1035	.1038	.1041	.1044	.1047	.1050	.1053	.1056	.1059
3.5	1.1032	1.1035	1.1038	1.1041	1.1044	1.1047	1.1050	1.1053	1.1056

TABLE 15.—*Continued*

%/wt. HCHO

%/wt. CH₃OH	36.8	36.9	37.0	37.1	37.2	37.3	37.4	37.5	37.6
3.6	.1030	.1033	.1036	.1039	.1042	.1045	.1048	.1051	.1054
3.7	.1028	.1031	.1034	.1037	.1040	.1043	.1046	.1049	.1052
3.8	.1025	.1028	.1031	.1034	.1037	.1040	.1043	.1046	.1049
3.9	.1022	.1025	.1028	.1031	.1034	.1037	.1040	.1043	.1046
4.0	1.1020	1.1023	1.1026	1.1029	1.1032	1.1035	1.1038	1.1041	1.1034
4.1	.1018	.1021	.1024	.1027	.1030	.1033	.1036	.1039	.1042
4.2	.1015	.1018	.1021	.1024	.1027	.1030	.1033	.1036	.1039
4.3	.1012	.1015	.1018	.1021	.1024	.1027	.1030	.1033	.1036
4.4	.1010	.1013	.1016	.1019	.1022	.1025	.1028	.1031	.1034
4.5	1.1008	1.1011	1.1014	1.1017	1.1020	1.1023	1.1026	1.1029	1.1032
4.6	.1005	.1008	.1011	.1014	.1017	.1020	.1023	.1026	.1029
4.7	.1002	.1005	.1008	.1011	.1014	.1017	.1020	.1023	.1026
4.8	.1000	.1003	.1006	.1009	.1012	.1015	.1018	.1021	.1024
4.9	.0998	.1001	.1004	.1007	.1010	.1013	.1016	.1019	.1022
5.0	1.0995	1.0998	1.1001	1.1004	1.1007	1.1010	1.1013	1.1016	1.1019
5.1	.0992	.0995	.0998	.1001	.1004	.1007	.1010	.1013	.1016
5.2	.0989	.0992	.0995	.0998	.1001	.1004	.1007	.1010	.1013
5.3	.0987	.0990	.0993	.0996	.0999	.1002	.1005	.1008	.1011
5.4	.0984	.0987	.0990	.0993	.0996	.0999	.1002	.1005	.1008
5.5	1.0982	1.0985	1.0988	1.0991	1.0994	1.0997	1.1000	1.1003	1.1006
5.6	.0979	.0982	.0985	.0988	.0991	.0994	.0997	.0000	.1003
5.7	.0977	.0980	.0985	.0986	.0989	.0992	.0995	.0998	.1001
5.8	.0974	.0977	.0980	.0983	.0986	.0989	.0992	.0995	.0998
5.9	.0972	.0975	.0978	.0981	.0984	.0987	.0990	.0993	.0996
6.0	1.0969	1.0972	1.0975	1.0978	1.0981	1.0984	1.0987	1.0990	1.0995
6.1	.0966	.0969	.0972	.0975	.0978	.0982	.0985	.0988	.0991
6.2	.0964	.0967	.0970	.0973	.0976	.0979	.0982	.0985	.0988
6.3	.0961	.0964	.0967	.0970	.0973	.0976	.0979	.0982	.0985
6.4	.0959	.0962	.0965	.0968	.0971	.0974	.0977	.0980	.0983
6.5	1.0957	1.0960	1.0963	1.0966	1.0969	1.0972	1.0975	1.0978	1.0981
6.6	.0954	.0957	.0960	.0963	.0966	.0969	.0972	.0975	.0978
6.7	.0951	.0954	.0957	.0960	.0963	.0966	.0969	.0972	.0975
6.8	.0949	.0952	.0955	.0958	.0961	.0964	.0967	.0970	.0973
6.9	.0947	.0950	.0952	.0955	.0958	.0961	.0964	.0967	.0970
7.0	1.0944	1.0947	1.0950	1.0953	1.0956	1.0959	1.0962	1.0965	1.0968

TABLE 15.—*Continued*

%/wt. HCHO

%/wt. CH₃OH	36.8	36.9	37.0	37.1	37.2	37.3	37.4	37.5	37.6
7.1	.0942	.0945	.0948	.0951	.0954	.0957	.0960	.0963	.0966
7.2	.0939	.0942	.0945	.0948	.0951	.0954	.0957	.0960	.0963
7.3	.0936	.0939	.0942	.0945	.0948	.0951	.0954	.0957	.0960
7.4	.0934	.0937	.0940	.0943	.0946	.0949	.0952	0.955	.0958
7.5	1.0952	1.0935	1.0938	.0941	1.0944	1.0947	1.0950	1.0953	.0956
7.6	.0929	.0932	.0935	.0938	.0941	.0944	.0947	.0950	.0953
7.7	.0926	.0929	.0932	.0935	.0938	.0941	.0944	.0947	.0950
7.8	.0924	.0927	.0930	.0933	.0936	.0939	.0942	.0945	.0948
7.9	.0922	.0925	.0928	.0931	.0934	.0937	.0940	.0945	.0946
8.0	1.0919	1.0922	1.0925	1.0928	1.0931	1.0934	1.0937	1.0940	1.0943
8.1	.0916	.0919	.0922	.0925	.0928	.0931	.0934	.0957	.0940
8.2	.0914	.0917	.0920	.0923	.0926	.0929	.0932	.0933	.0938
8.3	.0912	.0915	.0918	.0921	.0924	.0927	.0929	.0932	.0935
8.4	.0909	.0912	.0915	.0918	.0921	.0924	.0927	.0930	.0933
8.5	1.0906	1.0909	1.0912	.0915	0.0918	1.0921	1.0924	1.0927	.0930
8.6	.0904	.0907	.0910	.0913	.0916	.0919	.0922	.0925	.0928
8.7	.0902	.0905	.0908	.0911	.0914	.0917	.0919	.0922	.0925
8.8	.0899	.0902	.0905	.0908	.0911	.0914	.0917	.0920	.0923
8.9	.0896	.0899	.0902	.0905	.0908	.0911	.0914	.0917	.0920
9.0	1.0894	1.0897	1.0900	1.0903	1.0906	1.0909	1.0912	1.0915	1.0918
9.1	.0892	.0895	.0898	.0901	.0904	.0907	.0910	.0913	.0916
9.2	.0889	.0892	.0895	.0898	.0901	.0904	.0907	.0910	.0913
9.3	.0886	.0889	.0892	.0895	.0898	.0901	.0904	.0907	.0910
9.4	.0884	.0887	.0890	.0893	.0896	.0899	.0902	.0905	.0908
9.5	1.0882	1.0885	1.0888	1.0891	1.0894	1.0897	1.0900	.0903	0.906
9.6	.0879	.0882	.0885	.0888	.0891	.0894	.0897	.0900	.0903
9.7	.0876	.0879	.0882	.0885	.0888	.0891	.0894	.0897	.0900
9.8	.0874	.0877	.0880	.0883	.0886	.0889	.0892	.0895	.0898
9.9	.0872	.0875	.0878	.0881	.0884	.0887	.0890	.0893	.0896
10.0	1.0869	1.0872	1.0875	1.0878	1.0881	1.0884	1.0887	1.0890	1.0893
10.1	.0866	.0869	.0872	.0875	.0878	.0881	.0884	.0887	.0890
10.2	.0864	.0867	.0870	.0873	.0876	.0879	.0882	.0885	.0888
10.3	.0862	.0865	.0868	.0871	.0874	.0877	.0880	.0883	.0886
10.4	.0859	.0862	.0865	.0868	.0871	.0874	.0877	.0880	.0883
10.5	1.0856	1.0859	1.0862	1.0865	1.0868	1.0871	1.0874	1.0877	1.0880

TABLE 15.—Continued

%/wt. HCHO

%/w CH₃OH	36.8	36.9	37.0	37.1	37.2	37.3	37.4	37.5	37.6
10.6	.0854	.0857	.0860	.0863	.0866	.0869	.0872	.0875	.0878
10.7	.0852	.0855	.0858	.0861	.0864	.0867	.0870	.0873	.0875
10.8	.0849	.0852	.0855	.0858	.0861	.0864	.0867	.0870	.0873
10.9	.0846	.0849	.0852	.0855	.0858	.0861	.0864	.0867	.0870
11.0	1.0844	1.0847	1.0850	1.0853	1.0856	1.0859	1.0862	1.0865	1.0868
11.1	.0842	.0845	.0848	.0851	.0854	.0857	.0860	.0863	.0866
11.2	.0839	.0842	.0845	.0848	.0851	.0854	.0857	.0860	.0863
11.3	.0836	.0839	.0842	.0845	.0848	.0851	.0854	.0857	.0860
11.4	.0834	.0837	.0840	.0843	.0846	.0849	.0852	.0855	.0858
11.5	1.0832	1.0835	1.0838	1.0841	1.0844	1.0849	1.0846	1.0852	1.0855
11.6	.0829	.0832	.0835	.0838	.0841	.0844	.0847	.0850	.0853
11.7	.0826	.0829	.0832	.0835	.0838	.0441	.0844	.0847	.0850
11.8	.0824	.0827	.0830	.0833	.0836	.0838	.0841	.0844	.0847
11.9	.0822	.0825	.0828	.0831	.0834	.0837	.0840	.0843	.0846
12.0	1.0819	1.0822	1.0825	1.0828	1.0831	1.0834	1.0837	1.0840	1.0843
12.1	.0816	.0819	.0822	.0825	.0828	.0831	.0834	.0837	.0840
12.2	.0814	.0817	.0820	.0823	.0826	.0829	.0832	.0835	.0838
12.3	.0812	.0815	.0818	.0821	.0824	.0827	.0830	.0833	.0836
12.4	.0809	.0812	.0815	.0818	.0821	.0824	.0826	.0829	.0832
12.5	1.0806	1.0809	1.0812	1.0815	1.0818	1.0821	1.0824	1.0827	1.0830
12.6	.0804	.0807	.0810	.0813	.0816	.0819	.0821	.0824	.0827
12.7	.0802	.0805	.0808	.0811	.0814	.0817	.0818	.0822	.0824
12.8	.0799	.0802	.9805	.0808	.0811	.0814	.0816	.0819	.0822
12.9	.0796	.0799	.0802	.0805	.0808	.0811	.0814	.0817	.0820
13.0	1.0794	1.0797	1.0800	1.0803	1.0806	1.0809	1.0812	1.0815	1.0818
13.1	.0792	.0795	.0798	.0801	.0804	.0807	.0810	.0813	.0817
13.2	.0789	.0792	.0795	.0798	.0801	.0804	.0807	.0810	.0813
13.3	.0786	.0789	.0792	.0795	.0798	.0801	.0804	.0807	.0810
13.4	.0784	.0787	.0790	.0793	.0796	.0799	.0802	.0805	.0808
13.5	1.0782	1.0785	1.0788	1.0791	1.0794	1.0797	1.0800	1.0803	.0806
13.6	.0779	.0782	.0785	.0788	.0791	.0794	.0797	.0800	.0803
13.7	.0776	.0779	.0782	.0786	.0789	.0792	.0795	.0798	.0801
13.8	.0774	.0777	.0780	.0783	.0786	.0789	.0792	.0795	.0798
13.9	.0772	.0775	.0778	.0780	.0783	.0786	.0789	.0792	.0795
14.0	1.0769	1.0772	1.0775	1.0778	1.0781	1.0784	1.0787	1.0790	1.0793

TABLE 15.—*Concluded*

%/wt. HCHO

%/wt. CH₃OH	36.8	36.9	37.0	37.1	37.2	37.3	37.4	37.5	37.6
14.1	.0766	.0769	.0772	.0776	.0779	.0782	.0785	.0788	.0791
14.2	.0764	.0767	.0770	.0773	.0776	.0779	.0782	.0785	.0788
14.3	.0762	.0765	.0768	.0770	.0773	.0776	.0779	.0782	.0785
14.4	.0759	.0762	.0765	.0768	.0771	.0774	.0777	.0780	.0783
14.5	1.0756	1.0759	1.0762	1.0766	1.0769	1.0772	1.0775	1.0778	.0781
14.6	.0754	.0757	.0760	.0763	.0766	.0769	.0772	.0775	.0778
14.7	.0752	.0755	.0758	.0760	.0763	.0766	.0769	.0772	.0775
14.8	.0749	.0752	.0755	.0758	.0761	.0764	.0767	.0770	.0773
14.9	.0746	.0749	.0752	.0756	.0759	.0762	.0765	.0768	.0771
15.0	1.0743	1.0746	1.0749	1.0752	1.0755	1.0758	1.0761	1.0764	1.0767

(*Courtesy of Heyden Chemical Corporation*)

Values for the density of 45 and 50 per cent low methanol formaldehyde based on measurements made in Du Pont laboratories are listed in Table 16.

Although neither density nor refractivity alone affords a criterion of formaldehyde concentration, both formaldehyde and methanol content can be estimated when both of these figures have been measured. The influence of formaldehyde and methanol concentrations on density and refractive index for the ternary system formaldehyde-methanol-water was accurately determined by Natta and Baccaredda. A ternary diagram constructed from their data can be employed for the analysis of solutions containing 0 to 50 per cent formaldehyde and 0 to 100 per cent methanol (Figure 30, page 380).

The fact that some of the early investigators, such as Lüttke[17] and Davis[2], appear to have been unaware of the presence of varying concentrations of methanol in commercial formaldehyde accounts for the apparent lack of agreement in their data. This situation was clarified by Maue[19] and Gradenwitz[5, 6] in 1918. Data showing the density of solutions containing various concentrations of formaldehyde and methanol were published by Gradenwitz at that time.

Dielectric Constant. Dobrosserdov[3] reports the dielectric constant of "Formalin" as 45.0. The solution tested had the specific gravity $D_{15°}^{23.7°} = 1.0775$, indicating a methanol content of about 15 per cent. Acidity was not reported.

Expansion Coefficient. Approximate values for the expansion coefficients for various grades of commercial formaldehyde as calculated from density data are shown in Table 17.

Flash Point. Formaldehyde vapors are combustible within certain limits and explosions can occur when they are mixed with the proper amount of air. Formaldehyde solutions have a definite flash point, which is lowered by the presence of methanol. The flash point of commercial formaldehyde

TABLE 16. DENSITY OF 45 AND 50 PER CENT FORMALDEHYDE CONTAINING APPROXIMATELY 1 PER CENT METHANOL

Temperature (°C)	Formaldehyde (45%)	Concentration (50%)
35	1.129	1.142
45	1.123	1.136
55	1.116	1.129
65	1.109	1.121

TABLE 17. EXPANSION COEFFICIENTS OF COMMERCIAL FORMALDEHYDE SOLUTIONS

Formaldehyde Concentration (%)	Methanol Concentration (%)	Coefficient of Expansion	Temperature Range (°C)
37	1	0.00049	20 to 70
37	8	0.00047	6 to 36
37	10	0.00057	10 to 44
37	12	0.00057	−8 to 44
50	1	0.00062	35 to 65

TABLE 18. FLASH POINT OF COMMERCIAL FORMALDEHYDE SOLUTIONS*

Formaldehyde Content (Wt. %)	Methanol Content (Wt. %)	Flash Point (°C)	(°F)
37.2	0.5	85	185
37.2	4.1	75	167
37.1	8.0	67	152
37.2	10.1	64	147
37.1	11.9	56	133
37.5	14.0	56	132

* Data from Underwriter's Laboratories.

solutions containing various methanol concentrations as determined by the Underwriters' Laboratories using the Tag Closed Tester is shown in Table 18. Flash points for 45 and 50 per cent low methanol formaldehyde measured in the same equipment are approximately 176 and 174°F, respectively.

Partial Pressure of Formaldehyde. The partial pressure of formaldehyde vapor over commercial solutions is increased by the presence of methanol[16]. According to Ledbury and Blair[14], a 10.4 per cent formaldehyde solution containing 61.5 per cent methanol has a formaldehyde pressure of

1.16 mm at 20°C, whereas a 10.4 per cent solution containing no methanol has a partial pressure of approximately 0.37 mm. The work of these investigators indicates that this increase in formaldehyde partial pressure due to methanol grows less with decreasing temperature and becomes practically nil at 0°C[15]. Formaldehyde partial pressures for solutions containing various volume percentages of formaldehyde in which methanol is present in the molar ratio $CH_3OH/CH_2O = 0.13$ are reported[14, 15]. Unfortunately, weight per cent figures are not given.

According to our measurements, the partial pressure of formaldehyde over 37 per cent solution containing 9 per cent methanol is 4.2 mm at 35°C, whereas a 37 per cent solution containing 1 per cent methanol has a partial pressure of 2.7 mm under the same conditions. Partial pressure values for low methanol formaldehyde solutions at various concentrations and temperatures are approximately the same as those reported for pure formaldehyde in Chapter 5 (pages 91–99).

Resistivity. Resistivity of commercial formaldehyde solutions, as measured by means of a glass cell with platinum electrodes at 30°C, was found by the writer to be 35,000 to 37,000 ohms, measured with an A.C. bridge. The solutions examined contained 37 per cent formaldehyde and 8 to 10 per cent methanol. The presence of abnormal concentrations of inorganic salts and formic acid causes the resistivity to drop, whereas high methanol concentrations increase resistivity.

Viscosity. Viscosity values for representative commercial formaldehyde solutions as measured with the Hoeppler Viscosimeter (Du Pont Laboratories) are shown in Table 19. The presence of methanol causes an increase in solution viscosity which increases with the concentration of alcohol in the solution.

Storage of Commercial Formaldehyde

The principal changes which may take place in formaldehyde on storage are as follows:

(a) Polymerization and precipitation of polymer.

(b) The Cannizzaro reaction, involving oxidation of one molecule of formaldehyde to formic acid and reduction of another to methanol.

(c) Methylal formation: $CH_2O + 2CH_3OH \rightarrow CH_2(OCH_3)_2 + H_2O$.

(d) Oxidation to formic acid: $CH_2O + \frac{1}{2}O_2 \rightarrow HCOOH$.

(e) Condensation to hydroxyaldehydes and sugars.

These items are listed in their relative order of importance from a practical standpoint. The changes are all detrimental to product quality but may be avoided or kept at a minimum by maintenance of proper storage conditions. With optimum conditions of storage, commercial formaldehyde will remain unimpaired for long periods of time. In general, proper storage involves the avoidance of temperature extremes and the use of storage tanks constructed of materials which are substantially inert to corrosion by the mildly acidic solution. Low temperatures favor polymer precipitation, high temperatures accelerate the reactions leading to chemical loss of formaldehyde (b, c, d, and e) and improper materials of tank construction result in contamination with foreign materials, some of which catalyze the undesirable chemical reactions.

Precipitation of Polymer. The major factors which determine the precipitation of polymer from formaldehyde solutions are: (1) formaldehyde concentration, (2) concentration of solution stabilizers, if present, (3) temperature, (4) time, (5) hydrogen ion concentration, i.e., pH.

At improper storage temperatures, a formaldehyde solution gradually becomes cloudy and eventually solid hydrated polymer separates as a pre-

TABLE 19. VISCOSITY VALUES FOR COMMERCIAL FORMALDEHYDE

Formaldehyde Concentration (%)	Methanol Concentration (%)	Temperature (°C)	Absolute Viscosity Centipoises
30	—	25	1.87
30	—	60	1.04
37	—	60	1.21
50	—	60	1.82
37	6	25	2.45
37	8	25	2.56
37	10	25	2.58
37	12	25	2.69

cipitate. When this occurs the available formaldehyde in the container does not decrease, but if the polymer is permitted to settle out, the upper layer will be under-strength and the lower layer will be over-strength.

Polymer precipitation can be prevented by maintaining the solution above the minimum temperature at which precipitation takes place. Since this temperature is a function of the formaldehyde concentration as well as the concentration of any solution stabilizer (usually methanol), the solution composition must be known. Minimum temperatures for the prevention of polymer precipitation for representative solutions on long storage are listed in Table 20. However, since the polymerization reactions, which lead to the precipitation of polymers from formaldehyde solutions are slow under normal conditions of pH, exposure to temperatures below the values indicated in the table will not lead to precipitation of polymer if the exposure is not prolonged. For example, 37 per cent low methanol formaldehyde will not precipitate polymer if kept at 70°F for one day, at 80°F for 25 days or at 90°F for sixty days. A 50 per cent solution will remain clear for approximately 25 days at 131°F.

In general, solutions having a pH in the range 2.8 to 4.5 are the most stable with respect to polymer precipitation since polymerization reactions proceed at minimal rates in this range. Solutions in which polymer has precipitated can be clarified by warming if exposure to unfavorable temperatures has been short; after long exposure, clarification is practically impossible. The fundamental chemistry involved in polymer precipitation is reviewed in Chapter 3 (pages 57–58).

Solution Stabilizers. The action of methanol in preventing polymer precipitation in formaldehyde solutions is probably due to the formation of hemiacetals which exist in a state of chemical equilibrium with the hydrated

TABLE 20. MINIMUM STORAGE TEMPERATURES FOR COMMERCIAL FORMALDEHYDE

(Temperatures requisite for long storage periods)

Formaldehyde Concentration (%)	Methanol Concentration (%)	Temperature (°C)	(°F)
30	—	7	45
37	<1	35	95
37	7	16	60
37	10	7	45
37	12	6	43
45	<1	55	131
50	<1	65	149

formaldehyde (methylene glycols) in solutions to which it has been added. These equilibria are illustrated in the following equations:

$$HO-CH_2-OH \; + \; CH_3OH \; \rightleftharpoons \; HO-CH_2-OCH_3 \; + \; H_2O$$

Methylene glycol   *Methanol*   *Formaldehyde*   *Water*
(Formaldehyde hydrate)   *hemiformal*

$$HO-CH_2-O-CH_2-OH \; + \; CH_3OH \; \rightleftharpoons$$

Dioxymethylene glycol   *Methanol*
(Diformaldehyde monohydrate)

$$HO-CH_2-O-CH_2-OCH_3 \; + \; H_2O$$
Diformaldehyde hemiformal   *Water*

The fact that commercial formaldehyde manufactured from methanol usually contained a certain amount of unconverted methanol accounted for the discovery of the value of methanol as a stabilizing agent against polymer precipitation. Its utility, however, was apparently not fully realized until after Auerbach's study of pure formaldehyde solutions in 1905. Gradenwitz[6] reported with an air of novelty in 1918 that too low a methanol content rendered formaldehyde solutions unstable and stated that the German pharmaceutical requirement for density (1.079 to 1.081) assumed a 13 per cent methanol content—a concentration sufficient to maintain a clear solution indefinitely. Ethanol, propanol, isopropanol, glycols, and glycerol also act as stabilizing agents for formaldehyde solutions.

The temperature at which cloudiness begins is influenced by the strength of the solution and the amount of methanol which it contains, as well as by the temperature and the length of exposure. For this reason the amount

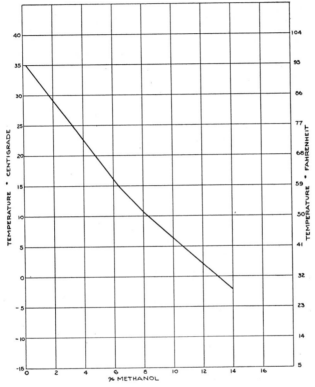

Figure 18. Effect of methanol on storage temperatures at which polymer is precipitated from 37 per cent formaldehyde solutions after 10 to 12 days.

of methanol in commercial formaldehyde solutions is usually controlled by the manufacturer with respect to the probable temperature of exposure. According to Ullmann[28], the methanol content for summer exposure is 7 to 10 per cent and for winter exposure 10 to 13 per cent. In the United States, summer production is usually adjusted to 6 to 10 per cent methanol and winter production to 10 to 12 and sometimes even 15 per cent. Solution containing 10 per cent methanol and 37 per cent formaldehyde will remain perfectly clear for about 4 days at 32°F, about 8 days at 39°F, and almost indefinitely at 45°F. Figure 18 shows the approximate storage temperatures at which polymer is precipitated after 10 to 12 days from 37 per cent solutions containing varying amounts of methanol.

Agents other than alcohols which will prevent polymer precipitation have also been discovered. Neutral or mildly acidic nitrogen compounds such as urea[31] and melamine[26] act as solution stabilizers and have been patented for this purpose. The use of hydrogen sulfide as a solution stabilizer has also been patented[11]. However, to date, methanol is the principal stabilizer in commercial use.

Reactions Leading to the Chemical Loss of Formaldehyde on Storage.— These reactions as previously listed on pages 74–75 include the Cannizzaro reaction, methylal formation, oxidation to formic acid and condensation to hydroxyaldehydes and sugars.

Increase of acidity of formaldehyde solution is due principally to the Cannizzaro reaction. Direct oxidation by reaction with oxygen is slow at ordinary temperatures and is probably negligible. Although the Cannizzaro reaction proceeds most rapidly in alkaline formaldehyde, it can also take place under acidic conditions. Its occurrence in hot formaldehyde acidified with hydrochloric acid has been reported[24]. This reaction explains the increase in formaldehyde acidity on storage when oxidation is obviated by the absence of substantial amounts of oxygen.

Formation of acid and consequent loss of strength due to the Cannizzaro reaction is accelerated by high storage temperatures and the catalytic effect of trace metallic impurities such as iron and aluminum. Concentrated formaldehyde solutions, which must be kept at elevated temperatures to prevent polymer precipitation, suffer appreciably from this cause on protracted storage. For example, 50 per cent formaldehyde solutions having an acidity in the range 0.02 to 0.04 per cent, calculated as formic acid, may develop an acid content in the range 0.07 to 0.11 after 6 weeks at 65°C[32]. Yates has demonstrated that this increase of acidity can be substantially reduced by addition of small concentrations (0.0001 to 0.1 per cent) phosphoric acid and water-soluble phosphates, alone or in the presence of aliphatic amines including hexamethylenetetramine[32]. Similar results are also claimed for mixtures of aliphatic hydroxy, amino and hydroxamino acids with hexamethylenetetramine[33].

Methylal formation is encountered only with N.F. formaldehyde solutions containing appreciable methanol as a solution stabilizer. The reaction is catalyzed by acidic conditions and the presence of metallic salts such as iron, zinc, and aluminum formates. Normally the extent of this reaction is negligible with pure solutions and proper storage conditions.

Condensation to hydroxyaldehydes and sugars may also occur in some instances. Although this reaction takes place chiefly under alkaline conditions, it is apparently catalyzed by very small concentrations of metallic ions such as lead, tin, magnesium and calcium[23], and its occurence cannot be ruled out in commercial formaldehyde solutions even though these solutions are slightly acidic.

Materials of Construction for Formaldehyde Storage. Due to its slight degree of acidity, formaldehyde corrodes some metals, and metallic impurities are introduced on storage in containers constructed from such metals. Recommended materials for formaldehyde storage are glass, stoneware, acid-resistant enamel, stainless steel, rubber and aluminum. When aluminum is first exposed to formaldehyde solution some corrosion takes place, but the metal surface is soon covered with a resistant film of corrosion products. However, aluminum should not be used with hot formaldehyde solutions since the increased temperature accelerates the corrosive effect of formic acid.

Stainless steel (18 per cent chromium, 8 per cent nickel) is the preferred material for most equipment used in handling formaldehyde. Stainless steels #302, 304, 316, and 347 are approximately equivalent if properly welded and normalized. Stainless steel #347 has the advantage of not requiring a heat treatment after welding. Since formaldehyde causes a slight degree of corrosion and is itself discolored by a number of common metals—in particular iron, copper, nickel, and zinc alloys—these materials should be avoided. Data by Teeple[26a] on the effect of metals in formaldehyde storage is tabulated in Table 21.

Fairly resistant storage vessels may also be obtained with the use of coating materials, such as certain asphalt-base paints, phenolformaldehyde resin varnishes, or rubber preparations. However, such storage vessels are not as satisfactory as those constructed from the materials recommended above.

According to Homer[12] reinforced concrete tanks lined with asphalt and acid-resistant brick are highly satisfactory for the bulk storage of formaldehyde. Concrete which has been coated with molten paraffin after thorough drying is also suitable, and will last without relining for a considerable time.

Wood shrinks on exposure to formaldehyde and for this reason is not entirely suitable for use in bulk storage, although properly constructed barrels are used. Formaldehyde solution in contact with wood may extract a small amount of resinous material and thus become discolored. If the solution is also in contact with iron or copper, deep colorations ranging from blue to black may result. This does not impair solution strength nor render it unsuitable for many ordinary uses. If discoloration is objectionable, exposure to these materials should be avoided.

Toxicity: Physiological Hazards and Precautions

Formaldehyde solutions and vapors are highly irritating to the mucous membranes of the eyes, nose and upper respiratory tracts and may also

TABLE 21. EFFECTS OF METALS ON STORAGE STABILITY OF COMMERCIAL FORMALDEHYDE[26a]

Material	Color After Seven Days	Color After Thirty Days	Color After Sixty Days	Acidity After Sixty Days (%)	Indicated Corrosion Rate, Ipy
"Monel".........	Very slight yellow	Slight yellow	Slight yellow	0.055	Less than 0.0001
Nickel...........	OK	Very slight yellow	Very slight yellow	0.037	Less than 0.0001
"Inconel".......	OK	OK	Very slight yellow	0.039	Less than 0.0001
Type 302 S.S.	OK	OK	OK	0.034	Slight gain in weight
Type 304 S.S...	OK	OK	OK	0.037	Slight gain in weight
Type 316 S.S...	OK	OK	OK	0.037	Slight gain in weight
Aluminum 2S...	Very slight yellow	Slight yellow	Slight yellow	0.170	0.0010
Copper..........	OK	Green tint	Greenish-blue sludge	0.083	0.0002
Mild steel.......	Yellow in two days	Yellow	Yellow	0.221	0.0015
Wrought iron...	Dark—8 hrs	Dark yellow	Dark yellow sludge	0.282	0.0027
Blank...........	OK	OK	OK	0.034	

* Initial acidity = 0.034 per cent as formic acid.

(Courtesy, National Association of Corrosion Engineers)

produce cutaneous irritations. However, if reasonable precautions are taken, no difficulties are encountered in handling commercial formaldehyde and the health hazards under proper working conditions are not serious. Formaldehyde is classed as an "economic poison" by the laws of several states, but at the present time, it is not considered a poison in Interstate Commerce Commission regulations. Regular periodic medical examinations are recommended for all workers who handle formaldehyde.

Since formaldehyde is often handled in the form of its solid polymer, paraformaldehyde (pages 119–129), and this product presents the same potential hazards as formaldehyde, the following discussion is equally applicable to the safe handling of this polymer. Paraformaldehyde vaporizes to give monomeric formaldehyde gas and dissolves in water to give formaldehyde solutions. However, since its rate of solution and vaporization are both slow at ordinary temperatures, it is recognized as less hazardous. Nevertheless, it should be remembered that the rate of solution and vaporization are accelerated on heating and both solution and vapor are identical with commercial formaldehyde at the concentrations involved.

Formaldehyde toxicity is manifested on inhalation, contact with the skin or mucous membrane and oral imbibition.

Fortunately, formaldehyde acts as its own warning agent and harmful concentrations in the air are generally made evident by its irritating action on the eyes and nose. The maximum allowable concentration in the air, as generally accepted, has been set by the American Standards Association[1] at 10 ppm by volume or 0.012 mg per liter at 25°C and atmospheric (760 mm) pressure. This concentration is conceded as allowable for exposures not exceeding 8 hours daily[29].

The least detectable odor of formaldehyde is reported at 0.8 ppm and the lowest concentration causing throat irritation at 5.0 ppm[27]. Concentrations of 20 ppm are reported by Haggard[7] to become unendurable, but Flury and Zernik[4] believe that this figure is probably too low. The first symptoms noticed on exposure to small concentrations of formaldehyde vapor are burning of the eyes, weeping, and irritation of the upper respiratory passages. Stronger concentrations produce coughing, constriction in the chest, and a sense of pressure in the head[18]. Inhaling a large quantity of gas may cause sleeplessness, a feeling of weakness, and palpitation of the heart[8]. In some cases, inhalation of formaldehyde may so affect the nervous system as to cause a condition similar to alcoholic intoxication[25].

Formaldehyde acts on the proteins of the body cells both as an irritant and tanning agent. This causes hardening of the skin, diminished secretion, and in some cases a dermatitis usually described as a moist eczema[9, 10]. Although, in general, dermatitis develops only after prolonged contact, some individuals have or develop hypersensitivity. A person who has been previously resistant may become acutely sensitive after one severe exposure. Formaldehyde dermatitis should be treated by a competent physician.

In general, the effect of formaldehyde on the skin may be described as a reddening and infiltration which may include formation of vesicles with superficial necroses or nodules as well as cracking of the hardened surface. In addition, the nails may become soft, fibrous or brownish in appearance and a painful inflammation of the nail bed may result[29].

Formaldehyde solution taken internally severely irritates the mouth, throat, and stomach. The symptoms are intense pain, vomiting, and sometimes diarrhea which may be followed by vertigo, stupor, convulsions, and unconsciousness. In 1925, Kline[13] reported that a search of the literature revealed only 27 cases of formaldehyde poisoning of which 12 proved fatal. In general, fatal doses described in the literature range from 1 to 3 ounces. However, in one instance a few drops apparently caused the death of a child of 3 years and, in another case, recovery attended prompt treatment following ingestion of 4 ounces of 37 per cent solution. Known cases of internal poisoning resulted either from taking formaldehyde by mistake or with deliberate suicidal intent. The severity of the poisoning naturally depends on the amount and concentration of the formaldehyde taken into the body. Dilute solutions cause only slight inflammation, whereas concentrated formaldehyde results in coagulation of the tissue as well as necroses. Injury of the kidneys and gastrointestinal disturbances complicate recovery from poisoning.

Safe Handling and First Aid

Exposure to undesirable concentrations of formaldehyde in the air can usually be avoided by adequate ventilation of working areas. Care must be taken to avoid concentrations exceeding 10 ppm. In cases of spills, only workers who are properly protected should remain in the area, and the spilled formaldehyde should be washed away with large quantities of water or neutralized with dilute ammonia and then washed up with water. In the case of paraformaldehyde, inhalation of the dust should be avoided and spilled polymer should be washed away since it will eventually build up undesirable concentrations of formaldehyde gas.

Operations which require the pouring or handling of formaldehyde or paraformaldehyde in open containers should be carried out under forced-draft hoods which will draw the contaminated air away from the workman. Large quantities of formaldehyde are best handled in completely closed equipment.

Air-line masks, hose masks with externally lubricated hand-operated blowers, self-contained breathing apparatus or canister-type gas masks should be used when men are required to enter an area in which the formaldehyde concentration exceeds permissible limits or where the air is polluted with paraformaldehyde dust. Safety practices recommended by the Manufacturing Chemists' Association for formaldehyde[18] and paraformaldehyde[18] should be consulted in choosing the proper equipment for the conditions of exposure involved. In cases involving formaldehyde concentrations of not more than 2 per cent by volume in atmospheres containing at least 16 per cent by volume of oxygen, canister masks may be used for exposure periods not exceeding 30 minutes. The canister should be changed as soon as the odor of formaldehyde is detected. Bureau of Mines Type B canister (for organic vapors) or Type AB (for atmospheres in which acid gases are also present) should be used for protection. Bureau of Mines Type N canisters should be used in atmospheres containing paraformaldehyde dust. Approved equivalents of these canisters are also acceptable.

Protective clothing such as waterproof boots, gloves, aprons, etc., should be employed as required by the conditions involved. Safety goggles with or without transparent face masks are essential when eye or face protection is necessary. Exposure of skin to formaldehyde solution or polymer should be avoided. Lanolin or similar protective agents rubbed well into the skin are helpful for individuals who are subject to occasional exposure. However, when formaldehyde does get on the skin, the part affected should be washed thoroughly with clean cold water as soon as possible. If formaldehyde solution is splashed in the eyes or if paraformaldehyde dust gets into the eyes, they should be gently flushed or washed with large quantities of water for at least 15 minutes. An eye specialist should be called promptly in such cases.

If formaldehyde has been swallowed, no time should be lost in putting the patient under the care of a competent physician. First aid measures include giving milk freely or one glass of water containing a tablespoonful of ammonium acetate. Following this, the patient should induce vomiting by sticking his finger down his throat. This procedure should be repeated at least three times[18]. Additional milk and raw eggs should be taken following the above treatment in cases where paraformaldehyde has been swallowed. The Manufacturing Chemists' Association[18] recommends production of vomiting with a pint or more of warm salt water (one cupful of salt per quart of water) or warm soapy water. Although gastric lavage is usually desirable, it should be noted that only a physician should use a stomach tube.

Following exposure to formaldehyde gas, the patient should be removed to fresh air and a physician summoned if the condition appears serious. If conscious, the patient should lie down without a pillow. He should be kept quiet and warm and should be given strong coffee or tea. In addition, smelling salts or aromatic spirits of ammonia may be inhaled.

In cases of formaldehyde poisoning in which the patient is unconscious, no attempt should be made to give anything by mouth. The patient should be laid down, preferably on the left side with head low, removing foreign objects such as false teeth, chewing gum, tobacco, etc., from the mouth. Artificial respiration should be employed if the patient is not breathing. In case of cyanosis or shallow breathing, oxygen with carbon dioxide or commercial oxygen should be administered, using artificial respiration if required.

References

1. American Standards Association, "Allowable Concentration of Formaldehyde," Standard, No. 37.16—1944, New York (1944).
2. Davis, W. A., *J. Soc. Chem. Ind.*, **16**, 502 (1897).
3. Dobrosserdov (Dobrosserdoff), Dm., *Chem. Zentr.*, **1911**, I. 956.
4. Flury, F., and Zernik, F., "Schädliche Gase," Berlin (1931).
5. Gradenwitz, H., *Chem. Ztg.*, **42**, 221 (1918).
6. Gradenwitz, H., *Pharm. Ztg.*, **63**, 241 (1918).
7. Haggard, H. W., *J. Ind. Hyg.*, **5**, 390 (1923–4).
8. Hamilton, A., "Industrial Toxicology," p. 197, New York, Harper and Brothers (1934).
9. *Ibid.*, p. 198.
10. Hamilton, A., "Industrial Poisons in the United States," p. 433, New York, Macmillan Co. (1925).
11. Hinegardner, W. F., U. S. Patent 2,002,243 (1935).
12. Homer, H. W., *J. Soc. Chem. Ind.*, **1941**, 213T.
13. Kline, B. S., *Arch. Int. Med.*, **36**, 220 (1925).
14. Ledbury, W., and Blair, E. W., *J. Chem. Soc.*, **127**, 2835–8 (1925).
15. *Ibid.*, p. 33–7.
16. Ledbury, W., and Blair, E. W., "The Production of Formaldehyde by Oxidation of Hydrocarbons," Special Report No. 1, pp. 31–4, London, Dept. of Scientific and Industrial Research, Published under the Authority of His Majesty's Stationery Office, 1927
17. Lüttke, H., *Fischer's Jahresbericht*, **1893**, 512.
18. Manufacturing Chemists' Association, Chemical Safety Data Sheets: "Formaldehyde"-SD-1, "Paraformaldehyde"-SD-6, Washington, D. C. (1950).
19. Maue, G., *Pharm. Ztg.*, **63**, 197 (1918).
20. National Formulary, Ninth Ed. pp. 221–2, Town State Printer (1950).
21. Natta, G., and Baccaredda, M., *Giorn. chim. ind. applicata*, **15**, 273–281 (1933).
22. Neidig, C. P., *Chem. Ind.*, **61**, 214–7 (1947).
23. Orloff, J. E., "Formaldehyde," p. 45, Leipzig, Barth (1909).
24. *Ibid.*, p. 32.
25. Patterson, F., Haines, W. S., and Webster, R. W., "Legal Medicine and Toxicology," Vol. II, pp. 626–7, Philadelphia, W. B. Saunders Co. (1926).
26. Swain, R. C., and Adams, P., (to American Cyanamid Co.), U. S. Patent 2,237,092 (1942).
26a. Teeple, H. O., *Corrosion*, **8**, No. 1, 14–28 (1952).

27. United States Dept. of Labor, Div. of Labor Standards, "Formaldehyde-Controlling Chemical Hazards", Series No. 3, U. S. Government Printing Office, Washington, D. C., 1945.

28. Ullmann, Fr., "Enzyklopädie der Technischen Chemie," 2nd Ed., Vol. 5, p. 415, Berlin, Urban & Schwarzenberg (1928).

29. United States Public Health Service, "Formaldehyde—Its Toxicity and Potential Dangers," Supplement No. 181 to Public Health Reports, Industrial Hygiene Research Laboratory, U. S. Printing Office, Washington, D. C. (1945).

30. "U. S. Pharmacopoeia," 11th Ed., p. 210–211, Easton, Pa., Mack Printing Co. (1936).

31. Walker, J. F., (to E. I. du Pont de Nemours & Co., Inc.), U. S. Patent 2,000,152 (1935).

32. Yates, E. S., (to E. I. du Pont de Nemours & Co., Inc.), U. S. Patent 2,488,363 (1949).

33. *Ibid.*, U. S. Patent 2,492,453 (1949).

CHAPTER 5

PHYSICAL PROPERTIES OF AQUEOUS FORMALDEHYDE

In considering the properties of water solutions of formaldehyde, care must be taken to differentiate between solutions containing only formaldehyde and water and commercial solutions of the type which also contain methanol or other solution stabilizers. Since in some cases investigators have not been aware of the methanol content of commercial formaldehyde, physical data must be critically examined to ascertain the type of solution investigated. The present chapter deals only with aqueous solutions which do not contain solution stabilizers.

In the chapter dealing with the state of dissolved formaldehyde (Chapter 3), it has been pointed out that formaldehyde is hydrated and partially polymerized in aqueous solutions, being present as an equilibrium mixture of the monohydrate, methylene glycol, and polymeric hydrates, polyoxymethylene glycols. The physical properties of formaldehyde solutions are such as would be expected in the light of this situation. They behave like solutions of a comparatively nonvolatile glycol; they do not behave like solutions of a volatile gas.

In the following pages, the known properties of substantially pure aqueous formaldehyde will be reviewed in alphabetical order, with the exception of properties so closely related that they are most readily treated together.

Acidity. Even the purest formaldehyde solutions are slightly acid. In general, the pH of pure aqueous formaldehyde lies in the range 2.5 to 3.5. Wadano[35] has demonstrated that this acidity is due to the presence of traces of formic acid. The titration curves of formaldehyde solutions are the result of the superposition of the curve for formic acid neutralization on the curve for formaldehyde neutralization. However, formaldehyde is so weakly acidic that formic acid can be readily determined by simple titration methods.

The dissociation constant of dissolved formaldehyde as an acid has been determined by Euler[13], Euler and Lovgren[12], Sauterey[31a], and Wadano[35] at various temperatures as shown below:

Temperature (°C)	Dissociation Constant $10^{13} K$	Ref.
0	0.1	13
15	1.2	31a
23	1.62	35
50	3.3	12

Neutral or basic formaldehyde solutions can be obtained only by the addition of a buffer such as borax. Even solutions distilled from magnesium carbonate are reported to have a pH of 4.4[1]. Solutions neutralized with caustic alkalies slowly become acid on standing. This change is due to the Cannizzaro reaction:

$$2CH_2O + H_2O \rightarrow HCOOH + CH_3OH$$

Appearance. Pure formaldehyde solutions are clear and colorless. Cloudiness or opalescence in formaldehyde is caused by polymer precipitation and is discussed in detail under this heading (page 75).

Boiling and Freezing Points. The boiling point of water is comparatively little affected by dissolved formaldehyde at concentrations below 50 per cent. According to Auerbach[3], the boiling point, which is approximately 100°C for a 3 per cent solution, falls off gradually with increasing formaldehyde concentration. The work of the Russian investigators, Korzhev and Rossinskaya[19], indicates that the boiling point may go through a minimum at 11 to 12 per cent concentration, since isotherms for the total vapor pressure of solutions at 97 and 98°C, when plotted against concentration, fall to a minimum at this concentration range and thereafter show a constant increase up to at least 30 per cent concentration. More recent data by Piret and Hall[27] indicate a minimum in the 20 to 30 per cent range. These boiling points as measured with a Swietoslawski-type apparatus[33] designed by Davis[8] are shown in Table 22.

Boiling point values determined by the conventional procedure, in which the vapor-liquid equilibrium is secured by pumping boiling solution over the thermometer bulb, differ considerably from still pot temperatures as measured by Walker and Mooney[36] (Table 29). The latter values were obtained by measuring temperatures below the surface of a gently boiling solution in a flask immersed in an oil bath heated to a temperature which would just maintain ebullition. As will be seen, these figures are much higher than the equilibrium data. Boiling temperatures range from 103.2°C for 50 per cent solution to 112.5°C for 80 per cent formaldehyde. These values are sufficiently reproducible to serve as a rough measure of formaldehyde concentration. For example, the liquid temperature of boiling 47 per cent formaldehyde in Piret and Hall's equilibrium still was 103.1°C at 760 mm[27].

The discrepancy between these figures and the equilibrium boiling point values may be due to a constant degree of superheat. However, since the gas-liquid relations of formaldehyde are complicated by reactions involving the depolymerization and dehydration of solvates, the two procedures may be measuring entirely different phenomena. The still pot boiling point can be regarded as the decomposition temperature of the formaldehyde hydrate composition that is being distilled.

The freezing point of water is lowered by formaldehyde. In the case of dilute solutions this lowering is as would be expected for the dissolved hydrate, methylene glycol, but decreases at higher concentrations because of the formation of the hydrated polymers (page 51). Freezing point measurements on 10 to 25 per cent formaldehyde are given in a PB report[26] covering an investigation of the possibility of concentrating formaldehyde by partial freezing. Solutions were prepared by dilution of 30 per cent formaldehyde and results demonstrated that the freezing point decreased on standing following dilution (Table 23). This effect is caused by depolymerization of

TABLE 22. BOILING POINTS OF AQUEOUS FORMALDEHYDE AS INDICATED BY VAPOR-LIQUID EQUILIBRIUM TEMPERATURE[27]

Concentration of CH_2O	Pressure 740 mm	760 mm
zero	99.3	100.0
2.57	98.95	99.75
4.46	98.85	99.60
7.98	98.65	99.35
12.62	98.45	99.10
15.5	98.35	99.00
19.8	98.35	98.95
22.65	98.35	98.95
25.35	98.35	98.95
31.1	98.35	98.95
41.5	98.40	99.00
43.8	98.45	99.05
45.8	98.55	99.15
47.0	98.60	99.20
47.2	98.65	99.25
49.3	98.80	99.40
51.5	99.10	99.70
53.1	99.40	100.00

polymeric hydrates. The ice separating from these solutions was substantially free of formaldehyde. Information on more concentrated solutions is lacking, since polymer precipitation takes place on chilling, before the true freezing point is attained.

Density and Refractivity. The density and refractivity of pure formaldehyde solutions bear a simple linear relation to formaldehyde concentration. Differences caused by variations in the relative proportions of simple and polymeric formaldehyde hydrates in solutions of the same concentration have practically no effect on these figures as ordinarily measured. Such variations can be measured only by the extremely sensitive methods of dilatometry and interferometry.

The relation of density to formaldehyde content for solutions containing up to 37 per cent CH_2O was determined by Auerbach and Barschall[2] for pure solutions prepared by dissolving gaseous formaldehyde in distilled water. Their results are shown in Table 24. Formaldehyde concentration is reported by volume as well as weight. It will be noted that the weight figures are lower than the volume figures. This disparity becomes even greater in the case of solutions containing methanol.

Earlier density data, such as those obtained by Lüttke[22] and Davis[9], were apparently made with solutions containing small percentages of methyl alcohol and are consequently misleading. Values for the density of pure

TABLE 23. FREEZING POINT OF 10 TO 25 PER CENT FORMALDEHYDE PREPARED BY DILUTION OF 30 PER CENT SOLUTION[26]

Solution Concn. After Dilution	Freezing Point (Time in Hours After Dilution)					
	0	1	2	3	6	24
25	−15.45	−16.15	−16.30	−16.40	−16.45	−16.45
20	−11.10	−12.20	−12.50	−12.70	−12.90	−12.95
15	−7.45	−8.70	−9.05	−9.25	−9.45	−9.50
10	−4.60	−5.60	−6.05	−6.25	−6.25	−6.30

TABLE 24. DENSITY OF AQUEOUS FORMALDEHYDE*

Formaldehyde Concentration		Density at 18°C
Weight Per Cent g CH₂O/100 g Solution	Volume Per Cent g CH₂O/100 cc Solution	
2.23	2.24	1.0054
4.60	4.66	1.0126
10.74	11.08	1.0311
13.59	14.15	1.0410
18.82	19.89	1.0568
23.73	25.44	1.0719
27.80	30.17	1.0853
34.11	37.72	1.1057
37.53	41.87	1.1158

* Data from Auerbach and Barschall[2].

formaldehyde solutions for various volume per cent concentrations up to 41 per cent and temperatures from 20 to 40°C are also reported by Datar[7].

A recent study of the density and refractivity of formaldehyde solutions by Natta and Baccaredda[24] confirms Auerbach's density measurements, but disagrees with the earlier refraction data of Reicher and Jansen[28], and Stutterheim[32]. Density and refraction data for solutions containing up to 50 per cent formaldehyde as determined by Natta and Baccaredda are shown in Table 25.

The density of formaldehyde solutions in the neighborhood of 18°C can be calculated with a fair degree of accuracy by the following formula:

$$D = 1.000 + 3\,\frac{W}{1000}$$

in which W is equivalent to the formaldehyde concentration in weight per cent. The following formulas based on the above equation are useful for converting volume per cent figures (V) to weight per cent (W) and *vice versa*. Their accuracy is naturally limited to temperatures around 18°C.

$$W = \frac{\sqrt{1 + 0.012V} - 1}{0.006}$$

$$V = W + 0.003W^2$$

The temperature coefficient (dD/dt) in the ordinary temperature range (15 to 30°C) is approximately 0.0002 for a 15 per cent solution and 0.0004 for a 47 per cent solution. Change of refractive index with temperature (dN/dt) for the same conditions is approximately 0.00015 for a 17 per cent

TABLE 25. DENSITY AND REFRACTION OF AQUEOUS FORMALDEHYDE*

Formaldehyde Concentration (g/100 g solution)	Density at 18°C	Refractive Index $n_D^{18°}$
0	0.9986	1.3330
5	1.0141	1.3388
10	1.0299	1.3445
15	1.0449	1.3504
20	1.0600	1.3559
25	1.0757	1.3617
30	1.0910	1.3676
35	1.1066	1.3735
40	1.1220	1.3795
45	1.1382	1.3857
50	1.1570	1.3925

* Data from Natta and Baccaredda[24].

solution and 0.00020 for a 38 per cent solution[24]. Values for these coefficients at higher temperatures are also reported by Natta and Baccaredda[24].

Freezing Points. See page 88.

Heat of Dilution. According to Delépine[11], when formaldehyde solutions are diluted, heat is evolved immediately and then gradually absorbed. On diluting 30 per cent and 15 per cent solutions to 3 per cent, 0.45 and 0.33 kcal per mol CH_2O, respectively, are liberated.

Heat of Formation and Free Energy. The heat of formation of aqueous formaldehyde is equivalent to the heat of formation of the anhydrous gas plus its heat of solution in water. As has been previously pointed out (page 37), the most acceptable value for the heat of formation of anhydrous formaldehyde at 18°C is 28 kcal per mol. According to Delépine[10], the heat of solution of formaldehyde gas in water is 15 kcal per mol. Since the thermal energy of dissolved formaldehyde is reported to be independent of concentration up to approximately 30 per cent (page 61), the heat of formation of dissolved formaldehyde at concentrations of 30 per cent or less may be taken as approximately 43 kcal per mol.

It is not possible to assign an exact value to the free energy of dissolved formaldehyde on the basis of data now available. The free energy change for the solution of monomeric formaldehyde gas in water at 25°C was estimated by Van de Griendt[25] as equivalent to -4.92 kcal per gram mol for a molal solution (approximately 3 per cent CH_2O). This calculation is based on partial pressure figures derived from Ledbury and Blair's data, with the assumption that Henry's law is followed by solutions containing up to 6 per cent formaldehyde. Since slight deviations from Henry's law become apparent at a 4 per cent concentration (page 53) and increase with increasing concentrations, Van de Griendt's value can be regarded only as an approximation. However, if we employ this figure in conjunction with the -27 kcal figure for the free energy of monomeric formaldehyde gas at 25°C, an approximate value of -32 kcal per mol is obtained for the free energy of dissolved formaldehyde at a concentration of 3 per cent and a temperature of 25°C.

Magnetic Properties. According to Good[16], the molecular magnetic susceptibility of dissolved formaldehyde varies with concentration and is not in agreement with the calculated value for CH_2O. These values serve as additional evidence for the theory that formaldehyde solution contains the monohydrate, methylene glycol, $CH_2(OH)_2$, in chemical equilibrium with polymeric hydrates. Meslin[23] reports a specific magnetic susceptibility (x) of -0.62×10^{-6} cgsm for dissolved formaldehyde solution but does not report the exact conditions of this determination. More recent measurements by Sauterey[31] gave a value of -0.50×10^{-6} cgsm for methanol-free solutions containing 3.7 to 32.2 g dissolved formaldehyde per 100 cc. These solutions were prepared by the hydrolysis of paraformaldehyde.

Partial Pressure. The partial pressure of formaldehyde over its pure aqueous solutions was studied by Auerbach and Barschall[4, 5], Ledbury and Blair[21], Korzhev and Rossinskaya[19], Piret and Hall[27], and Walker and Mooney[36]. As previously pointed out (pages 53–54), deviations of formaldehyde partial pressures from those which would be expected from Raoult's and Henry's laws may be explained by the hydration of dissolved formaldehyde and the increasing proportion of polymeric hydrates at low temperatures and high concentrations.

The partial pressure data of Ledbury and Blair[21] cover solutions containing up to 37 per cent formaldehyde at temperatures of 0 to 45°C. They were determined by measuring the formaldehyde content of air which had been passed through scrubbers containing solutions of definite temperature and concentration at equilibrium conditions. Formaldehyde concentrations were reported by these investigators in volume per cent. To conform with present American practice, we have converted their figures to the weight per cent basis in Table 26. The formaldehyde content of air in equilibrium with solution is given in the same table in terms of mg

TABLE 26. PARTIAL PRESSURE AND AIR SATURATION DATA
FOR AQUEOUS FORMALDEHYDE*

Formaldehyde Concentration (g CH₂O/100 g)	(g CH₂O/100 cc)	Partial Pressure of Formaldehyde (mm)	Mg Formaldehyde per Liter Saturated Air
Temperature = 0°C			
7.97	8.09	0.056	0.095
15.0	15.68	0.102	0.166
19.4	20.63	0.118	0.201
28.6	31.25	0.157	2.265
Temperature = 20°C			
9.25	9.52	0.340	0.59
18.6	19.7	0.575	1.01
27.2	29.5	0.780	1.39
28.6	31.1	0.795	1.40
36.2	40.2	1.025	1.75
Temperature = 35°C			
1.08	1.09	0.166	0.27
5.10	5.15	0.695	1.13
11.4	11.8	1.29	2.06
18.3	18.6	1.80	2.87
19.7	20.8	1.94	3.17
28.6	31.0	2.48	4.27
35.9	39.5	2.81	4.58
Temperature = 45°C			
10.5	10.8	2.30	3.77
19.4	20.4	3.79	6.17
27.1	28.75	4.72	7.70
35.5	39.2	5.60	9.12

* Data of Ledbury and Blair.

of formaldehyde per liter. The relation of these partial pressure values to concentration is shown graphically in Figure 19.

The following formula developed by Ledbury and Blair may be employed for expressing their experimental results for partial pressures of formaldehyde over 10 to 40 per cent solutions at various temperatures. Although presumably accurate only within the range of their measurements, it gives the correct order of magnitude for values at temperatures up to 100°C.

$$\log P_{CH_2O}^{mm} = \alpha - 2905/T$$

In this formula, T is equivalent to absolute temperature. Alpha (α) varies with concentration as shown below:

Formaldehyde Concn. (g/100 g Solution)	α (Values Calculated for Weight Per Cent)	Formaldehyde Concn. (g/100 cc Solution)	α (Values given by Ledbury and Blair)
10	9.48	10	9.47
20	9.71	20	9.70
30	9.83	30	9.81
40	9.89	40	9.87

Lacy[20] has recently demonstrated that partial pressure values for 10 to 40 per cent formaldehyde solutions can also be expressed as an empirical

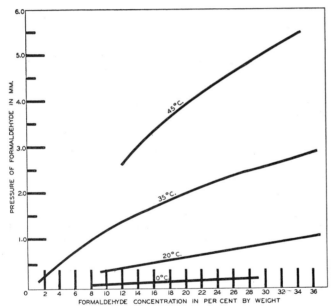

Figure 19. Partial pressure of formaldehyde over aqueous solutions at 0°–45°C. Based on data of Ledbury and Blair.

function of formaldehyde concentration in per cent by weight, W, and absolute temperature, T, by a modification of the above formula. This saves interpolation for alpha value at odd concentrations. Variants of the Lacy equation for calculating partial pressure in millimeters of mercury or atmospheres are shown below:

$$\log P_{CH_2O}^{mm} = 9.942 - 0.953 \, (0.488)^{W/10} - 2905/T$$

$$\log P_{CH_2O}^{atm} = 7.061 - 0.953 \, (0.488)^{W/10} - 2905/T$$

This equation appears to give satisfactory values for solutions of all concentrations below 60°C. For temperatures up to 100°C, its accuracy is limited to concentrations below about 20 per cent.

Auerbach[3] determined partial pressures for formaldehyde solutions at 18°C by the same method employed by Ledbury and Blair. However, these data are limited in scope, show some internal ambiguities, and are lower than other reported values.

Korzhev and Rossinskaya[19] measured partial pressure values for formaldehyde solutions at ordinary temperatures by analyzing the formaldehyde content of the condensate obtained by distilling solutions of known concentrations at reduced pressures. The values they obtained at these temperatures are higher than those of Ledbury and Blair.

Partial pressure measurements for boiling formaldehyde solution were determined by Auerbach and Barschall[4], Korzhev and Rossinskaya[19] and Piret and Hall[27] from distillation data. These figures are shown together with those obtained from the Lacy equation in Table 27.

Approximate values for the partial pressure of formaldehyde and water vapor over solutions containing up to 80 per cent formaldehyde are shown

TABLE 27. PARTIAL PRESSURE OF FORMALDEHYDE SOLUTIONS AT BOILING POINT
UNDER ORDINARY PRESSURE

Auerbach and Barschall[4]				Korzhev and Rossinskaya[19]			
CH_2O Concn.	Total Pressure	Temp. (°C)	P_{CH_2O}	CH_2O Concn.	Total Pressure	Temp. (°C)	P_{CH_2O}
2.4	734	99.1	8.9	4.55	735	99.0	23.4
3.1	765	100.1	13.2	4.71	753	99.0	24.6
5.1	736	99.0	19.9	7.06	735	98.5	39.0
5.9	765	100.0	23.1	7.03	753	98.8	43.6
8.2	737	99.0	28.1	11.74	753	98.5	54.4
9.6	765	99.7	34.0	17.34	735	97.5	81.1
10.0	740	98.9	38.5	17.50	753	98.5	85.1
15.2	740	98.8	53.7	19.51	753	98.5	88.9
16.8	740	98.8	61.5	19.54	735	97.5	87.2
19.6	763	99.3	76.5	25.90	735	97.5	109.5
21.6	763	99.2	86.5	25.90	753	98.5	114.5
24.7	767	99.2	101.3				
28.4	763	99.1	106.7	Calculated from Lacy Equation			
29.3	764	99.1	100.0				
32.2	752	99.0	90.3	10.0		99	47
32.7	764	99.1	95.3	20.0		99	81
34.4	763	99.1	97.2	30.0		99	105
36.4	752	98.9	109.9	40.0		99	120
36.1	763	99.1	112.6				
40.9	763	99.1	117.1				

TABLE 27—*Continued.*

Piret and Hall[27]

CH₂O Concn.	Total Pressure	Wt. % CH₂O in Vapor	P_{CH_2O}	
			Calc'd. for Indicated Pressure	Corrected for 760 mm Tot. Pressure
3.88	740	3.66	16.5	17.5
3.89	760	3.84	17.8	17.8
3.95	740	3.68	16.6	17.5
4.46	733	4.14	18.5	20.5
7.93	736	7.30	33.2	35.5
7.95	743	7.40	34.0	35.5
7.98	760	7.69	36.2	36.2
8.00	740	7.30	33.5	35.5
12.1	741	10.6	49.3	52.5
15.3	745	13.2	62.3	65.0
19.5	740	16.4	78.0	81.5
19.75	742	16.6	79.4	82.5
20.1	737	16.95	80.3	84.0
22.45	760	19.2	95.0	95.0
22.55	744	19.0	92.0	95.0
25.85	731	21.45	103.0	108.5
29.9	738	23.85	116.5	120.5
30.75	760	24.9	126.0	126.0
31.1	737	24.75	122.0	127.0
31.95	750	25.3	126.5	129.0
35.65	745	27.4	137.5	141.0
38.8	750	29.1	148.0	150.5
41.6	760	30.75	160.0	160.0
42.0	733	30.5	152.5	160.5
47.25	760	33.3	175.0	175.0
47.5	743	33.1	170.0	175.5
49.8	736	34.0	174.0	181.0

in Table 28. These figures are based on measurements made by Walker and Mooney[36] plus Ledbury and Blair's data for the lower concentrations. The tabular values were obtained by plotting logarithms of partial pressure values against the reciprocal of the absolute temperature and reading figures from the approximately linear curves obtained in this way. The accuracy for the formaldehyde partial pressure values at the higher concentrations is probably approximately ±2 per cent. The partial pressure values for water are presented only as rough approximations.

In view of the difficulties encountered in handling solutions containing over 50 per cent formaldehyde, the accuracy of the values for solutions in this range is limited. Equilibrium conditions are not easily attained for exact measurements and the usual equilibrium stills are not readily adaptable for handling solutions which solidify on cooling and are, in some cases, not readily reliquefied on heating. Still pot temperatures and partial pressure values for boiling formaldehyde were obtained by slow distillation

TABLE 28. APPROXIMATE PARTIAL PRESSURE VALUES FOR FORMALDEHYDE AND WATER OVER AQUEOUS FORMALDEHYDE SOLUTIONS

°C. Temp.	P_{H_2O}	FORMALDEHYDE CONCENTRATION IN PERCENT BY WEIGHT							
		10%	20%	30%	40%	50%	60%	70%	80%
20	18	0.4 / 17.6 / 18.0	0.7 / 17.3 / 18.0	0.9 / 17.1 / 18.0	1 / 17 / 18	1 / 17 / 18			
25	24	1 / 23 / 24	1 / 23 / 24	1 / 23 / 24	1 / 23 / 24	1 / 23 / 24			
30	32	1 / 31 / 32	1 / 31 / 32	2 / 30 / 32	2 / 30 / 32	2 / 29 / 31			
35	42	1 / 41 / 42	2 / 40 / 42	3 / 39 / 42	3 / 37 / 40	3 / 36 / 39	4 / 34 / 38	4 / 33 / 37	
40	55	2 / 53 / 55	3 / 52 / 55	4 / 50 / 54	4 / 49 / 53	5 / 47 / 52	5 / 45 / 50	6 / 42 / 48	
45	72	2 / 70 / 72	4 / 67 / 71	5 / 65 / 70	6 / 62 / 68	7 / 59 / 66	8 / 56 / 64	8 / 54 / 62	
50	93	3 / 90 / 93	6 / 86 / 92	7 / 83 / 90	8 / 80 / 88	9 / 76 / 85	11 / 71 / 82	11 / 67 / 78	
55	118	4 / 112 / 116	7 / 108 / 115	10 / 103 / 113	11 / 100 / 111	12 / 96 / 108	14 / 90 / 104	15 / 84 / 99	
60	149	6 / 141 / 147	10 / 135 / 145	13 / 130 / 143	15 / 125 / 140	17 / 119 / 136	19 / 112 / 131	21 / 105 / 126	
65	188	8 / 175 / 183	14 / 167 / 181	17 / 161 / 178	20 / 154 / 174	23 / 147 / 170	26 / 137 / 163	27 / 128 / 155	
70	234	10 / 222 / 232	18 / 212 / 230	24 / 201 / 225	27 / 193 / 220	32 / 183 / 215	34 / 166 / 200	37 / 153 / 190	
75	289	13 / 272 / 285	24 / 259 / 283	33 / 247 / 280	36 / 239 / 275	40 / 230 / 270	46 / 214 / 260	48 / 202 / 250	
80	355	18 / 337 / 355	33 / 307 / 340	43 / 292 / 335	51 / 279 / 330	58 / 262 / 320	62 / 253 / 315	63 / 237 / 300	
85	434	24 / 406 / 430	43 / 377 / 420	57 / 353 / 410	61 / 339 / 400	70 / 320 / 390	78 / 302 / 380	82 / 288 / 370	
90	526	30 / 480 / 510	52 / 453 / 505	68 / 432 / 500	80 / 410 / 490	92 / 388 / 480	95 / 365 / 460	103 / 337 / 440	
95	634	38 / 577 / 615	66 / 544 / 610	90 / 510 / 600	102 / 483 / 585	120 / 445 / 565	130 / 410 / 540	133 / 367 / 500	133 / 277 / 410
100	760	49 / 700 / 749	87 / 653 / 740	113 / 607 / 720	130 / 585 / 715	150 / 540 / 690	165 / 505 / 670	168 / 432 / 600	168 / 337 / 505
105	906	62 / 838 / 900	108 / 767 / 875	150 / 700 / 850	163 / 667 / 830	195 / 610 / 805	210 / 560 / 770	210 / 510 / 720	210 / 390 / 600
110	1075	77 / 928 / 1008	138 / 866 / 1004	183 / 819 / 1002	209 / 791 / 1000	240 / 730 / 970	250 / 680 / 930	255 / 585 / 840	260 / 435 / 695

HEAVY LINE MARKS REGION IN WHICH SOLUTIONS SOLIDIFY OR PRECIPITATE POLYMER.

All figures in millimeters of mercury.
Top Figure = P_{CH_2O}
2nd Figure = P_{H_2O}
Bottom Figure = Total Vapor Pressure

of 10 to 20 g of solution from approximately 500 g of various concentrates in a flask with a short still head insulated so that refluxing was substantially nil. Results are shown in Table 29. Partial pressure values for temperatures below the boiling point were measured by the kinetic method employed by Ledbury and Blair (Table 30). Total pressures were also ap-proximated at temperatures below the atmospheric pressure boiling point

TABLE 29. STILL POT TEMPERATURES OF BOILING FORMALDEHYDE AND PARTIAL PRESSURES CALCULATED FROM THESE VALUES

Approx. CH_2O Concn.	Concn. Range During Distillation	Total Press.	Still Pot Temp.	Concn. of Distillate	P_{CH_2O}
(%)	(%)	(mm)	(°C)	(%)	(mm)
20	19.3–19.4	756.5	100.9	17.9	
	19.4–19.4	756.5	100.9	18.2	
Adjusted average	20%	760	100.9	18	90
40	37.9–38.7	759	101.6	27.6	
	38.7–39.4	759	102.0	27.7	
Adjusted average	40%	760	101.8	28	142
50	49.9–51.2	745	102.7	34.0	
	51.2–51.6	745	102.9	34.5	
Adjusted average	50%	760	103.2	34	175
60	59.3–60.2	739	103.2	38.1	
	60.2–61.3	739	103.1	38.2	
Adjusted average	60%	769	103.9	38	203
70	68.3–69.9	748	107.0	41.5	
	69.9–70.2	748	107.3	42.4	
Adjusted average	70%	760	107.5	42	229
80	81.5–82.5	742	113.0	48.1	
	82.5–83.0	742	114.5	49.6	
Adjusted average	80%	760	112.5	48	267

TABLE 30. FORMALDEHYDE PARTIAL PRESSURE DATA FOR CONCENTRATED SOLUTIONS OBTAINED BY KINETIC METHOD

CH_2O Concn. (%)	Temp. (°C)	Formaldehyde Partial Pressure Experimental Values (mm)		Average Value (mm)
37	20	1.3	1.2	1.3
37	47	7.9	7.7	7.8
50	55	13.3	11.5	12
50	80	58.3	60.6	60
60	60	18.9	20.5	20
60	80	62.4	60.5	62
70	90	102	100	101

by measuring the pressure at which boiling occurred when the pressure over a solution sample held at the desired temperature was gradually reduced. The sample was held in a test tube containing an inverted capillary melting point tube to prevent superheating. Measurements were repeated until constant values were secured. This gave approximate values for total pressure from which partial pressures of water could be estimated.

Partial pressure data for boiling formaldehyde together with those obtained with the Lacy equation are shown graphically in Figures 20 and 21.

Accurate figures are not available for the partial pressure of water vapor over formaldehyde solutions. Korzhev and Rossinskaya[19] measured values

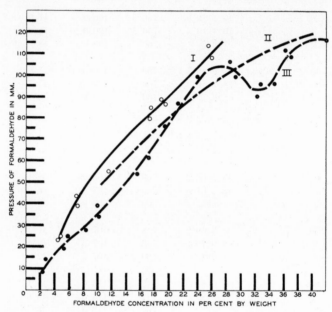

Figure 20. Partial pressure of formaldehyde over aqueous solutions at approximately 98–100°C.

        Curve I—Data of Korzhev and Rossinskaya
        Curve II—Lacy Equation
        Curve III—Data of Auerbach

for solutions containing 7, 17, and 28 per cent formaldehyde at various temperatures from 20 to 90°C, but their data show considerable variation. According to their findings, the partial pressure of water over formaldehyde solutions can be calculated by the formula shown below, in which A^1 is a constant varying with concentration.

$$\log P^{mm}_{CH_2O} = A^1 - 2168/T$$

$A^1 = 8.695$ for 7 per cent formaldehyde; 8.677 for 17 per cent formaldehyde; and 8.672 for 28 per cent formaldehyde.

Polymer Precipitation. Pure formaldehyde solutions containing up to

30 per cent formaldehyde remain clear and colorless on storage at ordinary temperatures. Solutions containing more than 30 per cent formaldehyde gradually become cloudy and precipitate polymer unless kept at elevated temperatures. For example, a 37 per cent solution must be maintained at a temperature of approximately 37°C to avoid polymer precipitation. Below room temperature, solutions containing 30 per cent or even less formaldehyde gradually become cloudy with polymer. However, since the reactions taking place in formaldehyde solution which lead to polymer

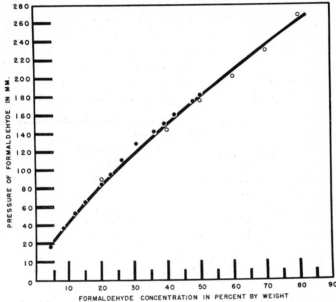

Figure 21. Partial pressure of boiling formaldehyde solutions.

● = Hall and Piret
○ = Walker and Mooney

precipitation are comparatively slow, solutions containing up to about 50 per cent formaldehyde remain clear for a short time on cooling to room temperature.

Proximate temperatures necessary to prevent polymer precipitation from solutions ranging from 58 to 90 per cent formaldehyde are shown in Table 31 as measured by Walker and Mooney[36]. These figures were obtained by placing solutions in ampoules sealed with an approximately 1-centimeter layer of white mineral oil in a heating bath whose temperature was reduced at the rate of 2°C per day. The 84 and 90 per cent figures were obtained on gradual cooling of a flask of hot solution in the air since this solution solidified too readily for employing the ampoule technique.

Refractivity. See page 66.

Solvent Properties. Although formaldehyde solutions are somewhat similar to water in solvent properties, they differ in some respects, bearing a definite resemblance to ethylene glycol and glycerol. As would be expected, these differences are intensified with increasing formaldehyde concentration. Solutions containing 30 to 40 per cent formaldehyde are reported to dissolve pyridine and quinoline bases, which are insoluble in 25 per cent or more dilute solutions[15]. Some organic liquids which are only partially miscible with water (*e.g.*, methyl ethyl ketone) are often infinitely compatible with strong formaldehyde. Inorganic salts are in general less soluble in formaldehyde solutions than in water[30]. Acidic or strongly basic materials are readily soluble in formaldehyde solution but cause polymer precipitation when added to strong solutions. Materials forming water-soluble formaldehyde addition compounds do not cause precipitation, even though they may be mildly acidic, if added in sufficient proportion to reduce the formaldehyde concentration substantially. Sul-

TABLE 31. POLYMER SEPARATION TEMPERATURES FOR CONCENTRATED AQUEOUS FORMALDEHYDE

Concn. of CH$_2$O (%)	Temperature of Polymer Separation (°C)
58	62–4
66	72–4
72	84–6
84	Approx. 107
90	Approx. 116

fur dioxide, for example, is readily soluble in strong formaldehyde, giving stable saturated solutions with 30 to 37 per cent formaldehyde[6, 29]. Such a solution probably contains formaldehyde sulfurous acid ($HOCH_2SO_3H$). Although this product is not sufficiently stable to be isolated, its sodium salt is the well known sodium bisulfite compound of formaldehyde.

Distribution constants for formaldehyde between water and various water-immiscible solvents are reported by Johnson and Piret (Chapter 3, pp. 62–63). A study of the system formaldehyde-chloroform-water is reported by Herz and Lewy[18].

The adsorption of formaldehyde from its aqueous solution has been studied by Flumiani and Corubolo[14], who state the adsorption isotherm holds only for solutions containing 0.15 to 2.42 per cent formaldehyde.

Surface Tension. According to Traube[34], the surface tension of a 0.75 per cent (0.25N) formaldehyde solution is approximately identical with that of water. At 20°C, this value is 73 dynes per cm.

Viscosity. The viscosity of aqueous formaldehyde decreases with increasing temperature. Recent data obtained with the Hoeppler viscosimeter will be found on page 75.

Data previously reported by Heiduschka and Zirkel[17] were apparently obtained with solutions containing methanol, since the reported density values for the concentrations studied are definitely lower than known values for pure formaldehyde solutions.

References

1. Atkins, H., *J. Marine Biol. Assoc. United Kingdom*, **12**, 717–771 (1922).
2. Auerbach, F., and Barschall, H., "Studien uber Formaldehyd, Part I, Formaldehyd in wässriger, Lösung," pp. 11–12, Julius Springer, Berlin (1905). (Reprint from *Arb. kaiserl-Gesundh.*, **22**, (3) (1905).
3. *Ibid.*, p. 31.
4. *Ibid.*, pp. 34–37.
5. *Ibid.*, pp. 40–44.
6. Cushman, A. S., U. S. Patent 1,399,007 (1921).
7. Datar, S. N., *Current Science*, **3**, 483–4 (1935).
8. Davis, H. L., *J. Chem. Ed.*, **10**, 47 (1933).
9. Davis, W. A., *J. Soc. Chem. Ind.*, **16**, 502 (1897).
10. Delépine, M., *Compt. rend.*, **124**, 816 (1897).
11. *Ibid.*, 1454 (1897).
12. Euler, H. V., and Lovgren, T., *Z. anorg. allgem. Chem.*, **147**, 123 (1925).
13. Euler, T., *Ber.*, **38**, 255 (1905).
14. Flumiani, G., and Corubolo, I., *Rad. Jugoslav Akad. Znanosti Umjetnosti*, **241**, 251–62; *C. A.*, **33**, 7170 (1939).
15. Goldschmidt, C., *J. prakt. Chem.* (*2*), **72**, 536 (1905).
16. Good, W., *J. Roy. Tech. Coll. Glasgow*, **2**, 401–409 (1931), *C. A.*, **25**, 2888 (1931).
17. Heiduschka, A., and Zirkel, H., *Arch. Pharm.*, **254**, 482–3 (1916).
18. Herz, W., and Lewy, M., *Jahresber. Schles. Ges. Vaterl. Kultur Naturw. Sekt.*, **1906**, 1–9; "International Critical Tables," **3**, 422, 434, New York, National Research Council (McGraw-Hill Book Co., Inc.), 1929.
19. Korzhev, P. P., and Rossinskaya, I. M., *J. Chem. Ind. (U. S. S. R.)*, **12**, 610–14 (1935).
20. Lacy, B. S., (Electrochemicals Department, E. I. du Pont de Nemours & Co., Inc.), unpublished communication.
21. Ledbury, W., and Blair, E. W., *J. Chem. Soc.*, **127**, 33–37, 2834–5 (1925).
22. Lüttke, H., *Jahresber.*, **1893**, 512.
23. Meslin, G., *Ann. Chim. Phys.* (*8*), **7**, 145 (1906); "International Critical Tables," **6**, 361, New York, National Research Council (McGraw-Hill Book Co., Inc.), 1929.
24. Natta, G., and Baccaredda, M., *Giorn. chim. ind. applicata*, **15**, 273–281 (1933).
25. Parks, G. S., and Huffman, H. M., "The Free Energy of Some Organic Compounds," p. 159, New York, Chemical Catalog Co., Inc. (Reinhold Publishing Corp.), 1932.
26. PB Report 70169, pp. 744–52 (1939).
27. Piret, E. L., and Hall, M. W., *Ind. Eng. Chem.*, **40**, 661–72 (1948).
28. Reicher, L. T., and Jansen, F. C. M., *Chem. Weekblad*, **9**, 104 (1911).
29. Reinking, K., Dehnel, E., and Labhardt, H., *Ber.*, **38**, 1075–6 (1905).
30. Rothmund, V., *Z. physik. Chem.*, **69**, 531–9 (1909).
31. Sauterey, R., *Compt. rend.*, **229**, 884–6 (1949).
31a. Sauterey, R., *Ann. de chimie* (*12*), **7**, 1–12 (1952).

32. Stutterheim, G. A., *Pharm. Weekblad*, **54,** 686 (1917).
33. Swietoslawski, W., "Ebulliometric Measurements," pp. 23, 37, 59, New York, Reinhold Publishing Corp. (1945).
34. Traube, I., *Ber.*, **42,** 2168 (1909).
35. Wadano, M., *Ber.*, **67,** 191 (1934).
36. Walker, J. F., and Mooney, T. J., unpublished communication, Electrochemicals Dept., E. I. du Pont de Nemours & Co., Niagara Falls, N. Y.

CHAPTER 6

DISTILLATION OF FORMALDEHYDE SOLUTIONS

The behavior of formaldehyde solutions on distillation under various conditions of temperature and pressure indicates that the partial pressure of formaldehyde over these solutions is in reality the decomposition pressure of dissolved formaldehyde hydrate[29]. With the aid of this concept it is possible to predict the behavior of aqueous formaldehyde solutions under various conditions and devise methods for the recovery or removal of formaldehyde from these systems. Solutions containing components other than formaldehyde and water can probably be handled in a similar fashion, provided these components do not react irreversibly with formaldehyde under the conditions stipulated.

Vacuum Distillation

Since the partial pressure of formaldehyde over its aqueous solutions is extremely low at ordinary temperatures, vacuum distillation affords a method for concentrating formaldehyde. This fact was first observed by Butlerov in 1859[6]. During vacuum distillation, formaldehyde hydrates concentrate in the still with eventual formation of the solid mixture of polymeric hydrates known as paraformaldehyde. This procedure has long been used commercially for the preparation of paraformaldehyde[11], and special process modifications have been described in patents[7, 22].

The formaldehyde content of the vapors obtained from formaldehyde solutions of various concentrations at 20-mm pressure (boiling point = approximately 20°C), as determined by Korzhev and Rossinskaya[18], is shown in Table 32.

Since the polymerization and depolymerization reactions which take place when the concentration of dissolved formaldehyde is changed at low temperatures are slow, it is highly probable they do not keep pace with the increase in formaldehyde content which occurs in the still residue during vacuum concentration. As a result, the formaldehyde partial pressures prevailing during vacuum distillation are probably anomalous, since the monohydrate or methylene glycol content for the concentrating solution may become higher than the equilibrium value. If this happens, the formaldehyde partial pressure will become higher than the normal value, since it is proportional to the methylene glycol concentration[27]. This means that the distillate will be somewhat richer in formaldehyde than would be predicted from partial pressure values reported for solutions having the concentration of the solution being distilled at any given moment. If distillation is rapid, the rate of dehydration is the controlling factor and partial pressure of formaldehyde will be lower than would be predicted. As a result, Trümpler and Bieber's polarographic studies of the dehydration reaction (pages 58–60) have practical connotations in this connection. These considerations must be kept in mind when dealing with the distillation of formaldehyde solution at low temperatures.

Conditions equivalent to those of a vacuum distillation are also encountered when formaldehyde solutions are distilled with solvents which form azeotropes with water. A German process[9, 10] for paraformaldehyde production is based on this technique. A combination of vacuum distillation and azeotropic distillation has been patented by Hasche[14].

TABLE 32. CONCENTRATIONS OF FORMALDEHYDE IN LIQUID AND GAS PHASES FOR SOLUTIONS BOILING AT 20 mm PRESSURE*

Concentration of Formaldehyde in Solution (%)	Concentration of Formaldehyde in Vapor (%)
6.83	0.46
16.9	1.19
28.2	4.66

* Data from Korzhev and Rossinskaya[18].

Pressure Distillation

When formaldehyde solutions are subjected to pressure distillation, the formaldehyde passes over in the first fractions distilled and distillates of higher concentration than the original solution are obtained. This effect is illustrated by the data of Ledbury and Blair[20], who employed pressure distillation for the recovery of formaldehyde from dilute aqueous solutions. Table 33 has been prepared from their data. Since the distillation temperatures of these formaldehyde solutions are approximately equivalent to those for pure water, the temperature figures based on water, which we have included in this table, are probably substantially correct. These data show that the concentration of formaldehyde in the distillate fraction increases with increasing pressure and temperature.

Ledbury and Blair explained this behavior by the assumption that the dissolved formaldehyde was polymerized and that the high temperatures caused the polymer to break up. Since Auerbach[1] showed that formaldehyde is practically all in the form of methylene glycol at concentrations of approximately 2 per cent, this hypothesis is untenable. Results are readily explained by the effect of temperature on the equilibrium constant of the dehydration reaction, $CH_2(OH_2) \rightleftharpoons H_2O + CH_2O$. When solutions containing over 2 per cent are involved, the effect of temperature on the depolymerization of polymeric hydrates is also involved. At high temperatures, formaldehyde solutions behave as a solution of a volatile gas would be expected to behave. The formaldehyde passes over and dissolves in the aqueous condensate. The results indicate that formaldehyde gas is comparatively insoluble in water at temperatures of about 100°C.

The utility of pressure distillation as a means of recovering formaldehyde from dilute solutions is limited by the fact that formaldehyde tends to undergo the Cannizzaro reaction, forming methanol and formic acid on heating with water. The rate at which this reaction takes place increases rapidly with rising temperatures. According to Ledbury and Blair, 9 per cent of the formaldehyde is lost because of this reaction in a distillation at 100-lbs pressure, whereas only 2 per cent is lost at 60 lbs.

TABLE 33. INFLUENCE OF PRESSURE ON THE DISTILLATION OF A 2.1 PER CENT FORMALDEHYDE SOLUTION*

Distillation Pressure (psi)	Approximate Temperature (°C)	Formaldehyde Concentration in First 200-cc Distillate from One Liter Still Charge (%)
0	100	3.2
20	126	4.6
40	142	5.4
60	153	6.2
80	162	7.4
100	170	7.6

* Data from Ledbury and Blair[20].

Conditions somewhat equivalent to those of pressure distillation are also obtained by dissolving a salt, such as calcium chloride, in the solution to be distilled. The boiling point of the solution is elevated and a concentrated formaldehyde solution is distilled. According to a Russian patent[17] describing a process of this sort, 36.7 per cent formaldehyde can be obtained by distilling a 23.6 per cent solution containing 20.3 per cent of calcium chloride. Hasche obtained an American patent covering a somewhat similar process[13]. Combinations of pressure distillation and azeotropic distillation have been patented by Wong[31].

Atmospheric Pressure Distillation

As might be expected, when formaldehyde is distilled at atmospheric pressure the results are intermediate between those of vacuum and pressure distillation. However, distillation studies as reported by various investigators show wide variations (Table 34). These differences are caused by the unique properties of the formaldehyde-water system whose behavior in a given distillation is controlled by the polymerization, depolymerization, hydration and dehydration reactions indicated below:

$$1/n \text{ HO}(CH_2O)_nH \rightleftharpoons HOCH_2OH \rightleftharpoons CH_2O + H_2O.$$

TABLE 34. DISTILLATION OF FORMALDEHYDE AT ATMOSPHERIC PRESSURE
(Results Reported by Various Investigators)

Investigators	Year	Results	Apparatus or Technique
Auerbach and Barschall[2]	1905	CH$_2$O content of still residue always stronger than distillate	No still head; no appreciable reflux
Wilkinson and Gibson[30]	1921	Constant boiling mixture containing 8% CH$_2$O; distillate always weaker than residue for solutions containing over 8%	Hempel column
Blair and Taylor[4]	1926	Constant boiling mixture containing 30% CH$_2$O	
Zimmerli[31]	1927	Distillate always more concentrated than residue; distillate concentrations of up to 55% obtainable from more dilute solutions	5 ft packed column surmounted by an efficient dephlegmator; slow distillation
Ledbury and Blair[20]	1927	Constant boiling mixture containing 30% CH$_2$O	5-section Young and Thomas column
Walker[23]	1932	Distillate always more concentrated than residue	Efficient dephlegmator with high reflux ratio, no packed column
Bond[5]	1933	Constant boiling mixture containing 30 to 33% formaldehyde	Packed column
Korzhev and Rossinskaya[18]	1935	Constant boiling mixture in the 9 to 12% concentration range	Equilibrium still
Reynolds[5]	1937	Constant boiling mixture containing 23% CH$_2$O	
Piret and Hall[23]	1948	CH$_2$O content of still residue always stronger than distillate	Equilibrium still
Pyle and Lane[24]	1950	Constant boiling mixture containing 20 to 21% CH$_2$O	Equilibrium data

Differences in the kinetics of the various reactions are significant factors whose influence is modified by the type and size of the distillation equipment, the rate of distillation and the reflux ratio. Equilibrium values are also affected by variations in the prevailing temperatures and pressures.

When formaldehyde solutions are distilled slowly and without fractionation, the results are determined by the equilibrium values for the partial pressures of formaldehyde and water. Vapor-liquid equilibrium concentrations, as reported by Piret and Hall[23] (Table 35), indicate that under these conditions the distillate is always less concentrated than the residue and

TABLE 35. LIQUID-VAPOR EQUILIBRIA FOR BOILING FORMALDEHYDE AT APPROXIMATELY ATMOSPHERIC PRESSURE[23]

In Liquid	Wt. Per Cent CH₂O In Vapor	Mol Fraction of CH₂O in Vapor	Pressure (mm)
3.88	3.66	0.0223	740
3.89	3.84	0.0234	760
3.95	3.68	0.0224	740
4.46	4.14	0.0253	733
7.93	7.30	0.0452	736
7.95	7.40	0.0458	743
7.98	7.69	0.0476	760
8.00	7.30	0.0452	740
12.1	10.6	0.0655	741
15.3	13.2	0.0837	745
19.5	16.4	0.1055	740
19.75	16.6	0.107	742
20.1	16.95	0.109	737
22.45	19.2	0.125	760
22.55	19.0	0.1235	744
25.85	21.45	0.141	731
29.9	23.85	0.158	738
30.75	24.9	0.166	760
31.1	24.75	0.165	737
31.95	25.3	0.1685	750
35.65	27.4	0.1845	745
38.8	29.1	0.1975	750
41.6	30.75	0.211	760
42.0	30.5	0.208	733
47.25	33.3	0.230	760
47.5	33.1	0.229	743
49.8	34.0	0.236	736

there are no apparent constant boiling mixtures. These findings are in qualitative agreement with the earlier findings of Auerbach and Barschall[2].

Actual distillations of formaldehyde solutions are not normally carried out under the above conditions but involve fractionation, fractional condensation of vapors and rates of distillation and condensation which are often too rapid to permit the attainment of liquid-vapor equilibria. Accordingly, in dealing with practical distillation processes, the effect of these factors must be taken into account. In this connection, one of the most important and unique factors is the fractional condensation phenomenon.

Fractional Condensation

When the mixture of formaldehyde and water vapor obtained by boiling a formaldehyde solution is subjected to partial condensation, the formaldehyde content of the uncondensed vapor is greater than that of the original gas mixture. In other words, the vapors behave as a mixture of water and a volatile gas would be expected to behave. By making use of this phenomenon, concentrated solutions of formaldehyde can be obtained from dilute solutions by a process involving distillation and partial condensation. A formaldehyde concentration process of this type was claimed by the writer in 1932[28]. Vapors of boiling formaldehyde were partially condensed with a water-cooled, bayonet-type condenser in such a manner that the condensate could be returned rapidly to the still with minimum exposure to the hot ascending vapors. The uncondensed vapors leaving the still head were then liquified with a conventional condenser. By this procedure, a substantial portion of the formaldehyde in a 28 per cent solution was collected as a 53 per cent distillate. Results were equivalent to those reported by Zimmerli[32], who used a five-foot packed distilling column surmounted by an efficient reflux condenser. This indicates that fractional condensation rather than fractionation accounted for the results obtained by Zimmerli.

Because of the fractional condensation effect, some vapor enrichment is encountered whenever vapors are partially condensed in the distillation of formaldehyde solutions. Piret and Hall[23] have studied the influence of this effect in distillation processes and explain the apparent constant boiling mixtures reported by some investigators as a result of this phenomenon. Their experiments show that as the degree of partial condensation is increased in the distillation of formaldehyde, the liquid-vapor composition curve is raised and constant boiling mixtures having higher and higher formaldehyde contents are obtained.

The fractional condensation process gives the best results when the condensate is chilled to a low temperature and removed promptly from further exposure to the hot gases. Apparently the rate at which formaldehyde dissolves in the condensate decreases with temperature and the liquid is removed before it has had the opportunity to come to equilibrium with the gas phase. Although the efficiency of the fractional condensation process falls off considerably when an attempt is made to carry it to its extreme limit, German investigators[16] have succeeded in producing paraformaldehyde by polymerization of the end gases obtained in a thoroughgoing process of this sort. Paraformaldehyde obtained in their process was found to contain as high as 93 per cent CH_2O plus 7 per cent combined water. Since vapor density measurements[3] have shown that paraformaldehyde is dissociated in the gas phase, it is evident that monomeric formaldehyde is the end-product of the fractional condensation process. Apparently, formaldehyde and water are not associated to any appreciable extent in the gaseous state. Additional evidence on this is supplied by recent vapor density studies by Hall and Piret[12]. Results indicate that the formaldehyde in vapors obtained from boiling aqueous formaldehyde solutions at 100°C is 95 per cent monomeric CH_2O and 5 per cent methylene glycol vapor. At approximately 110°C, the methylene glycol is completely dissociated. The vapor equilibrium constant, Kp, for aqueous formaldehyde compositions in the temperature range 40 to 160°C can be calculated from the formula, $\log_{10} Kp = -3200/T + 9.8$. The approximate heat of reaction for gaseous formaldehyde and steam as calculated from the above expression is 14.8 kcal per g mol CH_2O. This is approximately the same as the heat of solution of formaldehyde gas in water.

The fractional condensation phenomenon may be readily observed when a solution of formaldehyde is refluxed. A thin film of formaldehyde polymer is formed at the top of the condenser and a small percentage of formaldehyde gas escapes in the air. The fact that fractional condensation can be combined with other procedures for formaldehyde distillation is indicated by a patent which covers a procedure involving both fractional condensation and pressure distillation[10].

Fractionation

Adiabatic fractionation of formaldehyde should be differentiated from fractional condensation. In distillations of this type, an effective column is employed and ascending vapors are scrubbed thoroughly by hot descending liquor produced by returning a portion of the distillate obtained by total condensation of vapors at the top of the column. Conditions are the direct opposite of those favoring the fractional condensation process: exposure of gas to liquid is at a maximum to favor attainment of equilibrium and the column is insulated or heated to avoid temperature differentials.

Results obtained by fractional distillation of formaldehyde are in good agreement with vapor-liquid equilibrium curves presented by Pyle and Lane[24]. These curves (Figure 22) show apparent azeotropic compositions at atmospheric and higher pressures with no azeotrope indicated at 460-mm pressure. The concentrations of the apparent azeotropes at various pressures are as follows:

Pressure	CH_2O Concn. of Apparent Azeotrope
460 mm	—
760 mm	20 to 21 per cent
15 psi gage	41 per cent
60 psi gage	67.5 per cent

The term "apparent azeotrope" is employed in place of "azeotrope" since the latter term is normally warranted only for systems in which the chemical components are identical in the liquid and gaseous phase. In the case of formaldehyde-water systems, this is not the case.

The existence of the apparent azeotrope at atmospheric pressure is supported by data obtained by R. L. Craven[8] in which formaldehyde solutions of various concentrations were slowly distilled through an insulated five-plate glass bubble-plate column until an apparent state of equilibrium was obtained. This was determined by analysis of the liquid phase over the

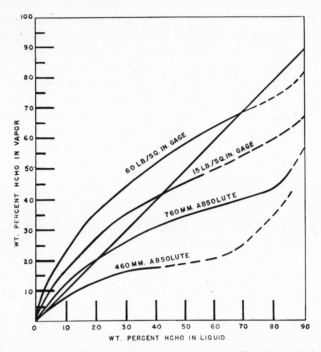

Figure 22. Vapor-liquid equilibrium curves for boiling formaldehyde.

five plates of the experimental column. The walls of the column were insulated by means of a glass jacket and heated electrically to prevent fractional condensation. The concentration of the still pot liquor was maintained at a constant value by continuous feed from a reservoir of solution at the starting concentration while running off liquor at a sufficient rate to avoid any substantial change of concentration. The apparent equilibrium values for the column sections with different still pot concentrations are shown in Table 36. It will be noted that the concentration at the first plate of the column is always lower than the still pot concentration and these values are in rough quantitative agreement with the equilibrium values of Piret and Hall[23]. However, the concentration changes in the column itself support the data of Pyle and Lane[24].

Although it is evident that the various results reported for formaldehyde fractionation (Table 34) show discrepancies attributable to the fractional condensation phenomenon, it is also evident that this is not the only factor involved.

Steam Distillation

The behavior of formaldehyde on steam distillation is approximately equivalent to its behavior on simple distillation at the same pressure. The volatility of various compounds with steam increases as hydration decreases. Virtanen and Pulkki[26] have calculated constants characteristic of the change in composition of solutions on steam distillation with the formula $(\log Y_1 - \log Y_2)/(\log X_1 - \log X_2) = H$, where X and Y refer respectively to the quantities of water and volatile organic compound, and subscripts 1 and 2 refer respectively to the quantities at the beginning and end of the distillation. According to their measurements the value of this constant for formaldehyde is 2.6, for acetaldehyde it is 40, and for benzaldehyde, 18.

TABLE 36. FRACTIONATION OF FORMALDEHYDE[8]

Equilibrium Concentrations of CH_2O in Per Cent at Plate Indicated

Still Pot	Plate 1	Plate 2	Plate 3	Plate 4	Plate 5
3.79	3.05	3.83	5.37	6.16	7.77
10.88	7.67	9.21	11.21	12.43	13.98
21.11	16.80	17.64	19.03	20.13	20.37
32.1	24.07	23.60	23.6	22.9	22.8
54.0	40.83	37.24	34.41	32.09	32.32

Steam distillation can be employed for the volatilization of concentrated formaldehyde according to a patent[21].

Distillation of Formaldehyde Containing Methanol

In general, the behavior of stabilized formaldehyde is similar to that of the methanol-free solutions. However, the initial boiling point of commercial solutions containing methanol is naturally lowered by the presence of methanol. On fractional distillation of such solutions, a methanol-rich fraction is first obtained. Removal of all the methanol by fractionation is somewhat difficult. This is probably due to the formation of the hemiacetal of formaldehyde in the solution. Ledbury and Blair[19] obtained methanol-free solution by refluxing dilute commercial formaldehyde in a packed column surmounted by a still head whose temperature was maintained at 66°C for 36 hours. Most of the methanol was removed in the first few hours of operation. According to Hirchberg[15], methanol can be removed from a gaseous mixture containing methanol, formaldehyde, and water by distilling the mixture in a rectifying column and withdrawing the aqueous condensate from the lower plates as soon as formed so that prolonged contact between condensate and vapors will be avoided.

Fractionation experiments by Ledbury and Blair[20] show that approximately 62 per cent of the methanol in a liter of formaldehyde solution containing 12 g of formaldehyde and 3.65 g of methanol per 100 cc is removed in the first 200 cc of distillate, whereas only 24 per cent of the formaldehyde is present in this fraction. A five-section Young and Thomas column was employed. On pressure distillation of a similar solution at 60-lb pressure, the first 200 cc distilled was found to contain 65 per cent of the methanol and 39 per cent of the formaldehyde present in the original solution. They report that in the absence of methanol a much larger proportion of formaldehyde would be carried over in the primary fraction.

References

1. Auerbach, Fr., "Studien über Formaldehyd, I, Formaldehyd in Wässeriger Lösung," pp. 10–27, Berlin, Julius Springer (1905).
2. *Ibid.*, p. 34.
3. Auerbach, Fr., and Barschall, H., "Studien über Formaldehyd II, Die festen Polymeren des Formaldehyds," p. 7, Berlin, Julius Springer (1907).
4. Blair, E., and Taylor, R., *J. Soc. Chem. Ind.*, **45**, 65T (1926).
5. Bond, H. A., (to E. I. du Pont de Nemours & Co., Inc.), U. S. Patent 1,905,033 (1933).
6. Butlerov (Butleroff), A., *Ann.*, **111**, 245, 247–8, (1859).
7. Consortium für elektrochemische Industrie, G. m. b. H., German Patent 489,644 (1930).
8. Craven, R. L., (Electrochemicals Dept., E. I. du Pont de Nemours & Co.), unpublished communication.
9. Deutsche Gold-und Silber-Scheideanstalt vorm. Roessler, German Patent 588,470 (1932); *Chem. Zentr.* **1932**, II, 2724.
10. Deutsche Gold- und Silber-Scheideanstalt vorm. Roessler. British Patent 479,255 (1938).
11. "Encyclopedia Britannica" (Fourteenth Ed.) Vol. 9, p. 512, New York, Encyclopedia Britannica, Inc.
12. Hall, M. W., and Piret, E. L., *Ind. Eng. Chem.*, **41**, 1277–86 (1949).
13. Hasche, R. L., U. S. Patent 2,015,180 (1935).
14. Hasche, R. L., and Brant, J. H., (to Eastman Kodak Co.), U. S. Patent 2,475,959 (1949).
15. Hirchberg, L. M., British Patent 199,759 (1923).
16. I. G. Farbenindustrie, A. G., German Patent 503,180 (1930).
17. Korzhev, P. P., Russian Patent 44,251 (1935).
18. Korzhev, P. P., and Rossinskaya, I. M., *J. Chem. Ind. (U. S. S. R.)*, **12**, 601–614 (1935).
19. Ledbury, W., and Blair, E. W., *J. Chem. Soc.*, **127**, 26 (1925).
20. Ledbury, W., and Blair, E. W., "The Production of Formaldehyde by Oxidation of Hydrocarbons," Special Report No. 1, Dept of Scientific and Industrial Research, pp. 40, 44–51, London, published under the Authority of His Majesty's Stationery Office, 1927.
21. N. V. Machinen Fabriek "de Hollandsche Ijssel," Dutch Patent 48,224, voorheen de Jough & Co. (1940).
22. Nasch, L., British Patent 420,993 (1934).

23. Piret, E. L., and Hall, M. W., *Ind. Eng. Chem.*, **40**, 661–72 (1948).
24. Pyle, C., and Lane, J. A., (to E. I. du Pont de Nemours & Co., Inc.), U. S. Patent 2,527,654 (1950).
25. Reynolds, B. M., (to Heyden Chemical Corp.), U. S. Patent 2,256,497 (1941).
26. Virtanen, A. I., and Pulkki, L., *Ann. acad. sci. Fennicae*, **29A**, 28 pp. (1927); *C. A.*, **22**, 4352 (1928).
27. Walker, J. F., *J. Phys. Chem.*, **35**, 1104–1131 (1931).
28. Walker, J. F., U. S. Patent 1,871,019 (1932).
29. Walker, J. F., *Ind. Eng. Chem.*, **32**, 1016 (1940).
30. Wilkinson, J. A., and Gibson, I. A., *J. Am. Chem. Soc.*, **43**, 695 (1921).
31. Wong, S. Y., (to Skelly Oil Co.), U. S. Patents 2,452,412–5 (1948).
32. Zimmerli, A., *Ind. Eng. Chem.*, **19**, 524 (1927); U. S. Patent 1,662,179 (1928).

CHAPTER 7

FORMALDEHYDE POLYMERS

Polymer formation is one of the most characteristic properties of the formaldehyde molecule. As indicated by the formulas shown below, two fundamentally different polymer types are theoretically possible:

$$\cdot CH_2 \cdot O \cdot CH_2 \cdot O \cdot CH_2 \cdot O \cdot CH_2 \cdot O \cdot CH_2 \cdot O \cdots \qquad (1)$$

$$
\begin{array}{c}
\text{H H H H H} \\
\cdot C \cdot C \cdot C \cdot C \cdot C \cdots \\
\text{O O O O O} \\
\text{H H H H H}
\end{array}
\qquad (2)
$$

Although both of these types are known, the term "formaldehyde polymer," as normally employed, refers chiefly to polymers of the first type which are known generically as polyoxymethylenes. They are reversible polymers and react chemically as solid forms of formaldehyde.

Representatives of the second type are encountered in the polyhydroxyaldehydes. They are substantially irreversible, and hexose sugars are apparently the highest members of the group. Reactions which result in the formation of these products are usually regarded as aldol-type condensations and will be discussed in connection with formaldehyde reactions.

In general, polyoxymethylene-type polymers fall in the two groups defined by Carothers as A- and C-polymers[15]. A-polymers are usually formed by the polymerization of anhydrous monomeric formaldehyde as indicated below:

$$H_2C:O + H_2C:O + H_2C:O + \cdots \rightarrow \cdot CH_2 \cdot O \cdot CH_2 \cdot O \cdot CH_2 \cdot O \cdots$$

C- or condensation polymers are derived from aqueous formaldehyde by polycondensation reactions involving formaldehyde hydrate or methylene glycol.

$$HO \cdot CH_2 \cdot OH + HO \cdot CH_2 \cdot OH + HO \cdot CH_2 \cdot OH + \cdots \rightarrow$$
$$HO \cdot CH_2 \cdot O \cdot CH_2 \cdot O \cdots CH_2 \cdot OH + (n-1) H_2O$$

Members of this group are known technically as polyoxymethylene glycols.

The cyclic polymers, trioxane and tetraoxymethylene, although true polyoxymethylenes, have apparently never been prepared by polymerization of pure monomeric formaldehyde.

Paraformaldehyde, the only formaldehyde polymer now being manu-

114

factured in large quantities for commercial use, is a mixture of polyoxymethylene glycols. As we have previously pointed out, paraformaldehyde is an important industrial chemical, finding use wherever formaldehyde is desired in a form that is substantially free of water.

The theory that the clouds on the planet Venus are due to solid polyoxymethylene glycols produced by the action of ultraviolet light on the water and carbon dioxide in the atmosphere of this planet has been withdrawn by its originator, Rupert Wildt[117]. Spectrographic studies show that formaldehyde gas, which would be present in equilibrium with polymer, is absent in the atmosphere of Venus.

The polymerization of formaldehyde is an exothermic reaction. Heats of polymerization, as calculated from available thermochemical data for various polyoxymethylene-type polymers, range from 12 to 15 kcal per mol of formaldehyde.

Formaldehyde polymers are classified by structure, molecular weight, and, in the case of polyoxymethylene glycols and their derivatives, by the type of end-groups attached to the linear molecules. The system of classification illustrated in Table 37 shows the relation of structure, molecular weight, and end-groups to characteristic properties and offers a logical order for the discussion of the major polymer groups which have received the attention of chemical investigators.

It will be noticed that "trioxymethylene" has been omitted from our list of formaldehyde polymers. This name has been so consistently misused as a designation for paraformaldehyde and the linear polyoxymethylenes that it can no longer be employed without confusion.

Linear Polymers

Linear formaldehyde polymers include the polyoxymethylene glycols, $HO \cdot (CH_2O)_n \cdot H$, polyoxymethylene glycol derivatives, and eu-polyoxymethylene, $(CH_2O)_n$.

As previously pointed out, the polyoxymethylene glycols are hydrated polymers chemically and structurally related to methylene glycol. This relationship is clearly indicated in the formulas shown below:

Name	Structural Formula
Methylene glycol	$HO \cdot CH_2 \cdot OH$
Dioxymethylene glycol	$HO \cdot CH_2 \cdot O \cdot CH_2 \cdot OH$
Trioxymethylene glycol	$HO \cdot CH_2 \cdot O \cdot CH_2 \cdot O \cdot CH_2 \cdot OH$
Tetraoxymethylene glycol	$HO \cdot CH_2O \cdot CH_2O \cdot CH_2O \cdot CH_2OH$
Higher polyoxymethylene glycols	$HO \cdot (CH_2O)_n \cdot H$

Although some of the polyoxymethylene glycols have been isolated in a comparatively pure state, they are usually encountered as mixtures, the formaldehyde content of which is a measure of the average degree of poly-

TABLE 37. FORMALDEHYDE POLYMERS

Type Formula	Approximate Range of Polymerization (n)	CH_2O Content (% by Weight)	Melting Range (°C)	Solubility*			
				Acetone	Water	Dilute Alkali	Dilute Acid
I. Linear Polymers (On vaporization these depolymerize to monomeric formaldehyde gas)							
Lower polyoxymethylene glycols	$HO(CH_2O)_n \cdot H$	2-8	77-93	80-120	s-i	vs	vs	vs
Paraformaldehyde	$HO(CH_2O)_n \cdot H$	6-100	91-99	120-170	s-i	ds	s	s
Alpha-polyoxymethylene	$HO(CH_2O)_n \cdot H$	>100	99.0-99.9	170-180	i	vds	ds	s
Beta-polyoxymethylene	$HO(CH_2O)_n \cdot H + H_2SO_4$ (trace)	>100	98-99	165-170	i	vds	ds	ds
Polyoxymethylene Glycol Derivatives†							
Polyoxymethylene diacetates	$CH_3CO \cdot O(CH_2O)_n \cdot COCH_3$	2-100	37-93	Up to ca. 165	i	i	ds	ds
Lower polyoxymethylene dimethyl ethers	$CH_3O \cdot (CH_2O)_n \cdot CH_3$	<100	72-93	Up to ca. 175	i for n >10	i for n >15	i for n >15	ds
Gamma-polyoxymethylene (Higher polyoxymethylene dimethyl ethers)	$CH_3O \cdot (CH_2O)_n \cdot CH_3$	>100	93-99	160 to ca. 180	i	i	i	ds
Delta-polyoxymethylene	$CH_3O \cdot (CH_2O)_n \cdot CH_2CH(OH)OCH_3$	>100	96-97	150-170	i	i	i	ds
Epsilon-polyoxymethylene	$(CH_2O)_n$(?)	Probably >100	99.7-99.9	195-200	i	i	—	—
Eu-polyoxymethylene	$(CH_2O)_n$	Approx. 5000	99.9-100	170-185	i	i	vds	vds
II. Cyclic Polymers (On vaporization these do not depolymerize)							
Trioxane (Alpha-trioxymethylene)	$(CH_2O)_3$	3	100	61-62	s	s	s	s
Tetraoxymethylene	$(CH_2O)_4$	4	100	112	s	s	s	s

Abbreviations: s = soluble, vs = very soluble, ds = difficulty soluble, vds = very difficulty soluble, i = insoluble.

*1. In the case of linear polymers, solution in water is accompanied by hydrolytic decomposition which is catalyzed by alkalies and acids. Solubility as ordinarily observed is a measure of this depolymerization process, since only the lower polyoxymethylene glycols are more than slightly soluble.

2. Cyclic polymers dissolve as polymers and give true solutions.

†Derivatives of low molecular weight do not depolymerize on vaporization.

merization. On the basis of molecular weight, physical properties, and methods of preparation, they may be classified in three groups: the lower polyoxymethylene glycols, paraformaldehyde, and alpha-polyoxymethylene. (The so-called beta-polyoxymethylene is apparently an alpha-polyoxymethylene containing a small percentage of sulfuric acid.) It must be remembered, however, that this classification is more or less arbitrary and is made only for purposes of convenience. The three groups merge into one another and absolute dividing lines cannot be drawn between them.

In general, the polyoxymethylene glycols have the appearance of colorless powders possessing the characteristic odor of formaldehyde. Their properties, such as melting point, solubility, chemical reactivity, etc., vary with their molecular weight. As the degree of polymerization, indicated by n in the type formula, $HO \cdot (CH_2O)_n \cdot H$, increases, their formaldehyde content approaches 100 per cent and the physical and chemical properties approach those of eu-polyoxymethylene, $(CH_2O)_n$.

Eu-polyoxymethylene is derived from anhydrous liquid formaldehyde and apparently does not contain combined water. However, since even a trace of water would be sufficient to hydrate the large molecules involved, it is possible that it may be a polyoxymethylene glycol of extremely high molecular weight. Conversely, it is also possible that some of the higher polyoxymethylene glycols may be true polyoxymethylenes contaminated with a trace of adsorbed water.

Lower Polyoxymethylene Glycols

When aqueous formaldehyde solutions containing from approximately 30 to 80 per cent formaldehyde are brought to room temperature or below, a precipitate consisting principally of the lower polyoxymethylene glycols is obtained. The point at which precipitation takes place is dependent on the concentration of the formaldehyde solution and the rate of cooling. As previously pointed out (pages 98–99), the lower members of this group are normally present in aqueous formaldehyde solutions. On standing, the products first precipitated undergo condensation, and polyoxymethylene glycols having a higher degree of polymerization are formed.

The lower polyoxymethylene glycols are colorless solids melting in the range 80 to 120°C. They differ from paraformaldehyde and other higher homologs in being soluble in acetone and ether, dissolving with little or no decomposition. They dissolve rapidly in warm water with hydrolysis and depolymerization to form formaldehyde solution. They are insoluble in petroleum ether.

A number of the lower polyoxymethylene glycols have been isolated by Staudinger[74] in a fair degree of chemical purity. In the case of the lowest homologs, this was accomplished by adding acetone to concentrated water solutions of formaldehyde, drying the resultant mixture with anhydrous sodium sulfate, and isolating the polyoxymethylene glycols in the acetone solution thus obtained by fractional precipitation with petroleum ether. Another procedure involved as a starting material the polyoxymethylene glycol gel obtained by an ether extraction of strong aqueous formaldehyde. Homologs containing 6 to 8 formaldehyde units per molecule were isolated when the wax-like solid, obtained on cooling a hot 90 per cent formaldehyde solution, was allowed to stand at room temperature for 2 to 3 weeks. This solid was first treated with a little acetone and filtered. The filter residue was then separated into fractions which were soluble in cold acetone, warm acetone, and boiling acetone. From the cold acetone fraction, hexa- and heptaoxymethylene glycols were precipitated with petroleum ether. Homologs containing 8 to 12 formaldehyde units were isolated from the other extracts by fractional crystallization.

The lowest fraction isolated was a mixture of di- and trioxymethylene glycols, $HOCH_2OCH_2OH$ and $HOCH_2OCH_2OCH_2OH$, which melted at 82 to 85°C and was highly soluble in cold acetone. Tetraoxymethylene glycol, $HO \cdot (CH_2O)_4 \cdot H$, was somewhat less soluble and melted at 95 to 105°C, with decomposition. Hexa- and heptaoxymethylene glycol fractions were fairly soluble in cold acetone. Octo-oxymethylene glycol was very soluble in hot acetone and melted at 115 to 120°C, with decomposition. Acetone solubility decreased progressively for higher polyoxymethylene glycols, dodecaoxymethylene glycol being only sparingly soluble in the boiling solvent. The purity of these fractions was indicated by the fact that further fractionation produced little or no change in physical properties or chemical composition.

Because of the chemical instability of the lower polyoxymethylene glycols, Staudinger was unable to obtain a complete structural identification of his purified fractions. Cryoscopic molecular weight determinations could not be made and attempts at conversion to the corresponding methyl ethers and acetates resulted in decomposition, so that a series of derivatives was obtained rather than a single product. However, since the homologous dimethyl ethers and diacetates, which are more stable, have been successfully isolated and identified, there is little reason for doubting their structure.

The octo-oxymethylene glycol fraction described by Staudinger is a typical representative of the group. This product, $HO \cdot (CH_2O)_8 \cdot H$, contained 93.0 per cent CH_2O and was apparently unchanged by further recrystallization from hot acetone. It dissolved readily in the hot solvent and crystallized on cooling in almost quantitative yield. It could also be recrystallized from chloroform, dioxane, and pyridine. By quick manipulation, it could even be recrystallized from water. Dilute acids and alkalies caused decomposition in a short time, even at low temperatures[74].

Octo-oxymethylene glycol crystallizes in large, well-formed needles that uniformly extinguish the rays of polarized light. It differs from crude mixtures of polyoxymethylene glycols which, even under the microscope, show no apparent crystalline form. It also differs from such products in having comparatively little formaldehyde odor. Staudinger is of the opinion that this is because it contains no formic acid, traces of which are present in crude polymers and catalyze decomposition.

The x-ray diagram of Staudinger's octo-oxymethylene dihydrate showed the characteristic lines of the higher polyoxymethylenes, but did not show interference rings from which the length of the molecule could be calculated[74]. It was concluded that this result might have been due either to the presence of small quantities of other polyoxymethylene glycols or to the possibility that the reflection planes containing the hydroxyl end groups did not occupy a regular position in a zone band. The ends of the molecules probably approach one another so closely that a chain of linear molecules has the net effect of a single molecule permeating the whole crystal.

Paraformaldehyde

Paraformaldehyde is defined as a mixture of polyoxymethylene glycols containing 93 to 99 per cent formaldehyde. However, commercial paraformaldehyde may also contain a small percentage of free or adsorbed water. Accordingly, the analyzed formaldehyde content of various grades may range from 91 to 98 per cent CH_2O. Flaked polymer, also known as flaked formaldehyde may contain as little as 91 per cent CH_2O, whereas special polymers of high molecular weight may have a formaldehyde content of over 98 per cent. The 95 to 96 per cent product is probably the most familiar grade. Finely ground, medium, granular and flake forms are available.

Tariff Commission figures show that U. S. production of paraformaldehyde reached 4,792,000 pounds in 1950. The corresponding figure for 1949 was 2,369,000 pounds.

Early History. Paraformaldehyde was first prepared by Butlerov[12] who obtained it by the vacuum distillation of formaldehyde solution. However, due to an erroneous determination of its vapor density, he concluded that it was dioxymethylene, $(CH_2O)_2$. This mistake was corrected by Hofmann[32] in 1869. Hofmann found that the polymer gave monomeric formaldehyde on vaporization, but made the incorrect assumption that it was a trimer and called it "trioxymethylene." The assumption was based on the apparent analogy of formaldehyde with thioformaldehyde, CH_2S, which was known to give a trimeric polymer. As a result of Hofmann's error, the term "trioxymethylene" has been applied in numerous instances to paraformaldehyde and the polyoxymethylenes and is often accompanied by the formula, $(CH_2O)_3$, when thus misused. Around the turn of the century, investigators such as Delépine[19] reserved "trioxymethylene" as a designation for polymers which were believed to be substantially free of combined water. However, in recent years, it has been employed chiefly as a synonym for paraformaldehyde.

The name *paraformaldehyde* was first employed in 1888 by Tollens and Mayer[106] who applied it to the polymeric residue which remains when formaldehyde solutions are evaporated. In 1890, Lösekann[44] discovered that this polymer contained combined water, and reported that it was a polymer hydrate having the formula $(CH_2O)_6 \cdot H_2O$. Following an exhaustive study of formaldehyde polymers, Delépine[18] concluded in 1897 that paraformaldehyde was a mixture of polymeric hydrates having the average formula, $(CH_2O)_8 \cdot H_2O$, and was formed by the condensation of methylene glycol as indicated by the equation

$$nCH_2(OH)_2 \rightarrow (CH_2O)_n \cdot H_2O + (n - 1)H_2O$$

In its essentials, this statement is in good agreement with the conclusions of modern chemists.

Composition and Structure. The chemical composition of paraformaldehyde is best expressed by the type formula, $HO \cdot (CH_2O)_n \cdot H$. Depending on conditions of preparation, it may contain, as previously stated, 91 to 99 per cent formaldehyde. Commercial paraformaldehyde is often specified to contain not less than 95 per cent formaldehyde by weight. According to the findings of Staudinger and his co-workers[97], paraformaldehyde is a mixture of polyoxymethylene glycols containing from 8 to 100 formaldehyde units per molecule. Although, as this would indicate, it usually contains small quantities of the lower polyoxymethylene glycols, this fraction is generally quite low, as shown by the fact that most samples contain only a small portion of acetone-solubles[75]. The majority of the polyoxymethylene glycols in paraformaldehyde must accordingly contain over 12 formaldehyde units per molecule. The average degree of polymerization is only roughly indicated by the formaldehyde content, since samples which are identical in this respect have been found by the writer to differ widely when compared by other indices of polymerization. This can be explained by the hypothesis that paraformaldehyde may in some instances contain varying amounts of adsorbed water or dilute formaldehyde solution.

Although isolation and rigid structural identification of the higher polyoxymethylene glycols in paraformaldehyde is not possible by known methods, the evidence offered by the chemical composition and physical characteristics of this polymer, plus its close relationship to the lower polyoxymethylene glycols, affords a sound basis for present conclusions concerning its structure and chemical composition. The information offered by the related polyoxymethylene glycol ethers and esters furnishes additional corroborative evidence

X-ray studies of paraformaldehyde indicate that it has a crystalline structure[78]. This is illustrated by the x-ray diagram shown in Figure 23. On the basis of x-ray studies, Ott[61] concludes that the structural unit of paraformaldehyde is $(CH_2O)_4$. Samples of the polymer examined by this investigator were reported to contain 8 such units, or 32 CH_2O groups per molecule[60, 61] These findings have been disputed by Staudinger[85], Mie and Hengstenberg[50], Sauter[63], and Sauterey[63a].

The formaldehyde content of paraformaldehyde is readily obtained by the ordinary procedures for formaldehyde analysis (pages 397–98).

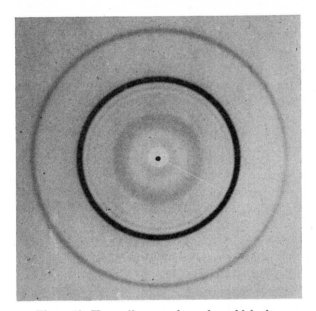

Figure 23. X-ray diagram of paraformaldehyde.

Thermochemistry. Heats of combustion for formaldehyde in the form of paraformaldehyde were determined by Delépine[18], von Wartenberg, Muchlinski and Riedler[107], and most recently by Delépine and Badoche[20]. The more recent Delépine value was measured at the Collège de France in 1942 using the calorimetric installation of Moureu, Dufraisse and Landieu with the object of securing a more accurate measurement than was possible in 1897. Both values were determined for a 93 per cent paraformaldehyde, $(CH_2O)_8 \cdot H_2O$. The more recent figure, 120.05 kcal per mol equivalent of CH_2O, compares with 120.8 kcal obtained previously. Von Wartenberg *et al.* report a value of 122.1 kcal per mol for CH_2O for a 94.6 per cent polymer. Heats of formation calculated from these data using Rossini's figures for carbon dioxide (beta-graphite standard) and water (page 37) are 42.3

kcal per mol CH_2O for the 93 per cent polymer (Delépine and Badoche) and 40.2 for von Wartenberg's 94.6 per cent polymer.

Using the value 28 kcal for the heat of formation of monomeric formaldehyde, the heat of polymerization for the conversion of one mol of CH_2O to paraformaldehyde is 12 to 14 kcal per mol formaldehyde.

As a result of a thermochemical study of the heat evolved when paraformaldehyde and aqueous formaldehyde are mixed with dilute sodium hydroxide solutions, Delépine[18] concludes that the heat of solution of paraformaldehyde is -2.5 kcal per mol CH_2O.

Properties. *Appearance and Odor.* Paraformaldehyde has the appearance of a colorless solid. Commercial grades have different degrees of subdivision, ranging from granular to a fine powder. Its odor is the characteristic pungent odor of monomeric formaldehyde.

Melting Range. On heating in a sealed tube, paraformaldehyde melts in a range the limits of which vary from approximately 120 to 170°C. The melting range is an index of the degree of polymerization. The lower limit of this range is only slightly higher than that for octo-oxymethylene glycol, $HO \cdot (CH_2O)_8 \cdot H$, whereas the higher limit approaches that of alpha-polyoxymethylene where n in the type formula is 100 or greater. The approximate degree of polymerization of different paraformaldehyde polymers may be roughly estimated and compared by the range in which they melt.

Decomposition Pressure and Thermal Decomposition. At ordinary temperatures paraformaldehyde gradually vaporizes, and on long exposure to the atmosphere complete volatilization eventually takes place. Nordgren[58] has demonstrated that the vapor obtained under these conditions consists principally of monomeric formaldehyde gas, probably accompanied by water vapor. This depolymerization is greatly accelerated by heat. Auerbach and Barschall[1] have shown that the gas obtained when a 93 per cent paraformaldehyde is completely vaporized at 224°C has a vapor density of 29.3, which is in fair agreement with the value 28.7, calculated for a mixture of 93 per cent CH_2O and 7 per cent water vapor.

Szalay[105] reports that exposure to ultrasonic waves causes partial depolymerization of paraformaldehyde but the effect is apparently of a minor order.

The partial pressure of formaldehyde over paraformaldehyde is in reality the decomposition pressure of the polyoxymethylene glycols of which it is composed. Exact measurement of this pressure is extremely difficult, since rates of decomposition are slow and equilibrium conditions are not readily attained. Variations in the composition of different samples of paraformaldehyde may also have an appreciable effect on the values obtained. Paraformaldehyde decomposition pressures reported by Nordgren[57] for temperatures from 10 to 58°C are reproduced in Table 38. Decomposition pressures for paraformaldehyde at higher temperatures as reported by Nielsen and Ebers[57] are also included, but these are admittedly rough measurements which were made merely for purposes of approximation.

It is of interest to note that the Lacy equation (page 93) for the partial pressure of formaldehyde over water solutions gives values approximating Nordgren's decomposition pressures when values of 80 to 100 are substi-

TABLE 38. DECOMPOSITION PRESSURE OF PARAFORMALDEHYDE TYPES

Temperature (°C)	Decomposition Pressure P_{CH_2O}(mm)
Data of Nordgren[58]	
10	0.84
21	1.24
25	1.45
33	3.06
37	5.00
43	7.07
47	8.22
51	10.32
58	13.56
Data of Nielsen and Ebers[57]	
65	156
80	246
90	331
100	372
110	401
120	587

tuted for W (weight per cent formaldehyde). The extrapolated equation obtained in this way is shown below:

$$\log P_{CH_2O} = 9.941 - \frac{2905}{T}$$

This equation does not reproduce the approximate values of Nielsen and Ebers at higher temperatures, giving instead much lower pressure figures.

Flash Point. Paraformaldehyde is a combustible solid whose flash point is reported as about 71°C (160°F) by the Tag closed cup tester and about 93°C (199°F) with the open cup[47].

Solubility. Paraformaldehyde dissolves slowly in cold water, more rapidly in hot water, hydrolyzing and depolymerizing as it dissolves. Formaldehyde solutions obtained in this way are identical with those obtained by dissolving gaseous formaldehyde in water[9]. According to Auerbach, solutions containing approximately 28 per cent formaldehyde may be obtained by agitating paraformaldehyde with water at 18°C for 5 weeks. At reflux temperatures almost any desired concentration of formaldehyde can be obtained in one or two hours. However, solutions obtained in this way are often cloudy due to trace impurities (*e.g.*, alumina and possibly high polymers) and must be filtered.

The effect of temperature and other variables on the solution of paraformaldehyde and the higher polyoxymethylene glycols in water has been studied in considerable detail by Löbering and various co-workers[39-43]. According to their findings[42], the temperature coefficient for the rate of paraformaldehyde solution at 12 to 28°C is 2.53 per 10°C temperature rise. This coefficient increases rapidly with rising temperature, *e.g.*, in the range 32.2 to 37.0°C it has a value of 8.39. The activation heat calculated for the range 21 to 28°C is 14.5 kcal per mol CH_2O, and for the range 32.2 to 37°C, 54.2 kcal.

Dilute alkalies and acids markedly accelerate the rate of paraformaldehyde solution, which varies with pH in the same manner as the solution reactions studied by Wadano, Trogus, and Hess (pages 56–57). The rate of solution as found by Löbering goes through a minimum between the pH values 2 to 5, increasing rapidly on either side of these limits, as illustrated in Figures 24 and 25[39]. Similar results have also been reported by Sauterey[68a].

According to Löbering and his co-workers[39, 42], measurements of the rate of solution of paraformaldehyde in water indicate that at low concentrations the reactions involved are monomolecular, as would be expected for a depolymerization reaction. However, as the formaldehyde concentration increases, the picture is complicated by reverse reactions of a higher order, *viz.*, condensation of dissolved polyoxymethylene glycols to higher homologs. As would be expected, the kinetics of paraformaldehyde solution are almost identical with those of the depolymerization reactions which take place when formaldehyde solutions are diluted.

According to Staudinger, the mechanism of the hydrolytic depolymerization reactions attending the solution of formaldehyde polymers differs for alkaline and acidic media. Under alkaline conditions, the hydroxyl end-groups are attacked and degradation proceeds in a step-wise fashion with successive splitting of formaldehyde units from the ends of the linear molecules. Under acidic conditions, the oxygen linkages within the chains may also be attacked, with splitting of the large molecules into smaller fragments. This theory is supported by the fact that the ethers of polyoxymethylene glycols are not dissolved by aqueous alkalies although they are readily attacked by dilute acids.

Löbering and Jung[42] have found that, other conditions being identical, the rate at which paraformaldehyde and other polyoxymethylene glycols dissolve in water is a good index of their degree of polymerization. Rate measurements accordingly afford a useful method for comparing samples of paraformaldehyde, alpha-polyoxymethylene, and other polymers of this type.

Toxicity. The toxicology of paraformaldehyde is essentially identical with that of formaldehyde[47]. Its activity depends on the fact that it de-

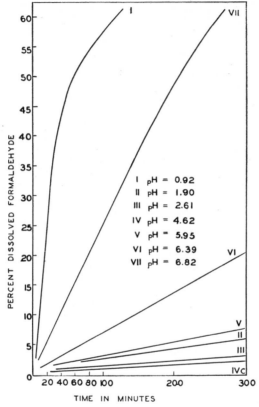

Figure 24. Effect of pH on solubility of paraformaldehyde. (From Löbering, J., *Ber.*, **69**, 1846 (1936).

polymerizes to give formaldehyde gas and dissolves to produce formaldehyde solutions. The fact that it dissolves and vaporizes slowly makes it somewhat less hazardous than concentrated formaldehyde solutions. However, it should be remembered that it can build up toxic vapor concentrations in a closed space and that solution and vaporization are accelerated by heat. Prolonged exposure, inhalation and oral intake must be avoided. Handling problems are treated in connection with formaldehyde solution (pages 79–84).

Paraformaldehyde Manufacture. In general, commercial paraformaldehyde is prepared from aqueous formaldehyde solutions by processes involving distillation and concentration to a point at which solidification
or precipitation of polymer takes place. The process is so controlled that
a product containing 91 per cent or more of formaldehyde is obtained.

Formaldehyde solutions may be readily concentrated by vacuum distillation, and this procedure has long been used commercially for the produc-

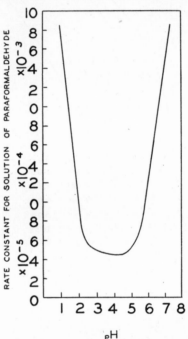

Figure 25. Effect of pH on paraformaldehyde solubility constant. (From Löbering,
J., *Ber.*, **69**, 1846 (1936).

tion of paraformaldehyde[22]. Dilute formaldehyde solution is distilled and
paraformaldehyde is left behind as a still residue.

Special variants of the vacuum distillation procedure are described in
patents.* Nasch[55] claims that a highly soluble polymer is obtained when
the vacuum concentration is carried out at 40°C or lower, especially when
colloidal substances such as egg albumen are present in the formaldehyde
solution. Another patent[53] describes the use of a fractionating column in
connection with the vacuum process. Formaldehyde solutions of low acidity
which contain 0.005 to 0.006 per cent formic acid are stated to be of special
value in the preparation of readily soluble products.[24]

* MacLean and Heinz describe a two-stage vacuum distillation procedure[45a].

According to Naujoks[56], an extremely soluble paraformaldehyde is precipitated when a water solution containing 55 to 65 per cent formaldehyde and 10 to 15 per cent methanol is gradually cooled from 65 to 15°C. The dried product, which is described as crystalline, is reported to contain 93.2 per cent formaldehyde. It is possible that this product is an octo-oxymethylene glycol, $HO \cdot (CH_2O)_8 \cdot H$ (calculated CH_2O content 93.0 per cent), of a comparatively high degree of purity.

Greenwald and Cohen[29] have patented a modified paraformaldehyde process in which the solution is aged for 8 to 20 hours after reaching the 55 to 65 per cent concentration while maintaining the temperature at about 55 to 65°C and then completing the vacuum distillation at a temperature below 50°C.

Production of paraformaldehyde by distillation processes involving the use of fractionating columns has been patented by Pyle and Lane[66].

Kuss[38] obtained paraformaldehyde by direct condensation of the vapors obtained in manufacturing formaldehyde from methanol. This gives a highly concentrated formaldehyde which is readily converted to paraformaldehyde by vacuum evaporation. Fuchs and Naujoks[27] concentrate formaldehyde by distillation with an inert organic liquid, such as ethyl acetate, which forms a low-boiling azeotrope with water, and thus obtain paraformaldehyde as an end-product of distillation. It has been previously pointed out that this is tantamount to distillation at reduced pressure. The use of fractional condensation technique for removal of water from mixtures of water vapor and gaseous formaldehyde until a gas capable of condensation to a solid paraformaldehyde has been obtained is also described in the patent literature[67]. This technique is also employed by Michael[49] who partially dehydrates a mixture of formaldehyde and water vapor by contacting it with a water-immiscible liquid at a temperature below its dew point and separating solid polymer from the enriched vapor by cooling. A screw conveyer is used to remove the polymer from the cylindrical tube in which it forms. A process patented by Walker[111] involves contacting a formaldehyde-water vapor with paraformaldehyde at a temperature above the dew point but well below the decomposition pressure of paraformaldehyde.

Modification of the polymerization process in the production of paraformaldehyde by vacuum concentration can be effected by addition of acids[16, 112], alkalies[17, 45b] and other additives[113].

Fuller[28] makes use of the catalytic effect of hydroxyl ions in accelerating paraformaldehyde solution. According to his claims, an extremely soluble polymer is obtained by mixing paraformaldehyde with a small quantity of a material which gives a strongly basic reaction when the mixture is treated with water. A composition containing up to 0.6 per cent sodium sulfite is described.

Peterson[63] obtains a soluble composition by adding hexamethylenetetramine to a difficultly soluble formaldehyde polymer. Soluble polymeric gels are obtained by adding sodium silicate solution to aqueous formaldehyde[31].

Paraformaldehyde of low reactivity and solubility is preferred for some uses and may be obtained by heating regular paraformaldehyde at 100°C[10].

In general, the solubility of paraformaldehyde is a measure of its reactivity[108].

Mechanism of Paraformaldehyde Production. When the concentration of formaldehyde in aqueous solution is increased by evaporation or distillation, the concentration and average molecular weight of the dissolved polyoxymethylene glycols increases until the saturation concentration of the less soluble products is exceeded and precipitation takes place. On standing or on further heating, the lower polyoxymethylene glycols primarily formed undergo further reaction with loss of combined water, and paraformaldehyde is produced.

The primary reactions are undoubtedly polycondensations of the type which has been previously shown to occur in formaldehyde solution (pages 47–55), *e.g.*:

$$HOCH_2OCH_2OCH_2OH + HOCH_2OH \rightarrow HOCH_2O(CH_2O)_2CH_2OH + H_2O$$

$$HOCH_2O \cdot (CH_2O)_2CH_2OH + HOCH_2OH \rightarrow HOCH_2O(CH_2O)_3CH_2OH + H_2O, \text{ etc.}$$

Conversion of the solid polymeric glycols to higher homologs may have another mechanism. Staudinger[103] believes that monomeric formaldehyde, formed by partial decomposition, undergoes a topochemical reaction with other polyoxymethylene glycol molecules as indicated below:

$$HOCH_2O(CH_2O)_3CH_2OH + CH_2O \rightarrow HOCH_2O(CH_2O)_4CH_2OH$$

The conversion of lower polyoxymethylene glycols to paraformaldehyde is demonstrated by the fact that the acetone-soluble polymers obtained by cooling hot 80 per cent aqueous formaldehyde gradually lose water on standing over phosphorus pentoxide, with a resulting increase in formaldehyde content. At the same time the solubility of the polymeric mixture falls off until it becomes completely insoluble in hot acetone, indicating the disappearance of molecules containing 12 or less formaldehyde units[77]. According to Staudinger[79], the reactions responsible for this conversion are catalyzed by the traces of formic acid usually present in crude polyoxymethylene glycols. A purified hepta-polyoxymethylene glycol fraction shows no change in formaldehyde content or acetone-solubility when subjected to desiccation under the above conditions. It loses weight due to volatilization, but does so at a slow though constant rate until completely evaporated. In the case of crude polyoxymethylene glycol mixtures, loss of weight by vaporization is at first rapid, but decreases constantly on storage.

Alpha-Polyoxymethylene

Alpha-polyoxymethylene is a solid formaldehyde polymer normally containing in the neighborhood of 99.7 to 99.9 per cent formaldehyde. It is generally prepared by the action of acids and alkalies on formaldehyde solutions. The term "trioxymethylene" was employed by Delépine and other early investigators to designate this polymer.

Composition and Structure. According to Staudinger[98], alpha-polyoxymethylene is a mixture of polyoxymethylene glycols in which the n value of the type formula, $HO \cdot (CH_2O)_n \cdot H$, is not less than 100 and probably much greater.

It is impossible to prove whether the small quantity of water (approximately 0.1 to 0.3 per cent) present in alpha-polyoxymethylene is chemically bound, as in the lower polyoxymethylene glycols, or physically adsorbed. However, the properties of this polymer are apparently those of a polyoxymethylene glycol. The determination of molecular weight is impossible by known methods, but analogy with related high molecular weight polyoxymethylene dimethyl ethers (gamma-polyoxymethylene) indicates that the degree of polymerization is at least 100. Estimates based on magnetic susceptibility measurement, as reported by Farquharson[23], indicate a much lower figure. However, this investigator's value for n in the type formula is 32 which would correspond to a formaldehyde content of 98.3 per cent for the polymer tested.

Properties. Alpha-polyoxymethylene has the appearance of a colorless powder. Its x-ray diagram indicates a definite crystal structure similar to that of paraformaldehyde. On heating in a sealed tube this polymer melts in the range 170 to 180°C. In general, its properties are similar to those of paraformaldehyde except that because of its high molecular weight, differences of degree are plainly manifest. According to Delépine[18], its heat of combustion is 122.9 kcal for each formaldehyde equivalent (CH_2O). Its heat of formation as calculated from this figure is 40.3 kcal per mol CH_2O.

Alpha-polyoxymethylene dissolves slowly in cold water, but the rate of solution is so slow that for ordinary purposes it may be regarded as practically insoluble. As in the case of paraformaldehyde, solution is accompanied by depolymerization and hydrolysis, and formaldehyde solution is obtained. On agitating a 5-g sample of alpha-polyoxymethylene with 25 cc water for one hour at room temperature (25°C), the writer[109] obtained a solution containing 0.02 per cent or less of formaldehyde. According to Auerbach[3], 3 g alpha-polyoxymethylene dissolved almost completely after agitation with the same quantity of water for 109 days at 15°C; solution is somewhat more rapid on boiling. Dilute alkali catalyzes solution and dissolves the polymer even at room temperature. On warming, solution takes place quickly and completely. Staudinger[76] reports that alpha-polyoxymethylene is soluble in hot formamide, and that although it depolymerizes in the hot solvent, polymers having a polymerization degree of $n = 50$ to 100 are apparently capable of existing for a short time in the dissolved state and can be crystallized out if the solution is chilled quickly.

Löbering and Hilber[41] have found that the solubility of this polymer can be increased by grinding in a vibratory mill and conclude that the polyoxymethylene chains are broken down by this treatment. A particle size of 1 to 3 μ was obtained on 20 hours' grinding, and although further grinding up to 90 hours produced no additional change in apparent particle size, the rate of solution increased continually with this treatment.

On standing, alpha-polyoxymethylene gradually evolves formaldehyde gas which can be recognized by its characteristic odor. Tensimeter measurements of the formaldehyde pressure over a polymer sample at 25°C showed gradually increasing values until an apparent equilibrium was attained at 6.8 mm after 6 to 7 months. At 37.5°C, tensimeter readings of 17 to 22 mm were obtained in approximately the same period. On heating at ordinary pressure, depolymerization takes place and almost pure monomeric formaldehyde is obtained[3].

Vapor density measurements on the gas obtained by vaporizing acid-precipitated alpha-polyoxymethylene in a Dumas bulb immersed in hot oil, as made by the writer, indicated the following molecular weight values.

Vaporization Temp. (°C)	Average Mol. Wt.
136	31.3
146	31.0
165	31.9

These values are slightly over 30, a fact attributed to the presence of small concentrations of trioxane. A mixture of 97 per cent formaldehyde monomer and 3 per cent trioxane would give an average molecular weight of approximately 32.

The formaldehyde content of alpha-polyoxymethylene is readily obtained by standard methods of formaldehyde analysis, such as the alkaline-peroxide procedure, the sodium sulfite method, and the iodometric method of Romijn. The latter method of analysis as developed by Signer[99], possesses an accuracy of at least 0.2 per cent.

Preparation. Alpha-polyoxymethylene cannot be obtained by dehydrating formaldehyde solutions at low temperatures (below 100°C) by purely physical methods. More strenuous procedures involving the use of alkaline and acidic catalysts or dehydrating agents are required. Alpha-polyoxymethylene can also be obtained by the action of heat on paraformaldehyde[4, 10, 108].

The preparation of a formaldehyde polymer by addition of sulfuric acid to "Formalin" was first reported by Cambier and Brochet[14] in 1894. However, their polymer was probably contaminated with polyoxymethylene dimethyl ethers which are always formed by this procedure when methanol is present. According to Auerbach[3], alpha-polyoxymethylene is best prepared by adding one volume of concentrated sulfuric acid gradually and with constant agitation and cooling to 10 volumes of 37 per cent methanol-free formaldehyde solution. The solution is then cooled for several hours at 0 to 10°C, until precipitation is complete. The polymer is filtered off and washed successively with alcohol and ether, and then dried. According to Staudinger[101], this procedure gives 60 to 80 per cent yields of alpha-polyoxymethylene, depending on the temperatures employed. Washing with alcohol and ether is not necessary if the product is subjected to vacuum desiccation after washing. More concentrated formaldehyde solutions can be employed, but if this is done, paraformaldehyde may be precipitated. Auerbach states that for this reason the use of higher concentrations than 50 to 54 per cent is not advisable. In the first stages of the process strenuous cooling should be avoided, since a difficultly filterable product (probably contaminated with paraformaldehyde) is thus formed. The temperature should be kept in the range 30 to 50°C. Slow cooling of the acidified product gives a precipitate which is easily filtered and which, on desiccation, has a formaldehyde content of 99.6 to 99.8 per cent. Kohlschütter[37] has made a morphological study of the forms in which the polymer is precipitated from acidified formaldehyde.

Precipitation of formaldehyde polymer by the addition of alkali to formaldehyde solution was observed by Mannich[46] in 1919. Staudinger, Signer, and Schweitzer[104] studied this process in 1931 and concluded that the polymeric product was alpha-polyoxymethylene. Although polymer prepared in this way contains a trace of metal hydroxide (0.1 to 0.2 per cent), Staudinger's study of polymers obtained with varying proportions of metal hydroxides of the first two groups of the periodic table gives no evidence for believing that the base is chemically combined with the polymer. Addition of potassium hydroxide to 40 per cent methanol-free formaldehyde in the proportion of one mol KOH per 100 mols dissolved CH_2O gives a 58 per cent yield of alpha-polyoxymethylene. No polymer is precipitated when as much as $1KOH/2CH_2O$ or as little as $1KOH/10,000CH_2O$ is employed. The yield goes through a maximum between the ratios $1KOH/20CH_2O$ and $1KOH/500CH_2O$.

A process of producing polyoxymethylenes by adding sulfuric or phosphoric acid to formaldehyde solutions or paraformaldehyde-type polymers at temperatures not higher than 60°C has been patented by Schilt[69]. The polymers obtained differ from paraformaldehyde in that they will continue to burn once they have been ignited, and can be used as a solid fuel for specialty purposes. Pirlet[64] has developed a method for the preparation of polyoxymethylenes by adding alkaline catalysts to 50 to 60 per cent formaldehyde. These catalysts may include ammonia, ammonium salts, hexamethylenetetramine, quaternary ammonium bases, ethanolamine, urea, and pyrones, as well as ordinary caustic. The presence of methanol in the formaldehyde solution does not prevent polymerization, but the best yields are obtained when it is absent. The product is stated to be crystalline in nature.

Beta-Polyoxymethylene

The term *beta-polyoxymethylene* was applied by Auerbach and Barschall[5] to the formaldehyde polymer obtained by gradual addition of 4 volumes of concentrated sulfuric acid to 10 volumes of approximately 40 per cent formaldehyde solution. If the acidified mixture is allowed to cool gradually, the polymer is deposited in a definite crystalline form, and on examination under the microscope is found to consist of clear, transparent, short, thick, hexagonal prisms with flat pyramidal surfaces or six-cornered tablets. In water it is even less soluble than alpha-polyoxymethylene, but like this polymer dissolves readily in sodium sulfite solution and dilute alkalies. Staudinger and co-workers[100] conclude that like alpha-polyoxymethylene it consists of polyoxymethylene glycols of high molecular weight. The fact that it is less soluble than the alpha-product may be due to its smaller surface. The crystal form probably has much to do with its physical properties. It also contains a small proportion of sulfuric acid (approximately 0.2 to 0.4 per cent) which cannot be completely removed by washing. This acid may be chemically combined, but Staudinger is inclined to doubt this since there is no clear relation between sulfuric acid content and molecular weight. Furthermore, the lower polyoxymethylene glycol sulfates have not been isolated. It is more likely that the sulfuric acid may be adsorbed or occluded in the polymer crystals.

On long standing, beta-polyoxymethylene becomes partially insoluble in dilute alkali and sodium sulfite solution. This is apparently due to reactions catalyzed by sulfuric acid, resulting in the formation of small amounts of polyoxymethylene dimethyl ethers. Staudinger[102] has demonstrated that methyl alcohol is produced by the action of sulfuric acid on formaldehyde, and the methyl alcohol accountable for the dimethyl ether formation is probably formed in this way. Sulfuric acid catalyzes the Cannizzaro reaction as well as the ether formation.

On heating in a sealed tube, beta-polyoxymethylene melts with decomposition at 165 to 170°C. On heating at ordinary pressure, it tends to vaporize without melting and sublimes readily. Auerbach and Barschall found that the average molecular weight of the gas obtained by vaporizing the beta-polymer at temperatures of 184 to 224°C varied from 31 to 33, and that on standing at 198°C the molecular weight increased gradually. Vapor density measurements obtained by the writer when beta-polyoxymethylene was vaporized at 116 and 122°C indicated average molecular weights of 35.6 and 34.3 respectively. The fact that these values exceed 30 is probably caused by the formation of trioxane, $(CH_2O)_3$, which is catalyzed by acid.

The ease with which beta-polyoxymethylene may be sublimed is a specific characteristic of this polymer definitely attributable to its acid content. According to Kohlschütter[36], the fibers obtained on subliming polymers of this type are produced by the formation of polyoxymethylene on the fine fiber-like crystals formed by the trioxane which is always present in the vapors obtained from these polymers. The fiber is thus a pseudomorph of polyoxymethylene on a trioxane crystal. Also, according to Kohlschütter, trioxane can itself undergo a kind of lattice conversion by which it approximates in its properties a complex polyoxymethylene and then forms an integral part of the fiber structure. This change occurs under the influence of the polymerizing formaldehyde, even during crystal growth, and leads to particularly stable polyoxymethylene fibers of uniform inner structure. According to Staudinger and co-workers[81], the structure of the fibers obtained by subliming beta-polyoxymethylene are similar to those of cellulose.

Polyoxymethylene Glycol Derivatives

The functional derivatives of the polyoxymethylene glycols are chiefly of theoretical interest for their contribution to our understanding of the structure of paraformaldehyde and related polyoxymethylenes. Staudinger's studies of the polyoxymethylene diacetates and dimethyl ethers are the basis for the modern interpretation of the chemistry of these polymers.

Although first discovered at an earlier date, the higher polyoxymethylene dimethyl ethers, gamma-polyoxymethylene, and delta-polyoxymethylene were shown by Staudinger to be members of this group. The last is a partially rearranged polyoxymethylene.

Polyoxymethylene Diacetates

Structure. Polyoxymethylene diacetates are obtained by the action of acetic anhydride on paraformaldehyde and alpha-polyoxymethylenes. Their chemical structure, as indicated below, is readily expressed by the type formula,

$$CH_3 \cdot COO \cdot (CH_2O)_n \cdot OC \cdot CH_3 \quad \text{or} \quad CH_3 \cdot CO \cdot O \cdot CH_2 \cdot O \cdot CH_2 \cdot O \cdots CH_2 \cdot O \cdot CO \cdot CH_3 \cdot$$

These derivatives are more stable than the glycols from which they are obtained, and accordingly can be subjected in many instances to the standard methods of chemical manipulation. Individual members of the series containing up to approximately 20 formaldehyde units per molecule have been isolated, and chemical structures have been determined by measurement of molecular weight and analysis for formaldehyde and acetic acid[82]. Molecular refraction figures are in agreement with the structure postulated and do not support the theory that they may be molecular compounds of polyoxymethylene and acetic anhydride, $CH_3CO \cdot O \cdot OC \cdot CH_3 \cdot nCH_2O$. X-ray studies of the diacetates show that they are chain molecules crystallized in a molecular lattice and that they lie in parallel layers[86]. The molecular lengths for pure diacetates can be accurately determined by the same methods which have been employed by Müller and Shearer[54] for fatty acids. Owing to their lack of solubility, the higher diacetates cannot be isolated in a pure state. They decompose at or below their melting point and gradually split up when heated in high-boiling solvents. That they are higher members of the homologous series is demonstrated by their physical and chemical properties and by chemical analysis.

Properties. The physical properties of some of the lower polyoxymethylene diacetates, together with the results of molecular weight determination and chemical analysis, are illustrated in Table 39, prepared from the data of Staudinger and Lüthy[82, 87]. As will be seen, the physical properties change progressively with increasing molecular weight just as in the case of other homologous series of organic compounds. Solubility in water, alcohol, ether, acetone, and other solvents continually drops off with increase in the value of n in the type formula, while melting point, density, viscosity, and refractive index increase. A high molecular weight diacetate which could not be purified by recrystallization was found to contain 92.7 per cent formaldehyde and to melt with decomposition at 150 to 170°C. Since this material is free of soluble polymeric acetates containing less than 23 CH_2O units, it must consist of a mixture of polyoxymethylene diacetates containing from 23 to approximately 70 formaldehyde units[89].

Magnetic susceptibilities of polyoxymethylene diacetates have been measured by Farquharson[23] and the molecular values reported are shown in Table 40. Based on these figures the susceptibility per CH_2O unit is -14.87×10^{-6}.

Pure polyoxymethylene diacetates are stable indefinitely at room temperature. Lower members, for which n is less than 14, may be kept at 100°C for 50 days without decomposition. Higher homologs are less stable. Impure diacetates decompose spontaneously, with formation of polyoxymethylenes. This decomposition is apparently due to the action of catalytic impurities and moisture. The soluble diacetates give stable solutions in

TABLE 39. PROPERTIES OF POLYOXYMETHYLENE DIACETATES*

Type Formula: $CH_3COO(CH_2O)_nOC \cdot CH_3$

Degree of Polymerization (n)	M.P. (°C)	B.P. 0.1 mm (°C)	Density	n_D	Viscosity (Water = 1)	Molecular Length in Crystals (Å)†	Appearance	Solubility (g per 100 cc Solvent)	CH₂O Content Found	CH₂O Content Calc.	Molecular Weight Found§	Molecular Weight Calc.
1	−23	39–40	$1.128_{24}°$	$1.4025_{24}°$	2.04	—	Colorless fluid	Miscible with alcohol	22.9	22.7	131	132
2	−13	60–62	$1.158_{24}°$	$1.4125_{24}°$	3.30	—	"	" " ether	37.1	37.0	162	162
3	−3	84	$1.179_{24}°$	$1.4185_{24}°$	5.55	—	"	" " " ether	46.8	46.9	189	192
4	7	102–104	$1.195_{24}°$	$1.4233_{24}°$	8.68	14.6	"	" " " acetone	53.7	53.6	216	222
5	17	124–126	$1.204_{24}°$	$1.4255_{24}°$	11.8	16.9	"	Sol. in ether + acetone	59.4	59.5	248	252
6 and 7‡	Approx. 15	180–190	—	—	—	—	"	" " " "	—	—	—	—
8	32–34		$1.216_{36}°$	$1.4297_{36}°$	—	23.7	Wax-like solid	" " " "	69.5	70.2	345	342
9	40–43		$1.353_{15}°$	—	—	25.2	White micro-cryst. powder	" " acetone + chloroform	72.3	72.6	—	372
10	52–53.5		—	—	—	27.2	"	" " " + "	73.8	74.6	385	402
11	65.5–67		—	—	—	—	"	Ether 0.5 g	76.6	76.4	—	432
12	73–75		—	—	—	32.1	"	Acetone 2.0 g	77.8	77.9	440	462
14	84–86		$1.364_{15}°$	—	—	34.6	"	" 1.5 g	80.3	80.4	493	522
15	90.5–92		—	—	—	36.8	"	" 1.0 g	81.0	81.5	543	552
16	93–95		—	—	—	38.5	"	" 0.8 g	82.3	82.5	570	582
17	98.9–99.5		$1.370_{15}°$	•	—	40.4	"	" 0.6 g	83.1	83.3	590	612
19	107–109		$1.390_{15}°$	—	—	43.7	"	" 0.4 g	84.2	84.0	665	672
20	111–112		—	—	—	—	"	Methyl acetate 0.2 g	85.2	85.5	—	702
22	116–118		—	—	—	—	"	" 0.1 g	85.9	86.8	—	762
35¶ (av.)	150–157		—	—	—	—	"	Sol. in hot formamide	91.0	91.3	1170	1152

* From Staudinger, H. et al., Ann., **474**, 195–197 (1929) and Staudinger, H.. "Die Hochmolekularen Organischen Verbindungen," 1932, page 234.

† See also: Mie, G. and Hengstenberg, J., Z. physik. Chem., **126**, 425–448 (1927).

‡ The isolation of these diacetates is difficult since they decompose on vacuum fractionation and are not easy to crystallize because of their low melting points.

§ First five determined by cryoscopy in benzene, last by cryoscopy in camphor, remainder by ebullioscopy in methylene chloride.

¶ Crystallized from hot formamide.

135

organic solvents and do not decompose on boiling at temperatures up to 80°C. In higher-boiling solvents, such as pyridine, toluene, xylene, etc., decomposition proceeds rapidly at the boiling temperature, and the odor of formaldehyde split off in decomposition becomes apparent[87]. Pure polyoxymethylene diacetates do not smell of formaldehyde. In the case of the more volatile members, irritation of the type produced by inhaling formaldehyde appears as a delayed reaction following exposure to vapors. This is probably due to hydrolysis taking place on the mucous membranes of the nose and throat.

The chemical properties of the polyoxymethylene diacetates combine those of an ester and a formaldehyde polymer. They are readily saponified by dilute alkali and acid to polyoxymethylene glycols which are then hydrolyzed to give formaldehyde solution. In solution, they behave chemically as formaldehyde, differing only in that they react more slowly, since hy-

TABLE 40. MOLECULAR SUSCEPTIBILITIES OF POLYOXYMETHYLENE DIACETATES[23]

CH_2O Units n	Molecular Susceptibility $-\chi \times 10^6$
1	70.7
2	85.8
3	100.5
4	115.6
6	145.2
9	190.6
22	383.0

drolysis and depolymerization must precede reaction. In the cold a 14-polyoxymethylene diacetate, $CH_3COO(CH_2O)_{14}COCH_3$, does not darken ammoniacal silver nitrate for 12 hours, whereas paraformaldehyde does so in 5 minutes and formaldehyde solution acts instantaneously[87]. On heating at 200°C, they are decomposed to formaldehyde and acetic anhydride.

Preparation. Descudé[21] prepared methylene diacetate, $CH_2(OOC \cdot CH_3)_2$, and dioxymethylene acetate, $CH_3COO(CH_2)_2OCCH_3$, in 1903 by the action of paraformaldehyde on acetic anhydride in the presence of zinc chloride. Staudinger and Lüthy[82] confirmed this work in 1925, but found that zinc chloride caused decomposition of the higher acetates, which could be obtained in good yield only when the metal salt was absent.

The higher diacetates are best obtained when more than one mol equivalent of CH_2O is reacted with one mol acetic anhydride. Good results are obtained by heating one mol acetic anhydride with 5 equivalents of formaldehyde in the form of paraformaldehyde for ½ to 1 hour at 160 to 170°C in a sealed tube. The reaction product contains liquid diacetates ($n = 1$ to 5), soluble solid diacetates ($n = 7$ to 22), and insoluble products

($n = 23$ to 70). Separate fractions and individual products are isolated by processes involving solvent extraction, vacuum distillation, and crystallization[88].

Polyoxymethylene Dimethyl Ethers

Structure. The polyoxymethylene dimethyl ethers make up a homologous series of polyoxymethylene glycol derivatives having the structure indicated below:

$$CH_3 \cdot O \cdot CH_2 \cdot O \cdot CH_2 \cdot O \cdot CH_2 \cdots O \cdot CH_2 \cdot O \cdot CH_3$$

They may also be represented by the type formula, $CH_3O \cdot (CH_2O)_n \cdot CH_3$. Their structure is clearly demonstrated by chemical analysis, synthesis, molecular weight, and other physical and chemical properties. Chemically, they are acetals closely related to methylal, $CH_3OCH_2OCH_3$, which may be regarded as the parent member of the group in which n of the type formula equals 1. They are synthesized by the action of methanol on polyoxymethylene glycols in the presence of an acidic catalyst, just as methylal is synthesized by the action of methanol on aqueous formaldehyde (methylene glycol). On hydrolysis they are converted to formaldehyde and methanol in the quantities indicated by their molecular weight and chemical constitution. Like other acetals, they possess a high degree of chemical stability. These ethers are not hydrolyzed under neutral or alkaline conditions, but are readily attacked by dilute acid. They are more stable than the polyoxymethylene diacetates.

Properties. Due to the relatively small differences in the physical properties (melting points, boiling points, and solubility) of adjacent members in this series, individual homologs are not readily separated. However, fractions having various average molecular weight values have been isolated. The physical properties of fractions of this sort are indicated in Table 41, data for which have been gathered principally from the publications of Staudinger and his co-workers[90]. Dioxymethylene dimethyl ether, which is included, was prepared by Löbering and Fleischmann[40] by the reaction of sodium methylate with dichlorodimethyl ether.

The higher members of the group are solids which crystallize readily in small leaflets. The molecular lattice forces in these crystals are probably greater than in the less symmetrical polyoxymethylene diacetates, as evidenced by the fact that derivatives with corresponding degrees of polymerization have higher melting points and lower solubilities.

That the thermal stability of the polyoxymethylene diethers is greater than that of the diacetates is indicated by the fact that the hexa- and hepta-fractions can be distilled without decomposition. Thermal stability decreases as the degree of polymerization increases. The melting point

TABLE 41. PROPERTIES OF POLYOXYMETHYLENE DIMETHYL ETHERS
Type Formula: $CH_3O\cdot(CH_2O)_n\cdot CH_3$

Average n	M.P. (°C)	B.P. (°C)	Density	Appearance	Solubility	Molecular Wt. Found*	Molecular Wt. Calc.	Formaldehyde Content Found	Formaldehyde Content Calc.
2†	—	91–93 (1 atm)	—	Colorless liquid	Mis. w. benzene + water	—	106	56.9	56.6
6	31–34	150–170 (20 mm)	—	Crystalline solids	Sol. in benzene	233	226	79.7	79.6
15	109–111	—	—	"	Diff. sol. in water	—	496	90.5	90.7
23	140–143	—	—	"	Insol. in water	650	736	92.7	93.7
33	152–156	—	—	"	Sol. in formamide at approx. 90°C	1010	1030	94.8	95.5
50	161–163	—	—	"	Sol. in hot xylol	1610	1546	97.1	97.0
80	165–170	—	1.467	"	Insol. in hot xylol	2490	2446	98.0	98.1
90	170–180	—	—	"	Sol. in hot formamide	2830	2746	98.2	98.3
100	170–175	—	—	"	Sol. in formamide at approx. 150°C	2950	3046	98.4	98.5

* Determined cryoscopically with camphor as solvent.
† Löbering, J. and Fleischmann, A., *Ber.*, **70**, 1680–3 (1937).

gradually increases with increasing molecular weight. Polymeric ethers having an n value of approximately 150, $(CH_3O \cdot (CH_2O)_{150}CH_3)$, melt with decomposition at 180°C. For higher polymers, decomposition takes place below the melting point.

The acid-catalyzed saponification of the polyoxymethylene dimethyl ethers has been studied by Löbering and co-workers[40, 43]. Reaction rates have been determined and the mechanism of the hydrolytic process is discussed. The breaking down of the pure ether linkages at the start proceeds most slowly. The solubility of the ethers, which is dependent on the molecular weight, plays a predominant role, since it determines the concentration of dissolved undecomposed molecules and with it the relative velocity of the total process.

Preparation. The polyoxymethylene dimethyl ethers are prepared by heating polyoxymethylene glycols or paraformaldehyde with methanol in the presence of a trace of sulfuric or hydrochloric acid in a sealed tube for 15 hours at 150°C, or for a shorter time (12 hours) at 165 to 180°C. Considerable pressure is caused by decomposition reactions, which produce carbon oxides, and by formation of some dimethyl ether. The average molecular weight of the ether products increases with the ratio of paraformaldehyde or polyoxymethylene to methanol in the charge. A good proportion of both low and high polymer is obtained with a 6 to 1 ratio of formaldehyde (as polymer) to methanol. The products are purified by washing with sodium sulfite solution, which does not dissolve the true dimethyl ethers, and may then be fractionated by distillation, in the case of the lower polymers, or by fractional crystallization from various solvents.

Gresham and Brooks[29a] obtain good yields of polyoxymethylene dimethyl ethers containing 2 to 4 formaldehyde units per molecule. This procedure is carried out by heating methylal with paraformaldehyde or concentrated formaldehyde solutions in the presence of sulfuric acid. Related dialkyl ethers, such as di-, tri- and tetraoxymethylene diisobutyl ethers, can be obtained by reactions involving the corresponding higher dialkyl formals.

Reaction of methanol and polyoxymethylene glycols in the absence of acids results in the formation of polyoxymethylene hemiacetals, $HO \cdot CH_2O \cdot CH_2 \cdot O \cdots CH_3 \cdot O \cdot CH_3$, which are of low stability and are readily soluble in dilute alkalies and sodium sulfite solution.

Gamma-Polyoxymethylene (Higher Polyoxymethylene Dimethyl Ethers)

Structure. This polymer, which was prepared and named by Auerbach[6], has been shown by Staudinger[91] to consist essentially of high molecular weight polyoxymethylene dimethyl ethers, in which n of the type formula is greater than 100. It is probable that the majority of the polymer molecules have an n value of 300 to 500. The presence of methoxy groups is demonstrated by isolation of methanol from the products of its hydrolysis. This can be done by dissolving gamma-polyoxymethylene in boiling 0.1N hydrochloric acid, adding sulfanilic acid to combine with the formaldehyde thus set free, distilling a water-methanol fraction from this mixture, and determining its methanol content by titration with potassium permanganate. Identification of methanol in the distillate is possible by converting it to methyl p-nitrobenzoate. Formation of methanol by the Cannizzaro reaction during polymer hydrolysis is ruled out by the fact that alpha-polyoxymethylene, when subjected to a similar treatment with 0.1N hydrochloric acid, gives only traces of methanol. This is not the case if N acid is employed. This method of determining the combined methanol in gamma-polyoxymethylene and related polymers has an accuracy of 0.3 to 0.5 per cent[93].

Farquharson[23] states that the structures of both gamma- and delta-polyoxymethylene must be radically different from the alpha polymer, since their magnetic susceptibility is lower.

Properties. Gamma-polyoxymethylene is a colorless crystalline product which decomposes on heating in the temperature range 160 to 210°C. The exact temperature of decomposition depends on the method of preparation and purification. It melts to a clear liquid at approximately 160 to 180°C. On vaporization, it depolymerizes to monomeric formaldehyde gas. The polymer does not smell of formaldehyde and is insoluble in water and organic solvents. It is not affected by dilute alkalies, but dissolves with hydrolysis in acidic solutions when heat is applied.

Preparation. Auerbach and Barschall[6] first prepared gamma-polyoxymethylene by the addition of two volumes concentrated sulfuric acid to five volumes of "Formalin" with sufficient cooling to keep the temperature at approximately 20°C. On standing, a mixture of beta- and gamma-polyoxymethylene was precipitated from which the beta polymer could be removed by extraction with sodium sulfite solution. These investigators[6] concluded that the presence of methanol in the commercial formaldehyde solution was responsible for the formation of the gamma polymer, but were not aware of the presence of methoxyl groups in the product, believing it to have the formula $(CH_2O)_n$.

Staudinger and his co-workers[92] have since demonstrated that the methanol concentration of the formaldehyde employed determines the ratio of gamma to beta polymer in the crude precipitate. Results obtained in their work by employing approximately 34 per cent formaldehyde (reported as 38 per cent by volume) in the presence of various amounts of methanol are shown in Table 42. In these experiments two volumes of sulfuric acid were added for every 5 volumes of formaldehyde solution in the manner prescribed by Auerbach. It will be observed that, although the percentage of gamma polymer increases with increasing concentrations of methanol in the solution employed, the total yield of precipitated polymer decreases. When the solution used contains 20 per cent methanol no precipitate is obtained.

Diethyl and dipropyl polyoxymethylene ethers have been prepared by Staudinger and his co-workers[94] by adding sulfuric acid to formaldehyde solution containing ethyl and propyl alcohol, following the same technique employed in making gamma-polyoxymethylene.

A process for preparing gamma-polyoxymethylene and related polyoxymethylene ethers of other alcohols as developed by Londergan[45] involves addition of sulfuric acid to a substantially nonaqueous solution of paraformaldehyde or trioxane in the desired alcohol. To secure satisfactory results, this solution must contain 60 to 90 parts by weight of the formaldehyde polymer. Gamma-polyoxymethylene is prepared by gradual addition of 24

TABLE 42. YIELDS OF BETA-AND GAMMA-POLYOXYMETHYLENE OBTAINED
BY ADDING SULFURIC ACID TO 36 PER CENT FORMALDEHYDE
CONTAINING VARYING AMOUNTS OF METHANOL*

Methanol added in Per Cent Based on Formaldehyde Solution Charged	Total Per Cent Yield of Polymer	% Beta Polymer Soluble in Na_2SO_3 Solution	% Gamma Polymer Insoluble in Na_2SO_3 Solution	Total Per Cent Yield of Gamma Polymer
1	92.1	93.3	6.7	6.3
5	85.5	43.3	56.7	48.5
10	68.4	33.3	66.7	45.6
15	52.6	16.7	83.3	43.8
20	No precipitate	—	—	—

* Data from H. Staudinger *et al.*[92]

cc concentrated sulfuric acid to a solution of 117 g paraformaldehyde in 50 g methanol. The polymer solution is obtained by adding a saturated solution of 0.1 g caustic soda in methanol to the paraformaldehyde-methanol mixture at 70 to 80°C. The sulfuric acid is added at such a rate that the temperature does not exceed 80°C. The cooled reaction mixture is broken up and treated with 350 cc of 15 per cent caustic at 60°C to remove paraformaldehyde and is then washed with water, filtered and dried. The gamma polymer obtained in this way melts at 140 to 150°C. Related products can also be obtained with ethylene glycol and other saturated aliphatics containing one to three hydroxy radicals.

Delta-Polyoxymethylene

Structure. Delta-polyoxymethylene, prepared by heating gamma-polyoxymethylene in boiling water, has been shown by Staudinger and his co-workers[95] to be a partially rearranged polyoxymethylene dimethyl ether in which a small fraction of the oxymethylene chains have been converted to the carbohydrate grouping as shown by the following equation,

$$CH_3 \cdot O \cdot CH_2 \cdot O \cdots CH_2 \cdot O \cdot CH_2 \cdot O \cdot CH_2 \cdot O \cdot CH_3 \rightarrow$$

Gamma-polyoxymethylene

$$CH_3 \cdot O \cdot CH_2 \cdot O \cdots CH_2 \cdot O \cdot CH_2 \cdot CH(OH) \cdot O \cdot CH_3$$

Delta-polyoxymethylene-type polymer

Staudinger demonstrated the presence of polymers of this type in the delta polymer by identifying glycolaldehyde triacetate, $CH_3COO \cdot CH_2 \cdot CH(OOC \cdot CH_3)_2$, in the products obtained by reacting this ether with acetic anhydride. Other polyoxymethylenes give only methylene diacetate and dioxymethylene diacetate. According to Staudinger, delta-polyoxymethylene is a mixture of partially rearranged polymers and probably contains some molecules in which three or more carbon atoms are involved in the rearrangement.

Properties. Delta-polyoxymethylene is a dusty white powder which, when viewed under a microscope, appears to consist of crystalline fragments such as might be obtained by mechanical sub-division of the gamma polymer from which it is made. Its x-ray diagram is similar to that of alpha-polyoxymethylene. The powder smells only slightly of formaldehyde and has a very low partial pressure. This polymer, as prepared by Auerbach[8], melted sharply at 169 to 170°C in a sealed melting-point tube. Staudinger's product melted in the range 150 to 170°C with decomposition, evolving formaldehyde gas. On further heating it is converted to a yellow oil of caramel-like odor. On volatilization at 198 to 218°C, its vapors were found by Auerbach to have an average molecular weight of approximately 190 to 280. With continued heating, the molecular weight of these vapors gradually decreases.

Preparation. The delta polymer was named by Auerbach and Barschall[8], who first prepared it. They believed it to be a polyoxymethylene isomer having the formula, $(CH_2O)_n$. The preparation is carried out by long boiling of gamma-polyoxymethylene with water. Auerbach heated approximately 95 g of the gamma polymer in two liters of water, boiling the mixture 6 hours per day for 14 days. Each day the water previously employed was poured off and fresh water added. Twenty-six grams of product were obtained. Since the polymer never dissolves, Staudinger assumes that it is formed by a topochemical reaction. Boiling gamma-polyoxymethylene with alkali does not give the delta polymer.

Epsilon-Polyoxymethylene

On repeated sublimation of trioxane (alpha-trioxymethylene), Hammick and Boeree[30] obtained as a residue a small amount of a nonvolatile polymer which they called epsilon-polyoxymethylene. Its properties and method of synthesis indicate a close resemblance to the higher polyoxymethylenes. Staudinger[96] believes it to be a delta-polyoxymethylene.

Epsilon-polyoxymethylene is a white, silky, paper-like material having no apparent crystalline form. On heating, it melts with decomposition at 195 to 200°C. Analyses indicate a formaldehyde (CH_2O) content of 99.7 per cent or better. It is insoluble in water, cold sodium sulfite solution, and organic solvents. It dissolves in hot nitrobenzene but separates on cooling as an amorphous precipitate. On boiling with ammoniacal silver nitrate, the material darkens but the solution is not discolored. It cannot be acetylated and therefore probably does not contain hydroxyl groups.

Eu-Polyoxymethylene

Structure. Eu-polyoxymethylene is obtained by the polymerization of anhydrous liquid formaldehyde. Chemical analyses indicate that it possesses the empirical formula $(CH_2O)_n$. It possesses the film- and fiber-forming characteristics of a substance of extremely high molecular weight, much higher than that of alpha-polyoxymethylene or any of the other formaldehyde polymers previously discussed. This indicates, according to Staudinger, that its molecules contain 5000 or more formaldehyde units. A structural formula involving multi-membered polyoxymethylene rings is a possibility[108]. However, since Heuer's study[73] of the high molecular weight polystyrenes indicates that these polymers are made up of linear molecules and do not contain multi-membered rings, Staudinger believes that eu-polyoxymethylene is also made up of linear molecules. This conclusion carries with it the assumption that some type of end group must be present. How these end-groups are formed is not clear, but it is possible that traces of polar impurities, possibly produced by decomposition reactions or structural rearrangement, may be responsible. Since the polymer can be formed from anhydrous formaldehyde at temperatures as low as −80°C, it is difficult to envisage reactions of this sort, which usually take place only at higher temperatures. Although eu-polyoxymethylene is amorphous in appearance, just as polystyrene is, it differs from polystyrene in showing an x-ray diagram which indicates a crystalline structure[89]. This x-ray diagram shows the same interference bands as those of alpha-, beta-, and gamma-polyoxymethylenes, indicating an identical internal structure. However, there is considerable cloudiness between the innermost rings of the diagram, which indicates that the crystalline structure is not as complete as in the other polymers, and points to the possibility of amorphous zones. When the polymer is warmed to its softening point and then cooled, the diagram shows a definite increase in the crystallite. It is probable that the polymer is largely made up of parallel macromolecules consisting of oxymethylene chains. The marked resemblance of this polymer-type to cellulose points to a fundamental structural similarity.

The polymerization of liquid formaldehyde is a true polymerization, analogous to that of styrene or vinyl acetate. The reaction may be conceived as a chain addition of formaldehyde molecules to an activated molecule. Chain growth is eventually stopped by a foreign molecule capable of forming polymer end-groups, or by some other process. This type of polymerization differs radically from the polycondensation reactions which account for polyoxymethylene glycol formation.

Properties. Depending on the conditions under which polymerization takes place, eu-polyoxymethylene may be obtained as a transparent glass, a hard porcelain-like solid, or a colorless powder[80, 108]. Transparent polymers, which are obtained by slow polymerization in the neighborhood of −80°C, probably have a higher degree of polymerization than the opaque polymers normally produced at higher temperatures. Polymerization of anhydrous formaldehyde gas during condensation on the walls of a glass vessel cooled to −80°C results in the formation of transparent flexible films. According to Staudinger, this film-forming process takes place most readily under reduced pressure.

On warming to the temperature range 160 to 200°C, eu-polyoxymethylene softens and in this state shows both plastic and elastic properties. Since decomposition takes place at 170 to 220°C, the elasticity interval is very small. Eu-polyoxymethylene glasses obtained by Staudinger sintered at 175°C in an open melting-point tube and decomposed with evolution of monomeric gas at higher temperatures. Under these conditions, complete liquefaction of the polymers was not observed[80]. Polymer obtained in powdered form by warming a solution of liquid formaldehyde in dry ether melts to a viscous liquid at 170 to 172°C in a sealed tube[108]. The monomeric formaldehyde obtained by vaporizing eu-polyoxymethylene shows little tendency to polymerize in the gaseous state because of its high purity, and on chilling condenses to liquid formaldehyde[108].

Polyoxymethylene glass prepared at −80°C is somewhat brittle at ordinary temperatures, but in the plastic state is extremely tough. Polymer obtained at higher temperatures is soft and may be easily drawn to give long fibers or may be pressed into film. Staudinger reports that fine fibers obtained in this way have a reversible elasticity of 10 per cent[80].

Eu-polyoxymethylene does not smell of formaldehyde under ordinary conditions. However, on long storage in a sealed container the odor of formaldehyde is definitely detectable, indicating a gradual depolymerization to monomer. The polymer is insoluble in water but gradually dissolves in dilute boiling alkali or sodium sulfite. On exposure to the air, eu-polyoxymethylene powder absorbs up to 2 per cent water which cannot be subsequently removed, even by vacuum desiccation over phosphorus pentoxide. This is probably a case of physical adsorption, but might possibly involve chemical reaction with formation of polyoxymethylene glycols[108].

According to Delépine and Badoche[20], the heat of combustion of this polymer is 120.3 kcal per formaldehyde unit. Its heat of formation as calculated from this figure using Rossini's values for carbon dioxide and water is 42.0 kcal per mol of formaldehyde. Delépine's earlier value for the heat of combustion of polyoxymethylene prepared from liquid formaldehyde was 122.9 kcal per formaldehyde unit[18].

Preparation. The polymerization of pure liquid formaldehyde to a solid polymer was first reported in 1892 by August Kekulé[34], who observed the slow polymerization at $-20°C$ and the rapid, almost explosive, reaction at higher temperatures.

For the preparation of eu-polyoxymethylene, best results are obtained by use of pure, redistilled liquid formaldehyde (pages 41–44). This material polymerizes to an opaque product when kept for several hours in a nitrogen atmosphere at $-80°C$. Oxygen inhibits the rate of polymer formation, complete polymerization requiring up to several days when this gas is present. Under these conditions, a clear glass-like polymer may be obtained[80]. The inhibiting action of oxygen on polymerization reactions at low temperature has also been reported in the case of vinyl acetate[84]. At high temperatures polymerization is accelerated by oxygen.

Eu-polyoxymethylene can be obtained in powdered form by polymerization in a dry inert solvent. Good results are obtained by sealing 10-cc quantities of a solution of one volume liquid formaldehyde and two volumes anhydrous ether in a "Pyrex" tube, which is then allowed to come to room temperature. Since considerable pressure is developed during polymerization, explosive shattering of the sealed tube is a possibility, and suitable safety precautions should be observed. Immersion of the "Pyrex" tubes in an ice-and-water bath gives good results and reduces the likelihood of explosions. After four hours, the tubes are opened and the polymer paste is freed from ether by vacuum drying.

The polymerization of liquid formaldehyde is accelerated by catalysts, such as boron trichloride and trimethylamine[80]. Primary amines are more effective in this respect than tertiary ones, probably because they react readily to form polymer end-groups[108].

$$R \cdot NH_2 + nCH_2O \rightarrow RNH \cdot CH_2 \cdot O \cdot CH \cdots CH_2 \cdot OH$$

A 33 per cent solution of liquid formaldehyde in dry ether is quantitatively polymerized with great rapidity by the addition of one part n-butylamine per 60,000 parts of formaldehyde. Isoamylamine and ethylamine are equally effective, whereas tri-n-butylamine is much less active[108]. Polymers obtained with these catalysts do not have the plastic properties of eu-polyoxymethylene to any substantial extent and undoubtedly possess greatly inferior molecular weights.

CYCLIC POLYMERS

Trioxane

Trioxane or alpha-trioxymethylene, $(CH_2O)_3$, the cyclic trimer of formaldehyde, is a stable chemical individual possessing unique, well-defined properties. Because of preparational difficulties, it has in the past been studied by very few investigators. Its obscurity has also been further enhanced by the incorrect use of the term "trioxymethylene" as a designation for linear formaldehyde polymers.

Structure. The structure of trioxane as shown below is demonstrated by its analysis, molecular weight, chemical behavior, and physical characteristics.

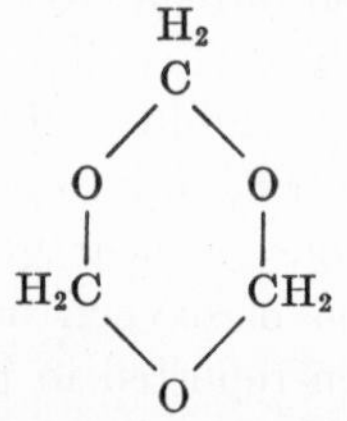

$$
\begin{array}{c}
H_2 \\
C \\
\diagup \quad \diagdown \\
O \qquad O \\
| \qquad\qquad | \\
H_2C \qquad CH_2 \\
\diagdown \quad \diagup \\
O
\end{array}
$$

Molecular-weight determinations in aqueous solution by cryoscopy or in the gaseous state by vapor density give the value 90, which is equivalent to three formaldehyde units. Studies of Raman spectra[33] and crystal structure[52] show that its cyclic molecule is a nonplanar hexagonal ring similar to cyclohexane. The x-ray diagram of trioxane (Figure 26) shows its crystalline nature and similarity to other polyoxymethylenes.

The dipole moment of trioxane has been determined by Maryott and Acree[48] and Calderbank and LeFèvre[13]. The experimental value in benzene is reported as 2.18×10^{-18} esu[48].

Physical Properties. Pure trioxane is a colorless, crystalline compound with a pleasant characteristic odor resembling that of chloroform. It melts at 61 to 62°C and boils without decomposition at approximately 115°C. It forms an azeotrope with water which distills at 91.3°C and contains 70 per cent trioxane[114]. It is combustible and burns readily when ignited. Determinations made by the writer show that its flash point is approximately 45°C (115°F).* The density of the molten compound is 1.17 at 65°C, according to our measurements.

---

* G. W. Jones, G. S. Scott and Irving Hartman of the U. S. Bureau of Mines have determined that the temperature limits between which air saturated with trioxane vapor is explosive are 38 to 75°C. The concentration limits are 3.57 and 28.7 per cent by volume. This would indicate that our approximate flash point measurement is high. This information was released for publication by the Bureau of Mines with the consent of the Office of the Quartermaster General.

Trioxane sublimes readily, forming fine colorless needles or refractive rhombohedral crystals. As these crystals have the remarkable property of being soft and pliant, they may be bent at will. They also possess a polar electric charge, which can be demonstrated by means of a gold-leaf electroscope. This electric charge is apparently built up when the crystals are forming or evaporating, since no electricity is produced when the air sur-

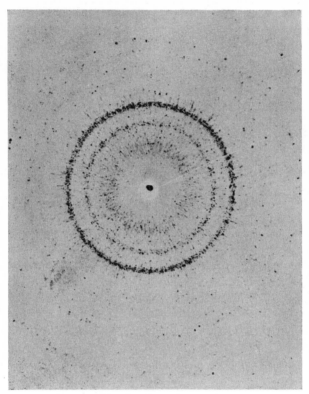

Figure 26. X-ray diagram of trioxane. This diagram was obtained with a sample of fine trioxane crystals precipitated from ether solution by addition of petroleum ether.

rounding the crystals becomes saturated with trioxane vapor. Our measurements indicate that at 20°C solid trioxane has a dielectric constant of 3.2–3.4 as measured at 1.6 megacycles. Under the same conditions, it has a power factor of approximately 3.4 per cent. Molten trioxane has a dielectric constant of 8 at 70°C[115].

The vapor pressure of trioxane at various temperatures from 25 to 129°C is shown in Table 43. Low-temperature measurements were made with a tensimeter by Auerbach and Barschall[9]. Measurements at higher temperatures are based principally on boiling points reported by Frank[26].

Trioxane is soluble in water and may be readily crystallized from this solvent. According to Auerbach and Barschall[9], saturated aqueous solutions at 18 and 25°C contain respectively 17.2 and 21.1 grams of trioxane per 100 cc. It is infinitely soluble in hot water, with which it forms a constant-boiling mixture that distills at approximately 91°C as previously pointed out (page 146).

Trioxane is soluble in alcohol, ketones, organic acids, ethers, esters, phenols, aromatic hydrocarbons, chlorinated hydrocarbons, etc. Solubilities at various temperatures in several organic solvents are shown in Table 44.

Figure 27. Trioxane crystals.

It is only slightly soluble in pentane and lower paraffin fractions, such as petroleum ether.

Like dioxane and other cyclic ethers, molten trioxane is an excellent solvent for numerous organic compounds, such as naphthalene, urea, camphor, dichlorobenzene, etc. Concentrated aqueous solutions of trioxane have properties not possessed by the trioxane itself, *e.g.*, a hot 80 per cent aqueous solution dissolves the protein zein.

When solid trioxane is mixed with two or three times its weight of phenol crystals at room temperature, solution takes place with considerable absorption of heat, and a clear liquid is produced. A solution of this type obtained with 30 grams trioxane and 94 grams phenol begins to crystallize at about 17°C and boils in the range 143 to 5°C.

Thermodynamic Properties. Heats of combustion reported for trioxane are shown below together with the corresponding heats of formation calculated with Rossini's figures for carbon dioxide (beta graphite standard) and water (see page 37). All figures are calculated per $\frac{1}{3}$ gram mol of trioxane or per CH_2O unit.

Investigator and Year	Heat of Combustion (per CH_2O Unit)	Heat of Formation (per CH_2O Unit)
	kcal	*kcal*
Palfrey[62] (1944)	118.9	43.4
Delépine and Badoche[20] (1942)	120.9	41.4
Von Wartenberg *et al.*[107] (1924)	109.5	52.8

Palfrey's figure is based on a heat of combustion of 3.96 ± 0.01 kcal per gram at 23°C and should have an accuracy of ±0.3 kcal for the $\frac{1}{3}$ mol figure.

TABLE 43. VAPOR PRESSURE OF TRIOXANE

Temperature (°C)	Vapor Pressure (mm Hg)	Authority
25	12.7	Auerbach and Barschall[9]
37.5	31.2	" " "
86	283	Frank[26]
87	296	"
90	330	"
114.5	759	Auerbach and Barschall[9]
129	1214	Frank[26]

TABLE 44. SOLUBILITY OF TRIOXANE IN ORGANIC SOLVENTS

Solvent	Trioxane Soly., G/100 G Soln.		
	25°C	35°C	45°C
Acetic acid	40	52	..
Benzene	37	49	62
Toluene	..	41	58
Trichloroethylene	23	35	47

A comparison of the three figures reported indicates that the von Wartenberg measurement may be low. Heats of formation as calculated from the Palfrey and Delépine data indicate that the heat of polymerization of monomeric formaldehyde to trioxane is 13 to 15 kcal.

$$CH_2O \text{ (g)} \rightarrow 1/3(CH_2O)_3\text{(s)} + 13 \text{ to } 15 \text{ kcal.}$$

The heat of vaporization of trioxane in the neighborhood of its boiling point, as calculated from vapor-pressure measurements, is approximately 9.8 kcal per mol, or 3.3 kcal per CH_2O unit. This figure gives a Trouton constant of 25, which is close to that of water.

Hydrolysis and Depolymerization. Trioxane has a relatively high degree of chemical stability. According to Auerbach and Barschall[9], pure samples show little decomposition even after heating for 6 hours in a sealed tube at 224°C. Only traces of formaldehyde and a slight amount of a substance with caramel-like odor are produced. Burnett and Bell[11] investigated the thermal decomposition of trioxane in the gaseous state at 270 to 345°C and found it to be a homogeneous reaction of the first order.

Trioxane is hydrolyzed by aqueous solutions of strong acids but, like the acetals, it is inert under neutral or alkaline conditions and for this reason gives negative results when subjected to the ordinary processes for formaldehyde analysis such as the sodium sulfite process or the alkaline peroxide procedure.

Measurement of the acid hydrolysis of trioxane in dilute solutions gives good first order rate constants as determined by the equation:

$$k = \frac{1}{t} \ln \frac{a}{a - x}$$

In this equation, a is the initial trioxane concentration, x indicates the amount depolymerized in unit time and k is the first order rate constant. This constant varies with acid strength and temperature. Values of k as determined by Chadwick and the writer[116] for varying sulfuric acid concentrations in a 3 per cent aqueous trioxane solution at various temperatures are shown graphically in Figure 28. Data obtained in a typical reaction are shown in Table 45. The time required for depolymerization of various fractions of trioxane at different temperatures and sulfuric acid normalities as calculated from the rate constants of Figure 28 is shown in Table 46. For solutions containing 12 and 20 per cent trioxane, the depolymerization rate was found to be approximately identical to the value measured for a 3 per cent solution but this rate increased rapidly for higher concentrations. At 75°C, a solution containing 20 per cent trioxane was 23 per cent depolymerized in one hour; 40 and 60 per cent solutions were 40 and 65 per cent depolymerized in the same time. Pure trioxane was more than 90 per cent depolymerized under the above conditions but in this case the monomeric formaldehyde polymerized to polyoxymethylene. These results indicate that there is a deviation from first order kinetics in highly concentrated aqueous trioxane solutions. A kinetic salt effect in the acid-catalyzed depolymerization has recently been reported by Paul[62a].

The acid hydrolysis constant of trioxane, as determined by Skrabal, Stockmair and Schreiner[71] at various temperatures in the presence of N p-toluene sulfonic acid, is shown below:

	Temperature (°C)				
	15	25	35	45	55
Acid hydrolysis constant (Min.$^{-1}$)					
$K_s \times 10^5$	0.0150	0.159	1.17	6.17	24.1

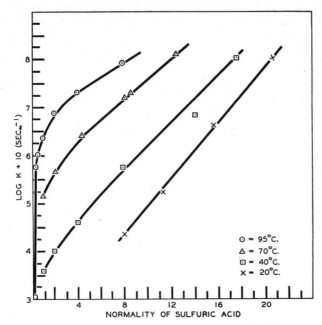

Figure 28. Effect of sulfuric acid concentration on rate of trioxane depolymerization.

TABLE 45. DEPOLYMERIZATION OF 3 PER CENT TRIOXANE IN 3.84*N* SULFURIC ACID AT 40°C

Time, Sec.	Concn., G/100 cc		$k \times 10^6$ for Interval 0 to t, Sec.$^{-1}$
	CH_2O, x	Trioxane, $a - x$	
0	0.000	3.000	
7,200	0.073	2.927	3.42
16,200	0.175	2.825	3.95
77,400	0.691	2.309	3.38
165,600	1.292	1.708	3.62
443,000	2.42	0.58	3.71
702,000	2.76	0.24	3.60

Trioxane hydrolysis may be readily compared with that of other acetals and ethers by means of the following hydrolysis constants, K_s^1, calculated on an equivalent ether oxygen basis[71].

$$K_s^1 (25°C)$$
(Calc. for One Ether Oxygen)

Diethyl ether	0.000,000,000,014,6
Trioxane	0.000,000,53
Methylal	0.000,76
Paraldehyde	0.002,73
Acetal	30.0

The rates of depolymerization of trioxane in acetic acid and various organic solvents have been reported by Chadwick and the writer[116]. Sulfuric acid was found to be a far more efficient depolymerization catalyst in acetic acid solutions of trioxane than in aqueous solutions, the reaction with

TABLE 46. RATE OF TRIOXANE DEPOLYMERIZATION AT VARIOUS TEMPERATURES WITH SULFURIC ACID AS CATALYST

Normality of H_2SO_4	Time of Liberation of Following Percentage Formaldehyde			
	10%	50%	75%	99%
	Depolymerization at 20°C			
8	12.3 hr.	3.4 days	6.8 days	22.5 days
12	55 min.	6.1 hr.	12.2 hr.	40 hr.
16	4.4 min.	29 min.	58 min.	190 min.
20	20 sec.	22 min.	4.4 min.	14.5 min.
	Depolymerization at 40°C			
1	3.0 days	21 days	42 days	137 days
2	29 hr.	8.0 days	16 days	53 days
4	6.5 hr.	1.78 days	3.6 days	11.8 days
8	31 min.	3.5 hr.	7.0 hr.	23 hr.
12	3.6 min.	24 min.	48 min.	160 min.
16	25 sec.	2.8 min.	5.6 min.	18.3 min.
	Depolymerization at 70°C			
0.5	41 hr.	1.13 days	2.3 days	7.5 days
1	20 hr.	13.4 hr.	27 hr.	3.8 days
2	7.9 hr.	5.3 hr.	10.6 hr.	35 hr.
4	10.1 min.	1.13 hr.	2.3 hr.	7.5 hr.
8	66 sec.	7.3 min.	14.6 min.	48 min.
12	10.7 sec.	71 sec.	2.4 min.	7.8 min.
	Depolymerization at 95°C			
0.1	11.3 days	75 days	150 days	500 days
0.25	33 min.	3.6 hr.	7.2 hr.	24 hr.
0.5	17.3 min.	1.8 hr.	3.7 hr.	12.8 hr.
1	7.4 min.	49 min.	1.6 hr.	5.4 hr.
2	2.4 min.	15.6 min.	31 min.	103 min.
4	45 sec.	5 min.	10 min.	33 min.
8	11.6 sec.	77 sec.	2.6 min.	8.5 min.

a given sulfuric acid concentration proceeding approximately one thousand times faster in the organic solvent. Depolymerization proceeds at a measurable rate in this solvent at 95°C even in the absence of an added catalyst.

When trioxane is heated in a substantially anhydrous system in the presence of strong acids, such as sulfuric, hydrochloric, and phosphoric acids, or acidic materials such as zinc chloride, ferric chloride, etc., it is readily depolymerized to monomeric formaldehyde which, when produced in this way, is extremely reactive and enters readily into combination when the depolymerization is carried out in the presence of a compound capable of reacting with formaldehyde[110]. In the absence of a formaldehyde acceptor, the monomer polymerizes to a high molecular weight polyoxymethylene. By use of this depolymerization reaction, trioxane may be employed as a special form of anhydrous formaldehyde. When used in this way, trioxane remains inert until the necessary catalyst has been added and then reacts at a rate determined by the concentration of catalyst. Addition of 0.01 per cent of a 2 per cent aqueous solution of sulfuric acid is sufficient to catalyze chemical reaction in a hot mixture of 1 gram mol of phenol and ⅓ gram mol of trioxane. Weak acids, such as acetic and boric, have no apparent action on boiling trioxane.

Physiological Properties. Physiological tests by the Haskell Laboratory of Industrial Toxicology (Du Pont) have shown that pure trioxane does not appear highly toxic to rats when inhaled or given orally[116]. Patch tests show that there does not appear to be much risk of its causing dermatitis in humans[116]. Knoefel[35] states that trioxane differs from paraldehyde in having no narcotic effect on rabbits.

Preparation. Trioxane was discovered in 1885 by Pratesi[65], who prepared it by heating paraformaldehyde in a sealed tube with a trace of sulfuric acid at 115°C. Several later workers mention unsuccessful attempts to repeat this synthesis. Seyewetz and Gibello[70] report an impure formaldehyde derivative melting at 69°C which probably contained trioxane. Auerbach and Barschall[9] succeeded in producing it by heating paraformaldehyde or a polyoxymethylene in a current of nitrogen and passing the resulting gases into a mixture of ice and water. A solution from which trioxane could be isolated was thus obtained.

Hammick and Boeree[30] described the optimum conditions for preparing trioxane by the methods of Pratesi, and Auerbach and Barschall. Yields of 10 and 20 per cent, respectively, were obtained by these two methods. Staudinger[83] mentions briefly a third method which involves heating moist paraform or an acid-precipitated polyoxymethylene in a vacuum at 100°C, the vapors being condensed in a receiver cooled to −80°C.

A practical process for trioxane preparation developed by Frank[25] involves distilling a 60 to 65 per cent formaldehyde solution in the presence of 2 per cent sulfuric acid and extracting trioxane from the distillate by means of a water-immiscible solvent such as methylene chloride[115]. Separation of trioxane in the distillation process is facilitated by fractionating so that only a low boiling mixture (b.p. 90 to 92°C) is removed. This distillate contains around 60 per cent trioxane[114]. Mikeska and co-workers[51] claim a trioxane process in which paraformaldehyde or a polyoxymethylene is heated with a metal halide catalyst. Sokol[72] describes a crystallization process for isolating trioxane from aqueous formaldehyde.

Tetraoxymethylene

Tetraoxymethylene is a cyclic tetramer of formaldehyde. It has been prepared by Staudinger[83], who obtained it by heating a water-insoluble, polyoxymethylene diacetate of high molecular weight. Obtained as a sublimate in the form of long needle-like crystals, it melts at 112°C. Vapor density measurements indicate a molecular weight corresponding to $(CH_2O)_4$. The vapor is stable at 200°C.

References

1. Auerbach, F., and Barschall, H., "Studien über Formaldehyde—Die festen Polymeren des Formaldehyds," p. 7, Berlin, Julius Springer (1907).
2. *Ibid.*, p. 9.
3. *Ibid.*, pp. 10–11.
4. *Ibid.*, p. 19.
5. *Ibid.*, pp. 20–26.
6. *Ibid.*, pp. 26–34.
7. *Ibid.*, pp. 27–29.
8. *Ibid.*, pp. 34–38.
9. *Ibid.*, pp. 38–45.
10. Brown, F. L., and Hrubesky, C. E., *Ind. Eng. Chem.*, **19**, 217 (1927).
11. Burnett, R. le G., and Bell, R. P., *Trans. Faraday Soc.*, **34**, 420 (1938).
12. Butlerov (Butlerow), A., *Ann.*, **111**, 245, 247–8 (1859).
13. Calderbank, K. E., and LeFèvre, R. J. W., *J. Chem. Soc.*, **1949**, 199–202.
14. Cambier, R., and Brochet, A., *Compt. rend.*, **119**, 607 (1894).
15. Carothers, W. H., *J. Am. Chem. Soc.*, **51**, 2548–59 (1929).
16. Craven, R. L., (to E. I. du Pont de Nemours & Co., Inc.), U. S. Patent 2,519,550 (1950).
17. Craven, R. L., (to E. I. du Pont de Nemours & Co., Inc.), U. S. Patent 2,551,365 (1949).
18. Delépine, M., *Compt. rend.*, **124**, 1528 (1897).
19. Delépine, M., *Ann. chim. phys.* (7), **15**, 530 (1898).
20. Delépine, M., and Badoche, M., *Compt. rend.*, **214**, 777–80 (1942).
21. Descudé, M., *Ann. chim. phys.* (7), **29**, 502 (1903).
22. "Encyclopaedia Britannica," 14th Ed., Vol. IX, p. 512 (1938).
23. Farquharson, J., *Trans. Faraday Soc.*, **33**, 824 (1937).
24. Finkenbeiner, H., and Schmache, W., (to Deutsche Gold- und Silber-Scheideanstalt vormals Roessler), U. S. Patent 2,116,783 (1938).
25. Frank, C. E., (to E. I. du Pont de Nemours & Co., Inc.), U. S. Patent 2,304,080 (1942).
26. Frank, C. E., (Experimental Station, E. I. du Pont de Nemours & Co., Inc.), unpublished communication.
27. Fuchs, O., and Naujoks, E., (to Deutsche Gold- und Silber-Scheideanstalt vormals Roessler), U. S. Patent 1,948,069 (1934).
28. Fuller, G. P., (to National Electrolytic Company), U. S. Patents 1,143,114 (1915), 1,170,624 (1916).
29. Greenwald, B. W., and Cohen, R. K., (to Cities Service Oil Co.), U. S. Patent 2,498,206 (1950).
29a. Gresham, W. F., and Brooks, R. E., (to E. I. du Pont de Nemours & Co., Inc.), U. S. Patent 2,449,469 (1948).

30. Hammick, D. L., and Boeree, A. R., *J. Chem. Soc.*, **121**, 2738–40 (1922).
31. Hinegardner, W. S., (to E. I. du Pont de Nemours & Co., Inc.), U. S. Patent 2,042,657 (1937).
32. Hofmann, A. W., *Ann.*, **145**, 357 (1868); *Ber.*, **2**, 156–60 (1869).
33. Kahovec, L., and Kohlrausch, K. W. F., *Z. physik. Chem.*, **B35**, 29 (1937).
34. Kekulé, A., *Ber.*, **25**, 2435 (1892).
35. Knoefel, P. K., *J. Pharmacol.*, **48**, 488 (1933).
36. Kohlschütter, H. W., *Ann.*, **482**, 75–104 (1930).
37. *Ibid.*, **484**, 155–178 (1930).
38. Kuss, E., (to I. G. Farbenindustrie A. G.), U. S. Patent 1,666,708 (1928).
39. Löbering, J., *Ber.*, **69**, 1844–54 (1936).
40. Löbering, J., and Fleischmann, A., *Ber.*, **70**, 1713–9 (1937).
41. Löbering, J., and Hilber, J., *Ber.*, **73**, 1382–8 (1940).
42. Löbering, J., and Jung, K. P., *Monatsh.*, **70**, 281–96 (1937).
43. Löbering, J., and Rank, V., *Ber.*, **70**, 2331–9 (1937).
44. Lösekann, G., *Chem. Ztg.*, **14**, 1408 (1890).
45. Londergan, T. E., (to E. I. du Pont de Nemours & Co., Inc.), U. S. Patent 2,512,950 (1950).
45a. MacLean, A. F., and Heinz, W. E., (to Celanese Corp. of America), U. S. Patents 2,568,016–7 (1951).
45b. MacLean, A. F., and Heinz, W. E., (to Celanese Corp. of America), U. S. Patent 2,568,018 (1951).
46. Mannich, C., *Ber.*, **52**, 160 (1919).
47. Manufacturing Chemists' Association, Inc., "Paraformaldehyde—Chemical Safety Data Sheet SD-6," Washington, D. C. (1950).
48. Maryott, A. A., and Acree, S. F., *J. Research, Natl. Bur. Standards*, **33**, No. 1, 71 (1944).
49. Michael, V. F., (to Stanolind Oil & Gas Co.), U. S. Patent 2,529,622 (1950).
50. Mie, G., and Hengstenberg, J., *Helv. Chim. Acta*, **11**, 1052 (1928),
51. Mikeska, L. A., *et al.*, (to Standard Oil Development Co.), U. S. Patent 2,270,135 (1942).
52. Moerman, N. F., *Rec. trav. chim.*, **56**, 161 (1937).
53. Mugdan, M., and Wimmer, J., (to Consortium fur elektrochemische Industrie G. m. b. H.), German Patent 489,644 (1925).
54. Müller, A., and Shearer, G., *J. Chem. Soc.*, **123**, 3156 (1923).
55. Nasch, L., British Patent 420,993 (1934).
56. Naujoks, E., (to Deutsche Gold- und Silber-Scheideanstalt formals Roessler) U. S. Patent 2,092,422 (1937).
57. Nielsen, H. H., and Ebers, E. S., *J. Chem. Phys.*, **5**, 824 (1937).
58. Nordgren, G., *Acta Path. Microbiol. Scand.*, (suppl.), **40**, 21–30 (1939).
59. *Ibid.*, p. 34.
60. Ott, E., *Science*, **71**, 465–6 (1930).
61. Ott, E., *Z. physik. Chem.*, **B9**, 378–400 (1930).
62. Palfrey, G. F., (E. I. du Pont de Nemours & Co., Inc.), unpublished communication.
62a. Paul, M. A., *J. Am. Chem. Soc.*, **74**, 141 (1952).
63. Peterson, O., (to E. I. du Pont de Nemours & Co., Inc.), U. S. Patent 2,373,777 (1945).
64. Pirlet, G., French Patent 746,540 (1934).
65. Pratesi, L., *Gazz. chim. ital.*, **14**, 139 (1885).

66. Pyle, C., and Lane, J. A., (to E. I. du Pont de Nemours & Co., Inc.), U. S. Patents 2,527,654–5 (1950); 2,581,881 (1952).

67. Sator, K., and Pfanmüller, W., (to I. G. Farbenindustrie A. G.), U. S. Patent 1,677,730 (1928).

68. Sauter, E., Z. physik. Chem., **B21**, 161, 186 (1933).

68a. Sauterey, R., Ann. chim. (*12*), **7**, 24–72 (1952).

69. Schilt, W., British Patent 342,668 (1929).

70. Seyewetz, A., and Gibello, Compt. rend., **138**, 1225 (1904).

71. Skrabal, A., Stockmair, W., and Schreiner, H., Z. physik. Chem., **A169**, 177 (1934).

72. Sokol, H., (to Heyden Chemical Corp.), U. S. Patent 2,465,489 (1949).

73. Staudinger, H., "Die Hochmolekularen Organischen Verbindungen," p. 216, Berlin, Julius Springer (1932).

74. Ibid., pp. 249–50.

75. Ibid., p. 250.

76. Ibid., p. 251.

77. Ibid., pp. 251–2.

78. Ibid., p. 252.

79. Ibid., p. 254.

80. Ibid., pp. 257–63.

81. Staudinger, H., Johner, H., Signer, R., Mie, G., and Hengstenberg, J., Z. physik. Chem., **126**, 425 (1927).

82. Staudinger, H., and Lüthy, M., Helv. Chim. Acta, **8**, 41 (1925).

83. Ibid., p. 65.

84. Staudinger, H., and Schwalbach, A., Ann., **488**, 8 (1931).

85. Staudinger, H., and Signer, R., Helv. Chim. Acta, **11**, 1047–51 (1928).

86. Staudinger, H., Signer, R., Johner, H., Lüthy, M., Kern, W., Russidis, D., and Schweitzer, O., Ann., **474**, 145–259 (1929).

87. Ibid., pp. 172–205.

88. Ibid., pp. 190–4.

89. Ibid., p. 199.

90. Ibid., pp. 205–229.

91. Ibid., pp. 218–29.

92. Ibid., p. 221.

93. Ibid., p. 225.

94. Ibid., pp. 230–1.

95. Ibid., pp. 232–7.

96. Ibid., p. 235.

97. Ibid., p. 241.

98. Ibid., pp. 243–5.

99. Ibid., p. 224.

100. Ibid., pp. 245, 248, 253–258.

101. Ibid., p. 251.

102. Ibid., pp. 254–5.

103. Ibid., p. 255.

104. Staudinger, H., Signer, R., and Schweitzer, O., Ber., **64**, 398–405 (1931).

105. Szalay, A., Z. physik. Chem., **A164**, 234–40 (1933).

106. Tollens, B., and Mayer, F., Ber., **21**, 1566, 2026, and 3503 (1888).

107. Von Wartenberg, H., Muchlinski, A., and Riedler, G., Z. angew. Chem., **37**, 457 (1924).

108. Walker, J. F., *J. Am. Chem. Soc.*, **55,** 2823–4 (1933).
109. *Ibid.*, pp. 2825–6.
110. Walker, J. F., (to E. I. du Pont de Nemours & Co., Inc.), U. S. Patent 2,304,431 (1942).
111. *Ibid.*, U. S. Patent 2,529,269 (1950).
112. *Ibid.*, U. S. Patent 2,369,504 (1945).
113. *Ibid.*, U. S. Patent 2,456,161 (1948).
114. *Ibid.*, U. S. Patent 2,347,447 (1944).
115. Walker, J. F., and Carlisle, P. J., *Chem. Eng. News*, **21,** 1250–1 (1943).
116. Walker, J. F., and Chadwick, A. F., *Ind. Eng. Chem.*, **39,** 974–7 (1947).
117. Wildt, Rupert, *Astrophys. J.*, **96,** 312-4 (1942).

CHAPTER 8

CHEMICAL PROPERTIES OF FORMALDEHYDE

Formaldehyde is one of the most reactive organic chemicals, differing both in reactivity and in many other respects from related compounds containing the carbonyl group. These differences are due to its unique chemical structure. In ketones the carbonyl group is attached to two carbon atoms; in other aldehydes it is attached to one carbon atom and one hydrogen atom; in formaldehyde it is attached to two hydrogen atoms. As a result, the characteristic properties of the carbonyl group in formaldehyde are not modified by the presence of other radicals. Alpha-hydrogen atoms and keto-enol tautomerism of the type which attends their presence are automatically ruled out.

The unusual reactivity of formaldehyde is reflected in the fact that although other aldehydes are normally handled in the pure monomeric state, formaldehyde does not appear commercially in this form. Instead, as we have previously pointed out, it is handled either as a solution in which it is chemically combined with water, or as a solid polymer. Fortunately, however, the net effect of reactions involving formaldehyde is usually what would be expected if monomeric formaldehyde were employed. The state of formaldehyde in its commercial forms is principally of importance in its effect on rates and mechanisms of reaction. Paraformaldehyde, for example, usually reacts more slowly than the aqueous solution, since it must depolymerize before it can react.

For purposes of simplicity in this book, formaldehyde is, in general, designated by the formula CH_2O in reaction equations, regardless of whether the monomeric gas, anhydrous liquid, aqueous solution, or polymer (paraformaldehyde or polyoxymethylene) is involved. The form intended, however, will be indicated as shown below:

CH_2O (g) Monomeric gas
CH_2O (l) Anhydrous liquid
CH_2O (aq) Aqueous solution
CH_2O (p) Polymer

Trioxane, the true cyclic trimer, will always be designated by the formula $(CH_2O)_3$.

It should be remembered that when the formula CH_2O (aq) or CH_2O (p) is employed, the actual reactant is not CH_2O but methylene glycol, $CH_2(OH)_2$, or a polyoxymethylene glycol, $HO \cdot (CH_2O)_n \cdot H$. When two or

158

more forms are equally effective the letter designations will be used together, separated only by commas. No designations will be employed in reactions which are more or less generally applicable to all forms of formaldehyde.

Chemical Stability and Decomposition

Aside from its tendency to form reversible polymers, formaldehyde possesses a surprising degree of chemical stability. Measurable decomposition of gaseous formaldehyde is practically negligible at temperatures below 300°C. However, Steacie and Calvert[16] have found that methanol and carbon monoxide are the major products of a slow heterogeneous decomposition reaction in the range 150 to 350°C. At high temperatures, formaldehyde decomposes almost exclusively to carbon monoxide and hydrogen with no separation of carbon[11], as indicated by the equation:

$$CH_2O \text{ (g)} \rightleftharpoons CO + H_2$$

Although the studies of Newton and Dodge (page 6) have shown that the equilibrium involved in this reaction favors the decomposition to such an extent that it may be regarded as almost irreversible under ordinary circumstances, reaction rates are almost infinitely slow below 300°C. Some idea of the rates at higher temperatures may be gained by the data of Medvedev and Robinson[59]. These investigators report that when a gaseous mixture containing 40 per cent formaldehyde and 60 per cent carbon dioxide is passed through a refractory glass tube (30-cm long) heated to various temperatures and under conditions such that the exposure time is 29 seconds in each case, the per cent decomposition is as follows: 10.2 at 450°, 38.6 at 500°, 44.7 at 550°, 87.5 at 600°, 94.9 at 650°, and 97.6 per cent at 700°C. Patry and Monceaux[69] report that low concentrations of formaldehyde (0.1 to 0.2 per cent) in the presence of nitrogen and air decompose slowly in a quartz tube at temperatures below 600°C but very rapidly at temperatures above 700°C. Decomposition at 700°C with 0.6 seconds exposure was 77 per cent in the presence of nitrogen and 81 per cent in air. This indicates that oxidation is a relatively minor reaction under these conditions. However, Snowdon and Style[75] report that traces of oxygen slowly induce the decomposition of formaldehyde at temperatures in the neighborhood of 300°C. The lengthy incubation period required for this reaction is completely eliminated by irradiation with the light of a mercury arc. This work indicates that the decomposition is catalyzed by oxidation intermediates whose slow formation is accelerated by radiant energy.

According to Fletcher[29], who studied formaldehyde decomposition between 510 and 607°C in a pressure range of 30 to 400 mm, the splitting of formaldehyde to carbon monoxide and hydrogen, although unquestionably the principal reaction, is also accompanied by condensation reactions which exert a disturbing influence on the rate of decomposition even though they eventually lead in part to carbon monoxide formation. An immediate pressure rise due to simple splitting is followed by a subsidiary rise which extends over several hours. Fletcher[29] also reports that the molecular heat of activation for the decomposition to carbon monoxide and hydrogen is 44.5 kcal. Patat and Sachsse[68] claim that free radicals are not produced when formaldehyde is decomposed.

Catalysts have considerable influence on both the rate and nature of formaldehyde decomposition. In the presence of finely divided platinum, decomposition is stated to occur at $150°C$[57], whereas copper shavings are reported to have no effect below $500°C$[47]. Various inorganic compounds, such as sodium carbonate, chromium oxide, alumina, etc., are stated to modify the decomposition with formation of small quantities of methanol, formic acid, carbon dioxide, methane, and complex organic products[82].

Ultraviolet light causes decomposition of formaldehyde to carbon monoxide and hydrogen at ordinary temperatures[8]. In the presence of water vapor these products are accompanied by methane and carbon dioxide[71]. Research on the mechanism of the photolysis at temperatures in the neighborhood of 100 to 300°C indicates that the following reactions may be involved:

$$CH_2O + h\nu = H + HCO \tag{1}$$

$$H + HCO = H_2 + CO \tag{2}$$

$$HCO = H + CO \tag{3}$$

$$HCO + CH_2O = H_2 + CO + HCO \tag{4}$$

$$2HCO = CO + CH_2O \tag{5}$$

Activation energies for reactions (2) and (3) are estimated as approximately 5 and 13 kcal respectively. Steacie and Calvert[16] suggest that reaction (5) may occur at the wall between two formyl radicals and also include the possibility of glyoxal being formed under these circumstances. Formyl radicals may also react with adsorbed hydrogen atoms to reform formaldehyde. The above investigators report that propylene and nitric acid inhibit photolysis. Style and Summers[79] suggest a mechanism based on reactions (1), (2), (4), and the wall reaction (5). However, they also indicate that reaction (3) may be substituted for (4). Since the formyl radical is apparently unstable at 100°C but has some degree of stability at lower temperatures, Axford and Norrish[4] conclude that it is probably due to this change of stability that the photolysis of formaldehyde becomes a chain reaction of constantly increasing quantum yield at temperatures above 100°C.

According to Löb[51], carbon monoxide, carbon dioxide, hydrogen, and methane are also produced when formaldehyde is subjected to the silent electric discharge in the presence of water. The reactions involved are as shown below:

$$CH_2O \ (g) \rightarrow CO + H_2$$

$$CO + H_2O \rightarrow CO_2 + H_2$$

$$CO + 3H_2 \rightarrow CH_4 + H_2O$$

Sugars and hydroxyaldehydes are produced as secondary products when formaldehyde is subjected to ultraviolet irradiation in the presence of water (page 165).

In pure aqueous or methanol-containing formaldehyde solutions decomposition to gaseous products is practically negligible under ordinary conditions. Loss of formaldehyde on heating such solution is due principally to the Cannizzaro reaction (pages 78, 164) which becomes fairly rapid when the solution is heated to 160°C and above (page 105).

Oxidation and Reduction

On oxidation under controlled conditions in the gaseous or dissolved state, formaldehyde may be converted to formic acid or under more energetic circumstances to carbon dioxide and water. Carbon monoxide also appears among the products of the oxidation, and peroxygen derivatives have been identified as reaction intermediates. According to Bowen and Tietz[12], performic acid is produced by the reaction of formaldehyde with oxygen at low temperatures. Style and Summers[79] suggest that hydroxymethyl peroxides are formed in the photochemical process.

Research on the thermally activated oxidation of formaldehyde gas indicates that the principal over-all reactions are:

$$CH_2O + O_2 = CO_2 + H_2O \tag{1}$$

$$CH_2O + \tfrac{1}{2}O_2 = CO + H_2O \tag{2}$$

Askey[2] and Fort and Hinshelwood[30] have demonstrated that the oxidation rate is highly dependent on the formaldehyde concentration but almost independent of the oxygen concentration. Inert gases including nitrogen and carbon oxides slow down the reaction but this effect is apparently merely the result of dilution[4]. Kinetic studies by Axford and Norrish[4] show that with conditions under which the oxidation is homogeneous at temperatures in the range 325 to 370°C, reactions (1) and (2) occur at the same time but (2) is the major reaction. Some decomposition to hydrogen and carbon monoxide is also induced but this reaction is of minor importance at high oxygen concentrations. The activation energy for the over-all oxidation reaction as calculated by the above investigators from their reaction rate data is 21.0 kcal. This is in good agreement with Fort and Hinshelwood's[30] 20.0 kcal figure. According to Axford and Norrish[4], the oxidation reaction may be best explained as following a chain mechanism involving hydrogen and oxygen atoms and OH and HO_2 radicals. This mechanism is analogous to that involved in the hydrogen-oxygen reaction. Surface effects are relatively minor and control the reaction only at high surface to volume ratios. Spence[77] reports that reaction (1) is almost complete in the presence of powdered glass and is apparently a heterogeneous reaction.

The photochemical oxidation of formaldehyde was first demonstrated by Patat[67] and has since been studied extensively by Norrish and Carruthers[63, 17], and Style and Summers[79]. Principal reaction products are carbon oxides, formic acid, hydrogen and water. Style and Summers[79] state that the HO_2 radical is an essential intermediate in the short chain reactions at about 100°C but that another mechanism becomes dominant at temperatures above 150°C. The photochemical reaction gradually merges into the thermal reaction at temperatures over 275°C.

The reaction of the free hydroxyl group with formaldehyde has been studied by Avramenko and Lorentso[3]. Carruthers and Norrish[17] report that ozone reacts with formaldehyde in the dark to give formic acid.

Formaldehyde solution is reported not to react with air or oxygen at temperatures up to 100°C. However, in the presence of platinum sponge, oxidation takes place even at room temperature with the formation of carbon dioxide[19, 22].

The results obtained when formaldehyde is subjected to the action of oxidizing agents depends on the specific agent involved and the conditions of treatment. With hydrogen peroxide either formic acid and hydrogen or carbon dioxide and water may be obtained (page 187). On heating with chromic acid solution, formaldehyde is quantitatively oxidized to carbon dioxide and water[10].

Sauterey's studies[73a] on the state of dissolved formaldehyde indicate that only the anion, $HOCH_2O^-$, can be directly oxidized by iodine to give formic acid. By use of this reaction, he determined the dissociation constant for dilute aqueous formaldehyde obtaining a value in good agreement with previous data (page 86).

On reduction, formaldehyde is converted to methanol.

$$CH_2O + H_2 \rightleftharpoons CH_3OH$$

The equilibrium relations of formaldehyde, hydrogen, and methanol have already been discussed in connection with the reverse reaction, *viz.*, the dehydrogenation of methanol (pages 8–9).

Under ordinary conditions the liquid-phase catalytic hydrogenation of formaldehyde is somewhat difficult, since excessive amounts of catalyst ranging from 50 to 100 per cent based on the weight of formaldehyde are required. This difficulty can be avoided if the hydrogenation is carried out in the presence of an alkaline buffer, as described in a process patented by Hanford and Schreiber[38]. Optimum results are obtained at pH values in the range 6 to 9 with temperatures of 50 to 150°C and pressures of 200 to 600 atmospheres. Approximately 1 to 10 per cent of the usual hydrogenation catalysts such as Raney nickel, finely divided cobalt, copper chromite, etc., are reported to give satisfactory results.

In alkaline solution, formaldehyde can be electrolytically oxidized to formic acid and carbon dioxide. When a specially prepared copper or silver anode is employed, Müller[62] reports that pure hydrogen is liberated in equal amounts from both cathode and anode. Under these circumstances sodium formate is produced in theoretical quantity and two equivalents of hydrogen are liberated for each faraday of electricity. The anode is best prepared by treating copper or silver with molten cuprous or silver chloride respectively and then subjecting to cathodic reduction in a solution of sodium hydroxide. Approximately equivalent results were obtained with 0.5 to 2N alkali containing about 17 per cent dissolved formaldehyde. The reactions taking place are shown below as postulated by. Müller.

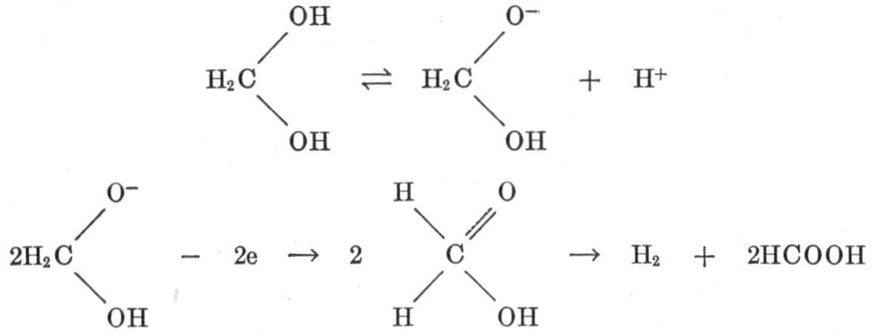

$$H_2C\begin{matrix}OH\\\\OH\end{matrix} \rightleftarrows H_2C\begin{matrix}O^-\\\\OH\end{matrix} + H^+$$

$$2H_2C\begin{matrix}O^-\\\\OH\end{matrix} - 2e \rightarrow 2 \begin{matrix}H\\C\\H\end{matrix}\begin{matrix}O\\\\OH\end{matrix} \rightarrow H_2 + 2HCOOH$$

When a platinum anode is employed, no hydrogen is evolved at this electrode. Hydrogen evolved in the anodic oxidation of formaldehyde apparently comes solely from the CH bond since Hoyer[42] has demonstrated that electrolysis of a formaldehyde solution containing NaOD liberates light hydrogen only at the anode. Hydroxymethyl peroxide, $HOCH_2OOH$, is suggested as the intermediate in the anode reaction.

Under some conditions, alkaline formaldehyde can be almost quantitatively reduced to methanol at a copper cathode[62]. Under other conditions small quantities of methane are obtained.

Reactions of Formaldehyde with Formaldehyde

Reactions of formaldehyde with formaldehyde include: (a) polymerization reactions, (b) the Cannizzaro reaction, (c) the Tischenko reaction and (d) aldol-type condensations.

Polymerization Reactions. The formation of formaldehyde polymers has been reviewed in detail in Chapter 7. It is only necessary to point out here that the formation of oxymethylene polymers is one of the most characteristic reactions of formaldehyde with formaldehyde.

Cannizzaro Reaction. This reaction involves the reduction of one molecule of formaldehyde with the oxidation of another.

$$2CH_2O \text{ (aq)} + H_2O \rightarrow CH_3OH + HCOOH$$

Although it is normally catalyzed by alkalies[53], it also takes place when formaldehyde is heated in the presence of acids[78]. Formaldehyde, benzaldehyde, and other aldehydes, which do not possess alpha-hydrogen atoms and hence cannot undergo ordinary aldol condensations, can be made to react almost quantitatively in this way. The reaction mechanism, at least under alkaline conditions, has been reported to involve the liberation of hydrogen, which reduces unreacted formaldehyde to methanol[31]:

$$CH_2O \text{ (aq)} + NaOH \rightarrow HCOONa + H_2$$

$$H_2 + CH_2O \text{ (aq)} \rightarrow CH_3OH$$

In this connection it is interesting to note that finely divided nickel is reported to accelerate the reaction[23]. (An alternative mechanism involves the Tischenko reaction with production of methyl formate, which is readily hydrolyzed in aqueous solution[73].)

An investigation of the effect of temperature on the reaction of equal volumes of $0.1N$ formaldehyde and $0.05N$ sodium hydroxide by Pajunen[65] indicated that the Cannizzaro reaction was predominant at temperatures of 40 to 60°C, whereas the aldol-type condensation leading to sugar formation assumed the major role at 70°C. This study also indicated that the Cannizzaro reaction was of the third order.

The production of methylal in reactions involving formaldehyde as the sole organic raw material is a sequel to the Cannizzaro reaction, being brought about by the action of formaldehyde on the methanol formed:

$$CH_2O \text{ (aq, p)} + 2CH_3OH \rightarrow CH_2(OCH_3)_2 + H_2O$$

Tischenko Reaction. Polymers of formaldehyde react with either aluminum or magnesium methylate to yield methyl formate[80]:

$$2CH_2O \text{ (p)} \rightarrow HCOOCH_3$$

Methyl formate is also reported among the products obtained on heating formaldehyde polymers, such as paraformaldehyde, alone[20] and in the presence of sulfuric acid[37].

Aldol-type Condensations. Although the formaldehyde molecule does not possess alpha-hydrogen atoms and hence cannot undergo orthodox aldol condensations, hydroxy aldehydes, hydroxy ketones, and sugars are formed by a closely analogous reaction under some conditions.

In aqueous formaldehyde this reaction is favored by alkaline agents; but with anhydrous solutions of formaldehyde in alcohols, glycols or glycerol it can apparently also be carried out in the presence of weak organic acids[55]. In addition, it is induced by ultraviolet light, and there is evidence that it is involved in the mechanism by which carbohydrates are formed in plants[6, 7, 26, 27, 52, 54]. In this connection, some investigators have concluded that excited formaldehyde molecules are produced by the photochemical reaction of carbon dioxide and water and that these molecules are promptly converted to carbohydrates. Although there is considerable doubt as to whether ordinary formaldehyde is produced under these circumstances, it does appear probable that it is formed in small concentrations in special cases, as claimed by Baly[7] and, more recently, by Yoe and Wingard[86] and Groth[36].

Although sugars are the predominant product when this condensation is allowed to proceed to completion, glycolic aldehyde has been isolated from reaction mixtures and is apparently the primary condensation product[25]:

$$2CH_2O \rightarrow CH_2OH \cdot CHO$$

Other relatively simple products whose presence has been demonstrated include: glyceraldehyde, $CH_2OH \cdot CHOH \cdot CHO$, dihydroxyacetone, $CH_2OH \cdot CO \cdot CH_2OH$; and erythrose, $CH_2OH \cdot CHOH \cdot CHOH \cdot CHO$[25, 64].

The fact that sugar-like products can be produced by formaldehyde condensation was first observed by Butlerov[15] in 1861, who designated the crude product obtained in his experiments as methylenitan. By careful handling of the crude product which decomposes to brown, tarry materials when heated in the presence of alkalies, Loew[52] succeeded in isolating a sweet syrup which reduced Fehling's solution but was optically inactive. From this syrup, Loew[52-54] obtained an approximately 75 per cent yield of a mixture of hexose sugars (formose) possessing the empirical formula $C_6H_{10}O_5$. An osazone isolated from formose was demonstrated to be phenyl-acrosazone, which is a derivative of inactive *dl*-fructose (alpha-acrose). A typical procedure for formose preparation involves agitating an approximately 4 per cent formaldehyde solution with an excess of calcium hydroxide for one half hour, filtering and allowing the filtrate to stand for 5 or 6 days. The solution is then neutralized, after which the formose is isolated by a process involving evaporation of water and removal of calcium by reaction with oxalic acid and extraction with alcohol. Prudhomme[72] has patented a photochemical process for the preparation of saccharose from formaldehyde in which calcium oxide is used as a catalyst. Balezin[5] claims that the role of calcium oxide in the condensation of formaldehyde to sugars does not consist in merely producing the proper degree of alkalinity but that it is an essential component of an intermediate complex formed in the condensation process.

Loew's work indicates that the aldol-type condensation proceeds best in the presence of the alkaline earth hydroxides and the hydroxides of the weakly basic metals such as lead and tin. The condensation is also characterized by a comparatively long incubation period. This usually indicates an autocatalytic process and there is a growing body of evidence that this is the case not only for this reaction but also for orthodox aldol condensations. Langenbeck[48] reports that the aldol-type formaldehyde condensation is definitely catalyzed by glycollic aldehyde, acetol or dihydroxyacetone ($CH_2OH \cdot CO \cdot CH_2OH$), fructose, glucose, etc. The most active catalysts are glycolic aldehyde and dihydroxyacetone. According to Langenbeck[48], no induction period is observed when these agents are added to alkaline formaldehyde. Studies by Katzschmann[45] indicate a probable mechanism of the autocatalytic reaction in which the slow primary formation of glycollic aldehyde is followed by the rapid reactions indicated:

$$CH_2O + CH_2O \rightarrow HOCH_2CHO$$

$$HOCH_2CHO + CH_2O \rightarrow HOCH_2CH(OH)CHO$$

$$HOCH_2CH(OH)CHO \rightleftharpoons HOCH_2COCH_2OH$$

$$HOCH_2COCH_2OH + 2CH_2O \rightarrow \underline{HOCH_2COC(OH)(CH_2OH)_2}$$

$$\underline{HOCH_2COC(OH)(CH_2OH)_2} + CH_2O \rightarrow HOCH_2COCH(OH)CH_2OH + HOCH_2CHO$$

The dimethylol derivative of acetol (bis(hydroxymethyl)acetol), underlined in the above equations, functions as the catalyst. This compound was isolated and identified by Katzschmann[45].

The formation of glycollic acid and related compounds from formaldehyde probably involves the formation of glycollic aldehyde as a primary step. Hammick and Boeree[37] obtained small yields of this acid by heating paraformaldehyde with 10 per cent its weight of concentrated sulfuric acid. Fuchs and Katscher[32] obtained monochloroacetic acid by heating trioxane, $(CH_2O)_3$, with sulfuryl chloride and zinc chloride. Green and Handley[35] claim the production of acetic acid by the action of heat and pressure on formaldehyde.

Type Reactions

Most of the chemical reactions of formaldehyde with other compounds may be broadly classified into three types: (a) *reduction reactions*, in which the aldehyde acts as a reducing agent, itself being oxidized to formic acid; (b) *addition or condensation reactions*, leading to the production of simple methylol or methylene derivatives; and (c) *polymerization reactions*, resulting in the formation of polymethylene derivatives.

Reduction Reactions. Formaldehyde is a reducing agent. In this capacity it is most effective in alkaline solutions in which it is converted to the formate ion. Latimer[49] gives the potential $(E°)$ for the half reaction shown below as 1.14 international volts.

$$CH_2O \text{ (aq)} + 3OH^- \rightarrow HCO_2^- \text{ (aq)} + 2H_2O + 2e^-$$

The potential for the acid reaction is reported to be 0.01.

$$CH_2O \text{ (aq)} + H_2O \rightarrow HCOOH \text{ (aq)} + 2H^+ + 2e^-$$

These figures are calculated from the free-energy values of Parks and Huffman[66].

In alkaline media formaldehyde has been found to bring about the precipitation of metals from solutions of salts of bismuth, copper, mercury silver, and gold[60]. The reduction of gold and platimun salts is reported even under strongly acid conditions (page 180). The reductions of silver oxide and ammoniacal silver nitrate are characteristic aldehyde reactions:

$$Ag_2O + CH_2O \text{ (aq)} \rightarrow 2Ag + HCOOH$$

In the case of cuprous oxide, reduction proceeds in a different manner, with evolution of hydrogen[39]:

$$Cu_2O + 2CH_2O \text{ (aq)} + 2NaOH \rightarrow 2Cu + H_2 + 2HCOONa + H_2O$$

The Cannizzaro reaction, previously discussed, involves the reduction of formaldehyde by formaldehyde. Since formaldehyde is apparently a more powerful reducing agent than most other aldehydes, this reaction can be utilized for the reduction of other aldehydes by a crossed Cannizzaro reaction[21].

$$C_6H_5CHO + CH_2O \text{ (aq)} + NaOH \rightarrow C_6H_5CH_2OH + HCOONa$$

The use of finely divided silver as a catalyst for this reaction is described in a patent by Pearl[70].

The apparent reduction of 9-formyl fluorene, containing a radioactive C^{14} carbon in the formyl group, with alkaline formaldehyde yields nonradioactive 9-fluorene methanol and radioactive formate[14]. This is explained by a mechanism involving an aldol condensation followed by decomposition of the unstable hydroxyaldehyde.

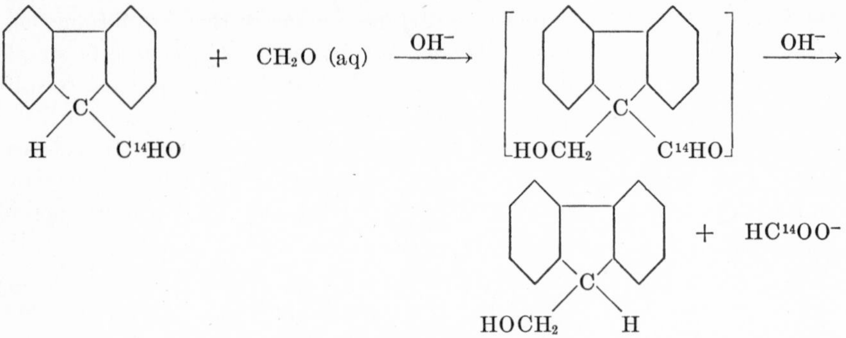

$$\text{(structure) } + \text{ CH}_2\text{O (aq) } \xrightarrow{\text{OH}^-} \left[\text{(structure)} \right] \xrightarrow{\text{OH}^-}$$

H   C^{14}HO   HOCH$_2$ C^{14}HO

$$\text{(structure)} + \text{HC}^{14}\text{OO}^-$$

HOCH$_2$ H

Reactions of this type should not be confused with the crossed Cannizzaro reaction. However, the tertiary aldehyde intermediate would normally be expected to undergo a true crossed Cannizzaro reaction with alkaline formaldehyde to yield a trimethylene glycol were it not unstable to alkali.

Methylations accomplished by heating formaldehyde or paraformaldehyde with ammonium hydrochlorides also involve the reducing action of formaldehyde[34].

$$6\text{CH}_2\text{O (aq, p)} + \text{NH}_4\text{Cl} \rightarrow \text{N(CH}_3)_3\cdot\text{HCl} + 3\text{HCOOH}$$

Addition or Condensation Reactions. *Formation of Methylol Derivatives.* Formaldehyde reacts readily with a wide variety of chemical compounds containing active hydrogen atoms with the formation of methylol derivatives in which one or more of the active hydrogens are substituted by the methylol (—CH$_2$OH) group. In general, these reactions are reversible. The production of methylene glycol is a representative example of this reaction type: $\text{HOH} + \text{CH}_2\text{O (g)} \rightleftharpoons \text{HOCH}_2\text{OH}$. Like methylene glycol, many of these methylol derivatives are unstable and decompose with regeneration of the original reactants or formation of methylene derivatives, polymers, etc., when their isolation in a pure state is attempted. However, other methylol derivatives are more stable and can be readily isolated. Illustrative of such is the well known bisulfite compound whose formation from aqueous formaldehyde and sodium bisulfite is indicated below[50]:

$$\text{CH}_2\text{O (aq)} + \text{NaHSO}_3 \rightarrow \text{HOCH}_2\cdot\text{SO}_3\text{Na}$$

As previously pointed out, the actual reactant in solution reactions is undoubtedly methylene glycol, and the true mechanism is probably more accurately represented by the condensation reaction shown in the following equation:

$$\text{HOCH}_2\text{OH} + \text{NaHSO}_3 \rightarrow \text{HOCH}_2\cdot\text{SO}_3\text{Na} + \text{H}_2\text{O}$$

Other important methylol derivatives include saligenin (methylolphenol) which is formed when alkaline formaldehyde reacts with phenol,

$$\text{C}_6\text{H}_5\text{OH} + \text{CH}_2\text{O (aq)} \rightarrow \text{C}_6\text{H}_4(\text{CH}_2\text{OH})\text{OH}$$

and dimethylolurea produced from neutral or alkaline formaldehyde and urea,

$$NH_2CONH_2 + 2CH_2O \text{ (aq)} \rightarrow HOCH_2 \cdot NHCONH \cdot CH_2OH$$

As a rule, the formation of methylol derivatives proceeds most readily under neutral or alkaline conditions, whereas under acidic conditions, or in the vapor phase, methylene derivatives are usually obtained. However, some compounds produce stable methylol derivatives even in the presence of acids, and others react to give methylene derivatives as the sole isolable product in alkaline media. Methylol derivatives are the primary formaldehyde reaction products, and it is probable that their formation is a part of the mechanism of all formaldehyde reactions.

Reactions of monomeric formaldehyde with metal-organic compounds lead to the production of metal substituted methylol compounds in direct analogy to the reactions shown above. On hydrolysis, these products give the corresponding methylol derivatives. The following reactions are illustrative:

Reaction with phenyl magnesium bromide[87]:

$$C_6H_5MgBr + CH_2O \text{ (g)} \rightarrow C_6H_5 \cdot CH_2 \cdot OMgBr$$

$$C_6H_5 \cdot CH_2 \cdot OMgBr + H_2O \rightarrow C_6H_5 \cdot CH_2OH + Mg(OH)Br$$

Reaction with sodium triphenylmethane[74]:

$$(C_6H_5)_3CNa + CH_2O \text{ (g)} \rightarrow (C_6H_5)_3C \cdot CH_2ONa$$

$$(C_6H_5)_3C \cdot CH_2ONa + H_2O \rightarrow (C_6H_5)_3C \cdot CH_2OH$$

Formation of Methylene Derivatives. Three types of simple methylene derivatives are formed by reactions of formaldehyde with other organic compounds: (a) compounds in which two similar or dissimilar radicals or groups are attached by means of a methylene group; (b) unsaturated compounds containing a double-bonded methylene group; and (c) compounds which may be regarded as simple cyclic polymers of the unsaturated methylene derivatives [type (b)]. In general, the methylenic compounds obtained with formaldehyde are more stable than the methylol derivatives and, in some cases, are formed by irreversible reactions. As previously pointed out, methylene derivatives are usually favored by acidic catalysts, although in some cases they are readily obtained in neutral or alkaline media.

Reactions leading to compounds of type (a), in which two similar groups are linked together, are found in the production of methylal from formaldehyde and methanol,

$$2CH_3OH + CH_2O \rightarrow CH_3OCH_2OCH_3 + H_2O$$

and diphenylmethane from formaldehyde and benzene,

$$2C_6H_5 + CH_2O \rightarrow C_6H_5 \cdot CH_2 \cdot C_6H_5 + H_2O.$$

Both these reactions take place in the presence of strongly acidic catalysts.

The linkage of dissimilar radicals is demonstrated by the chloromethylation of benzene[9],

$$C_6H_6 + HCl + CH_2O \rightarrow C_6H_5 \cdot CH_2Cl + H_2O$$

and the Mannich synthesis[56], illustrated by the reaction of acetone, dimethylamine (or dimethylamine hydrochloride) and formaldehyde.

$$CH_3COCH_3 + (CH_3)_2NH + CH_2O \rightarrow CH_3COCH_2 \cdot CH_2N(CH_3)_2 + H_2O$$

Formation of double-bonded methylene compounds (type b) occurs in reactions of formaldehyde solution with phenylhydrazine[83],

$$C_6H_5NHNH_2 + CH_2O \text{ (aq)} \rightarrow C_6H_5NHN:CH_2 + H_2O$$

and malonic ester[84].

$$CH_2(COOC_2H_5)_2 + CH_2O \text{ (aq)} \rightarrow CH_2:C(COOC_2H_5)_2 + H_2O$$

Double-bonded methylenic derivatives are also obtained by vapor-phase reactions of formaldehyde with carbonyl compounds such as acetaldehyde[33, 43],

$$CH_3CHO \text{ (g)} + CH_2O \text{ (g)} \rightarrow CH_2:CHCHO \text{ (g)} + H_2O \text{ (g)}$$

and acetone[13],

$$CH_3COCH_3 \text{ (g)} + CH_2O \text{ (g)} \rightarrow CH_2:CHCOCH_3 \text{ (g)} + H_2O \text{ (g)}$$

The malonic ester reaction shown previously can also be carried out in the gas phase.

Cyclic methylene derivatives (type c) are often formed in place of the double-bonded derivatives which might be expected. As a rule these compounds contain six-membered rings. They may be theoretically regarded as trimers of the simple unsaturate which may or may not be a reaction intermediate. The aniline-formaldehyde reaction[81, 85] is characteristic:

$$3C_6H_5NH_2 + 3CH_2O \text{ (aq)} \rightarrow (3C_6H_5N:CH_2 + 3H_2O) \rightarrow$$

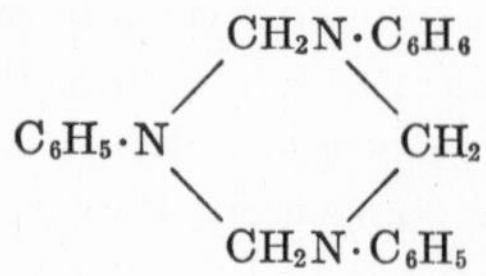

$$
\begin{array}{c}
CH_2N \cdot C_6H_5 \\
\diagup \qquad \diagdown \\
C_6H_5 \cdot N \qquad\qquad CH_2 \quad + \quad 3H_2O \\
\diagdown \qquad \diagup \\
CH_2N \cdot C_6H_5
\end{array}
$$

Trithiane rather than thioformaldehyde is produced when formaldehyde reacts with hydrogen sulfide in the presence of an acid[41]:

$$
\begin{array}{c}
S—CH_2 \\
\diagup \qquad \diagdown \\
3H_2S + 3CH_2O \text{ (aq)} \rightarrow H_2C \qquad\qquad S \quad + \quad 3H_2O \\
\diagdown \qquad \diagup \\
S—CH_2
\end{array}
$$

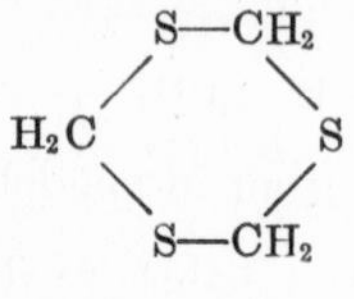

With ammonia, formaldehyde forms the tricyclic hexamethylenetetramine[24]:

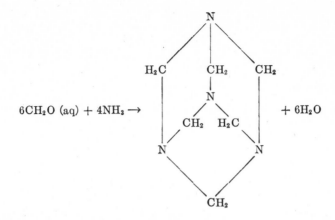

$$6CH_2O \text{ (aq)} + 4NH_3 \rightarrow$$

$$+ 6H_2O$$

It is highly probable that the mechanism of methylene compound formation always involves the primary production of methylol derivatives although in many cases these intermediates cannot be isolated. For example, there is excellent evidence that methylal formation is preceded by the production of a hemiacetal, as indicated by the following equations (page 202):

$$CH_3OH + CH_2O \text{ (aq, p)} \rightarrow CH_3OCH_2OH$$

$$CH_3OCH_2OH + CH_3OH \rightarrow CH_3OCH_2OCH_3 + H_2O$$

Polymerization Reactions. The property of forming resinous products on reaction with other chemicals is one of the most useful characteristics of formaldehyde and is the basis of its immense industrial importance in the synthetic-resin industry. Under suitable conditions, the molecules of many compounds are linked together by methylene groups when subjected to the action of formaldehyde. Phenol- and urea-formaldehyde resins are polymethylene compounds of this type.

Two distinct mechanisms are probably involved in resin-forming reactions: (1) the polycondensation of simple methylol derivatives and (2) the polymerization of double-bonded methylene compounds. Although in some cases the mechanism is definitely one or the other of the above, it is often not clear which is followed and both may play a part in some instances. Recent evidence indicates that the formation of urea-formaldehyde resins, which used to be regarded as a simple polycondensation of methylol ureas[43], may actually involve the primary formation of a methylene urea which then polymerizes to give a cyclic trimethylenetriamine whose methylol derivatives are finally cross-linked by condensation reactions[53] (see pages 299–300).

Thermoplastic resins are the result of simple linear condensations, whereas the production of thermosetting resins involves the formation of methylene cross-linkages between linear chains. Both of these resin types may be produced, sometimes from the same raw materials by variations in the relative amounts of formaldehyde employed, conditions of catalysis, and temperature. However, with compounds whose molecules present only two points of attack, or but two reactive hydrogen atoms, only thermoplastic resins can be obtained.

A diversified range of organic compounds, including hydroxy compounds, amines, amides, proteins, phenols, and hydrocarbons form resins with formaldehyde.

Reactions of Methylol Derivatives

A large portion of the chemical reactivity of formaldehyde is passed on to many of its methylol derivatives. Since the characteristics of these compounds are so closely related to those of formaldehyde, their chemical reactions form an important division of formaldehyde chemistry.

Of course, all reactions involving aqueous formaldehyde are logically reactions of methylol derivatives, since methylene glycols ($HO \cdot CH_2OH$, $HOCH_2 \cdot O \cdot CH_2OH$, etc.) are involved. However, since similar results could be expected with anhydrous formaldehyde, a needless degree of complexity is avoided by regarding them simply as reactions of formaldehyde (*i.e.*, when questions of mechanism are not of prime importance). Nevertheless, for some reactions this is not the case. Methylene glycol apparently adds to the ethylenic double-bond as a methylol derivative, $HO \cdot CH_2OH$, giving hydroxymethylol compounds in which —OH and —CH_2OH are added to adjacent carbon atoms. This reaction has accordingly been included in this section rather than in the sections dealing with simple formaldehyde reactions.

The following reaction types should be construed broadly to include polymethylol derivatives as well as the simple monomethylol compounds indicated in the type reactions.

Decomposition with Liberation of Formaldehyde. Type reaction: $RCH_2OH \rightarrow RH + CH_2O$. This is the reverse of the reaction by which methylol derivatives are formed. In aqueous solution, decomposition may take the form of a hydrolysis in which formaldehyde is liberated in the form of methylene glycol. Methylolformamide behaves in this way[44]:

$$HCONH \cdot CH_2OH + H_2O \rightarrow HOCH_2OH + HCONH_2$$

Under anhydrous conditions, formaldehyde gas is liberated. Thus, the methylol derivative of triphenylmethane is decomposed on heating[74]:

$$(C_6H_5)_3C \cdot CH_2OH \rightarrow (C_6H_5)_3CH + CH_2O \text{ (g)}$$

Some methylol derivatives are extremely stable and do not liberate formaldehyde even when heated to high temperatures.

Polycondensation. This type of reaction is analogous to the formation of polyoxymethylene glycols from methylene glycol in aqueous formaldehyde (page 47).

Condensation with Other Compounds. Type reaction: $R \cdot CH_2OH + R'H \rightarrow R \cdot CH_2 \cdot R' + H_2O$. This is illustrated by the formation of the dimethyl ether of dimethylolurea in the presence of a small concentration of acid[76]:

$$HOCH_2NHCONHCH_2OH + 2CH_3OH \rightarrow CH_3OCH_2NHCONHCH_2OCH_3 + 2H_2O$$

Under alkaline conditions formaldehyde cyanohydrin condenses with diethylamine[46]:

$$HOCH_2CN + HN(C_2H_5)_2 \rightarrow (C_2H_5)_2NCH_2CN + H_2O$$

Reactions of this type are common.

Dehydration to Double-Bonded Methylene Derivatives. Type reaction: $H \cdot R \cdot CH_2OH \rightarrow R{:}CH_2 + H_2O$. The preparation of methyl vinyl ketone by the distillation of methylolacetone (ketobutanol) in the presence of a trace of iodine or acidic catalyst is representative[61]:

$$CH_3COCH_2 \cdot CH_2OH \rightarrow CH_3COCH{:}CH_2 + H_2O$$

Cyclic trimers or higher polymers of the methylenic monomer are often the only isolable product of reactions of this type. Isolation of the pure unsaturated monomer can be realized in but a few cases.

Addition Reactions. Type reaction: $RCH_2OH + R'{:}R'' \rightarrow RCH_2 \cdot R' \cdot R''OH$. Under some conditions, methylol derivatives are capable of adding to ethylenic linkages. For example, methylene glycol apparently adds to propylene under pressure and in the presence of an acidic catalyst[28]:

$$HOCH_2OH + CH_2{:}CH \cdot CH_3 \rightarrow CH_2OH \cdot CH_2 \cdot CHOH \cdot CH_3$$

Reduction to Methyl Derivatives. Type reaction: $R \cdot CH_2OH + H_2 \rightarrow R \cdot CH_3 + H_2O$. A number of methylol derivatives may be reduced to the corresponding methyl derivatives by the action of hydrogen in the presence of an active nickel catalyst. Compounds which may be reduced in this way include dimethylolethanolamine[18],

$$HOCH_2CH_2N(CH_2OH)_2 + 2H_2 \rightarrow HOCH_2CH_2N(CH_3)_2 + 2H_2O$$

and dimethylolacetone[1],

$$CH_3COCH(CH_2OH)_2 + 2H_2 \rightarrow CH_3COCH(CH_3)_2 + 2H_2O.$$

Reductions of this type may be accomplished by a number of reducing agents including, in some cases, formaldehyde itself.

Reactions Involving Two or More Types

Formaldehyde reactions often involve two or more of the reaction types which have been described separately in the above discussion. For example, the formation of pentaerythritol by the action of formaldehyde on acetaldehyde in the presence of alkalies has been found to involve both methylol formation and a mixed Cannizzaro reaction (pages 220–221).

$$CH_3 \cdot CHO \quad + \quad 3CH_2O \text{ (aq)} \quad \rightarrow \quad C(CH_2OH)_3 \cdot CHO$$

$$\textit{Acetaldehyde} \qquad \textit{Formaldehyde} \qquad \textit{Trimethylolacetaldehyde or pentaerythrose}$$

$$C(CH_2OH)_3 \cdot CHO \quad + \quad CH_2O \text{ (aq)} \quad + \quad \tfrac{1}{2}Ca(OH)_2$$

$$\textit{Pentaerythrose} \qquad \textit{Formaldehyde} \qquad \textit{Lime}$$

$$\rightarrow \quad (CH_2OH)_4C \quad + \quad \tfrac{1}{2}Ca(OOCH)_2$$

$$\textit{Pentaerythritol} \qquad \textit{Calcium Formate}$$

References

1. Anon., *Chem. Age*, **32**, 501 (1935).
2. Askey, P. J., *J. Am. Chem. Soc.*, **52**, 974 (1930).
3. Avramenko, L. I., and Lorentso, R. V., *Doklady Akad. Navk.*, S. S. S. R., **69**, 205–7 (1949); *C. A.*, **44**, 2350 (1950).
4. Axford, D. W. E., and Norrish, R. G. W., *Proc. Roy. Soc.*, **A192**, 518–37 (1948).
5. Balezin, S. A., *Zhur. Obshchei Khim. (J. Gen. Chem.)*, **17**, 2288–91 (1947); *C. A.*, **43**, 148 (1949).
6. Baly, E. C. C., Heilbron, I. M., and Barker. W. F., *J. Chem. Soc.*, **119**, 1025–35 (1921).
7. *Ibid.*, **121**, 1078–88 (1922).
8. Berthelot, D., and Gaudechon, H., *Compt. rend.*, **150**, 1690–3 (1910); *C. A.*, **4**, 2408 (1910).
9. Blanc, G., *Bull. soc. chim. (4)*, **33**, 313 (1923).
10. Blank, O., and Finkenbeiner, H., *Ber.*, **39**, 1326 (1906).
11. Bone, W. A., and Smith, H. L., *J. Chem. Soc.*, **87**, 910 (1905).
12. Bowen, E. J., and Tietz, E. L., *J. Chem. Soc.*, **1930**, 234.
13. Brant, J. H., and Hasche, R. L., (to Eastman Kodak Company), U. S. Patent 2,245,567 (1941).
14. Burr, J. G., Jr., *J. Am. Chem. Soc.*, **73**, 823 (1951).
15. Butlerov (Butlerow), A. M., *Ann.*, **120**, 296 (1861).
16. Calvert, J. G., and Steacie, E. W. R., *J. Chem. Phys.*, **19**, 176–82 (1951).
17. Carruthers, J. E., and Norrish, R. G. W., *Trans. Faraday Soc.*, **32**, 195 (1936).
18. Cass, O. W., and K'Burg, R. T., (to E. I. du Pont de Nemours & Co., Inc.), U. S. Patent 2,194,294 (1940).
19. Comanducci, E., *Rend. accad. sci. Napoli*, **15**, 15–7 (1909); *Chem. Zentr.*, **1909**, I, 1530.
20. Contardi, A., *Gazz. chim. ital.*, **51**, I, 109–25 (1921); *C. A.*, **15**, 3072–3 (1921).

21. Davidson, D., and Bogert, M. T., *J. Am. Chem. Soc.*, **57**, 905 (1935).
22. Delépine, M., *Bull. soc. chim.*, (*3*) **17**, 938–39 (1897).
23. Delépine, M., and Horeau, A., *Compt. rend.*, **204**, 1605–8 (1937).
24. Duden, P., and Scharff, M., *Ann.*, **288**, 218 (1895).
25. Euler, H., and Euler, A., *Ber.*, **39**, 50 (1906).
26. Fischer, E., and Passmore, F., *Ber.*, **359**, 97 (1889).
27. Fischer, E., and Tafel, J., *Ber.*, **22**, 97 (1889).
28. Fitsky, W., (to I. G. Farbenindustrie, A. G.), U. S. Patent 2,143,370 (1939).
29. Fletcher, C. J. M., *Proc. Roy. Soc. (London)*, **A146**, 357 (1934).
30. Fort, R., and Hinshelwood, C. N., *Proc. Roy. Soc.*, **A129**, 284 (1930).
31. Fry, H. S., Uber, J. J., and Price, J. W., *Rec. trav. chim.*, **50**, 1060–5 (1931).
32. Fuchs, K., and Katscher. E., *Ber.*, **57**, 1256 (1924).
33. Gallagher, M., and Hasche, R. L., (to Eastman Kodak Company) U. S. Patent 2,246,037 (1941).
34. Gilman, H., "Organic Syntheses," Collective Vol. I, p. 514, New York, John Wiley & Sons, Inc. (1932).
35. Green, S. J., and Handley, R., (to Celanese Corporation of America) U. S. Patent 1,950,027 (1934).
36. Groth, W., *Z. Elektrochemie.* **45**, 262 (1939).
37. Hammick, D. L., and Boeree, A. R., *J. Chem. Soc.*, **123**, 2881 (1923).
38. Hanford, W. E., and Schreiber, R. S., (to E. I. du Pont de Nemours & Co., Inc.), U. S. Patent 2,276,192 (1942).
39. Harden, A., *Proc. Chem. Soc.*, **15**, 158–9 (1899); *Chem. Zentr.*, **1899**, II, 258.
40. Hodgins, T. S., and Hovey, A. G., *Ind. Eng. Chem.*, **31**, 673–7 (1939).
41. Hofmann, A. W., *Ann.*, **145**, 357 (1868).
42. Hoyer, H., *Z. Naturforsch.*, **4a**, 335–7 (1949).
43. I. G. Farbenindustri A. G., French Patent 847,370 (1939).
44. Kalle & Co., German Patent 164,610 (1905).
45. Katzschmann, E., *Ber.*, **77B**, 579–85 (1944).
46. Klages, A., *J. prakt. Chem.* (*2*), **65**, 193 (1902).
47. Kuznetzov, M. I., *J. Russ. Phys.-Chem. Soc.*, **45**, 557–68 (1913); *C. A.*, **7**, 3126 (1913).
48. Langenbeck, W., *Naturwiss*, **30**, 30–4 (1942); *Brit. C. A.*, **1942**, A, II, 299.
49. Latimer, W. M., "The Oxidation States of the Elements and their Aqueous Solutions," p. 119, New York, Prentice-Hall, Inc. (1938).
50. Lauer, W. M., and Langkammerer, C. M., *J. Am. Chem. Soc.*, **57**, 2360–2 (1935).
51. Löb, W., *Z. Elektrochemie*, **12**, 291, 301 (1906).
52. Loew, O., *J. prakt. Chem.*, **34**, 51 (1886).
53. Loew, O., *Ber.*, **20**, 144 (1887).
54. *Ibid.*, **22**, 475 (1889).
55. Lorand, E. J., (to Hercules Powder Co.), U. S. Patent 2,272,378 (1942).
56. Mannich, C., *Arch. Pharm.*, **255**, 261–76 (1917); *J. Chem. Soc.*, **112**, I, 634 (1917); *C. A.*, **12**, 684 (1918).
57. Marshall, M. J., and Stedman, D. F., *Trans. Roy Soc. Can.*, **17**, Sect. III, 53–61 (1923); *C.A.*, **18**, 968.
58. Marvel, C. S., Elliott, J. R., Boettner, F. E., and Yuska, H., *J. Am. Chem. Soc.*, **68**, 1681 (1946).
59. Medvedev, S. S., and Robinson, E. A., *Trans. Karpov. Inst. Chem. (Russian)*, (1925), No. 4, 117–25; *C. A.*, **20**, 2273.
60. Menzel, A., "Der Formaldehyd", p. 25, Vienna, Hartleben's Verlag, 1927.

61. Morgan, G., Megson, N. J. L., and Pepper, K. W., *Chemistry & Industry*, **57**, 885–891 (1938).
62. Müller, E., *Ann.*, **420**, 241 (1920).
63. Norrish, R. G. W., and Carruthers, J. E., *J. Chem. Soc.*, **1936**, 1036.
64. Orthner, L., and Gerisch, E., *Biochem. Z.*, **259**, 30 (1933).
65. Pajunen, V., *Suomen Kemistilehti*, **21**B, 21–4 (1948; *C. A.*, **42**, 8155 (1948).
66. Parks, G. S., and Huffmann, H. M., "Free Energies of Some Organic Compounds," New York, Chemical Catalog Co., Inc. (Reinhold Publishing Corp.) 1932.
67. Patat, F., *Z. physik. Chem.*, *Ber.*, **25**, 208 (1934).
68. Patat, F., and Sachsse, H., *Nachr. Ges Wis.*, *Göttingen.*, *Math.-physik. Klasse*, (*Fachgruppen III*), **1**, 41 (1935).
69. Patry, M., and Monceaux, P., *Compt. rend.*, **221**, 300–1 (1945).
70. Pearl, I. A., (to C. G. Parker), U. S. Patent 2,414,120 (1947).
71. Pribram, R., and Franke, A., *Monatsh.*, **33**, 415–39 (1912); *C. A.*, **6**, 2074 (1912).
72. Prudhomme, E. A., (to Carboxhyd Ltd.), U. S. Patent 2,121,981 (1938).
73. Rice, F. O., "The Mechanism of Homogeneous Organic Reactions from the Physical-Chemical Standpoint," p. 178, New York, Chemical Catalog Co., Inc. (Reinhold Publishing Corp.) (1928).
73a. Sauterey, R., *Ann. chim.* (*12*), **7**, 7–13 (1952).
74. Schlenk, W., and Ochs, R., *Ber.*, **49**, 608 (1916).
75. Snowdon, F. F., and Style, D. W. G., *Trans. Faraday Soc.*, **35**, 426 (1939).
76. Sorenson, B. E., (to E. I. du Pont de Nemours & Co., Inc.), U. S. Patent 2,201,927 (1940).
77. Spence, R., *J. Chem. Soc.*, **1936**, 649.
78. Staudinger, H., Signer, R., Johner, H., Schweitzer, O., and Kern, W., *Ann.*, **474**, 254–5 (1929).
79. Style, D. W. G. and Summers, D., *Trans. Faraday Sec.*, **42**, 388–95 (1946).
80. Tischenko (Tischtschenko), W., *J. Russ. Phys.-Chem. Soc.*, **38**, 355–418 (1906); *Chem. Zentr.*, **1906**, II, 1310.
81. Tollens, B., *Ber.*, **17**, 653 (1884).
82. Tropsch, H., and Roehlen, O., *Abhandl. Kenntnis Kohle*, **7**, 15–36 (1925); *Chem. Zentr.*, **1926**, I, 3298; *C. A.*, **21**, 3530 (1927).
83. Walker, J. W., *J. Chem. Soc.*, **69**, 1282 (1896).
84. Welch, K. N., *J. Chem. Soc.*, **1931**, 673–4.
85. Wellington, C., and Tollens, B., *Ber.*, **18**, 3309 (1885).
86. Yoe, J. H., and Wingard, R. E., *J. Chem. Phys.*, **1**, 886 (1933).
87. Ziegler, K., *Ber.*, **54**, 738 (1921).

CHAPTER 9

REACTIONS OF FORMALDEHYDE WITH INORGANIC AGENTS

Formaldehyde reacts chemically with a wide variety of inorganic agents. Representative reactions are summarized in this chapter beginning with those involving the strongly electropositive elements and compounds (*viz.*, alkalies, ammonia, and metals) and ending with those involving the electronegative mineral acids and halogens. This order has been followed because it emphasizes the fact that the chemical behavior of formaldehyde is determined to a very great extent by the polarity of the reactant or by the acidity or alkalinity of the reaction medium. As a result, related reaction types are grouped together and may be the more readily interpreted and compared.

Reactions of formaldehyde with simple carbon compounds which are often treated in inorganic texts will be found in this chapter. These compounds include carbon monoxide, cyanides, and ammonium thiocyanate.

Alkali Metals. Alkali metals apparently show little tendency to react with anhydrous formaldehyde. Elemental sodium does not act upon liquid formaldehyde at its boiling point[130]. However, according to Foelsing[36] hydrogen is liberated when sodium is heated with gaseous formaldehyde and a sodium derivative, possibly NaCHO, is produced.

Alkali and Alkaline-Earth Hydroxides. In the field of formaldehyde chemistry, alkalies are chiefly of importance as reaction catalysts. Since the utility of alkalies in this connection has already been pointed out, the subject need only be reviewed briefly in this place. When pure formaldehyde, paraformaldehyde, alpha-polyoxymethylene, or any of the polyoxymethylene glycols are treated with alkalies in aqueous systems the following reactions are catalyzed:

(1) Hydrolytic polymerization or depolymerization.

(2) Cannizzaro reaction.

(3) Aldol-type condensation.

With respect to hydrolytic polymerization and depolymerization, it is apparent that addition of small quantities of alkalies to aqueous systems containing dissolved formaldehyde or suspended polymers of the polyoxymethylene glycol type accelerates reactions leading to a state of equilibrium between solution and polymer. Alpha-polyoxymethylene is precipitated from solutions containing more than about 30 per cent formaldehyde at room temperature and paraformaldehyde or alpha-polyoxymethylene dissolve rapidly in approximately 3 times their weight of water under similar circumstances.

The effect of alkalies in catalyzing the Cannizzaro reaction (page 164) and the aldol-type condensation (pages 164–166) have been specifically discussed as indicated in the preceding chapter. In general, the weaker bases are the best catalysts for the latter reaction, stronger alkalies usually favoring the Cannizzaro reaction[87]. Calcium hydroxide gives better yields of aldolization products than strontium hydroxide, whereas sodium and potassium hydroxides are least effective.

The utility of alkalies as catalysts for the formation of methylol derivatives by reaction of formaldehyde with compounds containing active hydrogen atoms has been emphasized (page 169).

Hans and Astrid Euler[34] report that the primary reaction of dilute aqueous formaldehyde with alkalies is one of salt formation in which formaldehyde functions as a weak acid. According to their findings, 20 cc of a normal formaldehyde solution (3 per cent CH_2O) mixed with 20 cc water has a freezing point lowering of 0.93°C, a mixture of 20 cc N sodium hydroxide and 20 cc water has a lowering of 1.73°C, and a mixture of 20 cc N formaldehyde and 20 cc N alkali prepared at 0°C* has a lowering of 2.23°C, although it would have one of 2.66°C if no reaction took place. There is also a drop in the conductivity of the mixed solution which is less than half that of a 0.5 N sodium hydroxide solution containing no formaldehyde. This would indicate that about 50 per cent of the hydroxyl ions have been used up by reaction with formaldehyde, a fact which is further confirmed by comparison of the activity of sodium hydroxide as an esterification catalyst in the presence and absence of formaldehyde. In view of these facts, it is probable that the following reaction takes place (see also page 57):

$$H_2C \overset{\displaystyle OH}{\underset{\displaystyle OH}{\big\langle}} + Na^+ + OH^- \rightleftharpoons H_2C \overset{\displaystyle O^-}{\underset{\displaystyle OH}{\big\langle}} + Na^+ + H_2O$$

Franzen and Hauck[41] report the formation of formaldehyde salts of barium, calcium, strontium, and magnesium when the hydroxides of these alkaline earths are treated with formaldehyde solutions. It is their opinion that these compounds may act as intermediates in the formation of sugars from formaldehyde. The type of structural formula suggested for these

---

* The Cannizzaro reaction is not appreciable at this temperature.

products and the way in which they are believed to form is as follows for a calcium salt:

$$2CH_2O \text{ (aq)} + Ca(OH)_2 \rightarrow Ca(OCH_2OH)_2$$

When very dilute solutions are employed it is claimed that the compound HOCH$_2$OCaOH is produced. The length of chain is stated to vary with different alkaline earths involving more metallic atoms and oxymethylene groups. A strontium salt, believed to have the formula, CH$_2$(OSrOCH$_2$OH)$_2$ · 7H$_2$O, is a white powder which decomposes on storage, forming a brown material with a caramel odor. Staudinger's findings[117] make it appear somewhat unlikely that these products are definite compounds. It is possible that they are mixtures of alkaline-earth hydroxides and polymerized formaldehyde.

Metals, Metal Oxides, and Hydroxides. Metals apparently do not react with anyhydrous formaldehyde under ordinary conditions. The corrosive action of aqueous formaldehyde on some metals is caused by the small concentration of formic acid which is normally present in this solution (page 86). In solutions of strong mineral acids, formaldehyde acts as a corrosion inhibitor, reducing the attack of these acids on iron and steel[48] (see pages 493–496). The formation of sugars when strong formaldehyde is boiled with tin[87] and lead[88] is probably induced by the catalytic action of the metal hydroxides which are primarily formed. Similar results are also obtained with zinc dust[84]. These reactions illustrate the powerful catalytic action of the weakly basic metal hydroxides on the aldolization of formaldehyde. According to de Bruyn and van Ekenstein[17], a 70 per cent yield of formose is obtained when formaldehyde solution is warmed with freshly precipitated lead hydroxide.

Formaldehyde salts of metal hydroxides such as lead, zinc, cadmium, and copper are described by Franzen and Hauck[41]. These materials are obtained by adding sodium hydroxide to solutions containing formaldehyde and structures similar to those postulated for the compounds obtained with alkaline-earth hydroxides appear likely. Attempts to isolate aluminum, iron, nickel, and sodium salts were unsuccessful.

Under alkaline conditions, oxides and hydroxides of metals including silver, gold, copper (cuprous[51] and cupric[54] oxides), mercury[51], bismuth[52], and nickel[71] are reduced to metals by formaldehyde. The quantitative reduction of freshly precipitated bismuth hydroxide by excess formaldehyde in the presence of sodium hydroxide affords a method for the determination of bismuth[52].

$$2BiCl_3 + 3CH_2O + 9NaOH \rightarrow 2Bi + 3NaOOCH + 6NaCl + 6H_2O$$

Metallic Salts. Salts of metals whose oxides are reduced by formaldehyde are readily reduced when treated with formaldehyde and alkalies. However, gold and platinum chlorides are reported to undergo reduction even in strongly acid solution[2]. Under these conditions, the metals separate as crystalline precipitates. Colloidal gold is formed by the action of formaldehyde on gold chloride solution in the presence of potassium carbonate[12].

Silver chloride is rapidly and quantitatively reduced to silver when treated with formaldehyde in the presence of strong alkalies, whereas with silver bromide and iodide the reduction proceeds slowly and is quantitative only when special precautions are taken[125]. Under neutral or mildly acid conditions, formaldehyde is said to reduce iron from its salt solutions as a crystalline precipitate[2]. Kelber[71] obtained colloidal nickel by heating a solution of nickel formate and gelatin in glycerol with formaldehyde. Copper sulfate is reported not to be reduced by formaldehyde alone even on prolonged heating at $100°C$[94] (see also page 496).

Ammonia. Formaldehyde and ammonia react to form hexamethylenetetramine in quantitative yield. This reaction, which was first observed by Butlerov[19], was later studied by Duden and Scharf[28], Eschweiler[33], and many other investigators. Its formation and structure are indicated in the equation shown below:

$$6CH_2O + 4NH_3 \rightarrow$$

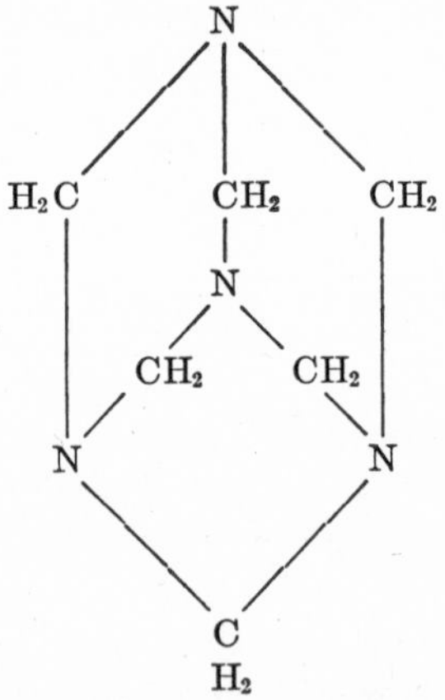

$$+ 6H_2O$$

Since hexamethylenetetramine is a formaldehyde product of considerable commercial importance and behaves chemically as a special form of anhydrous formaldehyde in many industrial applications, its formation, manufacture, and properties are treated specifically in Chapter 19 and will not be further discussed here.

Reaction products other than hexamethylenetetramine may apparently be obtained from ammonia and formaldehyde in special instances. Fosse's studies of the mechanism by which urea, hydrogen cyanide, and other nitrogenous compounds are formed in plants indicates that ammonia and formaldehyde may be the precursors of these products[37, 38]. On oxidation in the presence of ammoniacal silver and mercury salts, formaldehyde is partly converted to hydrogen cyanide, from which urea can be produced by heating with ammonium chloride[38, 39]. Small quantities of cyanamide have also been isolated from the products obtained by oxidation of polyoxymethylenes in the presence of ammonia. Inghilleri[65] reports the production of small amounts of an apparently alkaloidal product when formaldehyde, ammonia, and methanol are exposed to sunlight for seven months.

Ammonium Salts. Addition of formaldehyde to ammonium salt solutions at room temperature results in the liberation of acid. When this acid is neutralized or removed, hexamethylenetetramine is obtained. In the case of ammonium borate, the boric acid which is set free, precipitates in the same manner as if a strong acid had been added[111]. With ammonium carbonate, carbon dioxide is liberated as a gas and when the solution is evaporated under reduced pressure, a residue of hexamethylenetetramine remains[59]. When excess alkali is added to solutions containing ammonium salts of strong acids plus an excess of formaldehyde, the acid which was originally combined with ammonia may be accurately determined by measuring the amount of alkali consumed. A volumetric method for the determination of ammonium nitrate is based on the reaction shown below[49]:

$$6CH_2O \text{ (aq)} + 4NH_4NO_3 + 4NaOH \rightarrow C_6H_{12}N_4 + 4NaNO_3 + 10H_2O$$

A related analytical procedure employs ammonium chloride and caustic for determining formaldehyde[58].

Although formaldehyde-ammonium salt reactions of the above type can be explained by the assumption that hexamethylenetetramine, which is a weak monohydric base, is formed in the reaction solution, mechanism studies by Werner[132] indicate that this may not be the case since, according to his findings, a solution containing ammonium chloride and formaldehyde does not give the precipitates characteristic of polymethyleneamines when treated with picric acid. Werner concludes that weakly basic methyleneimine, whose salts are probably almost completely dissociated in water solution, is the true reaction product, and is formed by the mechanism:

$$NH_4Cl + CH_2O \text{ (aq)} \rightarrow NH_2CH_2OH \cdot HCl \rightarrow NH:CH_2(HCl) + H_2O$$

According to this hypothesis, methyleneimine is stable under acidic conditions and does not tend to polymerize. Hexamethylenetetramine is formed only after the acid has been neutralized or otherwise removed, e.g.,

$$6NH:CH_2(HCl) + 6NaOH \rightarrow (CH_2)_6N_4 + 2NH_3 + 6NaCl + 6H_2O$$

It should be noted that Werner's reaction mechanism is analogous to that by which formaldehyde destroys the basicity of amino acids in the Sorenson procedure for titrating these acids (pages 311, 461).

As the result of a comprehensive study of the reaction of formaldehyde with ammonium salts in aqueous solution and glacial acetic acid, Polley, Winkler and Nicholls[101] report that some hexamethylenetetramine is always formed in the reaction mixture as indicated by precipitation of its mercuric chloride complex (see page 424). However, although isolation of the hexamethylenetetramine is possible in good yield by vacuum evaporation at 50°C when ammonium carbonate, acetate, and propionate are employed; good yields are isolable with salts of stronger acids, such as ammonium nitrate, sulfate, phosphate and oxalate, only when the solution has been neutralized with alkali. The above investigators also demonstrated that the reaction rate, as well as the final yield of hexamethylenetetramine, increases with increasing pH. Furthermore, when various salt solutions are adjusted to the same pH with alkaline buffers, rates and yields are approximately identical. These findings are explained by the assumption that the active reagent with which formaldehyde reacts to produce hexamethylenetetramine in ammonium salt solutions is ammonia. Accordingly, the effect of pH on the reaction is explained by the equilibrium involving ammonia and the hydrogen ion.

$$NH_3 + H^+ \rightleftharpoons NH_4^+.$$

The above findings may also indicate that there is a pH-controlled equilibrium between the methyleneimine, postulated by Werner[132], and hexamethylenetetramine in aqueous solutions containing ammonium salts and formaldehyde.

On heating solutions containing formaldehyde and ammonium salts, salts of methylamines are formed (Plöchl's reaction[98]). According to Knudsen[73], salts of methylamine, dimethylamine, and trimethylamine can be produced by this reaction from ammonium chloride and sulfate. The reaction involves reduction with consequent oxidation of some of the formaldehyde to formic acid and carbon dioxide. When commercial formaldehyde containing methanol is employed, methyl formate and methylal distill as by-products from the acidic reaction mixture. Trimethylamine derivatives are not formed at temperatures of 104°C or below[132]. Werner's hypothesis[132] involving the primary formation of methyleneimine affords a possible interpretation of these reactions:

$$NH_3 \cdot HCl + CH_2O \ (aq) \rightarrow CH_2{:}NH(HCl) + H_2O$$

$$CH_2{:}NH(HCl) + CH_2O \ (aq) + H_2O \rightarrow CH_3NH_2 \cdot HCl + HCOOH$$

$$CH_3NH_2 \cdot HCl + CH_2O \ (aq) \rightarrow CH_3N{:}CH_2(HCl) + H_2O$$

$$CH_3N{:}CH_2(HCl) + CH_2O \ (aq) + H_2O \rightarrow (CH_3)_2NH \cdot HCl + HCOOH$$

$$2(CH_3)_2NH \cdot HCl + CH_2O \ (aq) \rightarrow CH_2(N(CH_3)_2)_2 \cdot 2HCl + H_2O$$

$$CH_2(N(CH_3)_2)_2 \cdot 2HCl + CH_2O \text{ (aq)} + H_2O \rightarrow$$

$$(CH_3)_3N \cdot HCl + (CH_3)_2NH \cdot HCl + HCOOH$$

It should, of course, be remembered that dissolved formaldehyde, CH_2O (aq), plus water in Werner's equations is probably methylene glycol, $CH_2(OH)_2$.

It is also possible that the intermediates in these reactions are methylol derivatives rather than methylene compounds, $i.e.$,

$$CH_2O \text{ (aq)} + NH_3 \cdot HCl \rightarrow HOCH_2NH_2 \cdot HCl$$

$$HOCH_2NH_2HCl + CH_2O \text{ (aq)} \rightarrow CH_3NH_2 \cdot HCl + HCOOH$$

This is not unlikely, since methylol compounds are known to give methyl derivatives on hydrogenation (page 173).

When ammonium salts are reacted with excess formaldehyde at ordinary temperatures, Polley, Winkler and Nicholls[101] report the formation of a by-product that cannot be converted to hexamethylenetetramine. The formation of this material increases as the temperature is raised from 20 to 40°C. It is probable that this product is methylamine or its formaldehyde derivative symmetrical trimethyl cyclotrimethylenetriamine (see page 283).

Emde and Hornemann[32] have asserted that the reduction reaction leading to the formation of methylamines from formaldehyde and ammonium salts is a Cannizzaro reaction, and support their hypothesis by demonstrating that almost exactly one mol of formic acid is produced per N-methyl radical formed in the reaction.

In connection with the interpretation of this reaction mechanism, it should be pointed out that methylamine can also be obtained by heating a solution of hexamethylenetetramine with hydrochloric acid[16].

$$C_6H_{12}N_4 + 4HCl + 4H_2O \rightarrow 4CH_3NH_2 \cdot HCl + 2CO_2$$

This does not indicate that hexamethylenetetramine is a reaction intermediate, since the formation of methylamine may be preceded by hydrolysis of hexamethylenetetramine to methylolamine or methyleneimine hydrochlorides and formaldehyde.

Although, in general, methylamines are the chief products obtained by heating formaldehyde with ammonium salts, other compounds are sometimes formed. For example, under special conditions it is claimed that resin-like materials can be prepared from ammonium sulfate and formaldehyde[27]. Salts containing active anions may also give unique reactions. When ammonium nitrate is heated with paraformaldehyde, nitrogen, formic acid and a small amount of nitrogen dioxide result[126].

When ammonium chloride and formaldehyde are added to hypochlorites, the methylene derivative of monochloramine ($CH_2:NCl$) is obtained in the form of needle-like crystals according to Cross, Bevan and Bacon[23]. Lindsay and Soper[82] claim that this product is identical with the cyclic symmetrical trichlorotrimethylenetriamine, $(CH_2:NCl)_3$, obtained by the action of hypochlorous acid on hexamethylenetetramine.

Ammonium thiocyanate gives resinous products with formaldehyde, behaving somewhat after the fashion of the isomeric thiourea. When reacted with equimolar quantities of formaldehyde in concentrated aqueous solution, ammonium thiocyanate gives yellow amorphous compounds which are insoluble in water and common solvents and have no definite melting points[112]. Acetone-soluble condensation products were obtained by Jacobson[66] on refluxing ammonium thiocyanate with approximately 5.5 mols formaldehyde as 37 per cent solution. Howald[64] produced a moldable product by reacting one mol ammonium thiocyanate with 1.5 to 2.0 mols formaldehyde, adding a plasticizer to the product, and mixing with thiourea. Resins from combinations of ammonium thiocyanate, urea, and formaldehyde were patented by Ellis[31].

Hydrazine. Hydrazine reacts readily with formaldehyde to give methylene derivatives. Pulvermacher[103], who first studied the reaction of these agents, obtained a product which he designated as formalazine. This compound is apparently a polymer of N,N'-bimethylene-hydrazine, $(CH_2:N \cdot N:CH_2)_x$. Although completely insoluble in water, it dissolves readily in acids giving unstable solutions which decompose with evolution of formaldehyde on being warmed. On heating in the dry state, formalazine chars but does not melt. By addition of a molecular proportion of formaldehyde to one mol of hydrazine hydrate, Stollé[118] obtained formalhydrazine, a soluble product believed to be a trimer of methylenehydrazine, $(CH_2:N \cdot NH_2)_3$. The same product was also formed by heating formaldehyde polymer with hydrazine hydrate in a sealed tube at 100°C. This compound is a colorless powder which explodes with a flash on strong heating. It reduces Fehling's solution and ammoniacal silver nitrate on warming. Aqueous solutions of formalhydrazine precipitate formalazine when heated. Another water-soluble formaldehyde derivative of hydrazine, tetraformaltrisazine was reported by Hofmann and Storm[61], who prepared it by gradual addition of commercial formaldehyde to cold hydrazine hydrate. Molecular weight measurements and elemental analyses of this product are in agreement with the bicyclic structure shown below:

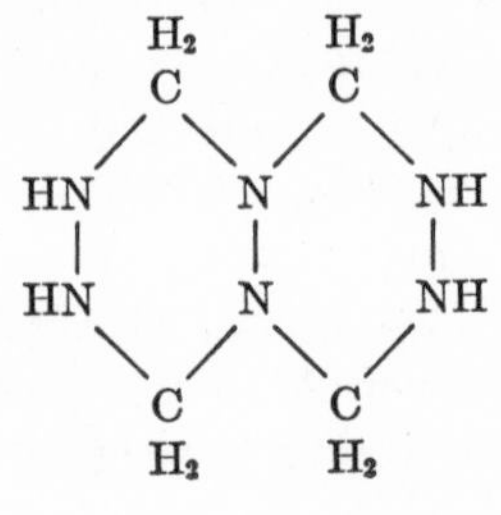

On heating, this compound begins to decompose at 225°C with evolution of gases smelling strongly of hydrogen cyanide. As in the case of formalhydrazine, aqueous solutions precipitate formalazine on warming, a reaction which is accelerated by the presence of traces of acid.

Hydroxylamine. When formaldehyde solution reacts with hydroxylamine, formaldoxime is obtained. The quantitative liberation of acid which accompanies this reaction forms the basis of Cambier and Brochet's[20] volumetric method for determining formaldehyde.

$$NH_2OH \cdot HCl + CH_2O \ (aq) \rightarrow CH_2{:}NOH + H_2O + HCl$$

Formaldoxime, isolated in a pure state by ether extraction of a reaction mixture of hydroxylamine hydrochloride and formaldehyde which has been neutralized with the calculated amount of sodium carbonate, polymerizes to give triformoxime, whose structure is indicated in the following equation[29, 114]:

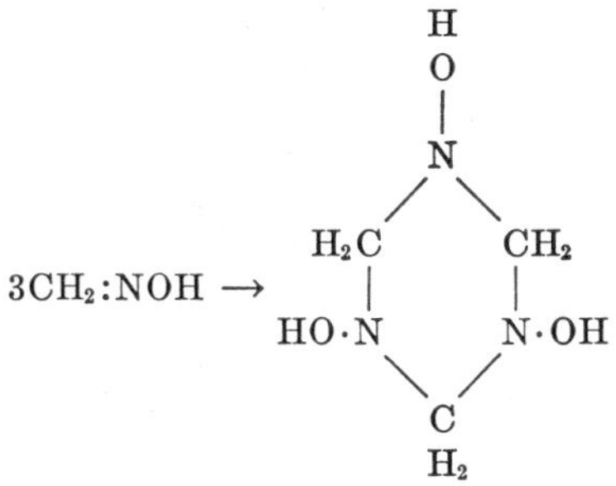

$$3CH_2{:}NOH \rightarrow$$

On heating with dilute acids, formamide is obtained, which is in turn converted to formic acid and ammonia. With sodium it is reported to form a derivative, $CH_2{:}NONa$, which explodes when heated and readily loses water to give sodium cyanide[29].

Formaldehyde, hydroxylamine hydrochloride, and concentrated hydrochloric acid give formic acid, ammonium chloride, and methylamine hydrochloride[76].

Hydrogen Cyanide and Cyanides. Hydrogen cyanide combines with formaldehyde producing glycolonitrile (formaldehyde cyanohydrin), the preparation of which in aqueous solution was first described by Henry[55] in 1890.

$$CH_2O \ (aq) + HCN \rightarrow HOCH_2CN$$

According to Lubs and Acree[89], it may be readily obtained in 70 to 80 per cent yield by mixing equivalent amounts of 90 per cent hydrogen cyanide and 35 per cent formaldehyde solution. Small quantities of ammonia act as a reaction catalyst and the product is isolated by vacuum distillation. Gaudry[43] reports a 76 to 80 per cent yield of glycolonitrile by reacting potassium cyanide with formaldehyde in an aqueous solution at 10°C, neutralizing with dilute sulfuric acid to a pH of 3 at the same temperature, and extracting the product with ether. In this procedure, the ether extract is dried with anhydrous calcium sulfate and a small quantity (approx. 5 ml/g mol) of ethanol is added before distillation. Alcohol is reported to prevent polymerization during the isolation process and to act as a stabilizer in the finished product after vacuum distillation.

Glycolonitrile is a colorless liquid possessing a cyanide odor. According to Lubs and Acree[89], pure, dry glycolonitrile can be distilled at atmospheric pressure with only slight decomposition. Its boiling point is 103°C at a pressure of 16 mm. However, on warming or storage in the presence of impurities, it sometimes decomposes spontaneously with formation of resinous products. It reacts with ammonia to give aminoacetonitrile[33, 72] (b.p. 58°C at 15 mm).

Alkali cyanides on reaction with formaldehyde form formaldehyde cyanohydrin as a primary product. Reaction is quantitative in dilute solution and serves as a basis for Romijn's[109] cyanide procedure (pp. 388–389) for formaldehyde analysis. By acidification and ether-extraction of the mixture obtained by reacting formaldehyde and potassium cyanide solutions at ice-water temperatures, Polstorff and Meyer[102] claim to have isolated formaldehyde cyanohydrin in satisfactory yield. On standing at room temperature cyanide-formaldehyde reaction mixtures undergo hydrolysis. Ammonia is evolved and the solution develops a yellow color. The products formed under these circumstances include alkali salts of glycolic acid, glycine, iminodiacetic acid and nitrilotriacetic acid[40, 102]. The probable mechanism by which these products are formed is indicated in the following equations:

$$KCN + CH_2O \text{ (aq)} + H_2O \rightarrow HOCH_2CN + KOH$$

$$CNCH_2OH + KOH + H_2O \rightarrow HOCH_2COOK + NH_3$$

$$CNCH_2OH + NH_3 \rightarrow NH_2CH_2CN + H_2O$$

$$NH_2CH_2CN + KOH + H_2O \rightarrow NH_2CH_2COOK + NH_3$$

$$NH_2CH_2CN + HOCH_2CN \rightarrow NH(CH_2CN)_2 + H_2O$$

$$NH(CH_2)CN_2 + 2KOH + 2H_2O \rightarrow NH(CH_2COOK)_2 + 2NH_3$$

$$NH_2CH_2CN + 2HOCH_2CN \rightarrow N(CH_2CN)_3 + 2H_2O$$

$$N(CH_2CN)_3 + 3KOH + 3H_2O \rightarrow N(CH_2COOK)_3 + 3NH_3$$

By reaction of 960 g of potassium cyanide in 4 liters water with 1200 cc 38 per cent formaldehyde at approximately 30°C, Franzen[40] obtained 130 g glycolic acid, 21 g glycine, 60 g iminodiacetic acid and 145 g nitrilotriacetic acid.

Reaction of formaldehyde, potassium cyanide, and ammonium chloride was found by Jay and Curtius[63] to give methyleneaminoacetonitrile (m.p. 129°C) as a crystalline product. This product is probably a trimer, $(CH_2:NCH_2CN)_3$, possessing the cyclic structure[69]:

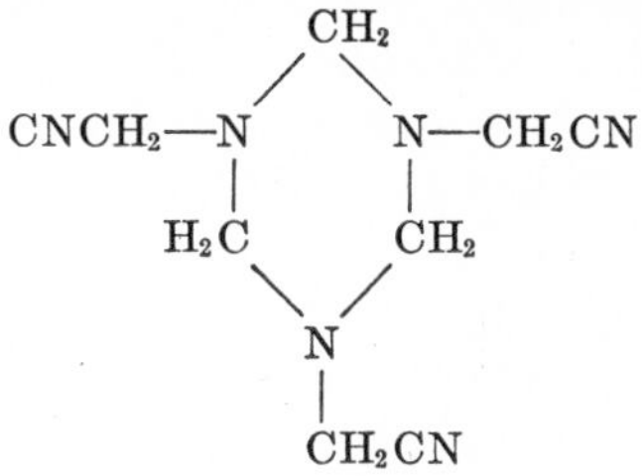

$$CNCH_2—N \underset{\underset{H_2C}{|}}{\overset{\overset{CH_2}{/\backslash}}{}} N—CH_2CN$$

$$H_2C \qquad CH_2$$

$$N$$

$$CH_2CN$$

Preparation of this product in 61 to 71 per cent yield from sodium cyanide, commercial formaldehyde, and ammonium chloride in the presence of glacial acetic acid is described in Gilman's "Organic Syntheses"[44].

Carbon Monoxide. Loder[85] and Larson[77] have shown that under high pressures (above 300 atm), carbon monoxide reacts with formaldehyde and water in the presence of an acid catalyst to produce glycolic (hydroxyacetic) acid in high yield.

$$CH_2O \text{ (aq)} + CO + H_2O \rightarrow CH_2OHCOOH$$

Similar processes involving reactions of carbon monoxide and formaldehyde with alcohols[86] or aliphatic acids[86a] yield alkoxy- and acyloxy-acetic acids respectively. The formation of chloroacetic acid by the joint reaction of carbon monoxide and hydrogen chloride with formaldehyde at 180°C and 900 atmospheres is described by Green and Handley[47].

Hydrogen Peroxide and Peroxides. Under alkaline conditions formaldehyde is rapidly and quantitatively oxidized by hydrogen peroxide with production of sodium formate and evolution of hydrogen.

$$2CH_2O \text{ (aq)} + H_2O_2 + 2NaOH \rightarrow 2NaOOCH + 2H_2O + H_2$$

This reaction is the basis of Blank and Finkenbeiner's[13] method of formaldehyde analysis. In studying the mechanism of this reaction, Wirtz and Bonhoeffer[134] employed hydrogen peroxide in heavy water. The hydrogen liberated was found to contain no deuterium, showing that formaldehyde has no exchangeable hydrogen atoms. Similar results are also obtained under acidic conditions. According to Hatcher and Holden[53], the reaction of formaldehyde with hydrogen peroxide in the absence of alkali is essentially the same as when alkali is present. However, according to Fry and Payne[42], all the following reactions take place:

$$2CH_2O \text{ (aq)} + H_2O_2 \rightarrow 2HCOOH + H_2 \qquad \text{(a)}$$

$$CH_2O \text{ (aq)} + H_2O_2 \rightarrow HCOOH + H_2O \qquad \text{(b)}$$

$$HCOOH + H_2O_2 \rightarrow H_2CO_3 + H_2O \qquad \text{(c)}$$

The extent of these reactions varies with the concentration of hydrogen peroxide in the reaction mixture, reaction (a) being favored by low concentrations. When barium peroxide reacts with formaldehyde, barium formate is the principal product. However, small quantities of barium carbonate are also obtained[90]. Sodium peroxide reacts violently with formaldehyde. Vanino[124] reports that a detonation may occur when this peroxide is thrown into commercial formaldehyde solution, and that it ignites when allowed to come into contact with solid paraformaldehyde.

Thus far, we have discussed only the final products of the reaction of hydrogen peroxide and formaldehyde. Under non-alkaline conditions, methylol peroxides have been successfully isolated. Since these products are probably formed as intermediates whenever peroxides and formaldehyde react, their formation casts considerable light on the study of reaction mechanism. The primary product methylol hydrogen peroxide, was obtained by Rieche and Meister[108] by the reaction of anhydrous formaldehyde in dry ether with anhydrous hydrogen peroxide.

$$H_2O_2 + CH_2O \rightarrow HOCH_2OOH$$

This product is a relatively stable oil of medium consistency having an index of refraction (n_D) of 1.4205 at 16°C. It is not sensitive to friction but explodes with extreme violence when heated in a flame. It reacts vigorously with alkali with evolution of hydrogen and oxygen in the ratio of 2 to 1 and formation of formic acid.

Dimethylol peroxide, $HOCH_2OOCH_2OH$, is a crystalline product melting at 62 to 65°C, which explodes on warming at about 70°C. Fenton[35] obtained this compound by evaporating a mixture of approximately equal volumes of concentrated formaldehyde and hydrogen peroxide solutions at a low temperature. Wieland and Wingler[133] obtained it in 80 to 90 per cent yield from an ether solution of the reactants. It takes fire on contact with iron, platinum black or warm copper oxide but is only slowly decomposed by exposure to sunlight. With alkali, it evolves hydrogen and is converted to sodium formate.

$$HOCH_2OOCH_2OH + 2NaOH \rightarrow 2HCOONa + 2H_2O + H_2$$

This product is apparently identical with the compound isolated by Legler[81] in 1881 from the products of the slow oxidation of ether.

On reaction of two mols of formaldehyde with dimethylol peroxide in ether solution, Rieche and Meister[107] obtained a dimethylol derivative of this product.

$$2CH_2O + HOCH_2OOCH_2OH \rightarrow HOCH_2OCH_2OOCH_2OCH_2OH$$

This compound is a viscous, not particularly explosive oil which is converted upon storage to a crystalline product melting at 152°C. The latter is probably a polyoxymethylene glycol or a polyoxymethylene glycol peroxide, *e.g.*, $HOCH_2O(CH_2O)_xOO(CH_2O)_yCH_2OH$.

By the action of phosphorus pentoxide, methylol peroxide derivatives are converted to peroxides of trioxane and tetraoxymethylene[107]. The following have been isolated in a pure state.

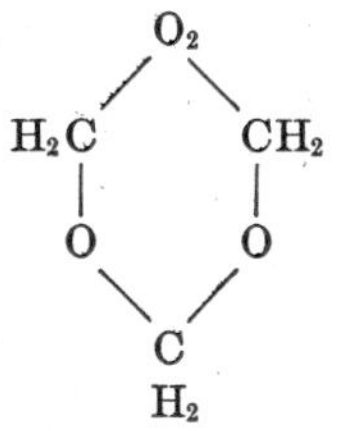

Pertrioxane
(Pertrioxymethylene)

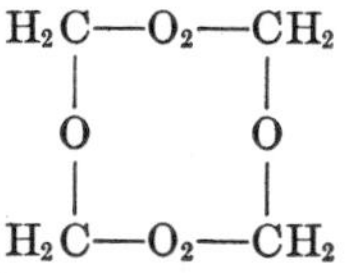

Tetraoxymethylene
Diperoxide

Pertrioxane is an oily liquid reported to boil at 35 to 36°C at 12 mm. It is insoluble in water but readily soluble in alcohol. Its odor is at first pleasant, then irritating, probably because it is hydrolyzed to formaldehyde on the mucous membranes. It is highly explosive and detonates on being warmed, with formation of a white sublimate of tetraoxymethylene. Tetraoxymethylene diperoxide is a crystalline product which detonates without melting on heating to 94°C. It is also sensitive to friction and shock. It is insoluble in water and only difficultly soluble in organic solvents.

Formaldehyde peroxides are also formed by the reaction of paraformaldehyde and hydrogen peroxide. Bamberger and Nussbaum[8] report that colorless crystals, which evolve hydrogen with alkalies, are deposited from a solution of paraformaldehyde in hydrogen peroxide. According to these investigators[9], mixtures of 60 per cent hydrogen peroxide with powdered paraformaldehyde are brisant explosives which detonate on heating or under the influence of a blasting cap. This mixture also detonates spontaneously when left in contact with lead for a short time, presumably from the heat evolved by the oxidation of the lead. Explosive crystals melting at 50°C were separated from the paraformaldehyde-peroxide mixtures.

Explosive peroxides are also obtained by joint reactions of hydrogen peroxide, formaldehyde, and nitrogen compounds. Formaldehyde, ammonium sulfate, and 3 per cent hydrogen peroxide give hexamethylene-triperoxidediamine, $N(CH_2O_2CH_2)_3N^5$. This product may also be obtained from hexamethylenetetramine and hydrogen peroxide[45, 93]. Von Girsewald[46] obtained an explosive peroxygen derivative by the reaction of formaldehyde, hydrazine sulfate, and peroxide at 40 to 50°C.

$$3CH_2O(aq) + H_2O_2 + N_2H_4 \cdot H_2SO_4 \rightarrow CH_2:N \cdot N \begin{matrix} CH_2\!-\!O \\ \diagup \quad \diagdown \\ \diagdown \quad \diagup \\ CH_2\!-\!O \end{matrix} + H_2SO_4 + 3H_2O$$

Hydrogen Sulfide and Sulfides. Hydrogen sulfide dissolves readily in aqueous formaldehyde forming partially sulfuretted analogs of methylene and polyoxymethylene glycols as indicated in the following equations:

$$HSH + HOCH_2OH \rightarrow HOCH_2SH + H_2O$$

$$2HOCH_2SH + HSCH_2OCH_2SH$$

$$HOCH_2SH + HSCH_2OCH_2SH \rightarrow HSCH_2SCH_2OCH_2SH + H_2O$$

Baumann[10] isolated mixed products of the above type by ether-extraction of commercial formaldehyde which had been saturated with hydrogen sulfide and maintained at a temperature of 10 to 12°C for 14 days. The products were viscous oils which gradually hardened and possessed a garlic odor. In the presence of dilute hydrochloric acid, Baumann obtained a solid, crystalline product which melted at 97 to 103°C after crystallization from ether. Analysis of this product indicated the empirical formula, $C_4H_8OS_3$. All these products are soluble in alcohol, ether, and benzene. They exhibit the chemical characteristics of mercaptans, producing disulfides on reaction with iodine. Water solutions give yellow precipitates with lead acetate which decompose on warming with formation of lead sulfide. They dissolve readily in dilute alkali, from which they are precipitated on acidification.

Crystalline hydrogen sulfide-formaldehyde reaction products known as formthionals are also obtained by saturating 37 per cent formaldehyde with hydrogen sulfide at temperatures in the neighborhood of 40 to 50°C[92]. Heat is evolved by the reaction and product is precipitated, almost completely solidifying the reaction mixture. A product of this type obtained from neutral formaldehyde melted at 80°C after crystallization from chloroform and contained 51.5 per cent sulfur. When pure, these compounds are colorless and have little odor.

Under the strongly alkaline conditions that prevail when formaldehyde is reacted with alkali metal sulfides and polysulfides, polymethylene sulfides are obtained. These polymers range from hard brittle resins to rubberlike elastomers. Rubber-like properties are apparently characteristic of the higher polymethylene polysulfides. The structure of the methylene sulfides is apparently analogous to the polyoxymethylenes.

$$\begin{matrix} & H_2 & \quad & H_2 & \quad & H_2 \\ HS\!-\!\!&C\!&-\!S\!-\!\!&C\!&-\!S\cdots\cdots&C\!&-\!SH \end{matrix}$$

Polymethylene Sulfide

$$\begin{matrix} & & S & & & S & & \\ & H_2 & \| & H_2 & \| & H_2 & \\ HSC\!-\!\!&S\!&-\!C\!&-\!S\cdots\cdots&CSH \end{matrix}$$

Polymethylene Disulfide

According to Kohno[74], methylene polysulfide obtained by the reaction of formaldehyde and sodium sulfide is a colorless amorphous powder which becomes plastic and rubbery on heating and is insoluble in organic solvents. Elastomeric resins having a composition corresponding to methylene disulfide, (CH_2S_2), are reported to harden and become brittle on aging. Kohno reports that these polymers gradually liberate free sulfur on storage and concludes that this is a causative factor for the slow hardening and embrittlement. Methylene trisulfide, $(CH_2S_3)_x$, can be converted progressively to the disulfide and monosulfide by removal of sulfur. Rubber-like polymers which show little or no change on storage were obtained by the writer[131] by reacting aqueous formaldehyde with alkali polysulfide solutions, in which the dissolved sulfide had a composition equivalent to sodium tetrasulfide or sodium pentasulfide. The reaction is carried out at 20 to 40°C with gradual addition of carbon dioxide. Methylene polysulfides are probably produced as indicated below:

$$Na_2S_4 + CH_2O + CO_2 \rightarrow 1/x(CH_2S_4)_x + Na_2CO_3$$

$$Na_2S_5 + CH_2O + CO_2 \rightarrow 1/x(CH_2S_5)_x + Na_2CO_3 .$$

These resins normally contain 90 to 93 per cent sulfur and range in color from yellow to dark olive. In dispersed form, they are yellow in color. They soften but do not melt at 120 to 140°C. They are soluble in carbon bisulfide but are substantially insoluble in other solvents.

The production of rubber-like resins from formaldehyde and alkali polysulfides appears to have been first described by Baer[4]. Ellis[30] also obtained plastic products by reaction of mixtures containing formaldehyde and acetaldehyde or other aldehydes with an alkali polysulfide. Resins suitable for molding were obtained by Patrick[96, 97] by heating formaldehyde with alkali-metal disulfides and tetrasulfides. Hills and Barnett[60] describe a light-colored, practically odorless powder, melting at approximately 140°C which is obtained by adding 37 per cent formaldehyde to a solution of sodium sulfide and sulfur at 90°C. This product, which is insoluble in common solvents, may be readily molded with fillers, pigments, etc. Kawano[70] reports a rubber-like polymer, "Thionac A", prepared by oxidizing the reaction product of sodium sulfide and formaldehyde.

Under strongly acidic conditions trithiane or trithioformaldehyde, $(CH_2S)_3$, is the principal reaction product of formaldehyde and hydrogen sulfide and is precipitated in practically quantitative yield[62]. Its preparation from commercial formaldehyde, hydrogen sulfide gas and hydrochloric acid is described in "Organic Syntheses" by Bost and Constable[14]. Sulfuretted methylene glycol derivatives of the type produced when hydrogen sulfide dissolves in aqueous formaldehyde are converted to trithiane on heating with concentrated hydrochloric acid[10]. Trithiane is also produced by reaction of 37 per cent formaldehyde, sodium thiosulfate, and hydrochloric acid[123].

Trithiane possesses a cyclic structure and is the sulfur analog of trioxane.

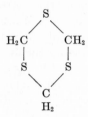

It is a stable, relatively inert, odorless compound melting at 215 to 218°C. It is insoluble in water but may be readily crystallized from hot benzene. Vapor density measurements show that it is trimeric in the gaseous state[63]. On oxidation, trithiane is converted to cyclic sulfoxides and sulfones[21].

Symmetrical dichloromethyl sulfide is formed when trithioformaldehyde is reacted with sulfur chlorides. When sulfur dichloride is employed, symmetrical dichloromethyl sulfide is obtained in practically quantitative yield according to the following equation:

$$(CH_2S)_3 + 2SCl_2 \rightarrow (CH_2Cl)_2S + CS_2 + 2HCl + 2S$$

According to Mann and Pope[91], who prepared the compound by the above reaction, it is a colorless liquid boiling at 57.5 to 58.5°C at 18 mm and has a density of 1.4144 at 14°C. These investigators report that it has no vesicant action on the skin.

Wohl's metathioformaldehyde, which he claimed to be a high molecular weight polymer of thioformaldehyde, produced by reaction of hydrogen sulfide and hexamethylenetetramine, is apparently a mixture of trithiane and products containing nitrogen. Repeated attempts to confirm his experiments gave Le Fevre and MacLeod[80] only mixed products of this type.

Reaction of formaldehyde with ammonium hydrosulfide yields a solid product melting at 198°C and possessing the empirical formula, $C_5H_{10}N_2S_2$. According to Delépine[25], this product is pentamethylenediamine disulfide:

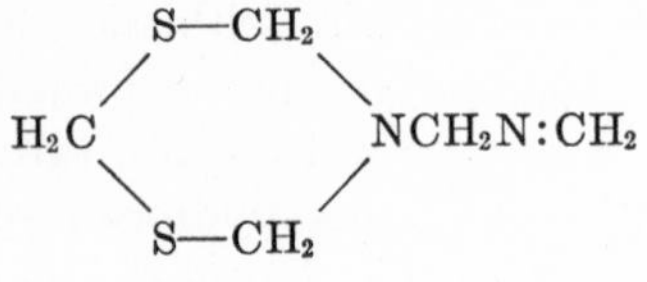

$$\begin{array}{c} S\!-\!CH_2 \\ H_2C \diagup \qquad \diagdown NCH_2N\!:\!CH_2 \\ \diagdown S\!-\!CH_2 \diagup \end{array}$$

A similar product melting at 200°C, obtained from formaldehyde and ammonium sulfide, was reported by Le Fevre and Le Fevre[79], who suggest the following structure:

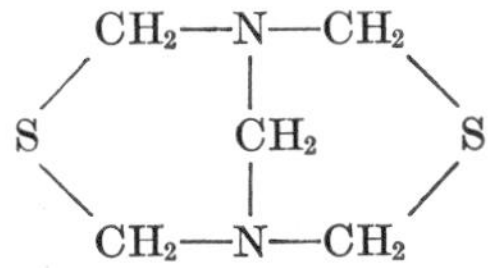

$$CH_2—N—CH_2$$
$$S \quad CH_2 \quad S$$
$$CH_2—N—CH_2$$

An isomeric product having the formula $C_5H_{10}N_2S_2$ and melting at 183°C was obtained by Jacobson[67] by reacting 37 per cent formaldehyde with ammonium chloride and sodium sulfide. It is readily purified by fractional crystallization from ethylene chloride.

Although reactions of hydrogen selenide and formaldehyde have received comparatively little study, it is probable that they are in many respects analogous to those obtained with hydrogen sulfide. Vanino and Schinner[128] report a solid selenoformaldehyde (m.p. 215°C) produced by the action of hydrogen selenide on acid formaldehyde. It is possible that this compound is a trimer analogous to trithiane.

Sulfur Dioxide and Sulfites. Sulfur dioxide dissolves readily in aqueous formaldehyde with evolution of considerable heat. According to Reinking and co-workers[106], an approximately 22 per cent formaldehyde solution dissolves about 2.6 times as much sulfur dioxide as an equivalent amount of water. Commercial 37 per cent formaldehyde dissolves even a larger ratio, giving a solution having a density of 1.4[24]. These solutions smell of both formaldehyde and sulfur dioxide. It is possible that they contain equilibrium concentrations of methylolsulfonic acid.

$$CH_2O(aq) + SO_2 + H_2O \rightleftharpoons HOCH_2SO_3H$$

Although this acid is unstable and cannot be isolated in a pure state, its sodium salt is the stable sodium bisulfite compound of formaldehyde which is readily isolated.

Sodium formaldehyde bisulfite is obtained by the action of sodium bisulfite on aqueous formaldehyde or by warming paraformaldehyde or polyoxymethylene with a solution of sodium bisulfite. It crystallizes from solution on addition of ethanol. Two crystalline forms containing water of crystallization have been described:

$$HOCH_2NaSO_3 \cdot H_2O \quad \text{and} \quad (HOCH_2NaSO_3)_2H_2O$$

Sodium formaldehyde bisulfite is soluble in water and methanol but only slightly soluble in ethanol[75]. Potassium formaldehyde bisulfite, $HOCH_2 \cdot KSO_3$, may be obtained in a manner similar to that employed for the sodium derivative. Normal alkali sulfites react with formaldehyde to produce bisulfite compounds with liberation of alkali hydroxide:

$$Na_2SO_3 + CH_2O (aq) + H_2O \rightarrow HOCH_2 \cdot NaSO_3 + NaOH$$

This reaction is the basis of the sodium sulfite method for formaldehyde analysis (pages 382–384).

For a considerable period following its discovery, sodium formaldehyde bisulfite was believed to possess the structure of a hydroxy sulfite ester, $HOCH_2OSO_2Na$. This formula was incorrect, and it has now been established that its structure is that of a hydroxysulfonic acid, $HOCH_2SO_3Na$. This structure is based on the studies of Raschig and Prahl[104, 105] in 1926–28 and Lauer and Langkammerer[78] in 1935.

On reaction of formaldehyde solution, sodium bisulfite, and ammonia at 70 to 75°C an amino derivative of the aldehyde bisulfite compound is produced. On cooling and acidifying, the parent acid crystallizes from the reaction mixture[104]. Backer and Mulder[3] have shown that this product is aminomethanesulfonic acid ($NH_2CH_2SO_3H$) and not a sulfurous ester, as was first supposed. This evidence serves as an additional check on the bisulfite compound structure. The formation of sodium aminomethane sulfonate is indicated in the following equation:

$$HOCH_2SO_3Na + NH_3 \rightarrow NH_2CH_2SO_3Na + H_2O$$

The above illustrates the so-called sulfomethylation reaction in which a hydrogen atom is replaced by an alkali sulfomethyl group, $-CH_2SO_3M$. The utilization of this reaction to introduce sulfomethyl groups in various organic molecules is reported by Suter, Bair and Bordwell[119]. Phenols, some ketones and malonic esters can be sulfomethylated by reaction with an aqueous formaldehyde solution to which an excess of sodium sulfite has been added.

Reduction of sodium formaldehyde bisulfite with zinc dust and acetic acid[104] or by hydrogenation in the presence of a nickel catalyst[115] yields the industrially important reducing agent, sodium formaldehyde sulfoxylate:

$$HOCH_2SO_3Na + H_2 \rightarrow HOCH_2SO_2Na + H_2O$$

Both sodium and zinc formaldehyde sulfoxylates are used commercially in large quantities for stripping and discharging dyed textiles. According to Wood[135], one process of manufacture involves reacting formaldehyde with sodium hydrosulfite in the presence of caustic soda.

$$Na_2S_2O_4 + CH_2O(aq) + NaOH \rightarrow HOCH_2 \cdot SO_2Na + Na_2SO_3$$

Sodium formaldehyde sulfoxylate and sodium formaldehyde bisulfite are said to be obtained in equimolar quantities when formaldehyde is added to sodium hydrosulfite in aqueous solution and may be separated by fractional crystallization. The so-called zinc and sodium formaldehyde hydrosulfites may be obtained in this way.

Other processes for commercial reducing compositions involve reaction of zinc, formaldehyde, and sulfur dioxide at 80°C, followed by double decomposition with caustic alkali if a sodium salt rather than a zinc salt is desired.

Various reducing compositions of the above types are sold under a variety of trade names.

Acids. *Sulfuric Acid.* When sulfuric acid is added to concentrated formaldehyde at ordinary temperatures, it functions as a polymerization catalyst and polyoxymethylenes are precipitated. On hot formaldehyde, it acts as a catalyst for the Cannizzaro reaction[116].

Paraformaldehyde is partially converted to trioxane on warming with sulfuric acid. On prolonged heating at 115 to 120°C, decomposition and condensation reactions take place with the formation of glycolic acid, methyl formate, carbon monoxide and water[50]. Paraformaldehyde reacts with fuming sulfuric acid (50 per cent oleum) at 60 to 70°C, producing methylene sulfate in 65 per cent yield. According to Baker[6, 7], this product is probably a dimer, $(CH_2SO_4)_2$, possessing the cyclic structure:

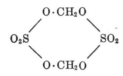

$$O_2S \underset{O \cdot CH_2O}{\overset{O \cdot CH_2O}{\diamondsuit}} SO_2$$

It is a colorless, microcrystalline powder, insoluble in water, alcohol, and other solvents, and melting at 155°C with decomposition. It reacts with potassium carbonate or potassium hydroxide and water to give potassium methylene bisulfate, $CH_2(OSO_2OK)_2$, and yields the corresponding ammonium salt with aqueous ammonia. In reporting the above salts, Baumgarten and Otto[11] state that potassium methylene bisulfate evolves formaldehyde when heated in the dry state and hydrolyzes rapidly in water at 100°C. The ammonium salt melts with decomposition at 110 to 115°C. Methylene sulfate also reacts with hydroxy compounds in the presence of alkalies, giving formals, and forms quaternary salts with tertiary amines. Salts of sulfamic acid, such as sodium and calcium sulfamate, react with formaldehyde to give methylol and methylene derivatives[23a, 130a].

Phosphoric Acid. On heating "trioxymethylene" (paraformaldehyde or polyoxymethylene) with anhydrous phosphoric acid at 140 to 145°C, Contardi[22] obtained methylene diphosphoric acid, $CH_2(H_2PO_4)_2$, in the form of a straw-colored syrup. The calcium and barium salts of this acid are only slightly soluble in cold water and less soluble in hot water. When heated to 400°C, these salts lose their formaldehyde quantitatively. On addition of phosphorus pentoxide to 37 per cent formaldehyde, heat is evolved and carbonization takes place; with more dilute solutions, formaldehyde polymer is formed[126].

Nitric Acid. At ordinary temperatures, nitric acid reacts violently with formaldehyde, oxidizing it to carbon dioxide and water, with formation of nitric oxide and a small amount of nitrogen. A mixture of pure nitric acid and formaldehyde may not react at once, but bursts into almost explosive action on being stimulated[129].

Travagli and Torboli[122] claim that formaldehyde solution can be nitrated with mixed acids at 5°C to give the dinitrate of methylene glycol:

$$CH_2(OH)_2 + 2HNO_3 \rightarrow CH_2(ONO_2)_2 + H_2O$$

If the temperature is allowed to reach 10°C, there is danger of violent decomposition. Methylene dinitrate separates from the nitration mixture as an upper layer and is fairly stable when purified by washing with water. It is a liquid, insoluble in water, but readily soluble in organic solvents. It is claimed to have value as an explosive.

Nitric oxide reacts slowly with formaldehyde at temperatures below 180°C to yield nitrous oxide, carbon oxides and water. This reaction becomes explosive at temperatures above 180°C. Reaction kinetics have been investigated by Pollard, Wyatt and Woodward[99, 100].

Nitrous Acid. Nitrous acid oxidizes formaldehyde to carbon dioxide and water, with liberation of nitrogen. This reaction is the basis for a method of nitrite analysis[127].

$$4HNO_2 + 3CH_2O(aq) \rightarrow 3CO_2 + 5H_2O + 2N_2$$

Hydrohalogen Acids. The primary reaction product of formaldehyde and hydrochloric acid is probably chloromethanol:

$$CH_2O(aq) + HCl \rightarrow ClCH_2OH$$

This product has never been isolated and its existence is purely hypothetical. When an excess of hydrogen chloride is reacted with formaldehyde solution, dichloromethyl ether is obtained[83]. The probable mechanism of this reaction is:

$$CH_2O(aq) + HCl \rightarrow ClCH_2OH$$
$$2ClCH_2OH \rightarrow ClCH_2OCH_2Cl + H_2O$$

This product is obtained in an 85 per cent yield when hydrogen chloride is passed into a solution of paraformaldehyde in cold concentrated sulfuric acid, the product separating as a clear, colorless upper layer[113]. When the reaction mixture is warmed, decomposition reactions take place and little or no dichloroether is produced. According to Schneider[113], traces of methylene chloride may be present in the products obtained under these circumstances.

Dichloromethyl ether is a colorless liquid boiling at approximately 100°C. It hydrolyzes on exposure to water or moist air, liberating hydrogen chloride and formaldehyde. It is not miscible with water, but is completely compatible with most organic solvents. It reacts with alcohols, producing formals.

Addition of small percentages of hydrochloric acid to concentrated formaldehyde solutions results in the precipitation of polyoxymethylenes. When commercial formaldehyde solution containing methanol is reacted with hydrogen chloride, monochloromethyl ether is obtained:

$$CH_3OH + CH_2O(aq) + HCl \rightarrow ClCH_2OCH_3 + H_2O$$

The symmetrical dibromo- and diiodomethyl ethers were prepared by Tischenko[120] by reacting polyoxymethylene with hydrogen bromide and iodide respectively. Symmetrical dibromomethyl ether, $(CH_2Br)_2O$, is reported to boil at 148.5 to 151.5°C; the diiodoether, $(CH_2I)_2O$, at 218 to 219°C. According to Henry[56], the dibromoether solidifies at -34°C and boils at 154 to 155°C.

Phosphine and Phosphorus Halides. Phosphine is readily taken up by warm aqueous formaldehyde acidified with hydrochloric acid. Hoffman[60a] reports the isolation of tetramethylol phosphonium chloride, $(HOCH_2)_4 PCl$, as a product of the reaction. Hydrolysis of this phosphorus compound yields dimethylol phosphinic acid, $(HOCH_2)_2PO_2H$, with formation of trimethylol phosphonium oxide as a reaction intermediate. Monomethylol phosphinic acid, $HOCH_2 \cdot PO_3H_2$, was obtained by Page[95] on hydrolysis of the primary reaction product of paraformaldehyde and phosphorus trichloride.

Methylene halides are obtained when the reaction of phosphorus halides and formaldehyde is pushed to its extreme limit. Phosphorus pentachloride and phosphorus pentabromide give methylene chloride and bromide respectively[57]. Phosphorus triiodide gives methylene iodide[18]. Phosphorus trichloride and polyoxymethylene react in the presence of zinc chloride to give *sym*-dichloromethyl ether and *sym*-dichloromethylal[26].

Phosphorus, bromine, water, and paraformaldehyde react to give a 60 per cent yield of *sym*-dibromomethyl ether[121]:

$$10Br + 2P + 3H_2O + 10CH_2O(p) \rightarrow 2H_3PO_4 + 5(CH_2Br)_2O$$

Halogens. In direct sunlight, paraformaldehyde reacts with chlorine in the cold, forming phosgene, carbon monoxide, and hydrogen chloride. In the absence of sunlight, the sole products are carbon monoxide and hydrogen chloride and gentle warming is required to initiate the reaction[15].

$$2Cl_2 + CH_2O(p) \rightarrow COCl_2 + 2HCl$$

$$Cl_2 + CH_2O(p) \rightarrow CO + 2HCl$$

Bromine reacts with aqueous formaldehyde to give carbon dioxide and formic acid[1]. The mechanism of this reaction has been studied by Scheffer and Van Went[110].

Iodine is stated not to act upon paraformaldehyde even in the presence of sunlight at ordinary temperatures. However, on heating in a sealed tube at 120 to 125°C for 7 to 8 hours, reaction takes place with formation of methyl iodide, acetaldehyde, diiodomethyl ether (ICH_2OCH_2I), carbon monoxide, and hydrogen iodide[120]. In the presence of alkali, solutions of iodine in potassium iodide oxidize formaldehyde quantitatively to formic acid.

References

1. Anderson, E., *Am. Chem. J.*, **49**, 182 (1913); *C. A.*, **7**, 1701 (1913).
2. Averkieff, N., *Z. anorg. allgem. Chem.*, **35**, 329–335 (1903).
3. Backer, H. J., and Mulder, H., *Rec. trav. chim.*, **52**, 454–468 (1933); **53**, 1120–1127 (1934).
4. Baer, J., U. S. Patent 2,039,206 (1936).
5. Baeyer, A., and Villiger, V., *Ber.*, **33**, 2486 (1900).
6. Baker, W., *J. Chem. Soc.*, **1931**, 1765.
7. Baker, W., and Field, F. B., *J. Chem. Soc.*, **1932**, 86.
8. Bamburger, M., and Nussbaum, J., *Monatsh.*, **40**, 411–416 (1920); *C. A.*, **14**, 1810 (1920).
9. Bamburger, M., and Nussbaum, J., *Z. ges. Schiess-u. Sprengstoffw.*, **22**, 125–128 (1927); *C. A.*, **21**, 4070 (1927).
10. Baumann, E., *Ber.*, **23**, 60–67 (1890).
11. Baumgarten, P., and Otto, T., *Ber.*, **75 B**, 1687–1694 (1942).
12. Black, J. H., *et al.*, *J. Am. Med. Assoc.*, **69**, 1855–1859 (1917).
13. Blank, O., and Finkenbeiner, H., *Ber.*, **31**, 2979 (1898).
14. Bost, R. W., and Constable, E. W., *Org. Syntheses*, **16**, 81–83 (1936).
15. Brochet, A., *Compt. rend.*, **121**, 1156–1159 (1895); *Chem. Zentr.*, **1896**, I, 362.
16. Brochet, A., and Cambier, R., *Bull. soc. chim. (3)*, **13**, 395, 534 (1895); *Chem. Zentr.*, **1895**, II, 30.
17. Bruyn, C. A. L. de, and Ekenstein, W. A. van, *Rec. trav. chim.*, **18**, 309 (1899).
18. Butlerov (Butlerow), A., *Ann.*, **111**, 242–252 (1859).
19. *Ibid.*, *Ann.*, **115**, 322 (1860).
20. Cambier, R., and Brochet, A., *Compt. rend.*, **120**, 449–452 (1895).
21. Camps, R., *Ber.*, **25**, 233, 248 (1892).
22. Contardi, A., *Gazz. chim. ital.*, **51**, I, 109–125 (1921).
23. Cross, C. F., Bevan, E. J., and Bacon, W., *J. Chem. Soc.*, **97**, 2404–2406 (1910); *Chem. Zentr.*, **1911**, I, 236.
23a. Cupery, M. E., (to E. I. du Pont de Nemours & Co., Inc.), U. S. Patent 2,369,-439 (1945).
24. Cushman, A. S., U. S. Patent 1,399,007 (1921).
25. Delépine, M., *Ann. chim. phys. (7)*, **15**, 570 (1898).
26. Descudé, M., *Bull. acad. roy. med. Belg.*, **1906**, 198–205; *Chem. Zentr.*, **1906**, II, 226.
27. Diesser, S., German Patent 246,038 (1910); *C. A.*, **6**, 2503 (1912).
28. Duden, P., and Scharf, M., *Ann.*, **288**, 218 (1895).
29. Dunstan, W. R., *Proc. Chem. Soc.*, **1894**, 55; *Chem. News*, **69**, 199–200 (1894).

30. Ellis, C., (to Ellis-Foster Co.), U. S. Patent 1,964,725 (1934).
31. *Ibid.*, U. S. Patent 2,011,573 (1935).
32. Emde, H., and Hornemann, T., *Helv. Chim. Acta*, **14**, 892–911 (1931).
33. Eschweiler, W., *Ber.*, **30**, 1001 (1897).
34. Euler, H., and Euler, A., *Ber.*, **38**, 2551 (1905).
35. Fenton, H. J. H., *Proc. Roy. Soc. (London) (A)*, **90**, 492–498 (1914); *J. Chem. Soc.*, **106**, I, 1121 (1914).
36. Foelsing, A., French Patent 328,425 (1903); *J. Soc. Chem. Ind.*, **22**, 962 (1903).
37. Fosse, R., *Ann. Inst. Pasteur*, **34**, 715–762 (1920); *C. A.*, **15**, 1142 (1921).
38. Fosse, R., *Compt. rend.*, **173**, 1370–1372 (1921).
39. Fosse, R., and Hieulle, A., *Compt. rend.*, **174**, 1021–1023 (1922); *C. A.*, **16**, 2111 (1922).
40. Franzen, H., *J. prakt. Chem. (2)*, **86**, 133–149 (1921); *C. A.*, **7**, 775 (1913).
41. Franzen, H., and Hauck, L., *J. prakt. Chem. (2)*, **91**, 261–284 (1915); *C. A.*, **9**, 2063 (1915).
42. Fry, H. S., and Payne, J. H., *J. Am. Chem. Soc.*, **53**, 1973–1980 (1931).
43. Gaudry, R., *Org. Syntheses*, **27**, 41–42 (1947).
44. Gilman, H., "Organic Syntheses," Coll. Vol. I, pp. 347–348, New York, John Wiley & Sons, Inc. (1932); Gilman, H. and Blatt, A. H., *Ibid.*, Second ed., pp. 355–357 (1941).
45. Girsewald, F. C. von, and Siegens, H., *Ber.*, **54**, 490 (1921).
46. *Ibid.*, p. 492.
47. Green, S., and Handley, R., (to Brit. Celanese Corp.), British Patent 334,207 (1930).
48. Griffin, R. C., *Ind. Eng. Chem.*, **12**, 1159–1160 (1920).
49. Grissom, J. T., *Ind. Eng. Chem.*, **12**, 172–173 (1920).
50. Hammick, D. L., and Boeree, A. R., *J. Chem. Soc.*, **123**, 2881 (1923).
51. Harden, A., *Proc. Chem. Soc.*, **15**, 158 (1899).
52. Hartwagner, F., *Z. anal. Chem.*, **52**, 17–20 (1913).
53. Hatcher, W. H., and Holden, G. W., *Trans. Roy. Soc. Can.*, **20**, 395–398 (1926).
54. Heimrod, G. W., and Levene, P. A., *Biochem. Z.*, **29**, 31–59 (1910); *C. A.*, **5**, 869–870 (1911).
55. Henry, L., *Compt. rend.*, **110**, 759–760 (1890).
56. Henry, L., *Bull. acad. roy. Belg. (3)*, **26**, 615–628; *Ber.*, **27**, Ref., 336 (1894)
57. Henry, L., *Bull. acad. roy. med. Belg.*, **1900**, 48–56; *Chem. Zentr.*, **1900**, I, 1122.
58. Herrmann, F., *Chem. Ztg.*, **35**, 25 (1911).
59. Herzog, W., *Angew. Chem.*, **33**, 48, 84, 156 (1920).
60. Hills, R. C., and Barnett, M. M., U. S. Patent 2,174,000 (1939).
60a. Hoffman, A., *J. Am. Chem. Soc.*, **43**, 1684 (1921); **52**, 2995 (1930).
61. Hofmann, K. A., and Storm, D., *Ber.*, **45**, 1725 (1912).
62. Hofmann, A. W., *Ann.*, **145**, 357–360 (1868).
63. Hofmann, A. W., *Ber.*, **3**, 588 (1870).
64. Howald, A. M., (to Toledo Synthetic Products, Inc.,), U. S. Patent 1,910,338 (1933).
65. Inghilleri, G., *Z. physiol. Chem.*, **80**, 64–72 (1912); *C. A.*, **7**, 95 (1913).
66. Jacobson, R. A., (to E. I. du Pont de Nemours & Co., Inc.), U. S. Patent 1,945,-315 (1934).
67. *Ibid.*, U. S. Patent 2,220,156 (1940).
68. Jay, R., and Curtius, T., *Ber.*, **27**, 59 (1894).
69. Johnson, T. B., and Rinehart, H. W., *J. Am. Chem. Soc.*, **46**, 768, 1653 (1924).

70. Kawano, T., *J. Soc. Rubber Ind.*, (*Japan*), **12**, 252–255 (1939); *C. A.*, **34**, 6482 (1940).
71. Kelber, C., *Ber.*, **50**, 1509–1512; *J. Chem. Soc.*, **114**, II, 19 (1918); *C. A.*, **12**, 2270 (1918).
72. Klages, A., *J. prakt. Chem.*, **65**, 188–197 (1902).
73. Knudsen, P., *Ber.*, **47**, 2694–2698 (1914).
74. Kohno, T., *J. Soc. Rubber Ind. Japan*, **14**, 436–441 (1941); *C. A.*, **42**, 7086 (1948).
75. Kraut, K., *Ann.*, **258**, 105–110 (1890).
76. Lapworth, A., *Proc. Chem. Soc.*, **23**, 108 (1907); *J. Chem. Soc.*, **91**, 1133–1138; *C. A.*, **2**, 101 (1908).
77. Larson, A. T., (to E. I. du Pont de Nemours & Co., Inc.) U. S. Patent 2,153,064 (1939).
78. Lauer, W. M., and Langkammerer, C. M., *J. Am. Chem. Soc.*, **57**, 2360–2362 (1935).
79. Le Fevre, C. G., and Le Fevre, R. J. W., *J. Chem. Soc.*, **1932**, 1142.
80. Le Fevre, R. J. W., and MacLeod, M., *J. Chem. Soc.*, **1931**, 474.
81. Legler, L., *Ann.*, **217**, 381 (1883).
82. Lindsay, M., and Soper, F. G., *J. Chem. Soc.*, **1946**, 791–792.
83. Litterscheid, F. M., and Thimme, K., *Ann.*, **334**, 13 (1904).
84. Löb, W., *Biochem. Z.*, **12**, 466–472 (1908); *Chem. Zentr.*, **1908**, II, 1017.
85. Loder, D. J., (to E. I. du Pont de Nemours & Co., Inc.), U. S. Patent 2,152,852 (1939).
86. *Ibid.*, U. S. Patent 2,211,625 (1940).
86a. Loder, D. J., and Bartlett, E. P., (to E. I. du Pont de Nemours & Co., Inc.), U. S. Patent 2,211,624 (1940).
87. Loew, O., *Ber.*, **21**, 270–275 (1888).
88. *Ibid.*, **22**, 470–478 (1889).
89. Lubs, H. A., and Acree, S. F., *J. Phys. Chem.*, **20**, 324 (1916).
90. Lyford, C. A., *J. Am. Chem. Soc.*, **29**, 1227–1236 (1907).
91. Mann, F. G., and Pope, W. J., *J. Chem. Soc.*, **123**, 1172–1178 (1923).
92. Marks, B. M., (to du Pont Viscoloid Co.), U. S. Patent 1,991,765 (1935).
93. Marotta, D., and Allessandrini, M. E., *Gazz. chim. ital.*, **59**, 942–946 (1929).
94. Nef, J. U., *Ann.*, **335**, 274 (1904).
95. Page, H. J., *J. Chem. Soc.*, **101**, 428–431 (1912).
96. Patrick, J. C., U. S. Patent 2,206,641 (1940).
97. Patrick, J. C., U. S. Patent 2,255,228 (1941).
98. Plöchl, J., *Ber.*, **21**, 2117 (1888).
99. Pollard, F. H., and Woodward, P., *Trans. Faraday Soc.*, **45**, 767–770 (1949).
100. Pollard, F. H., and Wyatt, R. M. H., *Trans. Faraday Soc.*, **45**, 760–767 (1949).
101. Polley, J. R., Winkler, C. A., and Nicholls, R. V. V., *Can. J. Research*, **25 B**, 525–534 (1947).
102. Polstorff, K., and Meyer, H., *Ber.*, **45**, 1905–1912 (1912).
103. Pulvermacher, G., *Ber.*, **26**, 2360 (1893).
104. Raschig, F., *Ber.*, **59**, 859–865 (1926).
105. Raschig, F., and Prahl, W., *Ber.*, **59**, 2025–2028 (1926); **61**, 179–189 (1928); *Ann.*, **448**, 265–312 (1926).
106. Reinking, K., Dehnel, E., and Labhardt, H., *Ber.*, **38**, 1069–1080 (1905).
107. Rieche, A., and Meister, R., *Ber.*, **66**, 718–727 (1933).
108. *Ibid.*, **68**, 1465–1473 (1935).
109. Romijn, G., *Z. anal. Chem.*, **36**, 18 (1897).
110. Scheffer, F. E. C., and van Went, N. B., *Rec. trav. chim.*, **47**, 406–414 (1928); *C. A.*, **22**, 1952 (1928).

111. Schiff, H., *Ann.*, **319**, 59–76 (1901); *Chem. Zentr.*, **1901**, II, 1333.
112. Schmerda, *Angew. Chem.*, **30**, 176 (1917); *J. Soc. Chem. Ind.*, **36**, 942; *C. A.*, **11**, 3377 (1917).
113. Schneider, H., *Angew. Chem.*, **51**, 274 (1938).
114. Scholl, R., *Ber.*, **24**, 573–579 (1891).
115. Schumann, C., Münch, E., Schlichting, O., and Christ, B., (to Grasselli Dyestuffs Corp.), U. S. Patents 1,714,636–7 (1929).
116. Staudinger, H., Signer, R., Johner, H., Schweitzer, O., and Kern, W., *Ann.*, **474**, 254–255 (1929).
117. Staudinger, H., Signer, R., and Schweitzer, O., *Ber.*, **64**, 398–405 (1931).
118. Stollé, R., *Ber.*, **40**, 1505 (1907).
119. Suter, C. M., Bair, R. K., and Bordwell, F. G., *J. Org. Chem.*, **10**, 470–478 (1945).
120. Tischenko, V. E., *J. Russ. Phys.-Chem. Soc.*, **19**, I, 479–483 (1887); *Chem. Zentr.*, **1887**, 1540–1541.
121. Tischenko, V. E., and Rabtzevich-Subkovskii, I. L., *J. Russ. Phys.-Chem. Soc.*, **46**, 705–708 (1914); *C. A.*, **9**, 1749 (1915).
122. Travagli, G., and Torboli, A., Italian Patent 333,080 (1935); *Chem. Zentr.*, **1937**, I, 3265.
123. Vanino, L., *Ber.*, **35**, 3251 (1902).
124. Vanino, L., *Z. anal. Chem.*, **41**, 619–620 (1902).
125. *Ibid.*, **79**, 369 (1930).
126. Vanino, L., and Seemann, L., *Pharm. Zentralhalle*, **45**, 733–735 (1904); *Chem. Zentr.*, **1904**, II, 1205.
127. Vanino, L., and Schinner, A., *Z. anal. Chem.*, **52**, 21–28 (1913); *C. A.*, **7**, 1461 (1913).
128. Vanino, L., and Schinner, A., *J. prakt. Chem.* (*2*), **91**, 116 (1915).
129. Verworn, M., *Arch. ges. Physiol.*, **167**, 289–308 (1917); *C. A.*, **12**, 492 (1918).
130. Walker, J. F., *J. Am. Chem. Soc.*, **55**, 2821–2822 (1933).
130a. Walker, J. F., (to E. I. du Pont de Nemours & Co., Inc.), U. S. Patent 2,321,-958 (1945).
131. *Ibid.*, U. S. Patent 2,429,859 (1947).
132. Werner, E. A., *J. Chem. Soc.*, **111**, 844–853 (1917).
133. Wieland, H., and Wingler, A., *Ann.*, **431**, 301–322 (1923).
134. Wirtz, K., and Bonhoeffer, K. F., *Z. physik. Chem.*, **32B**, 108–112 (1936).
135. Wood, H., *Chem. Age*, **38**, 85–86 (1938).

CHAPTER 10

REACTIONS OF FORMALDEHYDE WITH ALIPHATIC HYDROXY COMPOUNDS AND MERCAPTANS

The primary reaction products of formaldehyde and hydroxy compounds are hemiacetals. Since these compounds are unstable and have not been isolated in a pure state, evidence for their formation is based principally on physical measurements (see pages 61–63). The increased density and refractivity of alcoholic solutions of aldehydes are recognized indications of hemiacetal formation[81]. Hemiacetals are apparently formed in alcoholic solutions of many carbonyl compounds, as evidenced by abnormalities in the ultraviolet absorption spectra of these solutions. However, the results of such measurements indicate that the extent of compound formation decreases as the substituents surrounding the carbonyl group increase[49]. Formaldehyde has the property of forming such derivatives to an advanced degree.

Alcohols. In the case of the simpler aliphatic alcohols, the formation of formaldehyde hemiacetals takes place with the evolution of considerable heat: *e.g.*, 15 kcals are evolved when one gram molecule of formaldehyde gas is absorbed in methanol (page 61). The fact that a concentrated solution of anhydrous formaldehyde in methanol has a higher boiling point than either pure formaldehyde or methanol itself is definite proof of compound formation. It is probable that hemiacetals are present in alcoholic formaldehyde, just as polyoxymethylene glycols are present in aqueous solutions.

$$CH_3OH + CH_2O \text{ (g)} \rightarrow CH_3OCH_2OH$$
Methyl formaldehyde
hemiacetal

$$CH_3OH + 2CH_2O \text{ (g)} \rightarrow CH_3OCH_2OCH_2OH$$
Methyl dioxymethylene
hemiacetal

Chemical evidence for the existence of hemiacetals is found in the formation of stable alkoxymethoxy ($ROCH_2O$—) derivatives by alkali catalyzed reactions involving formaldehyde and alcohols. This is illustrated by Bruson and Riener's[15] process for the production of alkoxymethoxypropionitriles by the joint reaction of acrylonitrile, formaldehyde and an aliphatic alcohol. The exothermic formation of methoxymethoxypropionitrile (b.p. 200°C at 775 mm) when potassium hydroxide is added to a mixture of aqueous formaldehyde, methanol and acrylonitrile at ordinary temperatures probably involves the formation of the hemiacetal, CH_3OCH_2OH, as a reaction intermediate.

$$CH_3OCH_2OH + CH_2:CHCN \rightarrow CH_3OCH_2OCH_2CH_2CN$$

Since formaldehyde hemiacetals are comparatively unstable and are formed by reversible reactions, they may for most purposes be regarded as solutions or mixtures of formaldehyde and alcohols. Their existence explains the difficulties which are always encountered in separating formaldehyde from alcohols or other hydroxy compounds by purely physical methods. Under neutral or alkaline conditions, hemiacetals are apparently the sole reaction products of formaldehyde and alcohols. However, under acidic conditions the reaction proceeds further: water is liberated and formaldehyde acetals or formals are formed:

$$CH_3OCH_2OH + CH_3OH \rightarrow CH_3OCH_2OCH_2 + H_2O$$

Methyl	*Methanol*	*Methylal*	*Water*
formaldehyde			
hemiacetal			

In most cases, formal production takes place readily in the presence of acidic catalysts such as sulfuric acid, hydrochloric acid, ferric chloride, zinc chloride, and the like. As indicated above, the reaction is reversible and proceeds to a state of equilibrium. It is closely analogous to esterification. Since a minimum concentration of water in the reaction mixture is most favorable to a high conversion yield, paraformaldehyde is an excellent raw material for formal preparation.

In general, formals are ether-like compounds. They possess a high degree of chemical stability under neutral or alkaline conditions, but hydrolyze to formaldehyde and alcohol in the presence of acids. Formals of the lower alcohols are miscible with water, whereas the higher members of the group are insoluble or only partially soluble.

Methylal, $CH_2(OCH_3)_2$, known also as formaldehyde dimethyl acetal, formal and dimethoxymethane, is the dimethyl ether of methylene glycol. It is a colorless, flammable, ether-like liquid which boils at 42.3°C and freezes at −105°C. It has a density of 0.86 at 20°C and a refractive index n_D^{25} 1.3504. It dissolves in approximately three times its volume of water and is infinitely miscible with alcohol, ether and other organic solvents. It is converted to formaldehyde by vapor phase oxidation in the presence of a methanol oxidation catalyst[71].

According to Fischer and Giebe[34], methylal can be obtained in 75 per cent theoretical yield by mixing commercial formaldehyde solution with 1½ parts methanol containing 2 per cent hydrogen chloride, followed by the addition of calcium chloride equivalent to the weight of formaldehyde solution. Under these conditions methylal separates as an upper phase which can be removed and purified by distillation. Still better yields can be obtained by reacting methanol containing 1 per cent hydrogen chloride with paraformaldehyde. Adams and Adkins[1] report excellent yields of formals by heating alcohols with paraformaldehyde at 100°C in the presence of ferric chloride. These investigators claim that ferric chloride is the most efficient catalyst for formal production. The most practical procedure for methylal preparation is by distillation from the equilibrium mixture obtained by addition of an acid catalyst to an aqueous solution of formaldehyde and methanol.

The formals of tertiary alcohols are not readily produced. Conant, Webb, and Mendum[22] obtained tertiary butyl formal as a by-product in

TABLE 47. BOILING POINTS OF FORMALS

Formal	Boiling Point (°C)	Investigator	Ref.
Dimethyl	42	Timmermans	91
Methyl ethyl	67	Henry	48
Diethyl	89	Greene	39
Diisopropyl	119	Arnhold	2
Dipropyl	141	Favre	32
Diisobutyl	164	Arnhold	2
Ditertiary butyl	182–5	Conant, Webb and Mendum	22
Diisoamyl	207	Arnhold	2
Dioctyl	Approx. 360	Arnhold	2
Diallyl	138–139	Trillat and Cambier	94
Dibenzyl	280	Descudé	25

the formation of tertiary butyl carbinol when tertiary butyl magnesium chloride was reacted with anhydrous formaldehyde. This formal is recovered unchanged by steam distillation from 80 per cent sulfuric acid, but is hydrolyzed by heating with an equal weight of 95 per cent ethanol containing 2 cc of concentrated hydrochloric acid.

The boiling points of some representative formals are listed in Table 47. Formals of the higher alcohols are wax-like materials which may be obtained by heating the alcohol with paraformaldehyde in the presence of a small proportion of concentrated sulfuric acid[31]. Dicetyl formal is a solid; dilauryl formal a liquid.

Unsymmetrical mixed formals can be prepared by equilibrating paraformaldehyde with a mixture of alcohols in the presence of an acid catalyst, neutralizing and distilling[14a]. They can also be prepared by reacting chloromethyl ethers with alcohols (page 214).

Polyformals derived from the polyoxymethylene glycols are discussed in detail in Chapter 7 (see pages 137–141). Preparation of polymeric formals from simple dialkyl formals by reaction with formaldehyde in the presence of an acid catalyst is described by Gresham and Brooks[43].

Glycols and Glycerol. The reactions of glycols with formaldehyde are similar to those of the simple alcohols, with the exception that cyclic and polymeric formals are obtained. In the case of 1,2- and 1,3-glycols, cyclic formals are readily obtained on reaction in the presence of acidic catalysts. Hill and Carothers[52] also obtained a cyclic formal by heating tetramethylene glycol and paraformaldehyde in the presence of camphor sulfonic acid. The same investigators also obtained cyclic formals from tri- and tetramethylene glycols by a formal interchange carried out by heating dibutyl formal with the glycol in the presence of acid, distilling butanol as formed from the reaction mixture. Pentamethylene and higher glycols subjected to this procedure gave polyformals. From these products they were able to obtain, by low-pressure distillation, both large ring and linear polymers. These formals were obtained in the form of macrocrystalline solids, musk-like odorous liquids, and tough microcrystalline masses capable of being drawn into strong, pliable, highly oriented, silk-like filaments.

Dioxolane (glycol formal) is obtained by heating ethylene glycol with paraformaldehyde[93] or formaldehyde solution[95] in the presence of an acidic catalyst.

$$\begin{array}{l} CH_2OH \\ | \\ | \quad\quad + CH_2O \text{ (p, aq)} \rightarrow \\ CH_2OH \end{array} \quad\quad \begin{array}{l} CH_2O \\ | \quad\quad\diagdown \\ | \quad\quad\quad CH_2 + H_2O \\ | \quad\quad\diagup \\ CH_2O \end{array}$$

A process involving the distillation of this product as formed from a reaction mixture containing formaldehyde, ethylene glycol, and 5 to 10 per cent of an acid salt such as zinc chloride, ferric chloride, etc., has been patented[27].

Dioxolane is an ether-like liquid which boils at 76°C (755 mm) and has a density of 1.060 at 20°C[19]. It is infinitely compatible with water and has excellent solvent properties. The formals of trimethylene and tetramethylene glycol boil at 106°C[47], and 112 to 117°C[52] respectively.

1,3,5-Trioxepane, a cyclic glycol formal containing two oxymethylene units, is obtained by heating dioxolane with formaldehyde, paraformaldehyde or trioxane in the presence of an acid catalyst[42]. The reaction, which is carried out under pressure at 50 to 200°C, also yields polymeric formals of high molecular weight. High temperatures favor formation of the cyclic product. Trioxepane is best isolated by vacuum distillation of the acidic reaction mixture.

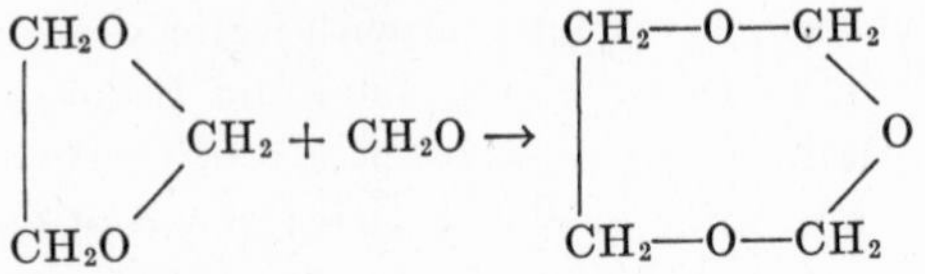

$$CH_2O \diagdown \atop \mid \atop CH_2O \diagup CH_2 + CH_2O \rightarrow \quad CH_2—O—CH_2 \atop \mid \quad\quad O \atop CH_2—O—CH_2$$

Trioxepane is a colorless liquid which boils at 132°C at atmospheric pressure. At 50 mm pressure, it distills at 57°C.

Monoformals of ethylene glycol can be prepared by reacting a formal with a glycol chlorohydrin in the presence of an acidic catalyst and then hydrolyzing the resultant chloroformal[64], *e.g.*:

$$CH_3OCH_2OCH_3 + HOCH_2CH_2Cl \rightarrow CH_3OCH_2OCH_2CH_2Cl + CH_3OH$$

$$CH_3OCH_2OCH_2CH_2Cl + H_2O \rightarrow CH_3OCH_2OCH_2CH_2OH + HCl$$

Sussman and Gresham[90a] obtain these monoformals by the acid catalyzed reaction of dioxolane with an aliphatic alcohol.

Polymeric glycol formals having the structure,

$$HOCH_2CH_2(OCH_2OCH_2CH_2)_xOCH_2OCH_2CH_2OH,$$

are prepared by reacting ethylene glycol and formaldehyde in the presence of an acidic catalyst, neutralizing the catalyst and isolating the resultant products[40]. High molecular weight polymers are obtained by the acid catalyzed reaction of formaldehyde and ethylene oxide under substantially anhydrous conditions[41].

Schulz and Tollens[86] succeeded in showing that glycerol reacts with formaldehyde to produce two simple formals probably having the structural formulas indicated below:

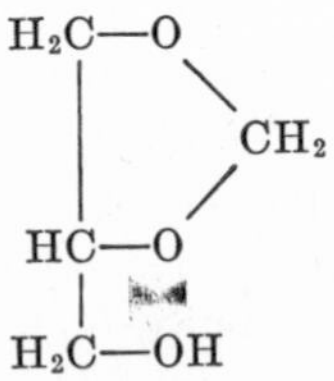

$$H_2C—O \diagdown \atop \mid \quad\quad CH_2 \atop HC—O \diagup \atop \mid \atop H_2C—OH$$
$$\text{I}$$

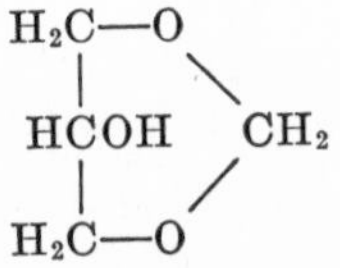

$$H_2C—O \diagdown \atop \mid \quad\quad CH_2 \atop HCOH \diagup \atop \mid \atop H_2C—O$$
$$\text{II}$$

Further studies of the preparation and properties of glycerol formals are reported by Fairbourne, Gibson and Stephens[30]. Glycerol formal obtained by heating formaldehyde and glycerol in the presence of hydrogen chloride boils at 195°C and is a good solvent for cellulose esters, resins, etc.

Polyhydroxy Compounds. Polycyclic acetals are obtained from polyhydroxy aliphatic compounds and formaldehyde in much the same way that the simpler monocyclic formals are produced. In general, these compounds are crystalline solids.

Dimethylene pentaerythritol (m.p. 50°C, b.p. 230°C) may be obtained by the action of pentaerythritol on formaldehyde.solution in the presence of zinc chloride[3]. It is reported to be more stable with respect to hydrolysis than diethyl ether[88]. Barth and Snow[5] obtained the cyclic monoformal of pentaerythritol (m.p. 55 to 60°C) by refluxing a dipentaerythritol formal with 10 per cent hydrochloric acid. The same investigators also describe mixed formal esters derived from pentaerythritol monoformal[5] and dipentaerythritol formals[4].

Tricyclic formals are obtained as crystalline precipitates when mannitol and sorbitol are heated with commercial formaldehyde and concentrated hydrochloric acid. As first reported by Schulz and Tollens[84, 85], trimethylene *d*-mannitol melts at 227°C and the corresponding *d*-sorbitol derivative at 206°C. Recent studies by Ness, Hann and Hudson[69, 70] indicate that the structures of these formals are as indicated below:

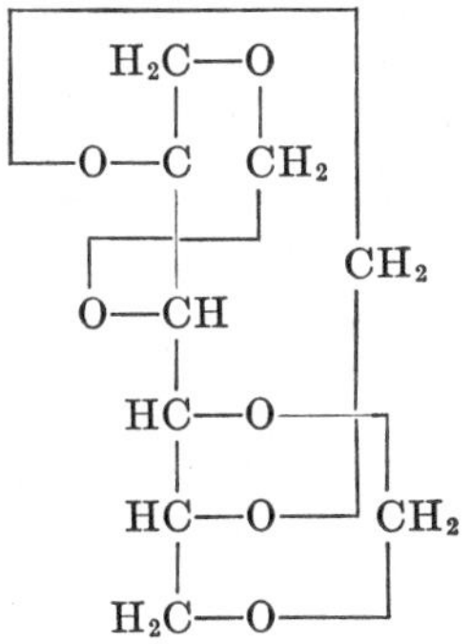

1,3:2,5:4,6-*Trimethylene*
d-Mannitol

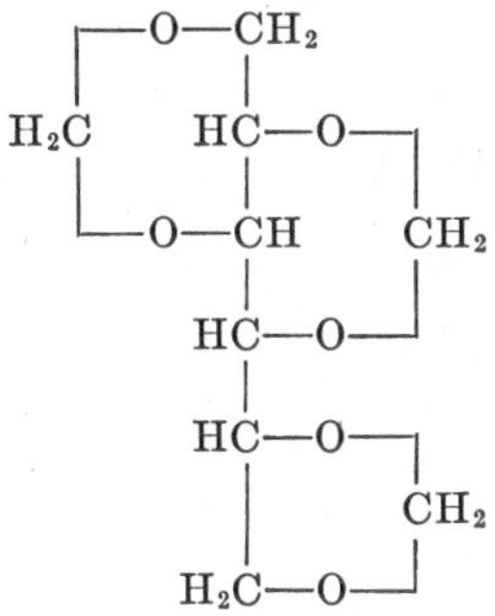

1,3:2,4:5,6-*Trimethylene*
d-Sorbitol

Ness, Hann and Hudson[70] have also obtained 1,3:2,4-dimethylene-*d*-sorbitol (m.p. 174 to 175°C) as a by-product of the hydrochloric acid catalyzed reaction of sorbitol with formaldehyde. Partial acetolysis and saponification of trimethylene-*d*-sorbitol yields 2,4-methylene-*d*-sorbitol, whereas the corresponding mannitol derivative[69] gives 2,5-methylene-*d*-mannitol when subjected to the same treatment. Results indicate that methylene linkages through primary hydroxyl groups are more easily ruptured by acetolysis than linkages through secondary hydroxyl groups.

Polyvinyl alcohol also reacts with formaldehyde in the presence of acids to give polycyclic formals which are industrially important resins[63]. Their structure is indicated below.

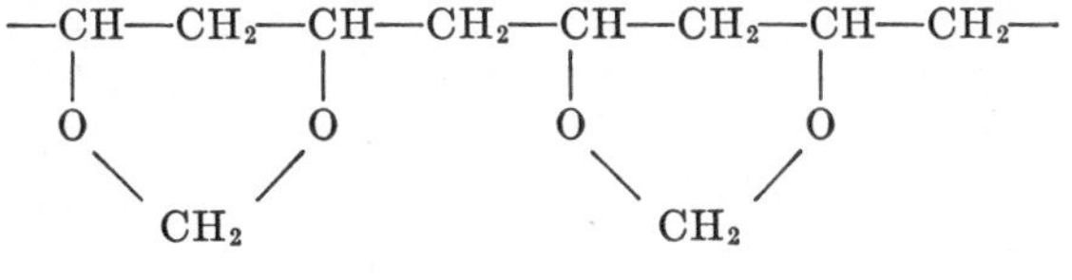

In addition to formals involving adjacent hydroxyl groups as shown above, groups on separate chains may also react, with the production of cross linkages. Gayler[36] claims that some cross-linking always takes place when the polyvinyl formal is formed from aqueous solutions and that formals which are not cross-linked can be obtained by passing hydrogen chloride into a suspension of polyvinyl alcohol and paraformaldehyde in dioxane. The product obtained by this technique is stated to give fibers which will orient on stretching, whereas cross-linked formals are brittle and inextensible. The polyvinyl formal resins are made both by the direct reaction of polyvinyl alcohol with formaldehyde and by the hydrolysis of polyvinyl acetate in the presence of formaldehyde and a suitable catalyst[68].

Sugars. Sugars, starch, and cellulose apparently react with formaldehyde in much the same manner as the simpler hydroxy compounds, with the formation of unstable hemiacetals and the more stable methylene ethers or formals.

When formaldehyde is heated with sugars or added to sugar solutions, loose compounds are formed. A gradual increase in the rotatory power of sugar solutions has been noted after the addition of formaldehyde and has been attributed to compound formation[57]. Lauch and Quade[58] describe the formation of compounds of formaldehyde and bioses such as maltose, sucrose, and lactose in an American patent. According to their findings, colorless crystalline compounds containing 1, 2, 3 or 4 mols of formaldehyde for each mol of biose may be obtained by dissolving the biose in a minimum amount of hot water, adding commercial formaldehyde, and evaporating at 70°C under reduced pressure. The products are stated to be without sharp odor in the dry state. They dissolve readily in water and are rapidly split up in the resulting solution, yielding formaldehyde and sugar. Other patents describe the preparation of biose-formaldehyde compounds containing up to 5 molecules of formaldehyde per molecule of sugar[74, 76]. Preparations involving heating the sugars with paraformaldehyde are also described[87]. The formaldehyde-sugar compositions have been used in pharmaceutical and numerous other products, and process variations are described in the patent literature. From a study of compounds of this general type, Heiduschka and Zirkel[45] have concluded that they are not definite chemical compounds but are apparently solid solutions of formaldehyde in sugar. On the basis of research covering the reactions of formaldehyde with a wide variety of polyhydroxy compounds, Contardi and Ciocca[23] conclude that these products are hemiacetals.

True acetals or formals are obtained when the reaction of sugars with formaldehyde is catalyzed with strongly acidic agents. Schwenkler[86a] reports that pentosans extracted from paper birch wood give stable methylene ethers under these conditions. According to Tollens[92], a stable formal of glucose and formaldehyde is obtained by the use of hydrochloric and acetic acids. The product, known as methylene glucose, is reported to have the formula $C_2H_{10}(CH_2)O_6 \cdot 1\frac{1}{2}H_2O$.

Starch. The reaction products of formaldehyde and starch are in general similar to those obtained with other hydroxy bodies. Under neutral and alkaline conditions unstable derivatives are obtained which are probably hemiacetals, whereas under acidic conditions more stable products are obtained which are apparently formals. At high temperatures complicated reactions take place, probably involving changes in the starch radicals, and lead to resinous products.

The temperature of gelatinization of potato starch is reported to decrease on treatment with formaldehyde. The magnitude of this effect increases with the concentration of formaldehyde employed in the treating solution and with the duration of its action[75]. A 38 per cent formaldehyde solution causes gelatinization after two days, even at as low a temperature as 15 to 16°C. On heating, the reaction takes place rapidly. Starch treated with formaldehyde in this way loses its characteristic property of giving a blue color with iodine[54]. Woker claims that formaldehyde acts like a diastase causing the hydrolysis of starch to sugar[66, 67, 104, 105, 106]. However, her conclusions have been contested by other investigators[53, 54, 78, 103], who have demonstrated that starch can be recovered unchanged from formaldehyde-starch reaction products on heating or on reacting with a formaldehyde acceptor, such as ammonia, ammonium acetate, or phenylhydrazine.

The simple formaldehyde-starch complexes of the type described in the above paragraph are colorless powders which are comparatively stable to heat. They may or may not form gels, depending on the method of preparation. Their production does not involve the use of acid catalysts, although in some cases[10, 65] dilute alkalies are employed. Their use in antiseptic surgical dressings has been suggested by Classen[20, 21] and others[38, 82]. A product of this type, which swells in cold water and forms short soft pastes, is produced by heating a partially converted starch with a minor proportion of formaldehyde under substantially nongelatinizing conditions at a temperature above about 300°F[6]. In most instances, these compounds are probably hemiacetals of formaldehyde.

Highly stable water-resistant formaldehyde-starch compounds can be obtained by the use of acidic catalysts. Stollé and Kope[90] have prepared a compound of this type in which the starch is treated with formaldehyde at a low temperature in the presence of a rather concentrated acid. Leuck[60] obtains a water-resistant product by dehydrating starch to remove all adsorbed water plus some water of hydration and reacting this material with formaldehyde and acid. Similar results are also achieved by treating the starch with alkali before reacting with formaldehyde[61]. Pierson[72] describes a process in which unburst starch granules are heated at 150 to 220°F with 3 to 10 per cent formaldehyde, to which sufficient acid has been added to give a pH in the range 2.5 to 4.0. The resultant product will not swell appreciably in either boiling water or aqueous alkali. A surgical dusting powder that will not swell or cause a foreign body reaction is obtained by Fenn[33] by heating starch with 2 per cent paraformaldehyde and an acid catalyst. The latter product is freed of unreacted formaldehyde by treatment with hot water and hydrogen peroxide. Kerr and Schink[55] produce a gelatinizable paste from tapioca or sweet potato starch by heating in an aqueous suspension at pH 1.6 to 2.5 with about 0.1 to 0.5 per cent formaldehyde. The heating temperature is kept below the gelatinization point and the final product is washed with bisulfite to remove unreacted formaldehyde.

Resins are obtained by heating formaldehyde and starch under various conditions. A rubber-like product is claimed on heating starch and formaldehyde at 150 to 175°F under pressure in the presence of metallic magnesium[9]. Alkali-treated starch that has been reacted with formaldehyde under acidic conditions acts as an adhesive when subjected to heat and pressure[24].

Cellulose. The reactions of formaldehyde with cellulose are fundamentally similar to its reactions with other hydroxyl compounds. However, as in the case of starch, the results obtained are modified by the effect of reaction conditions on the complex carbohydrate molecule involved and the various forms in which it occurs. Since formaldehyde treatment changes the physical and chemical characteristics of cellulose, its utilization for the modification of paper, rayon, cotton, wood and other cellulosic materials has been the subject of intensive industrial research, and special techniques for accomplishing various types of modification are covered by a large and growing volume of patent literature (cf. Chapter 21).

Formaldehyde-cellulose reactions may be summarized as including the formation of hemiacetals under neutral or alkaline conditions and acetal or formal linkages under acidic circumstances. The introduction of polymeric hemiacetal radicals, $-O(CH_2O)_nH$, and polyoxymethylene formal linkages, $-O(CH_2O)_n-$, is also indicated, as well as the simple deposition of formaldehyde polymer in or on the cellulosic fibers. Heuser[50] concludes that the formal linkages connect hydroxyl groups in separate cellulose chains. A ring closure involving neighboring hydroxyl groups in the same glucopyranose unit is believed to be unlikely since the resultant disposition of atoms would lead to strained and complex rings.

Reaction with formaldehyde in the presence of alkalies causes temporary alterations in the physical characteristics of cellulose; but it is apparent that no stable condensation product is formed, since material treated in this way can be freed of formaldehyde by boiling with ammonia water or sodium bisulfite solution or even by heating with water alone under pressure[107]. Modifications of this type are apparently caused by the formation of simple or polymeric hemiacetals or the deposition of polymer. Treatment of cellulose with formaldehyde under alkaline or even mildly acidic conditions produces no substantial modification in its chemical or physical properties.

Under acidic conditions, formaldehyde reacts with cellulose to produce stable methylene ethers or formals. The cross-linking of cellulose molecules, which results from this process, alters both the physical and chemical properties to a marked degree. Principal changes include reduction in swelling on exposure to water or aqueous alkalies, an increase in elastic recovery and stiffness, a decrease in alkali solubility and resistance to direct cotton dyes. Heuser[50] points out that only a small percentage of formal linkages is necessary to effect radical modifications. About 2 per cent combined formaldehyde renders a viscose rayon completely insoluble in cold 10 per cent caustic soda, whereas the untreated material may dissolve to the extent of 40 per cent. Goldthwait[37] reports complete dye resistance for cotton containing only 0.5 per cent combined formaldehyde. Methylene cellulose containing 7.5 per cent combined formaldehyde is so stiff that it is exceedingly brittle and is readily powdered[50]. Both Wood[107] and Schenk[79, 80] have noted that, depending upon the method of formation, products with both increased and decreased affinity for substantive dyes can be obtained. The use of acid condensation was apparently first described by Eschalier[28, 29] in a process for increasing the strength of artificial silk.

Reaction of cellulose with acid formaldehyde takes place only at relatively high temperatures (around 100 to 150°C) and is competitive with evaporation of formaldehyde when carried out at ordinary pressures[44, 50]. The extent of reaction varies with temperature, formaldehyde concentration, concentration and nature of catalyst and time of treatment. Since strong acids cause cellulose degradation, the mildest catalysts, which will permit the reaction to take place, give the best results. Ammonium chloride, aluminum chloride, zinc chloride and combinations of these with acetic and boric acids are reported to give satisfactory results. Wagner and Pacsu[95a] have demonstrated methylene cross-linking of cellulose by a vapor phase reaction at 110 to 150°C using boric acid as a catalyst. Stringent treatments necessary to obtain a high degree of methylenation are not normally employed for commercial modification of cellulose fibers. However, in this connection it must be pointed out that Dillenius[26] has demonstrated that physical strength lost by hydrolytic degradation of cellulose can also be regained by formation of methylene cross-linkages.

Saegusa[77] reports that the methylene ether of cellulose, prepared by reacting 35 per cent formaldehyde solution with viscose for 15 hours at a pH of 1.7 at room temperature followed by heating for 5 hours at 70°C, is insoluble in organic solvents or cuprammonium hydroxide, is hydrolyzed by acids but not by alkalies, and is decomposed on heating to 255°C. Increase in the degree of methylenation gives a progressively more brittle product. The degree of methylenation was found to increase with increasing concentration of formaldehyde vapor during baking, but did not exceed 17.3 per cent combined formaldehyde. This accounts for a reaction involving approximately two of the three potentially available hydroxyl radicals in the glucopyranose unit. Higher combined formaldehyde contents obtainable under some circumstances probably indicate formation of polyoxymethylene ether linkages[44].

Acid catalyzed reactions of cellulose with formaldehyde in the presence of alcohols, phenols, urea, and other polar agents lead to the formation of cellulose derivatives containing substituted oxymethyl groups ($-O-CH_2R$).

Organic derivatives of cellulose such as cellulose acetate can also be reacted with formaldehyde if they contain free hydroxyl groups. Charch and Cramer describe the preparation of a methoxymethyl derivative of cellulose acetate by reacting paraformaldehyde and methanol with a partially acetylated cellulose in an organic solvent using paratoluene sulfonic acid as the reaction catalyst[16].

The utilization of formaldehyde for modifying the properties of cotton, wood and various cellulosic products will be reviewed in the section dealing with formaldehyde uses (Chapter 21).

Joint Reactions of Formaldehyde with Alcohols and Hydrogen Halides. The joint reaction of formaldehyde and aliphatic or araliphatic alcohols with hydrogen halides results in the formation of alpha-halomethyl ethers. For example, when a cold (0 to 10°C) solution of formaldehyde containing an equimolecular proportion of methanol is saturated with hydrogen chloride, monochloromethyl ether is formed and comes to the top of the reaction mixture as a separate phase.

$$CH_2O \text{ (aq)} + CH_3OH + HCl \rightarrow CH_3OCH_2Cl + H_2O$$

This synthesis appears to have been first described by Henry[46]. Somewhat later Wedekind[100, 101] demonstrated that these ethers could also be obtained by passing the dry hydrogen halide into a chilled suspension of a formaldehyde polymer, e.g., paraformaldehyde, in an alcohol. This investigator also recommended the use of zinc chloride as a condensing agent. Better yields are obtained with polymer, since the reaction involves a chemical equilibrium and less water is present in the mixture than when solution is employed. Trioxane is readily utilized because of its solubility in alcohols[98].

An excess of hydrogen chloride is also desirable. The substituted ether is purified by separation from the aqueous phase, removing unreacted hydrogen halide with a stream of dry air or carbon dioxide, drying with calcium chloride and then subjecting to fractional distillation[17].

A large variety of chloro-, bromo- and iodoethers have been prepared, but to date fluoromethyl ethers apparently have not been made. Monochloromethyl ether distills at 59.5°C, whereas the bromo-[46] and iodomethyl[14] derivatives boil at 87° and 123 to 125°C, respectively.

In preparing chloromethyl ethers of the higher fatty alcohols, the use of an inert solvent such as benzene appears to be desirable[96]. Octadecylchloromethyl ether may be obtained in yields as high as 90 per cent by passing hydrogen chloride into a benzene solution of octadecyl alcohol in which paraformaldehyde is suspended, at a constant temperature of 5

TABLE 48. BOILING POINTS OF ALPHA-CHLOROMETHYL ETHERS ($ROCH_2Cl$)

R of Type Formula	Boiling Point (°C)	Ref.
Methyl	59.5	51
Ethyl	80.0	51
Propyl	112.5	51
n-Butyl	134	51
Isobutyl	120–1	51
Benzyl	125 (40 mm)	51
n-Octyl	217–8	56
Decyl	124–5 (10 mm)	56
Lauryl	119–21.5 (2 mm)	56
Myristyl	171–2.(5 mm)	56
Cetyl	158 (1–2 mm)	56
Octadecyl	184–6 (2 mm)	56

to 10°C. The aqueous phase which forms is separated, the benzene layer dried, and the dissolved product isolated by distillation of the solvent.

Table 48, prepared from the data of Hill and Keach[51], and Kursanov and Setkina[56] gives the boiling points of some representative chloromethyl ethers.

Ethylene glycol reacts with hydrogen chloride and formaldehyde or paraformaldehyde in the same manner as the simple aliphatic and araliphatic alcohols, with the formation of ethylene bis(chloromethyl ether)[97].

$$\begin{array}{ccc} CH_2{-}OH & & CH_2{-}OCH_2Cl \\ | & + 2CH_2O \text{ (aq, p)} + 2HCl \rightarrow & | \qquad + 2H_2O \\ CH_2{-}OH & & CH_2{-}OCH_2Cl \end{array}$$

This product is a viscous liquid, somewhat heavier than water, which distills at 97 to 99°C at approximately 13-mm pressure. Other polyhalomethyl ethers prepared by reactions of formaldehyde with polyhydroxyl compounds are described by Lichtenberger and Martin[62] and Lefevre[59].

In general, the alpha-chloromethyl ethers are colorless compounds possessing a good degree of chemical stability. However, the bromo- and iodomethyl ethers are less stable. All these ethers are highly reactive and hydrolyze readily on contact with moisture, liberating formaldehyde, hydrogen halide, and alcohol. For this reason their vapors are extremely irritating and they should be handled with caution.

The reactions of the alpha-chloromethyl ethers are of value in chemical syntheses. With alcohols, they react to give mixed formals[63], *e.g.*:

$$C_2H_5OH + ClCH_2OCH_3 \rightarrow C_2H_5OCH_2OCH_3 + HCl$$

With alkali phenates, they give phenyl formals[14]:

$$C_6H_5OK + ClCH_2OCH_3 \rightarrow C_6H_5OCH_2OCH_3 + KCl$$

Methylene ether-esters are formed with the salts of organic acid[18, 97, 102]:

$$CH_3COONa + ClCH_2OCH_3 \rightarrow CH_5COOCH_2OCH_3 + NaCl$$

The alpha-chloromethyl ethers apparently do not react with alkali-metal cyanides, but give good yields of nitriles with copper cyanide[35]:

$$C_2H_5OCH_2Cl + CuCN \rightarrow C_2H_5OCH_2CN + CuCl$$

They also react in a normal manner with Grignard reagents[13] and many active derivatives of alkali metals.

When alpha-chloromethyl ethers are subjected to the Friedel-Crafts reaction with benzene, the principal product is benzyl chloride rather than a benzyl ether[89]. In the case of alpha-chloromethyl propyl ether, the benzyl ether is produced in 30 per cent yield together with benzyl chloride:

$$C_6H_6 + C_3H_7OCH_2Cl \rightarrow C_6H_5CH_2OC_3H_7 + HCl$$

$$C_6H_6 + C_3H_7OCH_2Cl \rightarrow C_6H_5CH_2Cl + C_3H_7OH$$

With tertiary bases such as pyridine and triethylamine, the ethers unite to give quaternary salts[96]:

$$C_{18}H_{37}OCH_2Cl + (C_2H_5)_3N \rightarrow C_{18}H_{37}OCH_2 \cdot (C_2H_5)_3NCl$$

Octadecyl chloromethyl ether	*Triethyl- amine*	*Octadecyloxymethyl triethylammonium chloride*

Reactions of Formaldehyde with Mercaptans. Reactions of formaldehyde with thioalcohols or mercaptans are similar in many respects to those encountered with alcohols. According to Posner[73], mercaptals or methylene dithiols are formed in a two-step reaction, as indicated by the equations shown below in which the R of RSH stands for an alkyl, aralkyl or aryl radical.

$$CH_2O \text{ (aq)} + RSH \rightarrow RSCH_2OH \tag{a}$$

$$RSCH_2OH + RSH \rightarrow RSCH_2SR + H_2O \tag{b}$$

Reaction (a) is sufficiently definite to make possible the preparation of mixed mercaptals such as ethyl phenyl, ethyl benzyl and amyl phenyl. Posner[73] prepared these mixed mercaptals by treating one mol of mercaptan with one mol of formaldehyde in aqueous solution and then adding a mol of another mercaptan after approximately one hour and completing the condensation by the use of hydrogen chloride and zinc chloride as catalysts. Products obtained in his work were not isolated as mercaptals but were converted to sulfones by oxidation with permanganate.

Baumann[7] obtained diethylmethylene disulfone, $C_2H_5SO_2CH_2SO_2C_2H_5$ (m.p. 103°C), and diethyldimethylene trisulfone, $C_2H_5SO_2CH_2SO_2CH_2SO_2$-$C_2H_5$ (m.p. 149°C), by oxidation of the products obtained on reacting ethyl bromide with a solution of hydrogen sulfide in aqueous formaldehyde in the presence of caustic soda. These results indicate that mercaptals were probably formed as primary products. A possible reaction mechanism is:

$$CH_2O \text{ (aq)} + 2H_2S \rightarrow CH_2(SH)_2 + H_2O$$

$$CH_2(SH)_2 + 2C_2H_5Br + 2NaOH \rightarrow CH_2(SC_2H_5)_2 + 2NaBr + 2H_2O$$

$$2CH_2O \text{ (aq)} + 3H_2S \rightarrow HSCH_2SCH_2SH + 2H_2O$$

$$HSCH_2SCH_2SH + 2C_2H_5Br + 2NaOH \rightarrow C_2H_5SCH_2SCH_2SC_2H_5 + 2NaBr + 2H_2O$$

Mercaptals are stable compounds having a powerful mustard-like odor. Like formals, they are stable to alkali but are also said to have a fair resistance to acids. Methylene dimethyl mercaptal is prepared by the action of methyl mercaptan on commercial formaldehyde solution at 110°C. It is a colorless, oily fluid which boils at 147°C[33]. Methylene diethyl mercaptal is described as a liquid which boils at 180 to 185°C. Its preparation by the action of ethyl mercaptan on formaldehyde in the presence of hydrochloric acid is reported in a German patent[8].

On reaction with formadehyde and hydrogen chloride, mercaptans give alpha-chloromethyl thioethers analogous to the alpha-chloromethyl ethers obtained with alcohols.

Böhme[11] prepared chloromethyl ethyl thioether (or chloromethyl ethyl sulfide) in 60 per cent yield by saturating an ice-cold mixture of alpha-polyoxymethylene and ethyl mercaptan with hydrogen chloride until a clear liquid was obtained. After treatment with calcium chloride, the product separated as a separate phase with loss of unreacted hydrogen chloride, and was purified by distillation. It is a colorless liquid with an unpleasant odor which boils at 128 to 131°C. It is quite stable when pure. On treatment with water, it forms methylene diethyl mercaptal with liberation of formaldehyde and hydrogen chloride.

$$2C_2H_5SCH_2Cl + H_2O \rightarrow CH_2(SC_2H_5)_2 + CH_2O \text{ (aq)} + 2HCl$$

Most of its reactions are analogous to those of the alpha-chloromethyl ethers. However, like other sulfides, it gives a sulfone on oxidation.

Table 49 gives the boiling points of representative halomethyl thioethers prepared by Böhme, Fischer and Frank[12] and Walter[99]. The chloro- and bromo- derivatives were prepared by reaction of mercaptans with paraformaldehyde and hydrogen halides. The iodomethyl thioethers were prepared by reacting the chloromethyl thioethers with sodium iodide in purified acetone.

TABLE 49. BOILING POINTS OF HALOMETHYL THIOETHERS TYPE FORMULA = $RSCH_2X$

Alkyl or Aryl Group R	Halogen X	Boiling Point (°C)	Pressure (mm)	Ref.
Methyl	Chlorine	110–2	760	12
Methyl	Bromine	131–4	760	12
Methyl	Iodine	72	36	12
Ethyl	Bromine	67	45	12
Ethyl	Iodine	74	40	12
n-Propyl	Chlorine	58	60	12
isopropyl	Chlorine	56–8	42	12
n-Butyl	Chlorine	64–6	16	99
Isobutyl	Chlorine	160–1	760	99
sec-Butyl	Chlorine	58–9	11	99
t-Butyl	Chlorine	57–8	12	99
n-Amyl	Chlorine	170–6	760	99
n-Hexyl	Chlorine	105–6	22	99
Cyclohexyl	Chlorine	101–3	13.5	99
Benzyl	Chlorine	136–9	25	12
Phenyl	Chlorine	98	12	12
Allyl	Chlorine	52–5	15	99

References

1. Adams, E. W., and Adkins, H., *J. Am. Chem. Soc.*, **47**, 1358 (1925).
2. Arnhold, M., *Ann.*, **240**, 199 (1887).
3. Backer, H. J., and Schurink, H. B. J., *Rec. trav. chim.*, **50**, 1066–1068 (1931).
4. Barth, R. H., (to Heyden Chemical Corp.), U. S. Patent 2,446,257 (1948).
5. Barth, R. H., and Snow, J. E., (to Heyden Chemical Corp.), U. S. Patent 2,464,-430 (1949).
6. Bauer, H. F., (to Stein Hall and Co.), U. S. Patent 2,396,937 (1946).
7. Baumann, E., *Ber.*, **23**, 1875 (1890).
8. Bayer & Co., German Patent 97,207 (1897); *Chem. Zentr.*, **1898**, II, 524.
9. Beyer, R., (to Robert Beyer Corporation), U. S. Patent 1,983,732 (1934).
10. Blumer, E. R. L., German Patent 179,590 (1906); *Chem. Zentr.*, **1907**, I, 383.
11. Böhme, H., *Ber.*, **69**, 1610 (1936).
12. Böhme, H., Fischer, H., and Frank, R., *Ann.*, **563**, 54–72 (1949).
13. Braun, J. v., and Deutsch, H., *Ber.*, **45**, 2176 (1912).
14. Breslauer, J., and Pictet, A., *Ber.*, **40**, 3785 (1907).
14a. Brooks, R. E., (to E. I. du Pont de Nemours & Co., Inc.), U. S. Patent 2,321,542 (1943).

15. Bruson, H. A., and Riener, T. W., (to The Resinous Products and Chemical Co.), U. S. Patent 2,435,869 (1948).
16. Charch, W. H., and Cramer, F. B., (to E. I. du Pont de Nemours & Co., Inc.), U. S. Patent 2,430,911 (1947).
17. Clark, F. E., Cox, S. F., and Mack, E., *J. Am. Chem. Soc.*, **39**, 712 (1917).
18. *Ibid.*, p. 714.
19. Clarke, H. T., *J. Chem. Soc.*, **101**, 1804 (1912).
20. Classen, A., German Patent 92,259 (1896); *Chem. Zentr.*, **1897**, II, 456.
21. *Ibid.*, German Patent 94,282 (1896); *Chem. Zentr.*, **1898**, I, 229.
22. Conant, J. B., Webb, C. N., and Mendum, W. C., *J. Am. Chem. Soc.*, **51**, 1246–1255 (1929).
23. Contardi, A., and Ciocca, B., *Rend. ist. lombardo sci.*, **69**, 1057–1066 (1936); *Chem. Zentr.*, **1937**, II, 233–234; *C. A.*, **33**, 4583 (1939).
24. Corn Products Refining Co., British Patent 523,665 (1938); *Brit. Chem. Abs.*, **1940B**, 757.
25. Descudé, M., *Bull. soc. chim.* (*3*), **27**, 1218 (1902).
26. Dillenius, H., *Jentgen's Kunstseide und Zellwolle*, **24**, 520 (1942).
27. Dreyfus, H., Canadian Patent 355,557 (1936); U. S. Patent 2,095,320 (1937).
28. Eschalier, X., French Patent 374,724 (1906); *J. Soc. Chem. Ind.*, **26**, 821 (1907); British Patent 25,647 (1906).
29. Eschalier, X., *Rev. gen. mat. color.*, **12**, 249–251 (1908); *C. A.*, **2**, 3285 (1908).
30. Fairbourne, A., Gibson, G. P., and Stephens, D. W., *J. Soc. Chem. Ind.*, **49**, 1069 (1930).
31. Farb- and Gerbstoffwerke Carl Flesch Jr., French Patent 773,190 (1934); *Chem. Zentr.*, **1935**, I, 3476.
32. Favre, C., *Bull. soc. chim.* (*3*), **11**, 881 (1894).
33. Fenn, J. E., U. S. Patent 2,469,957 (1949).
34. Fischer, E., and Giebe, G., *Ber.*, **30**, 3054 (1897).
35. Gauthier, M. D., *Ann. chim. phys.*, **16**, 289–358 (1909).
36. Gayler, C. W., (to American Viscose Corp.), U. S. Patent 2,487,864 (1949).
37. Goldthwait, C. F., *Textile Research J.*, **21**, 55–62 (1951).
38. Gottstein, A., *Therap. Monatsh.*, **11**, 95 (1897); *Chem. Zentr.*, **1897**, I, 715.
39. Greene, W. H., *Compt. rend.*, **89**, 1077–1078 (1879); *Chem. Zentr.*, **1880**, 71.
40. Gresham, W. F., (to E. I. du Pont de Nemours & Co., Inc.), U. S. Patent 2,350,350 (1944).
41. *Ibid.*, U. S. Patent 2,395,265 (1946).
42. Gresham, W. F., and Bell, C. D., (to E. I. du Pont de Nemours & Co., Inc.), U. S. Patent 2,475,610 (1949).
43. Gresham, W. F., and Brooks, R. E., (to E. I. du Pont de Nemours & Co, Inc..), U. S. Patent 2,449,469 (1949).
44. Gruntfest, I. J., and Gagliardi, D. D., *Textile Research J.*, **18**, No. 11, 643–650 (1948).
45. Heiduschka, A., and Zirkel, H., *Arch. Pharm.*, **254**, 456–487 (1916); *J. Chem. Soc.*, **112**, I, 446 (1917).
46. Henry, L., *Bull. classe sci., Acad. roy. Belg.* (*3*), **25**, 439–440 (1893); *Ber.*, **26**, Ref. 933 (1893).
47. *Ibid.*, **1902**, 445–494; *Chem. Zentr.*, **1902**, II, 929.
48. *Ibid.*, **1908**, 6–17; *Chem. Zentr.*, **1908**, I, 2014.
49. Herold, W., *Z. Elektrochem.*, **39**, 566–571 (1933).
50. Heuser, E., *Paper Trade J.*, **122**, No. 3, 43–48 (Jan. 17, 1946).
51. Hill, A. J., and Keach, D. T., *J. Am. Chem. Soc.*, **48**, 259 (1926).

52. Hill, J. W., and Carothers, W. H., *J. Am. Chem. Soc.*, **57**, 925 (1935).
53. Jacoby, M., *Ber.*, **52B**, 558–562 (1919); *C. A.*, **13**, 2528 (1919).
54. Kauffmann, W. v., *Ber.*, **50**, 198–202 (1917).
55. Kerr, R. W., and Schink, N. F., (to Corn Products Refining Co.), U. S. Patent 2,438,855 (1948).
56. Kursanov, D. N., and Setkina, V. N., *J. Appl. Chem.* (*U.S.S.R.*), **16**, 36–46 (1943); *C. A.*, **38**, 3139 (1944).
57. Landini, G., *Atti accad. Lincei*, **16**, 52–58 (1908); *C. A.*, **2**, 933 (1908).
58. Lauch, R., and Quade, F., U. S. Patent 1,055,405 (1913).
59. Lefevre, G., *et al.*, *Compt. rend.*, **229**, 222–224 (1949).
60. Leuck, G. J., (to Corn Products Refining Co.), U. S. Patent 2,222,872 (1940).
61. *Ibid.*, U. S. Patent 2,222,873 (1940).
62. Lichtenberger, J., and Martin, L., *Bull. soc. chim. France*, **1947**, 468–476.
63. Litterscheid, F. M., and Thimme, K., *Ann.*, **334**, 26 (1904).
64. Loder, D. J., (to E. I. du Pont de Nemours & Co., Inc.), U. S. Patent 2,359,134 (1944).
65. Lolkema, J., and Van der Meer, W. A., (to N. V. W. A. Scholten's Chemische Fabriehen), Dutch Patent 60,861 (1948); *C. A.*, **42**, 3984 (1948).
66. Maggi, H., and Woker, G., *Ber.*, **50**, 1188–1189 (1917).
67. *Ibid.*, **51**, 790–793 (1918).
68. Morrison, G. O., *Chem. & Ind.*, **60**, 390–392 (1941).
69. Ness, A. T., Hann, R. M., and Hudson, C. S., *J. Am. Chem. Soc.*, **65**, 2215 (1943).
70. *Ibid.*, **66**, 665 (1944).
71. Payne, W., (to E. I. du Pont de Nemours & Co., Inc.), U. S. Patent 2,467,223 (1949).
72. Pierson, G. G., (to Perkins Glue Co.), U. S. Patent 2,417,611 (1947).
73. Posner, T., *Ber.*, **36**, 296 (1903).
74. Quade, F., U. S. Patent 1,062,501 (1913).
75. Reichard, A., *Z. ges. Brauw.*, **31**, 161–163 (1908); *C. A.*, **2**, 2881 (1908).
76. Rosenberg, P., German Patent 189,036 (1907).
77. Saegusa, H., *J. Cellulose Inst. Tokyo*, **17**, 81–89 (1941); *Brit. Chem. Abs.*, **1941**, B, II, 181.
78. Sallinger, H., *Ber.*, **52B**, 651–656 (1919); *C. A.*, **13**, 2529 (1919).
79. Schenk, M., *Helv. Chim. Acta.*, **14**, 520–541 (1931).
80. *Ibid.*, **15**, 1088–1102 (1932).
81. Schimmel & Co., *Ber. Schimmel & Co.*, **1933**, 78–81; *C. A.*, **28**, 5041–5042 (1934).
82. Schleich, C. L., *Therap. Monatsh.*, **11**, 97 (1897); *Chem. Zentr.*, **1897**, I, 715.
83. Schneider, W., *Ann.*, **386**, 349 (1912).
84. Schulz, M., and Tollens, B., *Ber.*, **27**, 1893 (1984).
85. Schulz, M., and Tollens, B., *Ann.*, **289**, 21 (1896).
86. *Ibid.*, p. 29.
86a. Schwenkler, J. J., *J. Forestry*, **45**, 798 (1947).
87. Sefton-Jones, H., British Patent 17,616 (1911); *C. A.*, **8**, 2032 (1914).
88. Skrabal, A., and Zlatewa, M., *Z. physik. Chem.*, **119**, 305 (1926).
89. Sommelet, M., *Compt. rend.*, **157**, 1444 (1913).
90. Stolle & Kopke, German Patent 201,436 (1908).
90a. Sussman, S., and Gresham, W. F., (to E. I. du Pont de Nemours & Co., Inc.), U. S. Patent 2,340,907 (1944).
91. Timmermans, J., *Bull. classe sci. Acad. roy. Belg.*, **24**, 244–269 (1910); *Chem. Zentr.*, **1910**, II, 442.
92. Tollens, B., *Ber.*, **32**, 2585–2588 (1899).

93. Trillat, A., and Cambier, R., *Bull. soc. chim.* (*3*), **11,** 752 (1894).

94. *Ibid.*, p. 757.

95. Verley, A., *Bull. soc. chim.* (*2*), **21,** 275–277 (1899); *Chem. Zentr.*, **1899,** I, 919–920.

95a. Wagner, R. E., and Pacsu, E., *Textile Research J.*, **22,** 12–20 (1952).

96. Wakelin, *Chem. Ind.*, **43,** 53 (1938).

97. Walker, (J.) F., *Plastic Products*, **9,** 187–188 (1933).

98. Walker, J. F., and Chadwick, A. F., *Ind. Eng. Chem.*, **39,** 977 (1947).

99. Walter, L. A., (to The Maltbie Chem. Co.), U. S. Patents 2,354,229 and 2,354,230 (1944).

100. Wedekind, E., German Patent 135,310 (1902).

101. Wedekind, E., *Ber.*, **36,** 1384 (1903).

102. *Ibid.*, p. 1385.

103. Wohlgemuth, J., *Biochem. Z.*, **94,** 213–224 (1919); *C. A.*, **13,** 2539 (1919).

104. Woker, G., *Ber.*, **49,** 2311–2318 (1916).

105. *Ibid.*, **50,** 679–692 (1917).

106. Woker, G., and Maggi, H., *Ber.*, **52,** 1594–1604 (1919).

107. Wood, F. C., *J. Soc. Chem. Ind.*, **50,** 411T (1931).

CHAPTER 11

REACTIONS OF FORMALDEHYDE WITH ALDEHYDES AND KETONES

Reactions of formaldehyde with other aldehydes and with ketones may be classified broadly as belonging to three generic types: (a) aldol condensations resulting in the production of methylol derivatives, (b) condensations leading to the formation of unsaturated methylene derivatives (vinyl compounds), and (c) mixed Cannizzaro reactions in which formaldehyde functions as a reducing agent. The first two types of reaction occur only with aldehydes or ketones containing alpha-hydrogen atoms, and are accordingly not encountered with aryl aldehydes, diaryl ketones, or substituted compounds such as trimethylacetaldehyde.

Reactions of Formaldehyde with Other Aldehydes

Acetaldehyde: Reactions in Water Solution. When calcium hydroxide is added to an aqueous solution containing one mol of acetaldehyde and at least four mols of formaldehyde, the tetrahydric alcohol pentaerythritol (or pentaerythrite) and calcium formate are produced. This reaction was apparently first observed by Tollens and Wiegand[74] in 1891.

$$CH_3CHO + 4CH_2O \text{ (aq)} + \tfrac{1}{2}Ca(OH)_2 \rightarrow C(CH_2OH)_4 + \tfrac{1}{2}Ca(OOCH)_2$$

The probable reaction mechanism is indicated in the following equations:

$$CH_3CHO + CH_2O \text{ (aq)} \rightarrow CH_2(CH_2OH)CHO \qquad (a)$$

$$CH_2(CH_2OH)CHO + CH_2O \text{ (aq)} \rightarrow CH(CH_2OH)_2CHO \qquad (b)$$

$$CH(CH_2OH)_2CHO + CH_2O \text{ (aq)} \rightarrow C(CH_2OH)_3CHO \qquad (c)$$

$$C(CH_2OH)_3CHO + CH_2O \text{ (aq)} + \tfrac{1}{2}Ca(OH)_2 \rightarrow C(CH_2OH)_4 + \tfrac{1}{2}Ca(OOCH)_2 \quad (d)$$

Reactions (a), (b), and (c), which lead to the formation of trimethylolacetaldehyde or pentaerythrose, are aldol condensations in which calcium hydroxide serves merely as a catalyst. Step (d) is a mixed Cannizzaro reaction in which alkali is consumed. Although isolation of the intermediates formed in the reaction sequence is difficult because of their instability and the rapidity with which further reaction takes place, the presence of mono- and dimethylolacetaldehyde has been demonstrated, and pentaerythrose itself has been prepared as a reaction product under special circumstances. In general, however, this is possible only when weak alkalies are employed as catalysts.

Stepanov and Shchukina[70] report that, when acetaldehyde is treated with two to three equivalents of formaldehyde in the presence of potassium carbonate at 12 to 18°C, a mixture of acetaldol and monomethylolacetaldehyde or beta-hydroxypropionaldehyde [product of reaction (a)] is obtained. Although the latter product could not be isolated because of polymerization at its boiling point (90°C at 18 mm), it was identified by the ease with which it was dehydrated to acrolein.

$$CH_2(CH_2OH)CHO \rightarrow CH_2{:}CHCHO + H_2O$$

The isolation of 2-hydroxymethyl propandiol-1,3 (m.p. 178°C) by Fujii[27] on hydrogenation of the reaction product of one mol acetaldehyde with 3 mols formaldehyde indicates the formation of dimethylolacetaldehyde by reaction (b).

According to McLeod[43], crude pentaerythrose, the product of reaction (c), may be prepared by reacting 3 mols of formaldehyde with one mol of acetaldehyde. This viscous, sugar-like, liquid product is identified by the fact that it can be converted to pentaerythritol by reduction with sodium amalgam. A process patented by Fitzky[24] covers the preparation of pentaerythrose by heating 30 per cent formaldehyde with acetaldehyde in the presence of sodium carbonate at 40°C.

Of interest in connection with the study of the mechanism of pentaerythritol formation are processes[15a, 86] for obtaining improved yields of pentaerythritol by the use of subsidiary catalysts such as finely divided copper oxide, copper and platinum which accelerate the Cannizzaro reaction [reaction (d)]. The writer[80] has demonstrated that yields can be appreciably improved by catalytic hydrogenation of the reaction mixture prior to product isolation. This technique also improves conversion of the intermediate product to pentaerythritol.

Pentaerythritol. Pentaerythritol has become an important chemical in recent years and its preparation has received intensive study. It is of considerable interest as a raw material for alkyd resins[2, 44, 58, 13] (alsop age 445), synthetic drying oils[5, 72], plasticizers[12], insecticides[68], etc.

Pure pentaerythritol is a colorless crystalline compound melting at 260.5°C. However, as normally obtained, it is almost always contaminated with the ether dipentaerythritol (m.p. 221°C) which gives it a lower melting point. The formation and structure of this ether is indicated thus:[22, 26]

$$2C(CH_2OH)_4 \rightarrow (CH_2OH)_3CCH_2OCH_2C(CH_2OH)_3 + H_2O$$

Wyler and Wernett[91] report that pentaerythritol forms a eutectic with 35 per cent dipentaerythritol melting at 190°C and that the crude product as usually produced melts at 230 to 244°C and contains 10 to 15 per cent of this ether. A process for isolating pure pentaerythritol from mixtures with dipentaerythritol developed by these investigators involves hot crystallization carried out by cooling aqueous solutions from approximately 108° to 70°C[91].

A recommended procedure for obtaining accurate melting points on samples of pentaerythritol involves adding 7 to 8 mm of the finely powdered product, which has been dried for 2 hours at 100 to 110°C, to a dry 16-cm tube (1.0 to 1.5 mm outside diameter) and immersing in a 200°C melting point bath. The temperature of the bath is raised one degree per minute up to 245°C and then one degree every three minutes. The melting range is measured from the formation of the first meniscus to complete liquefaction[92].

Pentaerythritol is highly soluble in hot water from which it crystallizes readily when the solution is cooled. It is only slightly soluble in methanol and is substantially insoluble in most organic solvents. On nitration it yields the explosive pentaerythritol tetranitrate ("Penthrite," PETN, see pages 472-473).

In general, pentaerythritol is manufactured by adding a suspension of hydrated lime to a dilute aqueous solution of commercial formaldehyde containing acetaldehyde in the proportion of one mol of the latter to four or more mols of formaldehyde. Since heat is evolved, the mixture must be cooled so that the temperature will remain in the neighborhood of 50°C until the reaction is complete. Following this, the reaction mixture is subjected to vacuum distillation and pentaerythritol is isolated by crystallization from the concentrated solution. In most methods of preparation, calcium is precipitated from the crude reaction mixture as calcium sulfate by addition of sulfuric acid before vacuum concentration[32]. According to Burke[14], better results are obtained by precipitating the lime as calcium carbonate by addition of potassium or sodium carbonate. Naujoks[59] recommends vacuum concentration of the filtered reaction mixture without removal of dissolved lime or calcium formate. The same investigator also claims that optimum yields are obtained when the reaction is followed by determining the concentration of reducing substances and stopped by adjusting to an acidic pH as soon as this figure falls to a minimum.

Although some investigators report that good results can be secured by employing sodium or barium hydroxide in place of calcium hydroxide, it appears that the best yields are obtained with calcium hydroxide[15, 53]. In a process patented by Peterson[61] a 50 per cent excess of calcium hydroxide is stated to give superior results. Spurlin[71] has patented the use of an alkaline ion exchange resin in a column through which the acetaldehyde-formaldehyde solution containing a small concentration of sodium chloride is slowly passed at a temperature of 30 to 32°C. Wyler[85] claims that the presence of methanol hinders pentaerythritol formation and advocates the use of formaldehyde that is substantially free of methanol. The same inventor[88] also claims that superior results are obtained by the use of a distilled formaldehyde solution with a high specific transmission for light having a wave length of 320 millimicrons. Jackson[38] reduces formation of polypentaerythritols by employing excess formaldehyde which is recovered by distillation.

Various subsidiary catalysts have been found of value in pentaerythritol preparation. Kuzin[41] reports that small proportions of sugars such as glucose or fructose act as catalysts for the aldol reaction and have a definite accelerating effect on product formation.

A process employing organic catalysts such as glycolaldehyde, hydroxymethyl acetaldehydes, glycol, glycerol and dihydroxy diethyl ketone is described by Cryer, Owens and Jui Fu Lee[16]. As previously pointed out, it is claimed[15a, 86] that catalysts such as copper oxide, copper, and platinum accelerate and improve the efficiency of pentaerythrose reduction. The use of quaternary ammonium compounds[84] and amides[87] as catalysts or subsidiary catalysts has also been patented by this investigator.

Various procedures for isolating pure pentaerythritol from formates, aldols and sugar-like products which occur as by-products in the crude reaction mixture have been developed and numerous methods of purification have been patented, e.g.[42, 66], including the use of hydrogen peroxide for dealing with undesirable aldehydic by-products[60, 90]. Bludworth[7] describes a procedure for obtaining pentaerythritol from a crude dilute aqueous formaldehyde produced by hydrocarbon oxidation. Special techniques favoring the formation of polypentaerythritols have also been patented[65, 89].

The patent literature on pentaerythritol is voluminous and a detailed discussion is beyond the scope of this book.

Gas-phase Reactions Involving Formaldehyde and Acetaldehyde. In the gaseous state at about 300°C, formaldehyde reacts with acetaldehyde in the presence of activated alumina, lead acetate on silica, and other catalysts to form acrolein and water[19, 29, 37, 79].

$$CH_3CHO + CH_2O \text{ (g)} \rightarrow CH_2{:}CHCHO + H_2O$$

Gaseous mixtures of formaldehyde, acetaldehyde, and ammonia, when subjected to temperatures of 200 to 400°C and pressures of approximately 50 atmospheres, yield mixtures of pyridines and picolines. When acetaldehyde and ammonia are subjected to similar treatment in the absence of formaldehyde only picolines are obtained. This reaction is catalyzed by tungsten carbide[8].

Higher Aliphatic Aldehydes. The higher aliphatic aldehydes react with formaldehyde in a manner analogous to that observed in the case of acetaldehyde. With the exception of the crossed Cannizzaro reaction, these reactions involve alpha-hydrogen atoms and differ only as a result of the lesser number of these atoms available. Gas-phase condensations to substituted acroleins are possible only with normal aldehydes containing two alpha-hydrogen atoms.

When potassium carbonate is used as a reaction catalyst, propionaldehyde combines with two mols of formaldehyde to give the dimethylol aldol known as pentaglycerose:

$$CH_3CH_2CHO + 2CH_2O \text{ (aq)} \rightarrow CH_3-\overset{\displaystyle CH_2OH}{\underset{\displaystyle CH_2OH}{\overset{|}{\underset{|}{C}}}}-CHO$$

According to Koch and Zerner[40], this product is a colorless, non-distillable syrup which decomposes on heating. On reduction with sodium amalgam it yields pentaglycerol, $[CH_3C(CH_2OH)_3]$, and on treatment with caustic undergoes the Cannizzaro reaction:

$$2CH_3C(CH_2OH)_2CHO + KOH \rightarrow CH_3C(CH_2OH)_3 + CH_3(CH_2OH)_2COOK$$

By the reaction of propionaldehyde with formaldehyde in the presence of alcoholic potash, the reduction product of a mixed aldol of propionaldehyde and formaldehyde is also obtained under some circumstances[40].

$$2CH_3CH_2CHO + 2CH_2O \text{ (aq)} + KOH \rightarrow$$
$$CH_3CH_2CH(OH)C(CH_2OH)_2CH_3 + HCOOK$$

Pentaglycerol (m.p. 199°C) appears to have been first obtained by Hosaeus[34] in 1893 by reaction of dilute formaldehyde with propionaldehyde and slaked lime.

$$CH_3CH_2CHO + 3CH_2O \text{ (aq)} + \tfrac{1}{2}Ca(OH)_2 \rightarrow$$

$$CH_2-\overset{\displaystyle CH_2OH}{\underset{\displaystyle CH_2OH}{\overset{|}{\underset{|}{C}}}}-CH_2OH + \tfrac{1}{2}Ca(OOCH)_2$$

A recent process involving the use of alkali-metal hydroxides and a special method of product isolation has been recently patented by Brubaker and Jacobson[11]. Condensation of propionaldehyde with two mols of formaldehyde followed by catalytic hydrogenation at room temperature is reported to give an approximately 80 per cent yield of pentaglycerol[80].

Gresham[33] obtains alpha-methylacrolein in good yield by the gradual addition of propionaldehyde to hot aqueous formaldehyde containing boron fluoride. The alpha-methylacrolein distills as produced in the form of a water azeotrope which boils at approximately 65°C.

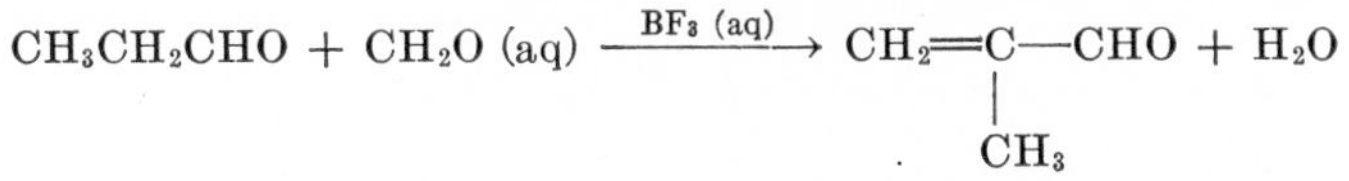

$$CH_3CH_2CHO + CH_2O \text{ (aq)} \xrightarrow{\text{BF}_3 \text{ (aq)}} CH_2{=}C{-}CHO + H_2O$$
$$\underset{CH_3}{|}$$

The methylol derivative of isobutyraldehyde, which is known as pentaldol, has a fair degree of stability and can be distilled without decomposition at 67 to 69°C at 14 mm. At 747 mm it distills with partial decomposition at 172 to 173°C. It is a crystalline solid melting at 89 to 90°C. The product is readily obtained by reaction of isobutyraldehyde and 37 per cent formaldehyde in the presence of potassium carbonate[82].

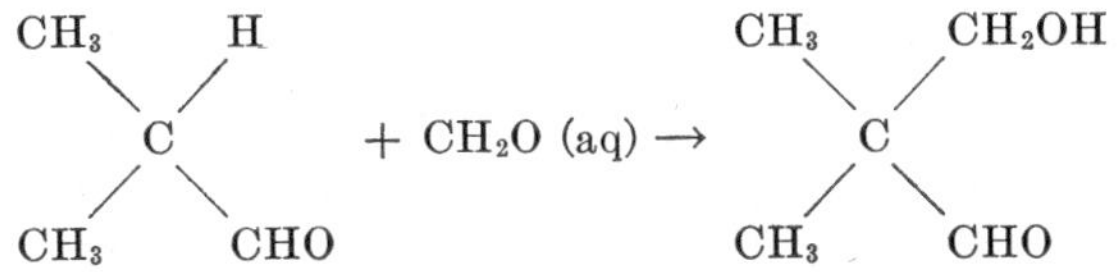

Like many hydroxy-aldehydes, pentaldol reduces ammoniacal silver nitrate but does not give a sodium bisulfite compound.

The dihydroxy compound, pentaglycol, was apparently first prepared by Apel and Tollens[3, 4] by reacting dilute formaldehyde with isobutyraldehyde and slaked lime.

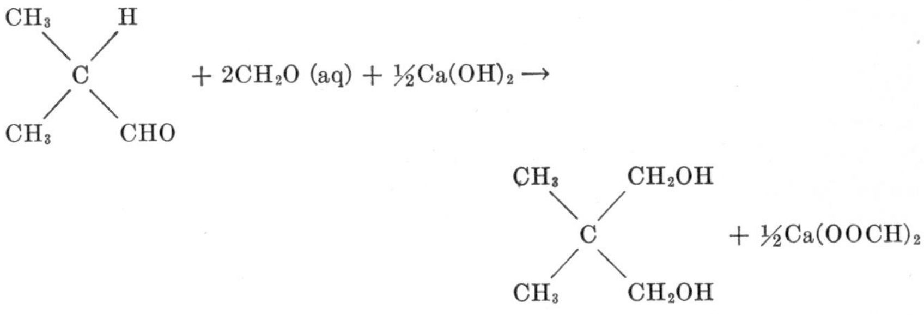

A more recent process for pentaglycol preparation developed by N. Turnbull and the writer[81] is carried out by adding approximately one mol of caustic soda to a mixture of approximately two equivalents of 37 per cent formaldehyde and one mol of isobutyraldehyde. A good yield of pentaglycol may be readily obtained by this procedure. Pentaglycol crystallizes in colorless needles which melt at 127 to 128°C. It boils at approximately 210°C. It is highly soluble in water and alcohol. It is also quite soluble in hot trichloroethylene and can be readily purified by crystallization from this solvent. Utilization of pentaldol formation as a step in the synthesis of pantothenic acid has been recently patented by Van House[76]. In this process, the aldol condensation is catalyzed by potassium cyanide with which the pentaldol is then reacted to form the corresponding cyanhydrin.

Mannich syntheses involving the joint reaction of isobutyraldehyde with primary and secondary amines, preferably as hydrochlorides, yield substituted aminomethyl derivatives[47, 49]:

$$CH_3NH_2 + CH_2O \text{ (aq)} + (CH_3)_2CHCHO \rightarrow CH_3NHCH_2 \cdot C(CH_3)_2CHO + H_2O$$

$$(CH_3)_2NH + CH_2O \text{ (aq)} + (CH_3)_2CHCHO \rightarrow (CH_3)_2NCH_2 \cdot C(CH_3)_2CHO + H_2O$$

Although comparatively little has been published concerning the reactions of aldehydes containing more than 4 carbon atoms with formaldehyde, reaction types appear to be quite similar to those obtained with the lower aldehydes. Isovaleraldehyde, for example, was found by van Marle and Tollens[77] to yield a trihydric alcohol when warmed for 24 to 36 hours with milk of lime and 37 per cent formaldehyde solution:

$$(CH_3)_2CH \cdot CH_2 \cdot CHO + 3CH_2O + \tfrac{1}{2}Ca(OH)_2 \rightarrow$$

$$(CH_3)_2CH \cdot C(CH_2OH)_3 + \tfrac{1}{2}Ca(OOCH)_2$$

The preparation of 2-ethyl-2-butyl-1,3-propanediol from 2-ethyl hexanal and formaldehyde is described by Tribit[75].

Unsaturated aldehydes such as acrolein and substituted acroleins do not react readily with formaldehyde and comparatively little information has been published on this subject. Bruson and Niederhauser[10] report production of unsaturated glycols, such as $CH_3CH_2CH:CHC(C_2H_5)(CH_2OH)_2$, by heating alpha, gamma-diethyl crotonaldehyde with an excess of formaldehyde in the presence of a strong alkali.

Aromatic Aldehydes. As previously stated, aromatic aldehydes, since they do not contain alpha-hydrogen atoms, cannot react with formaldehyde to form methylol derivatives or double-bonded methylene derivatives. On treatment with formaldehyde they are reduced to the corresponding alcohols by means of the crossed Cannizzaro reaction[17].

$$C_6H_5CHO + CH_2O \text{ (aq)} + NaOH \rightarrow C_6H_5CH_2OH + NaOOCH$$

Reactions of Formaldehyde with Ketones

The reactions of formaldehyde with aliphatic ketones are similar to those involving aldehydes; however, in general these reactions are more readily controlled so that fair yields of the intermediate methylol derivatives can be obtained. Since these methylol ketones are generally more stable than the methylol aldehydes, they can in many instances be isolated in a pure state.

Acetone. When acetone is reacted with calcium hydroxide and a large excess of formaldehyde in dilute aqueous solution, anhydroennea-heptitol (m.p. 156°C) is obtained[3, 67]. This is a cyclic pentahydroxy ether, apparently the dehydration product of the heptahydroxy alcohol, which would be normally expected.

$$7CH_2O \text{ (aq)} + CH_3COCH_3 + \frac{1}{2}Ca(OH)_2 \rightarrow$$

$$(CH_2OH)_3C \cdot CHOH \cdot C(CH_2OH)_3 + \frac{1}{2}Ca(OOCH)_2$$

$$(CH_2OH)_3C \cdot CHOH \cdot C(CH_2OH)_3 \rightarrow (CH_2OH)_2C \cdot CHOH \cdot C(CH_2OH)_2 + H_2O$$

$$\underset{H_2C \cdot O \cdot CH_2}{\overset{\mid \qquad\qquad\qquad\qquad \mid}{}}$$

Anhydroennea-heptitol

The formation of this product is analogous to that of dipentaerythritol as a by-product of the pentaerythritol reaction.

The primary condensation product of formaldehyde and acetone, monomethylolacetone or gamma-ketobutanol, is obtained by the reaction of equimolar quantities of formaldehyde and acetone in the presence of dilute alkali[52,56]. It may also be prepared by the reaction of paraformaldehyde with an excess of anhydrous acetone which has been made faintly alkaline by the addition of a small percentage of alcoholic caustic[62].

$$CH_2O \text{ (aq, p)} + CH_3COCH_3 \rightarrow CH_3COCH_2CH_2OH$$

Dreyfuss and Drewitt[21] state that good yields and low by-product formation are favored by a high acetone-formaldehyde ratio, the presence of water and a pH of 8.5 to 9.5. Monomethylolacetone can be isolated as a colorless oil, distilling at 109 to 110°C at 30 mm. On heating alone or, better, in the presence of catalysts, this product is dehydrated to methyl vinyl ketone, a mobile lachrymatory fluid which distills in the neighborhood of 80°C and is readily polymerized to light-colored, transparent thermoplastic resins. Iodine is stated to be an effective catalyst for its dehydration[55].

$$CH_3COCH_2CH_2OH \rightarrow CH_3COCH{:}CH_2 + H_2O$$

Methyl vinyl ketone may also be obtained by the vapor-phase reaction of acetone and formaldehyde in the presence of special catalysts in the neighborhood of 300°C[9].

In the presence of excess caustic, the sole reaction product of equimolar quantities of formaldehyde and acetone is an orange-colored amorphous resin whose empirical formula is C_4H_6O[57]. Balz[6] has obtained a molding resin which sets without gas evolution by a controlled alkali-catalyzed condensation of acetone and formaldehyde. Ellis[23] claims that good water-insoluble resins are obtained by refluxing commercial formaldehyde with three parts of acetone in the presence of potassium carbonate, neutralizing, and vacuum-distilling water and solvent. These resins are readily soluble in acetone and xylene and are said to be of value in quick-drying inks.

Liquid, water-soluble resins are obtained by alkaline condensation of two to four molecules of formaldehyde with acetone and are employed commercially for modification of textiles and other miscellaneous applications. Numerous patents cover various processes for the manufacture of these products.

Dimethylolacetone may be prepared by reacting acetone and formaldehyde solution in the presence of alkali at a pH of 10 or greater[25].

$$CH_3COCH_3 + 2CH_2O \text{ (aq, p)} \rightarrow CH_3COCH(CH_2OH)_2$$

When subjected to distillation it decomposes, yielding methylenebutanolone (methylene gamma-ketobutanol)[55]:

$$CH_3COCH(CH_2OH)_2 \rightarrow CH_3COC(CH_2OH):CH_2$$

On hydrogenation, it is reported to yield methyl isopropyl ketone[1]:

$$CH_3COCH(CH_2OH)_2 + 2H_2 \rightarrow CH_3COCH(CH_3)_2 + 2H_2O$$

Although trimethylolacetone has apparently not been isolated, its presence in acetone-formaldehyde reaction mixtures obtained under alkaline conditions is indicated by the isolation of ethyl isopropyl ketone in the products formed when such mixtures are subjected to strenuous hydrogenation[1]:

$$CH_2(CH_2OH)COCH(CH_2OH)_2 + 3H_2 \rightarrow C_2H_5COCH(CH_3)_2 + 3H_2O$$

The product obtained when four mols of formaldehyde are reacted with one of acetone is a cyclic ether boiling at 164 to 165°C, for which the following structures have been suggested:

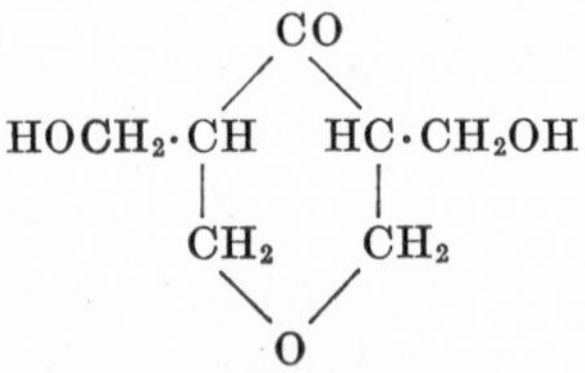

Tetrahydro-dimethylol
gamma-pyrone[56]

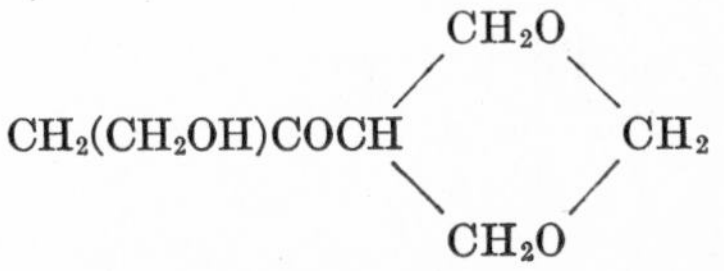

Trimethylol-acetone formal[54]

Joint reactions of formaldehyde with acetone and the salts of ammonia and amines have been investigated by Mannich and co-workers[45,48]. The principal products isolated when these compounds are refluxed together in aqueous solution and then treated with caustic are piperidine derivatives, such as 1,4-dimethyl-3-aceto-4-hydroxy-piperidine, which is obtained in fair yield from methylamine hydrochloride[45]. The reaction mechanism postulated involves the primary formation of an ammono analogue of methylolacetone.

$$CH_3NH_2 + 2CH_2O \text{ (aq)} + 2CH_3COCH_3 \rightarrow CH_3N(CH_2 \cdot CH_2COCH_3)_2 + 2H_2O$$

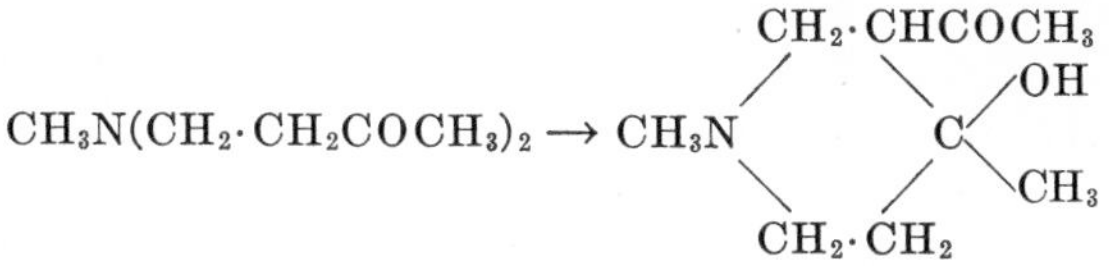

$$CH_3N(CH_2 \cdot CH_2COCH_3)_2 \rightarrow CH_3N \begin{array}{c} CH_2 \cdot CHCOCH_3 \\ \diagup \quad \diagdown \quad \diagup OH \\ \quad\quad C \\ \diagdown \quad \diagup \diagdown CH_3 \\ CH_2 \cdot CH_2 \end{array}$$

In the case of reactions involving ammonium chloride, Mannich and Ritsert[48] isolated the oxime of the primary product, $N(CH_2 \cdot CH_2COCH_3)_3$.

The condensation of monochloracetone with alkaline formaldehyde solution has been studied by Hurd, McPhee, and Morey[35]. The following products were identified:

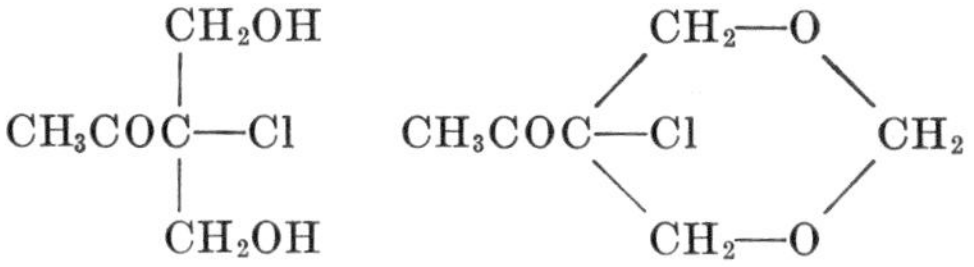

$$\begin{array}{ccc} CH_2OH & & CH_2\text{—}O \\ | & & \diagup \quad\quad \diagdown \\ CH_3COC\text{—}Cl \quad\quad CH_3COC\text{—}Cl & & CH_2 \\ | & & \diagdown \quad\quad \diagup \\ CH_2OH & & CH_2\text{—}O \end{array}$$

Higher Aliphatic Ketones. The reactions of the higher aliphatic ketones with formaldehyde under alkaline conditions are less complicated than those obtained with acetone and may, in general, be more readily controlled.

In the case of methyl ethyl ketone, mono- and dimethylol derivatives are readily formed[51]:

$$CH_3COCH_2CH_3 + CH_2O \text{ (aq)} \rightarrow CH_3COCH(CH_2OH)CH_3$$

Monomethylol Methyl Ethyl Ketone
(Beta-methyl-gamma-ketobutanol)

$$CH_3COCH_2CH_3 + 2CH_2O \text{ (aq)} \rightarrow CH_3COC(CH_2OH)_2CH_3$$

Dimethylol Methyl Ethyl Ketone

As will be observed, the alpha-hydrogen atoms of the ethyl group are attacked. The monomethylol derivative is a colorless liquid distilling at 193°C[55], whereas the dimethylol derivative is a crystalline solid melting at approximately 66°C.[18]

On distillation with a small proportion of iodine, oxalic acid, sulfuric acid, or other acidic catalysts, monomethylol methyl ethyl ketone is readily dehydrated to methyl isopropenyl ketone (b.p. 98°C) which yields light-colored resins on polymerization[55] and other resins by chemical reaction[78].

A detailed account of the effect of reaction variables on the preparation of monomethylol methyl ethyl ketone and methyl isopropenyl ketone from methyl ethyl ketone and formaldehyde is given by Landau and Irany[41a]. Methyl iospropenyl ketone may also be obtained by the gas-phase condensation of formaldehyde and methyl ethyl ketone[9].

High molecular weight ketones, such as 2-methyl-7-ethyl undecanone-4, although insoluble in water, react readily with paraformaldehyde in the presence of alcoholic caustic, giving monomethylol derivatives[62].

Unsaturated aliphatic ketones apparently react in the same manner as those which are fully saturated. Methyl vinyl ketone gives the mono-methylol derivative, methylenebutanolone, when reacted with formaldehyde solution in the presence of potassium carbonate[30]:

$$CH_3COCH:CH_2 + CH_2O \text{ (aq)} \rightarrow CH_3COC(CH_2OH):CH_2$$

Cyclic Ketones. The reactions of cyclopentanone and cyclohexanone with hydrated lime and excess formaldehyde lead to the production of polyhydroxy compounds analogous to pentaerythritol; in these, all the alpha-hydrogen atoms of the original ketones are replaced by methylol radicals and the carbonyl groups are reduced by the crossed Cannizzaro reaction[46,83]. The following products are obtained respectively:

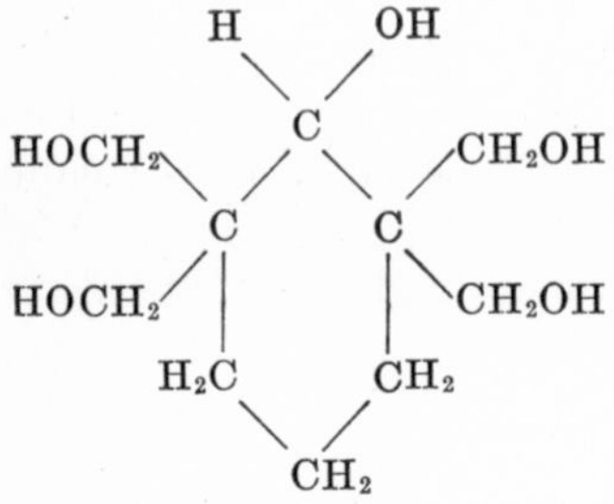

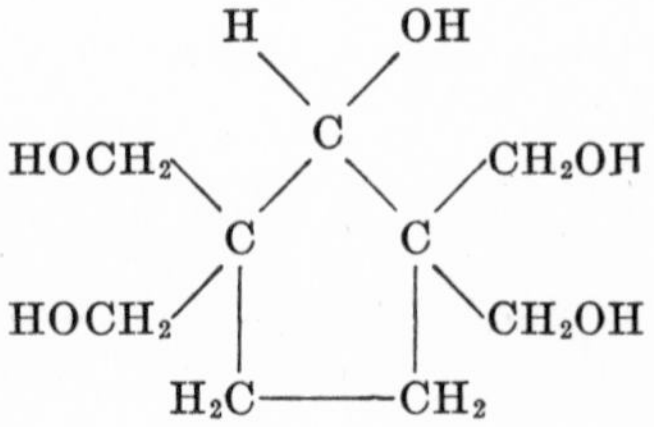

Gault and Skoda[31] describe the methylolation of cyclopentanone to produce the intermediate products, 2-methylolcyclopentanone, 2,2-di-methylolcyclopentanone and 2,2,5,5-tetramethylolcyclopentanone. Ray[64] patented a process for preparing tetramethylolcyclopentanone involving condensation of cyclopentanone with aqueous formaldehyde at a pH of 9 to 10. A temperature of 115 to 125°F is maintained with a liquid hydrocarbon coolant boiling in the range indicated.

Cyclohexanone-formaldehyde resins may be obtained by condensing 1.2 to 1.9 mols of formaldehyde with one mol cyclohexanone in aqueous solution using caustic alkali as a reaction catalyst[36]. These resins are claimed to have good color stability and may be combined with cellulose esters in coating compositions.

Other reaction products of cyclohexanone and formaldehyde isolated by Mannich and Brose[46] include the mono- and tetramethylol ketones. The reaction mechanisms are apparently identical with those involving acetone and acetaldehyde. Methylene cyclohexanone was obtained by Dimroth[20] by allowing one mol of formaldehyde to react with a slight excess of cyclohexanone in a dilute caustic solution for four days at room temperature.

Menthone, carvone, and camphor do not react with formaldehyde in the presence of lime[46].

DiKetones. Acetylacetone condenses with formaldehyde in aqueous solution to give methylenebis-acetylacetone[63]. This compound and its two methylol derivatives shown below[39, 69] are also obtained when the reaction is carried out in the presence of organic bases such as pyridine and diethylamine.

$$CH_3COCHCOCH_3 \qquad CH_3COC(CH_2OH)COCH_3 \qquad CH_3COC(CH_2OH)COCH_3$$
$$\mathord{|} \qquad\qquad\qquad \mathord{|} \qquad\qquad\qquad \mathord{|}$$
$$CH_2 \qquad\qquad\qquad CH_2 \qquad\qquad\qquad CH_2$$
$$\mathord{|} \qquad\qquad\qquad \mathord{|} \qquad\qquad\qquad \mathord{|}$$
$$CH_3COCHCOCH_3 \qquad CH_3COCHCOCH_3 \qquad CH_3COC(CH_2OH)COCH_3$$

Aryl Aliphatic Ketones. Reactions of formaldehyde with aromatic ketones appear to proceed with greater difficulty than those involving simple aliphatic ketones, and when pushed by the use of heat and strong alkalies give only complex products. However, according to Fuson, Ross and McKeever[28], the formal of dimethylolacetophenone can be obtained in good yield by the action of paraformaldehyde on acetophenone in methanol solution when potassium carbonate is employed as a catalyst if the reaction is allowed to proceed at room temperature for seven days:

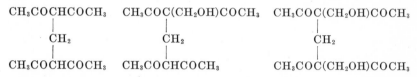

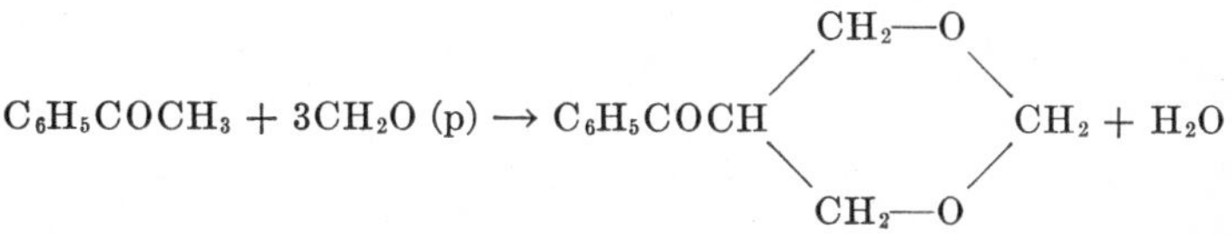

$$C_6H_5COCH_3 + 3CH_2O \text{ (p)} \rightarrow C_6H_5COCH \begin{smallmatrix} CH_2\!-\!O \\ \diagup \qquad \diagdown \\ \qquad\qquad CH_2 \\ \diagdown \qquad \diagup \\ CH_2\!-\!O \end{smallmatrix} + H_2O$$

This reaction is unusual in that formals are normally obtained only as a result of acid-catalyzed reactions. However, it is not entirely unique in this respect (pages 246 and 282).

On heating formaldehyde with acetophenone in the presence of ammonium chloride, Morgan, Megson and Pepper[55] obtained a complex product which they identified as trimethylolbis-acetophenone (m.p. 156°C) together with beta-benzoyl ethanol (monomethylolacetophenone).

With propiophenon, Fuson, Ross and McKeever[28] obtained beta-benzoyl propyl alcohol (monomethylolpropiophenone) by the same technique as that employed in the previously mentioned reaction of paraformaldehyde with acetophenone. This keto-alcohol shows no tendency to lose water under ordinary circumstances, but dehydrates on treatment with cold concentrated sulfuric acid to give the cyclic alpha-methyl hydrindone instead of the expected isopropenyl ketone.

$$\overset{\text{COCHCH}_3}{\underset{\text{CH}_2\text{OH}}{\bigcirc}} \rightarrow \quad \bigcirc \underset{\overset{|}{\text{C}}}{\overset{\text{CO}}{\underset{\text{H}_2}{\big|}}} \underset{\text{CHCH}_3}{} + \text{H}_2\text{O}$$

By heating propiophenone and formaldehyde in the presence of potassium carbonate, Manta[50] prepared *sym-β,β'*-dihydroxy-t-butyl benzyl alcohol in which the original carbonyl group has been reduced:

$$2\text{C}_6\text{H}_5\text{COCH}_2\text{CH}_3 + 6\text{CH}_2\text{O} \text{ (aq)} + \text{K}_2\text{CO}_3 + \text{H}_2\text{O} \rightarrow$$

$$2\text{C}_6\text{H}_5\text{CHOH(CH}_2\text{OH)}_2\text{CH}_3 + 2\text{KOOCH} + \text{CO}_2$$

With isobutyrophenone, Manta[50] obtained a molecular complex made up of two aliphatic hydroxy derivatives of benzyl alcohol.

Acetophenone, *m*-nitroacetophenone and propiophenone form sulfomethyl derivatives on reaction with aqueous formaldehyde and sodium sulfite[73].

References

1. Anon., *Chem. Age*, **32,** 501 (1935).
2. Anon., *J. Soc. Chem. Ind.*, **59,** 474 (1940).
3. Apel, M., and Tollens, B., *Ber.*, **27,** 1087 (1894).
4. Apel, M., and Tollens, B., *Ann.*, **289,** 38 (1896).
5. Arvin, J. A., (to E. I. du Pont de Nemours & Co., Inc.), U. S. Patent 2,029,851 (1936).
6. Balz, E. H., (to Plaskon Co.), U. S. Patent 2,237,325 (1941).
7. Bludworth, J. E., (to C. Dreyfuss), Canadian Patent 435,051 (1946).
8. Böhme, A. G., H. T., British Patent 280,511 (1928).
9. Brant, J. H., and Hasche, R. L., (to Eastman Kodak Co.,), U. S. Patent 2,245,567 (1941).
10. Bruson, H. A., and Niederhauser, W. D., (to Resinous Products and Chemical Co.), U. S. Patent 2,418,290 (1947).
11. Brubaker, M. M., and Jacobson, R. A., (to E. I. du Pont de Nemours & Co., Inc.), U. S. Patent 2,292,926 (1942).
12. Bruson, H. A., (to Rohm & Haas Co.), U. S. Patent 1,835,203 (1932).
13. Burrell, H., *Ind. Eng. Chem.*, **37,** 86–89 (1945).
14. Burke, C. E., (to E. I. du Pont de Nemours & Co., Inc.), U. S. Patent 1,716,110 (1929).
15. Corbellini, A., *et al.*, *Giorn. Chim. Ind. Appl.*, **15,** 53–56 (1933); *Brit. Chem. Abs., A,* **1933,** 486.
15a. Cox, R. F. B., (to Hercules Powder Co.), U. S. Patent 2,329,514 (1943).
16. Cryer, J., Owens, J. C., and Jui Fu Lee, R., (to the Sherwin-Williams Co.), U. S. Patent 2,534,191 (1950).
17. Davidson, D., and Bogert, M. T., *J. Am. Chem. Soc.*, **57,** 905 (1935).
18. Décombe, J., *Compt. rend.*, **203,** 1079 (1936).
19. Deutsche Gold-und Silber-Scheideanstalt vorm. Roessler, French Patent 835,834 (1939).

20. Dimroth, K., Resin, K., and Zetzsch, H., *Ber.*, **73**, 1399–1409 (1940).
21. Dreyfuss, H., and Drewitt, J. G. N., (to British Celanese Ltd.), U. S. Patent 2,387,933 (1945).
22. Ebert, L., *Ber.*, **64**, 114 (1931).
23. Ellis, C., (to Ellis Laboratories, Inc.), U. S. Patent 2,292,748 (1942).
24. Fitzky, W., (to General Aniline and Film Corp.), U. S. Patent 2,275,586 (1942).
25. Flemming, W., and von der Horst, H. D., (to I. G. Farbenindustrie A. G.), German Patent 544,887 (1930); U. S. Patent 1,955,060 (1934).
26. Friedrich, W., and Brün, W., *Ber.*, **63**, 2681 (1930).
27. Fujii, S., *Rev. Phys. Chem. Japan, Shinkichi Horiba Commen.*, **1946**, 153–157; *C. A.*, **44**, 1906 (1950).
28. Fuson, R. C., Ross, W. E., and McKeever, C. H., *J. Am. Chem. Soc.*, **60**, 2935 (1938).
29. Gallagher, M., and Hasche, R. L., (to Eastman Kodak Co.), U. S. Patent 2,246,-037 (1941).
30. Gault, H., and Germann, L. A., *Compt. rend.*, **203**, 514–516 (1936).
31. Gault, H., and Skoda, J., *Bull. Soc. Chim.*, **1946**, 308–316, 316–327.
32. Gilman, H., "Organic Syntheses," Collective *Vol. I*, pp. 417–419, New York, John Wiley & Sons, Inc. (1932).
33. Gresham, W. F., (to E. I. du Pont de Nemours & Co., Inc.), U. S. Patent 2,549,457 (1951).
34. Hosaeus, H., *Ann.*, **276**, 76 (1893).
35. Hurd, C. D., McPhee, W. D., and Morey, G. H., *J. Am. Chem. Soc.*, **70**, 329 (1948).
36. Hurst, D. A., and Miller, W. J., Jr., (to Allied Chem. and Dye Corp.), U. S. Patents 2,540,885–6 (1951).
37. I. G. Farbenindustrie A. G., French Patent 847,370 (1939).
38. Jackson, H. J., and Jones, G. G., (to Imperial Chemical Industries, Ltd.), U. S. Patent 2,562,102 (1951).
39. Knoevenagel, E., *Ber.*, **36**, 2136–2180 (1903).
40. Koch, K., and Zerner, T., *Monatsh.*, **22**, 443–459 (1901).
41. Kuzin, A., *J. Gen. Chem. (U.S.S.R.)*, **5**, 1527–1529 (1935).
41a. Landau, E. F., and Irany, E. P., *J. Org. Chem.*, **12**, 422–425 (1947).
42. Lonze Elektrizitätswerke und Chemische Fabrik A. G., British Patent 440,691 (1935).
43. McLeod, A. G., *Am. Chem. J.*, **37**, 20–50 (1907).
44. Malowan, S., *Plastic Products*, **10**, 147 (1934).
45. Mannich, C., and Ball, G., *Arch. pharm.*, **264**, 65–77 (1926); *C. A.*, **20**, 1808–1809 (1926).
46. Mannich, C., and Brose, W., *Ber.*, **56**, 833–844 (1923).
47. Mannich, C., Lesser, B., and Silten, F., *Ber.*, **65**, 378–385 (1932).
48. Mannich, C., and Ritsert, K., *Arch. pharm.*, **264**, 164–167 (1926); *C. A.*, **20**, 1808 (1926).
49. Mannich, C., and Wieder, H., *Ber.*, **65**, 385–390 (1932).
50. Manta, J., *J. prakt. Chem.*, **142**, 11 (1935).
51. Marling, G., and Kohler, H., (to Farbenfabriken vormals. F. Bayer & Co.), U. S. Patent 981,668 (1911).
52. *Ibid.*, U. S. Patent 989,993 (1911).
53. Molinari, H., *Giorn. Chim. Ind. Appl.*, **15**, 325–328 (1933); *C. A.*, **28**, 104 (1934).
54. Morgan, G., and Griffith, C. F., *J. Chem. Soc.*, **1937**, 841.
55. Morgan, G., and Megson, N. J. L., and Pepper, K. W., *Chem. & Ind.*, **57**, 885–891 (1938).

56. Morgan, G. T., and Holmes, E. L., *J. Chem. Soc.*, **1932**, 2667.
57. Müller, A., *Ber.*, **54**, 1142–1148 (1921).
58. Nagel, K., and Koenig, F., (to Chemical Marketing Co.), U. S. Patent 2,270,889 (1942).
59. Naujoks, E., (to Deutsche Gold- und Silber-Scheideanstalt vorm. Roessler), U. S. Patent 2,004,010 (1935).
60. Owens, G. R., (to Monsanto Chemical Co.), U. S. Patent 2,372,602 (1945).
61. Peterson, T. R., U. S. Patent 2,011,589 (1935).
62. Quattlebaum, W. F., Jr., (to Union Carbide and Carbon Corp.), U. S. Patent 2,064,564 (1936).
63. Rabe, P., and Elze, F., *Ann.*, **323**, 83–112 (1902).
64. Ray, Gardner C., (to Phillips Petroleum Co.), U. S. Patent 2,500,570 (1950).
65. Remensnyder, J. P., Bowman, P. E., and Barth, R. H., U. S. Patent 2,441,944 (1948).
66. Rheinische-Westfalische Springstoff, A. G., German Patent 298,932 (1919).
67. Roach, J. R., Wittcoff, H., and Miller, S. E., *J. Am. Chem. Soc.*, **69**, 2651 (1947).
68. Rose, W. G., and Haller, H. L. J., U. S. Patent 2,140,481 (1938).
69. Scholtz, M., *Ber.*, **30**, 2295–2299 (1897).
70. Stepanov, A. V., and Shchukina, M., *J. Russ. Phys. Chem. Soc.*, **58**, 840–848 (1926); *C. A.*, **21**, 1094 (1927).
71. Spurlin, H. M., (to Hercules Powder Co.), U. S. Patent 2,364,925 (1944).
72. Stingley, D. V., *Ind. Eng. Chem.*, **32**, 1217–1220 (1940).
73. Suter, C. M., Bair, R. K., and Bordwell, F. G., *J. Org. Chem.*, **10**, 470–478 (1945).
74. Tollens, B., and Wiegand, P., *Ann.*, **265**, 316 (1891).
75. Tribit, S. W., (to U. S. Industrial Chemicals), U. S. Patent 2,413,803 (1947).
76. Van House, R. W., (to Parke Davis and Co.), U. S. Patent 2,443,334 (1948).
77. van Marle, C. M., and Tollens, B., *Ber.*, **36**, 1342 (1903).
78. Vierling, K., and Hopff, H., (to I. G. Farbenindustrie A. G.), U. S. Patent 2,201,750 (1940).
79. Wagner, H., (to Chemical Marketing Co.), U. S. Patent 2,288,306 (1942).
80. Walker, J. F., (to E. I. du Pont de Nemours & Co., Inc.), U. S. Patent 2,400,724 (1946).
81. Walker, J. F., and Turnbull, N., (to E. I. du Pont de Nemours & Co., Inc.), U. S. Patent 2,135,063 (1938).
82. Wessely, L., *Monatsh.*, **21**, 216–234 (1900).
83. Wittcoff, H., (to General Mills Inc.), U. S. Patent 2,462,031 (1949).
84. Wyler, J. A., (to Trojan Powder Co.), U. S. Patent 2,152,371 (1939).
85. *Ibid.*, U. S. Patent 2,170,624 (1940).
86. *Ibid.*, U. S. Patent 2,206,379 (1940).
87. *Ibid.*, U. S. Patent 2,240,734 (1941).
88. *Ibid.*, U. S. Patent 2,486,774 (1949).
89. *Ibid.*, U. S. Patent 2,490,567 (1949).
90. *Ibid.*, U. S. Patent 2,532,252 (1950).
91. Wyler, J. A., and Wernett, E. A., (to Trojan Powder Co.), U. S. Patent 2,299,048 (1942).
92. Wyler, J. A., *Paint Ind. Mag.*, **61**, 235 (1946).

CHAPTER 12

REACTIONS OF FORMALDEHYDE WITH PHENOLS

Phenol-formaldehyde resins are the complex final products of many reactions involving formaldehyde and phenolic compounds. Because of their immense industrial importance, the art of preparing these bodies and converting them to useful articles of commerce has advanced rapidly since their potential value was first realized in the early part of the twentieth century. Exact scientific knowledge of phenol-formaldehyde reactions and the chemical structure of the products obtained has been a subject of slower growth. However, in recent years, more and more emphasis has been placed on fundamental studies. As a result, although many unsolved problems and controversial items still exist, worthwhile advances have been made.

Of particular importance in this field of chemistry are the simple reactions which are most amenable to exact study. These reactions are in some cases the precursors of more complex condensations. In all cases they illustrate the fundamental facts of phenol-formaldehyde chemistry and thus afford a practical approach to the more involved problems of resin structure. Since our purpose is to coordinate phenol-formaldehyde reactions with the broad field of formaldehyde chemistry, special emphasis will be placed on these less complicated reactions which are often neglected because of the importance of the intricate condensations that result in resin production.

Historical. The general character of phenol-aldehyde reactions was probably first recognized by Baeyer[7] in 1872. Baeyer apparently produced his first phenol-formaldehyde type resin by reacting phenol with methylene acetate in the presence of hydrochloric acid. This acid hydrolyzed the methylene ester, with liberation of formaldehyde, and catalyzed the subsequent condensation of this product with phenol. Following commercial production of formaldehyde in 1889, Kleeberg[51], at the suggestion of Emil Fisher, investigated its reactions with various phenols and published the results in 1891. Tollens also carried out studies in this field at approximately the same period, since in 1892 Hosaeus[46] refers to unpublished work of this investigator dealing with the action of formaldehyde on phenol, resorcinol, pyrogallol, and phloroglucinol in the presence of acid catalysts. Commercial applications of phenol-formaldehyde resins were developed in the first decade of the twentieth century. In the latter part of this decade, Baekeland discovered a technique for molding these resins.

This work, together with that of Redman and other investigators, led to the development of the present phenolic resin industry.

Fundamental Characteristics of Phenol-Formaldehyde Reactions

In general, phenols react readily with formaldehyde in the presence of both alkaline and acidic catalysts, producing a variety of substances ranging from simple methylol and methylene derivatives to the complex resins in which large numbers of phenolic molecules are linked together by methylene groups. These reactions are shared by formaldehyde polymers, such as paraformaldehyde and the linear polyoxymethylenes, which depolymerize under reaction conditions. Trioxane (symmetrical-trioxymethylene) is unique in that it does not react under neutral or alkaline conditions. When trioxane crystals are mixed with crystalline phenol at ordinary temperatures, liquefaction takes place, heat is absorbed, and a colorless solution is obtained. Reaction takes place only when a strongly acidic depolymerization catalyst is added to the phenol-trioxane solution[85, 86].

In the absence of added catalysts, anhydrous formaldehyde and paraformaldehyde dissolve in molten phenol without apparent reaction to give clear, colorless solutions which smell strongly of formaldehyde. In such solutions, it is probable that the dissolved formaldehyde is present in the form of hemiformals such as $C_6H_5OCH_2OH$ and $C_6H_5OCH_2OCH_2OH$. The analogy of these solutions to alcoholic formaldehyde solutions is clearly apparent (pages 61–63). According to Reychler[73], a small percentage of sodium phenate catalyzes the solution of linear formaldehyde polymers in phenol, just as sodium alcoholates catalyze solution in methanol, ethanol, and other alcohols. That hemiformals are produced is also indicated by the isolation of methyl phenyl formal from an acid-catalyzed reaction of phenol with formaldehyde solution containing methanol[20].

$$\text{\big\langle\ \big\rangle}-\text{OH} \xrightarrow{\ \text{CH}_2\text{O(aq)}\ } \text{\big\langle\ \big\rangle}-\text{OCH}_2\text{OH} \xrightarrow{\ \text{CH}_3\text{OH}\ }$$

$$\text{\big\langle\ \big\rangle}\text{OCH}_2\text{OCH}_3 + \text{H}_2\text{O}$$

The reactivity of the ortho- and para- hydrogen atoms in the phenolic nucleus is the predominant factor in phenol-formaldehyde reactions. However, the meta-hydrogen atoms may become involved when other positions are blocked (pages 257–258). Nuclear derivatives are the principal reaction products and formals, such as are obtained from alcohols and formaldehyde, are seldom isolated.

The two most important phenol-formaldehyde reactions are: (a) the formation of nuclear methylol phenols or phenol alcohols, as they are commonly designated, and (b) the formation of polynuclear methylene derivatives.

The production of phenol alcohols, which is the primary reaction, may be illustrated by the formation of ortho-methylolphenol (saligenin). As indicated, this reaction probably involves a rearrangement of the hemiformal, whose existence in solutions containing formaldehyde and phenol has been previously postulated.

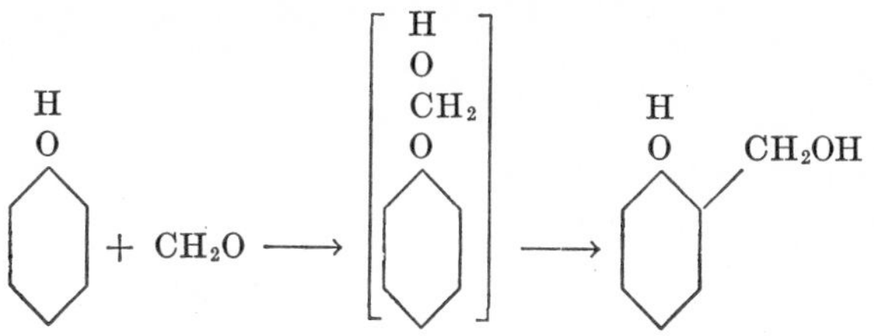

Nuclear methylene derivatives are apparently formed by a secondary reaction in which a nuclear methylol group reacts with an active hydrogen atom on the nucleus of another phenolic molecule, *e.g.*:

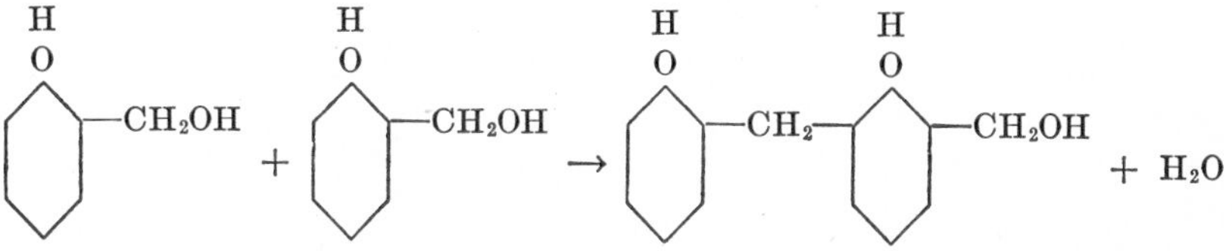

Resin formation results from reactions of this type in which a number of mono- or polymethylolphenol molecules condense to form complex polymethylene derivatives. Under some circumstances it is possible that resin production may also involve phenolic hydroxyl groups, giving rise to methylene ether bridges:

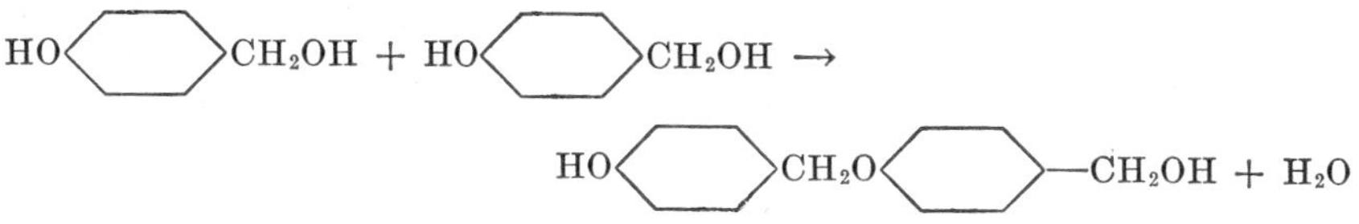

Cyclic formals in which only one phenolic nucleus is involved may be obtained in some instances:

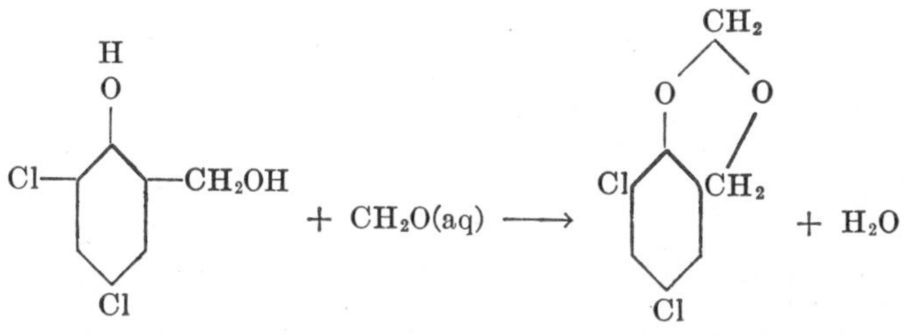

The individual characteristics of reactions involving various phenolic compounds are determined in large part by the structure of the phenol and the nature of the catalyst.

Influence of Phenolic Structure. The active hydrogen atoms in the phenolic nucleus occupy the 2, 4 and 6 positions with reference to the hydroxyl group. Substituent groups in these positions have a direct and restricting influence on reactions, since they limit the number of methylol and methylene groups which may enter the nucleus. Substituents also have an indirect effect, depending on the degree to which they activate or inactivate the remaining active hydrogen atoms. Indirect effects are also produced by substituents in the meta positions, 3 and 5.

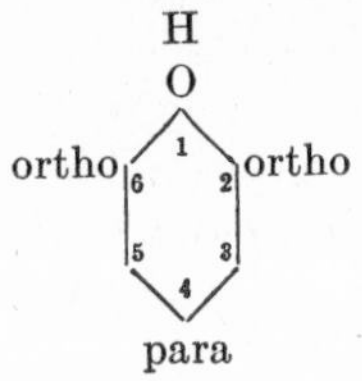

In general, phenols in which the 2, 4 and 6 positions are unblocked by substituents react most vehemently with formaldehyde. Phenols containing one substituent in an active position tend to be less reactive, and those containing two such substituents are still less reactive. However, when all three active positions are blocked, reaction may take place at the meta position under some circumstances. Finn and Musty[35] have demonstrated that the crystalline compound reported by Megson[58], when mesitol (2,4,6-trimethylphenol) is reacted with formaldehyde under strongly acidic conditions, is a meta-linked diphenyl methane.

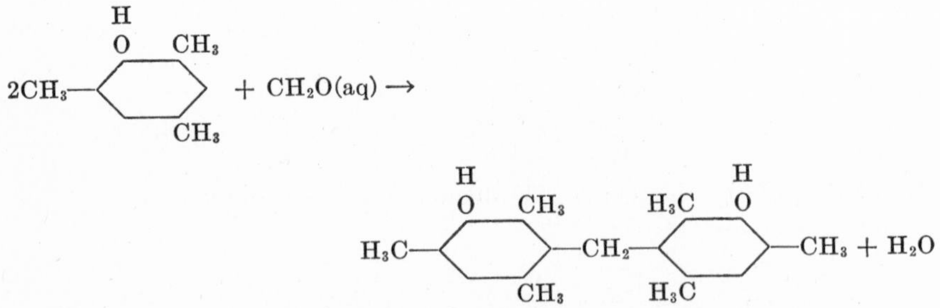

It is believed that substitution in the meta position is due to the cumulative effect of the ortho and para directive influence of the methyl groups. This reaction does not take place in alkaline solution.

The effect of substituent groups on reaction velocities is admirably illustrated by Sprung's study[76] of the rates at which phenol and methyl-substituted phenols react with paraformaldehyde in the presence of an alkaline catalyst (triethanolamine) at 98°C. Phenols included in this study are arranged below in the order of decreasing reactivity:

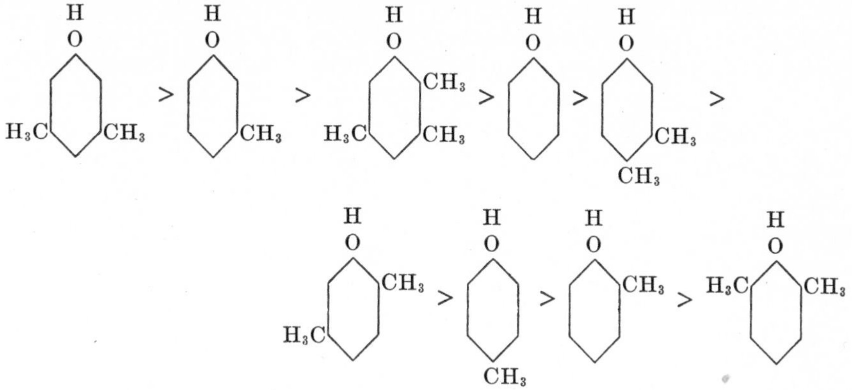

The total spread in reaction rates for this series was fifty-fold.

Relative reactivity values vary as the reaction progresses and the specific catalyst employed may be a critical factor in some instances. Accordingly, the results reported by various investigators differ in some respects. Sprung's data is based on initial rate constants. Holmes and Megson[45] obtained the following order of reactivity by comparing the time required for resin precipitation when commercial formaldehyde solution was employed with trimethylamine as catalyst: 3,5-xylenol, 3-cresol, 2,5-xylenol, 2,4-xylenol, 3,4-xylenol, 2-cresol, phenol and 4-cresol. The decreasing order of reactivity for five phenols studied by Muller and Muller[66] was 3,5-xylenol, 3-cresol, phenol, 2-cresol and 4-cresol.

Effect of Catalyst. From a behavioristic standpoint, reaction catalysts may be classified according to whether they are alkaline or acidic in nature. Although different acids and bases may vary somewhat in their action, such variations are usually subordinate. Hydrogen and hydroxyl ions may thus be regarded as the principal factors in reaction catalysis. As with other formaldehyde reactions, alkalies tend to favor the formation of methylol derivatives and acids tend to favor the formation of methylene derivatives.

Under alkaline conditions, the rate at which the primary mono- and polymethylol derivatives are formed is usually more rapid than the subsequent condensations which result in the formation of methylene derivatives. The rate at which available formaldehyde is consumed is also more rapid under these circumstances.

Under acid conditions, the situation is reversed in that the conversion of methylol derivatives to methylene derivatives is usually more rapid than their formation. Granger[39] suggests that this is because monomethylol derivatives are converted to methylene derivatives as soon as they are formed. As a result, the introduction of more than one methylol group in the simple phenol is prevented and the polynuclear phenols thus obtained, as would be expected, react less readily with additional formaldehyde.

A further difference between alkaline and acidic reactions, clearly demonstrated by Granger[39], lies in the fact that after a small concentration of alkali has been reached, additional alkali has little or no effect on reaction velocity, whereas when acid is employed the reaction rate increases continuously with acid concentration. As a result, acid condensations may attain a violence which cannot be equalled under alkaline conditions.

Although these generalizations furnish a satisfactory picture of the effect of alkali and acid on most phenol-formaldehyde reactions, the structure of the phenols involved is sometimes of paramount importance and may result in definite exceptions to general rules. Ortho-nitrophenol forms methylol derivatives under acidic conditions and beta-naphthol gives a quantitative yield of methylene bis-beta-naphthol in the presence of alkaline catalysts.

Other factors of importance in these reactions will be brought out in the following pages. For a detailed study of phenol-aldehyde resins, the reader is referred to "Phenoplasts" by Carswell[25], "Chemistry of Synthetic Plastics" by Wakeman[84], "The Chemistry of Synthetic Resins" by Ellis[30], and other texts devoted to resin technology.

Methylol Derivatives

Phenol Alcohols. As has been previously pointed out, the simplest definite reaction products of formaldehyde with phenols are the methylolphenols or phenol alcohols. Although some of these derivatives are too reactive to be isolated, a number have been obtained as pure crystalline products. In some respects they are analogous to the methylol derivatives of aldehydes and ketones, a similarity which is readily demonstrated when the keto- or ortho- and para-quinodal forms of the phenolic nucleus are designated in the structural formula. The mechanism of their formation from the primary phenolic hemiformals may involve tautomeric rearrangements of the sort indicated on page 241.

The preparation and isolation of phenol alcohols was first accomplished in 1894 by two independent investigators, Manasse[56] and Lederer[54], who employed processes involving basic catalysis. Although acid catalysts, as previously stated, usually accelerate the more complex condensations, it has been definitely demonstrated that phenol alcohols are also produced in the early stages of the normal acid reaction.[39] This fact supports the theory that methylolphenols are the primary products of all phenol-formaldehyde condensations.

The procedure of Lederer[54] involved the use of heat in the presence of a small amount of base, whereas Manasse[56] favored the employment of an approximately equimolar amount of strong alkali and allowed the reaction to take place at ordinary temperature. The latter method has been generally adopted as the standard procedure for phenol alcohol preparation.

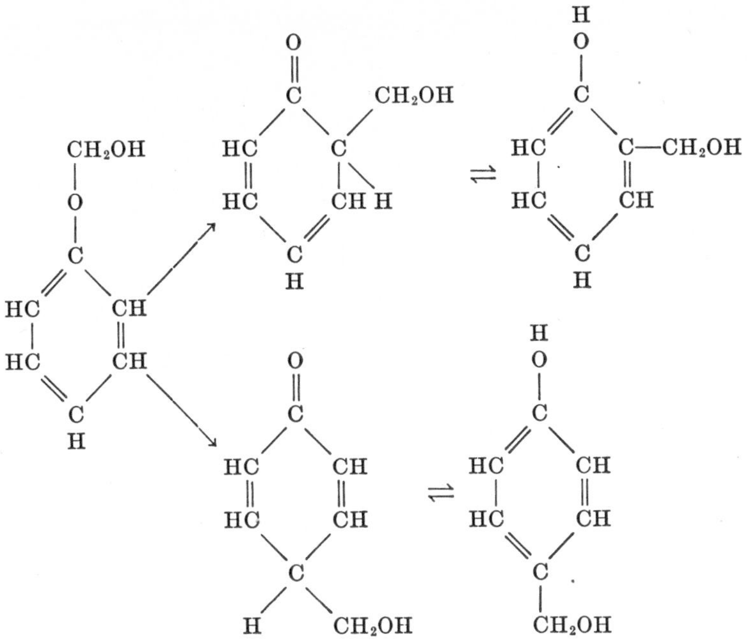

Saligenin and Methylolphenols. Saligenin (ortho-hydroxy benzyl alcohol, ortho-methylolphenol) and homosaligenin (para-hydroxy benzyl alcohol, para-methylolphenol) are the simplest phenol alcohols produced by the reaction of phenol with formaldehyde. Meta-methylolphenol is known but is apparently not obtained in phenol-formaldehyde reactions. Saligenin was first described in 1843 by Piria[69], who isolated it from the natural glucoside, salicin, long before the reactions of phenol and formaldehyde had been explored. On heating alone or in the presence of an acid it is converted to resinous products. Piria[70] obtained a resin of this sort by warming it with hydrochloric acid. This resin was named saliretin by Piria, a name which has since been applied generally to the primary condensation products of phenol alcohols. Homosaligenin was probably first prepared by Biedermann[16] in 1886 by reducing para-hydroxy benzaldehyde with sodium amalgam. The pure product is reported to melt at approximately 125°C[2]. It is apparently less stable than saligenin and resinifies readily on heating.

Saligenin is synthesized by the method of Manasse. Phenol is dissolved in somewhat more than an equivalent amount of dilute sodium hydroxide and a molecular quantity of commercial 37 per cent formaldehyde is added.

The reaction mixture is then allowed to stand at room temperature for one day or more until the odor of formaldehyde has disappeared. After this, the solution is neutralized with acetic acid and extracted with ether, which removes the phenol alcohols and unreacted phenol. The latter is distilled with steam, leaving a mixture consisting principally of saligenin and the isomeric para-methylolphenol. Saligenin may be isolated from an anhydrous mixture of these two alcohols by extraction with benzene at 50°C. According to Sprung[77], pure saligenin melts at approximately 84°C. Bender and co-workers report a melting point of 86°C.[15]

According to Granger[37], only two-thirds of the phenol reacts in the above preparation, although all the aldehyde is consumed. Polymethylolphenols are formed as by-products, and remain in the reaction mixture from which they are not extracted, since they are more soluble in water than in ether. By working up this mixture, Granger isolated three liquid phenol alcohol fractions believed to represent the three isomeric methylolphenols. Bender and co-workers[15] have recently reported the melting points of these products when isolated in the crystalline state.

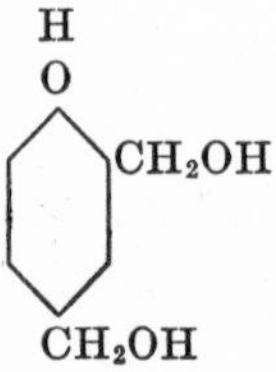

| 2,4-Dimeth-ylolphenol (m.p. 92–93°C, w. decompn.) | 2,6-Dimethylol-phenol (m.p. 96–98°C, w. decompn.) | 2,4,6-Trimethylol-phenol (m.p. approx. 80°C, very unstable) |

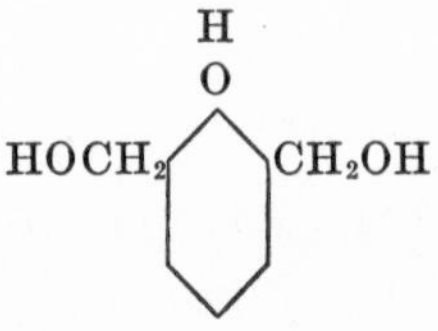

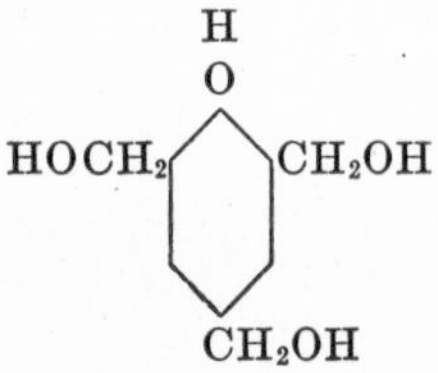

Martin[57] reports the isolation of the pure sodium salt of trimethylolphenol as a colorless granular precipitate when sodium phenate is reacted with a 40 volume per cent formaldehyde solution at 25 to 40°C and the reaction mixture is added to excess methanol. On neutralization, this product loses formaldehyde and water giving a tetramethylol dihydroxydiphenyl methane.

Cresol and Xylenol Derivatives. Cresols also give the expected methylol derivatives with formaldehyde. In the case of these compounds, pure crystalline dimethylol derivatives have been isolated by reacting ortho- and para-cresol with formaldehyde and alkali. The dialcohol, 2,6-dimethylol-p-cresol, melts at 135°C and was made by Ullmann and Brittner[82] by adding a solution of 108 g p-cresol in 200 cc water containing 50 g sodium hydroxide to 215 g 35 per cent formaldehyde. After four days the reaction mixture became crystalline and gave a 91 per cent yield of dimethylolcresol after acidification with dilute acetic acid. The 2,4-dimethylol-*o*-cresol was prepared by Granger[37], following a similar technique, and after recyrstallization from chloroform melted at 94°C.

The monoalcohols of ortho-cresol are difficult to prepare, since on reaction of equimolar quantities of ortho-cresol and formaldehyde in the presence of alkali, the dialcohol is obtained as the principal product and a considerable quantity of unreacted cresol is recovered[37]. A monoalcohol melting at 40°C is described by Lederer[54] and another melting at 87°C is described by Bayer & Co.[10]. Granger[37] believes the latter is probably the dialcohol.

Meta-cresol, which has three positions capable of methylolation, is highly reactive with formaldehyde, and the alcohols formed resinify rapidly even under alkaline conditions. As a result, it is extremely difficult to isolate pure phenol alcohols. However, crystalline products believed to be monoalcohols and melting at 105, 110, 117, and 122°C have been reported[10, 54, 56]. The difficulties of isolating pure products are considerable, since three monomethylol derivatives, three dimethylol derivatives and one trimethylol derivative are possible.

Xylenols readily form monoalcohols, as demonstrated by the work of Bamberger[9], and of Megson and Drummond[60]. However, these phenols are slow to react with the full quantity of formaldehyde, as determined theoretically by the number of free ortho- and para-positions available. This may be due to steric hindrance[37].

Miscellaneous Substituted Phenols. Phenols containing other substituents than methyl groups and containing free ortho and/or para positions form phenol alcohol derivatives with varying degrees of ease when subjected to reaction with formaldehyde in the presence of alkali. Numerous examples of isolable products obtained by this procedure, such as the 2,6-dimethylol derivative of *p*-benzylphenol[83] and the methylol derivatives of ortho-chlorophenol, *viz.*, 2,4-dimethylol-1-chlorophenol and 4-methylol-1-chlorophenol[41], etc., are described.

Ortho-nitrophenol, as previously pointed out, presents a notable exception to the general rules of reaction, in that its methylol derivatives can be prepared in acidic media. The production of small quantities of 2-nitro-4-methylolphenol by the reaction of formaldehyde with ortho-nitrophenol in the presence of hydrochloric acid was first noted by Behn and Stoermer[79]. Fishman[26] later demonstrated that in order to obtain the best yields of methylol derivatives: (a) a high concentration of hydrochloric acid is essential, (b) only theoretical amounts of formaldehyde should be used, and (c) the reaction mixture should not be heated longer than six hours. Under these conditions the nitrophenol reacts only partially, but a minimum of by-products is formed. By reaction of 200 g formaldehyde solution (37 per cent by weight) with 300 g *o*-nitrophenol in 1500 g concentrated hydrochloric acid, this investigator obtained 63 g crude isomeric mono-methylol derivatives containing 75 per cent 2-nitro-4-methylolphenol (m.p. 97°C) and 25 per cent 2-nitro-6-methylolphenol (m.p. 75°C). These can be separated by making use of the fact that the potassium derivative of the 2-4 isomer is soluble in alcohol, whereas the potassium derivative of the 2-6 isomer is insoluble.

According to Borsche and Berkhout[19] and other investigators[26], para-nitrophenol reacts with two molecular proportions of formaldehyde in the presence of an excess of approximately 80 per cent sulfuric acid to give the methylene ether or formal of the ortho-methylol derivative, which is prob-ably the primary reaction product[19, 26].

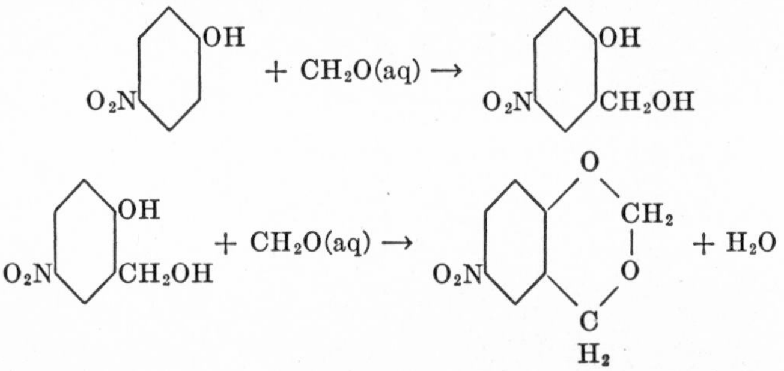

On oxidation, the above formal is converted to the methylene ether-ester of 5-nitrosalicyclic acid. Homologous methylene ethers or formals are also obtained with numerous nitrophenols, including *o*-nitro-*p*-cresol and 4-nitro-2-naphthol[19].

Apparently the nitro group, when present in the phenol nucleus, reduces the tendency of the primary products to condense further, with formation of methylene-linked products. The carboxyl group, as might be expected, appears to function in the same manner in some cases. Baeyer[8] obtained a monomethylol derivative of gallic acid by reacting this compound with formaldehyde in the presence of concentrated sulfuric acid[8].

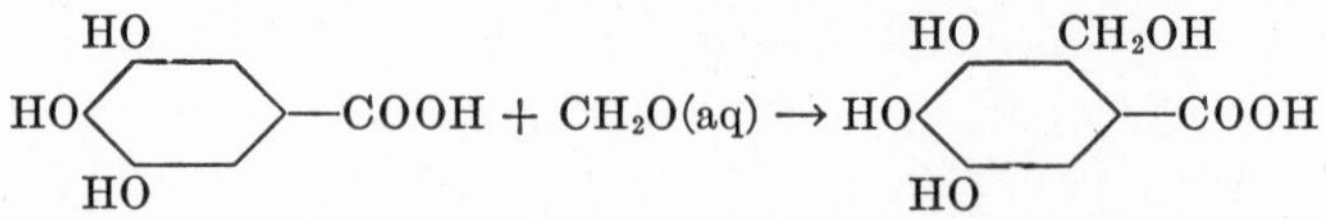

Results similar to those observed with nitrophenols are also obtained with 2,4-dichlorophenol. Ziegler and co-workers[89] obtained the methylene ether of 2,4-dichlorosaligenin on reaction of this phenol with formaldehyde polymer* at 0°C for three days in a medium consisting of glacial acetic acid to which a small amount of sulfuric acid had been added. The pre-

* Designated as "trioxymethylene" (?).

formed dichlorosaligenin was found to give the same formal when treated with cold concentrated hydrochloric acid and formaldehyde polymer, demonstrating the mechanism shown below:

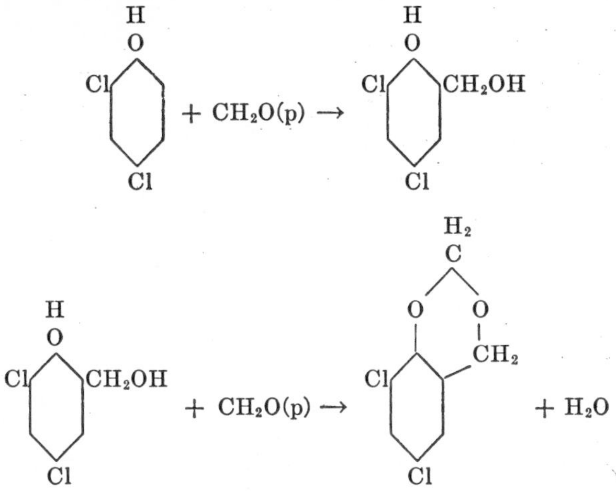

Polyhydric Phenols. Methylol derivatives of di- or polyhydroxy phenols are not readily isolated since they react with considerable rapidity to produce methylene derivatives. However, in the case of hydroquinone, Euler[32] has prepared both the 2,5-dimethylol and the tetramethylol derivatives by reaction with formaldehyde and alkali:

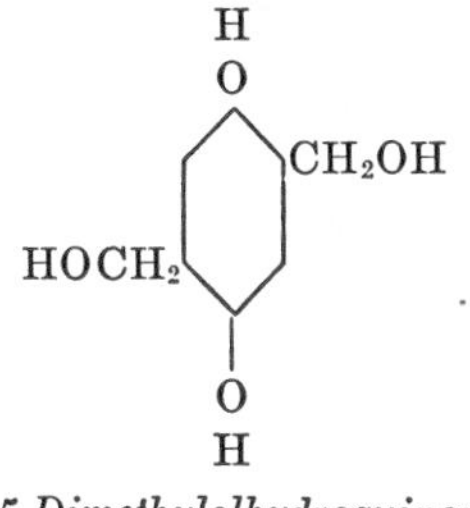

2,5-Dimethylolhydroquinone 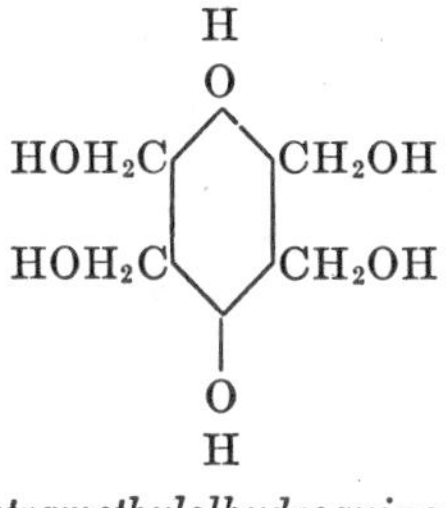   *Tetramethylolhydroquinone*

With resorcinol, Euler obtained a dialcohol of unknown constitution. Sen and Sarkar[75] report the preparation of a monomethylolresorcinol by reaction of resorcinol in alkaline solution with formaldehyde at 0 to 5°C. However, the fact that their product is insoluble in water casts considerable doubt on its structural identity. Boehm and Parlasca[18] synthesized 5-methylolresorcinol (1,3-dihydroxy-5-hydroxymethyl benzene) from the corresponding dicarbomethoxy aldehyde but the 4-methylol derivative proved unstable since similar attempts at synthesis gave a resinous product.

Derivatives of Methylolphenols. On heating in the absence of catalysts, methylolphenols tend to give hydroxybenzyl ethers as primary products. In the case of saligenin, Sprung[77] has demonstrated the formation of 2,2'-dihydroxybenzyl ether in excellent yield. With dimethylolphenols, long chain polyethers are obtained. Hanus and Fuchs[42] have demonstrated that the liberation of water, which accompanies ether formation when phenol alcohols are heated at a given temperature, is followed by liberation of formaldehyde with formation of diphenyl methane derivatives when higher temperatures are employed. Similar results also take place in the presence of alkaline or acidic catalysts.

$$RCH_2OCH_2R \rightarrow RCH_2R + CH_2O$$

Ethers may also be formed by the reaction of methylolphenols with alcohols. Durenol (2,3,5,6-tetramethyl phenol) gives a 4-ethoxymethyl derivative on reaction with formaldehyde in ethanol in the presence of sodium hydroxide as shown by Buraway and Chamberlain[22].

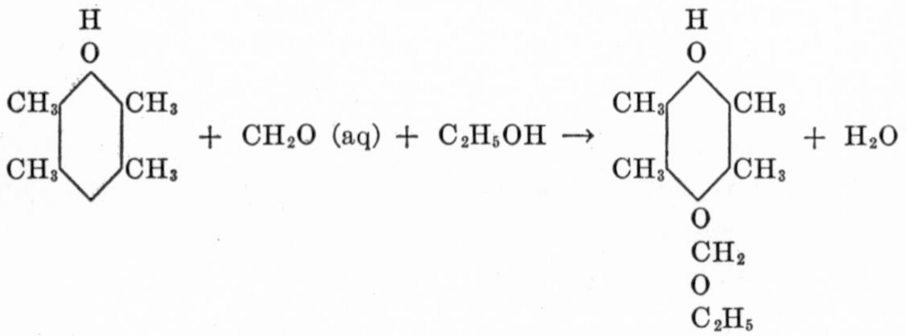

Derivatives of methylolphenols are also sometimes obtained by joint reactions involving a phenol, formaldehyde, and various polar compounds. In this way derivatives of phenol alcohols, which are not themselves readily obtained by the usual procedures, may in some cases be easily synthesized. The type reaction may be represented by the equation shown below in which HX represents a polar compound such as HCl, $H \cdot HSO_3$, $H \cdot N(CH_3)_2$, etc.:

$$C_6H_5OH + CH_2O + HX \rightarrow C_6H_4(OH)CH_2X + H_2O$$

Chloro-derivatives of methylolphenols are formed by the chloromethylation reaction (pages 337–342). Formation of resinous by-products is kept to a minimum when the phenol is gradually added to a cooled solution containing an excess of formaldehyde and saturated with hydrogen chloride[48]. For example, it is reported that 2,6-bis(chloromethyl)*p*-cresol may be obtained by adding 540 g (5 mols) molten *p*-cresol to a solution containing 600 g formaldehyde polymer (equivalent to approximately 20 mols CH_2O) in 3250 g concentrated hydrochloric acid, which has first been saturated with gaseous hydrogen chloride. The reaction mixture is maintained at 20°C, and approximately 8 hours should be allowed for the addition of *p*-cresol. The product, which separates as an oil, gradually crystallizes. It is then filtered, washed with water, and air-dried. After crystallization from hot petroleum ether, it melts at 86°C. Ortho-cresol gives 2,4-bis(chloromethyl)phenol when subjected to the above procedure. Phenol itself does not give the 2,4,6-trichloromethyl derivative which would at first be expected, but is converted to a bis(chloromethyl)derivative of the methylene ether of saligenin[48].

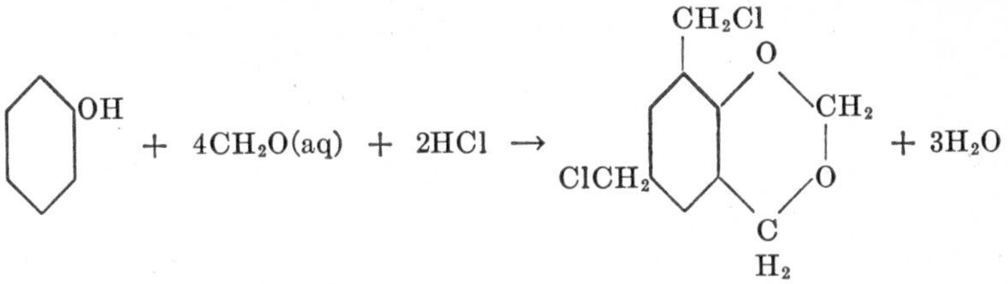

That the representation of chloromethylated phenols as phenol alcohol derivatives is not a paper relationship is demonstrated in one case by hydrolysis with boiling water to the corresponding methylolphenol[13].

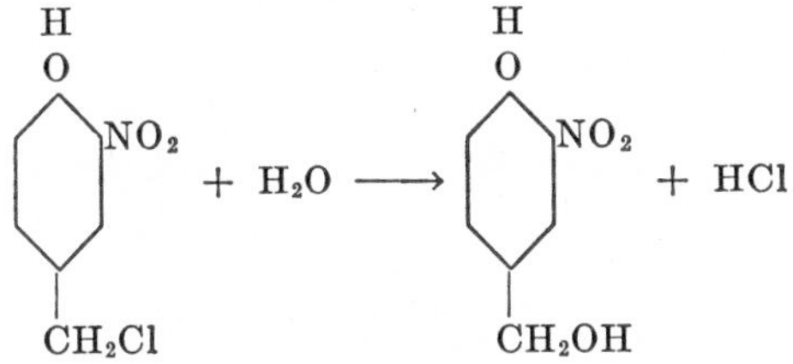

Finn and Musty[35] obtained the meta-methylol derivative of mesitol, which cannot be obtained by reaction with alkaline formaldehyde, by employing this technique and isolated it as a crystalline product melting at 130°C. On treatment with concentrated hydrochloric acid, it is almost quantitatively converted to the chloromethyl derivative.

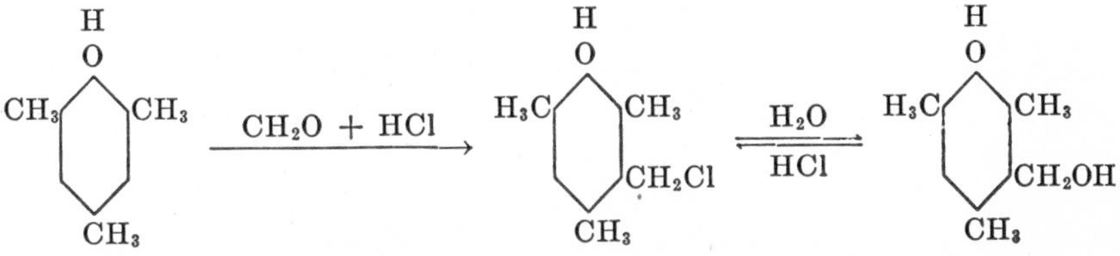

A sulfomethylation analogous to chloromethylation may be accomplished by reacting a phenol with formaldehyde and sodium sulfite or bisulfite. Sulfomethylation of beta-naphthol by this technique was disclosed in 1896 in a patent of Bayer and Co.[11]

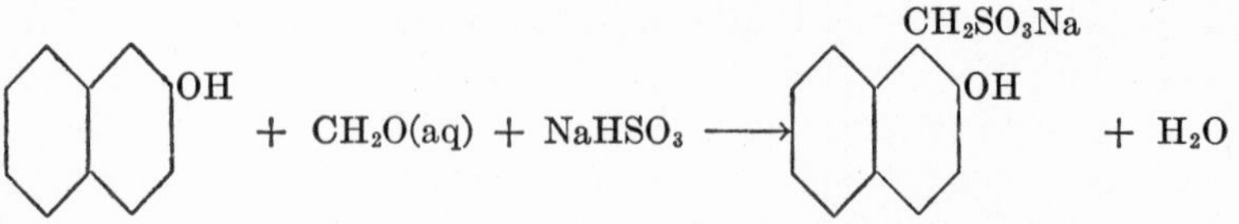

$$\text{(naphthol)OH} + CH_2O(aq) + NaHSO_3 \longrightarrow \text{(naphthol)}^{CH_2SO_3Na}_{OH} + H_2O$$

Suter, Bair, and Bordwell[80] recently confirmed this preparation using sodium sulfite and obtained a 75 per cent yield of sodium 2-hydroxy-1-naphthyl methane sulfonate. They also obtained similar condensation products from phenol and *p*-cresol but reported negative results with alpha-naphthol. The formation of the sulfonate of the beta-naphthol analogue of saligenin is of particular interest, since monomethylolnaphthols have never been isolated.

Nuclear aminomethyl derivatives are obtained by the Mannich synthesis when phenols or naphthols are reacted with secondary cyclic and acyclic alkyl amines. *p*-Hydroxydimethylbenzylamine (m.p. 200°C) is obtained from dimethylamine, formaldehyde, and phenol[12, 29]:

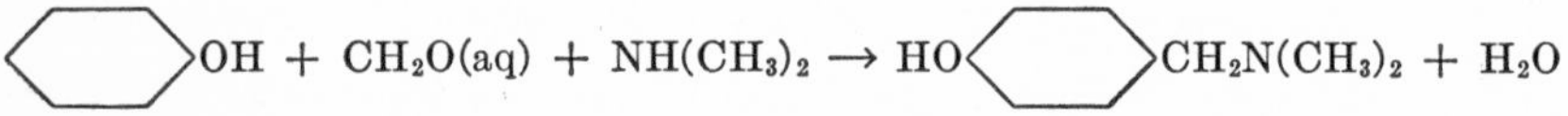

$$\text{(phenyl)OH} + CH_2O(aq) + NH(CH_3)_2 \rightarrow HO\text{(phenyl)}CH_2N(CH_3)_2 + H_2O$$

The ether, dimethylaminomethoxyphenol, is also formed as a product of this reaction:

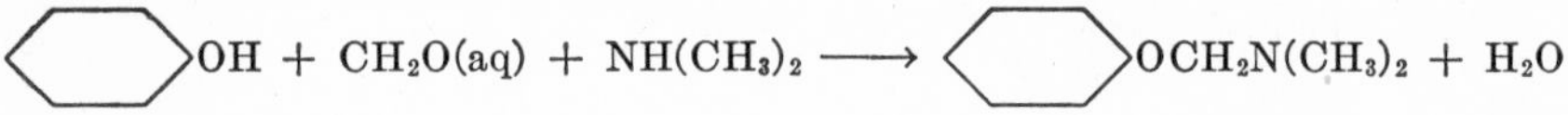

$$\text{(phenyl)OH} + CH_2O(aq) + NH(CH_3)_2 \longrightarrow \text{(phenyl)}OCH_2N(CH_3)_2 + H_2O$$

The phenolic product is readily separable from the ether by virtue of its solubility in aqueous alkali.

According to Caldwell and Thompson[23], dimethylaminomethylphenols are reduced to methylphenols when subjected to hydrogenation. By this procedure, they obtained 2,3,5-trimethylphenol (m.p. 93°C) on hydrogenating 2-dimethylaminomethyl-3,5-dimethylphenol at 165°C under a pressure of approximately 2500 lbs in the presence of a copper chromite catalyst. 2,6-Bis(dimethylaminomethyl)-hydroquinone (m.p. 190°C), which was readily obtained by the same investigators from hydroquinone, formaldehyde, and dimethylamine, gave 2,6-dimethylhydroquinone (m.p. 123 to 124°C) when subjected to the same procedure:

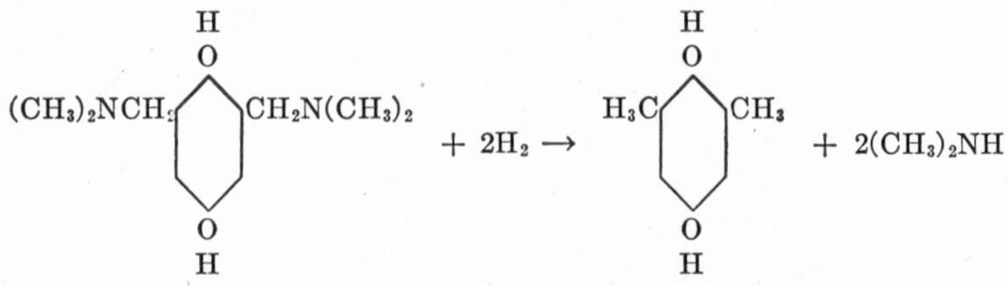

$$(CH_3)_2NCH_2 \overset{OH}{\underset{OH}{\text{(ring)}}} CH_2N(CH_3)_2 + 2H_2 \rightarrow H_3C \overset{OH}{\underset{OH}{\text{(ring)}}} CH_3 + 2(CH_3)_2NH$$

Bruson and MacMullen[21] prepared 2,4,6-tris(dimethylaminomethyl)-

phenol (b.p. 130 to 135°C at 1 mm) by the reaction:

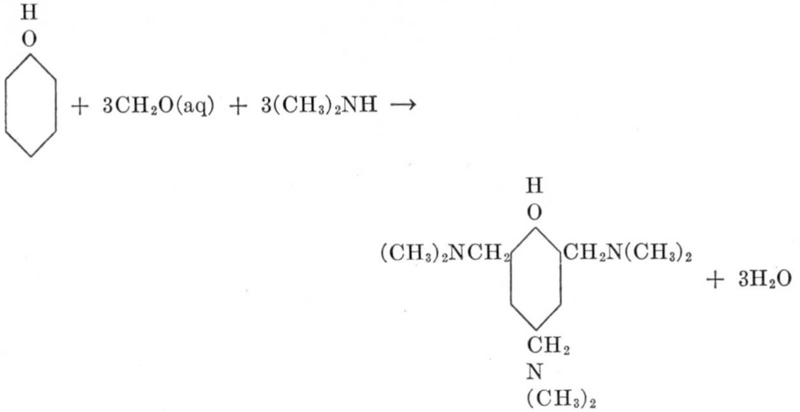

$$
\underset{\text{(phenol)}}{\text{C}_6\text{H}_5\text{OH}} + 3\text{CH}_2\text{O(aq)} + 3(\text{CH}_3)_2\text{NH} \rightarrow
$$

$$
(\text{CH}_3)_2\text{NCH}_2 \underset{\overset{|}{\text{CH}_2}}{\overset{\text{OH}}{\bigcirc}} \text{CH}_2\text{N(CH}_3)_2 + 3\text{H}_2\text{O}
$$

$$
\text{CH}_2\text{N(CH}_3)_2
$$

On acetylation with acetic anhydride, the above product is converted to the tetraacetate of trimethylolphenol (b.p. 200 to 205°C). This product has not previously been isolated. Its structure is demonstrated by the fact that it gives the acetate of mesitol, 2,4,6-trimethylphenol, on reaction with acetic anhydride. The latter is readily hydrolyzed to mesitol (m.p. 69°C). Bruson and MacMullen[21] also obtained morpholinomethylolphenols by substituting morpholine for dimethylamine in the Mannich reaction.

Production of phenols containing nuclear methyl groups by direct hydrogenation of methylolphenols has recently been demonstrated by Kennedy[50] by hydrogenation of the dimethylol derivatives of ortho- and para-cresol. Good yields of mesitol were obtained by this procedure. This hydrogenation was accomplished by heating the dimethylol derivatives in dioxane solution at 165°C under a hydrogen pressure of approximately 2500 lbs in the presence of copper chromite:

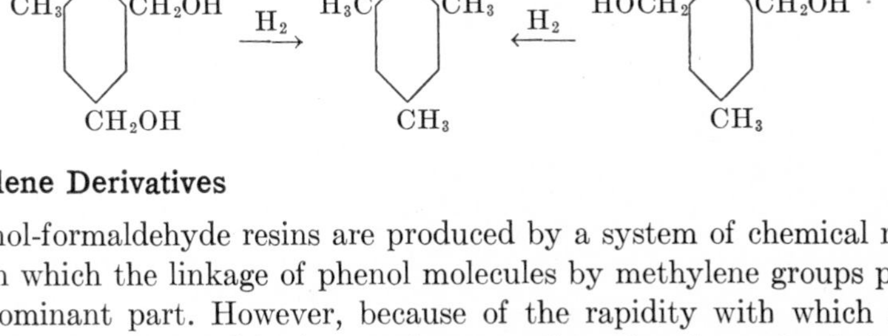

$$
\text{CH}_3 \underset{\text{CH}_2\text{OH}}{\overset{\text{OH}}{\bigcirc}} \text{CH}_2\text{OH} \xrightarrow{\text{H}_2} \quad \text{H}_3\text{C} \underset{\text{CH}_3}{\overset{\text{OH}}{\bigcirc}} \text{CH}_3 \xleftarrow{\text{H}_2} \quad \text{HOCH}_2 \underset{\text{CH}_3}{\overset{\text{OH}}{\bigcirc}} \text{CH}_2\text{OH} \cdot
$$

Methylene Derivatives

Phenol-formaldehyde resins are produced by a system of chemical reactions in which the linkage of phenol molecules by methylene groups plays a predominant part. However, because of the rapidity with which consecutive phenol-formaldehyde reactions usually take place, the isolation of simple methylene derivatives containing only two or three phenolic nuclei is often extremely difficult. This is particularly true of phenols in which the active ortho and para positions are unsubstituted and in which groups which tend to lower reactivity, such as nitro and carboxyl radicals, are absent. Since in these cases a large number of simple isomeric products can be formed, the separation of any one isomer in a pure state is difficult. The isolation of the simple methylene derivatives is noteworthy, since it clarifies our knowledge of the mechanism by which the more complicated products are produced.

Some phenols react readily to give simple methylene derivatives in good yield and show little tendency to form more complicated products. In many cases the nature and position of substituent groups in the phenol molecule plainly account for the results obtained. The importance of the reaction catalyst has already been emphasized, but it should be regarded as second in importance to the nature of the phenol involved. The molar ratio of formaldehyde to phenol is also an important factor. High ratios naturally favor the formation of polymethylene derivatives, whereas low ratios, as would be expected, favor the formation of the simpler methylene bis- and methylene tris-phenols.

Phenols and Cresols. The simpler methylene derivatives of phenol and the cresols having molecular weight values of under 1000 are prepared by the action of an acid catalyst on mixtures of the phenol with formaldehyde containing less than one mol of formaldehyde per mol phenol. Formation of products having minimal molecular weight is favored by use of a large excess of phenol in the reaction mixture. However, as a result of the high reactivity of phenol and its primary reaction products, the simple dinuclear methylene derivatives condense readily to polynuclear resins (novolacs) and are not readily isolated in good yield. Although methylol phenols are extremely unstable in the acidic reaction mixtures, their presence has been demonstrated in the early stages of the condensation by de Laire[34] and Baekeland[5]. Accordingly, it is believed that the mechanism of the production of methylene derivatives may involve intermediate formation of phenol alcohols. This is indicated below for the synthesis of 4,4'-dihydroxy diphenyl methane:

$$HO\langle\bigcirc\rangle + CH_2O(aq) \rightarrow HO\langle\bigcirc\rangle CH_2OH$$

$$HO\langle\bigcirc\rangle CH_2OH + \langle\bigcirc\rangle OH \rightarrow$$

$$HO\langle\bigcirc\rangle CH_2\langle\bigcirc\rangle OH + H_2O$$

However, although phenol alcohols have been isolated in the acid catalyzed phenol-formaldehyde condensation, this does not prove that they are the true intermediates. It is also possible that the acid condensation may follow a different mechanism. This mechanism may be analogous to the acid condensation of aromatic hydrocarbons and formaldehyde (pages 342–345), whereas the alkaline catalyzed formation of phenol alcohols resembles the alkaline condensation of formaldehyde and ketones (see pages 220–232). This is consistent with Finn and Musty's demonstration that mesitol gives a methylene-bis derivative involving the meta position of the phenolic nucleus in the presence of strong acids but does not react in the presence of alkalies. An alkaline catalyzed reaction in the meta position could not be explained if a quinoidal rearrangement of the phenolic structure were involved in the reaction mechanism.

Bender, Farnham and Guyer[14] have recently isolated three chemically pure isomeric dihydroxydiphenylmethanes from the reaction products of phenol and formaldehyde in the presence of hydrogen chloride and have rigorously demonstrated both their structure and chemical purity. These are:

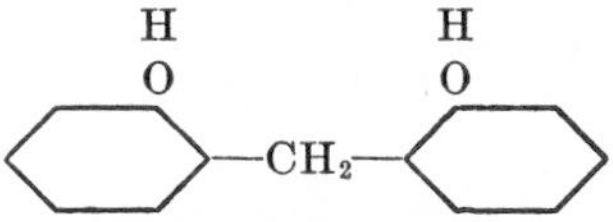

H   H
O   O
⬡—CH₂—⬡

2,2′-Dihydroxydiphenylmethane
(m.p. 118.5 to 119.5°C)

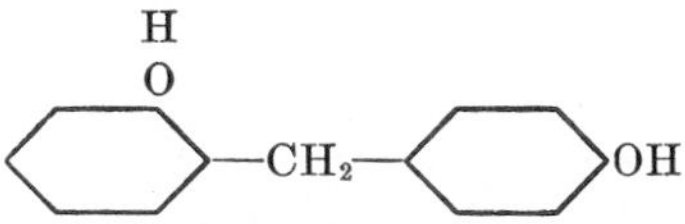

H
O
⬡—CH₂—⬡OH

2,4′-Dihydroxydiphenylmethane
(m.p. 119 to 120°C)

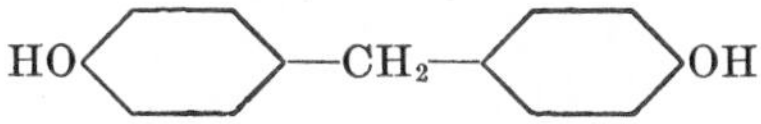

HO⬡—CH₂—⬡OH

4,4′-Dihydroxydiphenylmethane
(m.p. 162 to 163°C)

The 2,2′ isomer has not previously been isolated from phenol-formaldehyde reactions. However, in 1892, Noelting and Herzberg[67] reported the 4,4′ as a crystalline product melting at 148°C and, in 1923, Traubenberg[81] obtained an isomer melting at 115°C which Megson and Drummond[60] identified as the 2,4′ derivative in 1930. Bender and colleagues[14] have now demonstrated that mixtures of these isomers can be recrystallized from numerous solvents without change of melting point so that it is not surprising that low melting combinations have been mistaken for pure compounds.

The reactivity of the above diphenylmethanes has been compared by measuring the time required for gel formation on heating with 15 per cent hexamethylenetetramine at 160°C. The results as obtained by Bender *et al.*[14] are:

Isomer	Gel Time in Seconds
2,2′	60
4,4′	175
2,4′	240

The high reactivity of the 2,2′ isomer probably explains the fact that it has not previously been isolated from the phenol-formaldehyde reaction.

Bender and co-workers[14] have also synthesized the meta-dihydroxydiphenylmethane isomers by indirect methods since these isomers have not been obtainable, to date, by direct reaction of phenol with formaldehyde. Their melting points and reactivity with hexamethylenetetramine are:

Isomer	Melting Point (°C)	Gel Time in Seconds
3,3′	102–103	120
3,4′	116–116.5	130
2,3′	95.5–96	140

Trinuclear methylene phenols can be synthesized as crystalline products by reaction of methylol phenols with the dinuclear dihydroxy diphenyl methanes. The following were prepared in this way by Bender, Farnham, Apel and Gibb[15].

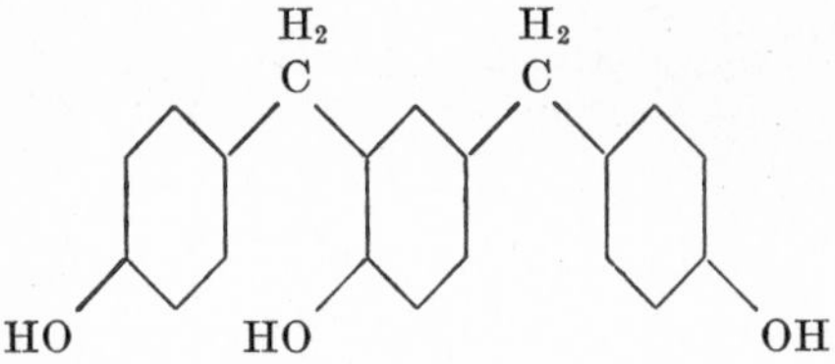

2,4-bis(4-hydroxybenzyl)phenol
(m.p. 149.5 to 150°C)

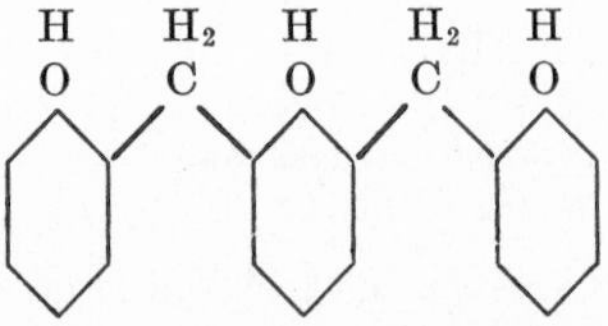

ex homosaligenin
and 4,4′-dihydroxy-
diphenylmethane

2,6-bis(2-hydroxybenzyl)phenol
(m.p. 161 to 162°C)

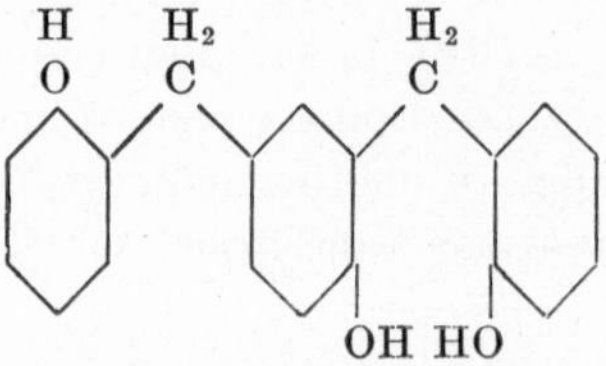

2,4-bis(2-hydroxybenzyl)phenol
(m.p. 122 to 123°C)

ex saligenin and
2,2′-dihydroxy-
diphenylmethane

The 2,6-bis(2-hydroxybenzyl)phenol has also been synthesized by Finn, Lewis and Megson[34a] as shown below:

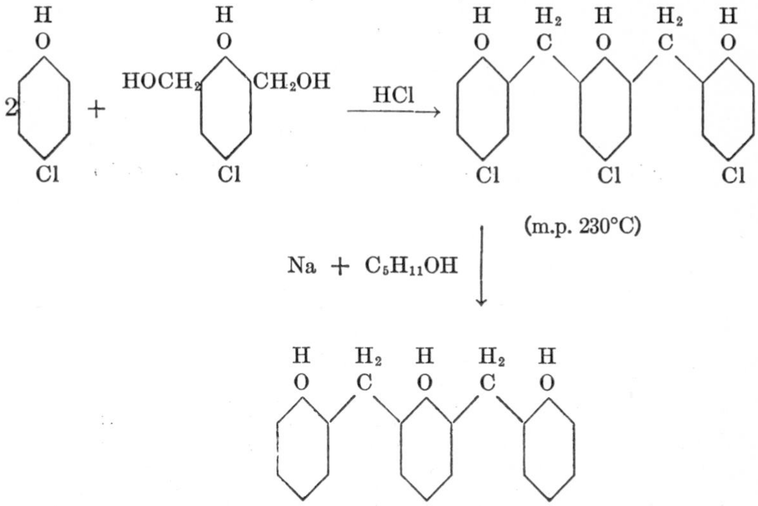

(m.p. 230°C)

$$Na + C_5H_{11}OH$$

(m.p. 158 to 159°C)

Simple methylenic derivatives of the isomeric cresols have also been isolated from reaction mixtures involving formaldehyde or paraformaldehyde and two or more molecular quantities of the pure cresols catalyzed with small percentages of hydrochloric acid[52, 60]. The structural formulas for these products are indicated below:

From ortho-cresol:

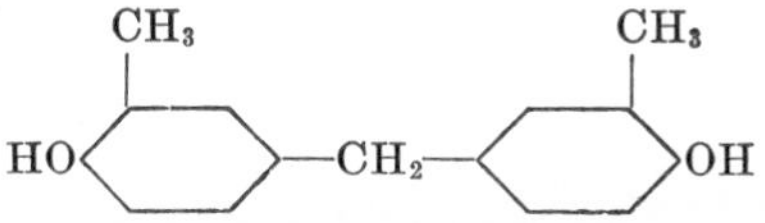

4,4'-Dihydroxy-3,3'-dimethyldiphenylmethane (m.p. 126 to 127°C)

From meta-cresol:

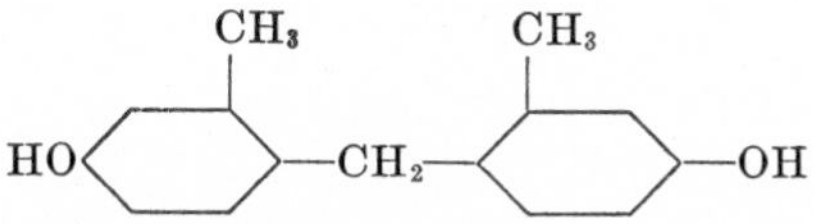

4,4'-Dihydroxy-2,2'-dimethyldiphenylmethane (m.p. 161°C) (or possibly
2,2'-Dihydroxy-4,4'-dimethyldiphenylmethane)

From para-cresol:

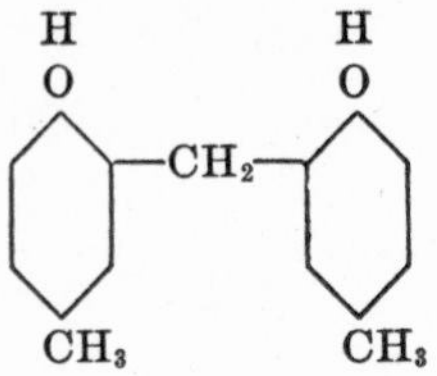

$$
\begin{array}{c}
\text{H} \qquad\qquad \text{H} \\
\text{O} \qquad\qquad \text{O} \\
\bigcirc\!-\!\text{CH}_2\!-\!\bigcirc \\
\text{CH}_3 \qquad\qquad \text{CH}_3
\end{array}
$$

6,6'-Dihydroxy-3,3'-dimethyldiphenylmethane (m.p. 126°C)

With para-cresol in which the active para position is blocked by a methyl group, a trinuclear methylenic derivative has been isolated from the condensation reaction[52, 60]. The structure of this product is indicated by its synthesis, carried out by reacting 2,6-dimethylol-*p*-cresol with excess *p*-cresol in the presence of hydrochloric acid.

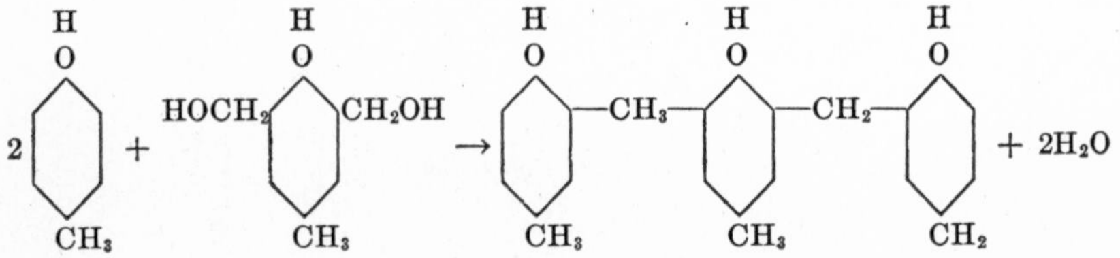

$$
2\,\bigcirc + \text{HOCH}_2\bigcirc\text{CH}_2\text{OH} \rightarrow \bigcirc\!-\!\text{CH}_3\!-\!\bigcirc\!-\!\text{CH}_2\!-\!\bigcirc + 2\text{H}_2\text{O}
$$

Di- and trinuclear products were obtained by Koebner[52] from *p*-cresol and formaldehyde in the following manner:

A mixture of 1000 g *p*-cresol (9.3 mols) and 250 g 30 per cent formaldehyde solution (2.5 mols) was warmed to 50°C and treated with 25 cc conc. hydrochloric acid. The violently reacting mixture was then cooled and mixed with 1500 cc benzene. On further cooling, approximately 100 g of the trinuclear derivative (m.p. 215°C) crystallized and was filtered off. The filtrate was then subjected to vacuum distillation. After removal of benzene, water, and unreacted cresol, 150 g of the dinuclear product was distilled at 240°C and 15-mm pressure, leaving a resinous residue which decomposed on heating to its boiling point. On cooling, this product crystallized and showed a melting point of 126°C.

By acid condensation of the di- and trinuclear methylene derivatives of *p*-cresol with formaldehyde and with the dimethylol derivatives of *p*-cresol and its polynuclear methylene derivatives, Koebner[52] claims to have synthesized pure crystalline polymethylene derivatives containing up to five *p*-cresol nuclei. The dimethylol derivatives of the di- and trinuclear compounds were prepared by alkaline condensation of the corresponding phenols with formaldehyde, using added methanol as a subsidiary reaction solvent. These various products, illustrating successive steps in the cresol-formaldehyde condensation, are shown below:

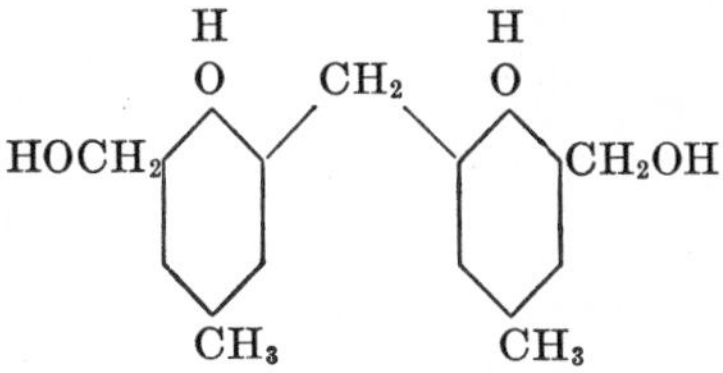

Dinuclear dialcohol
(m.p. 148.5°C)

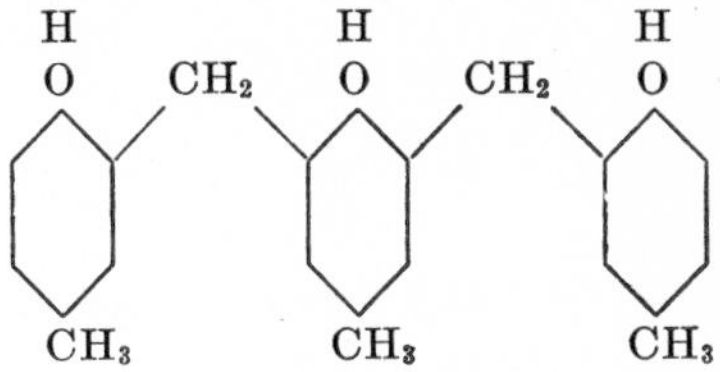

Trinuclear phenol
(m.p. 215°C)

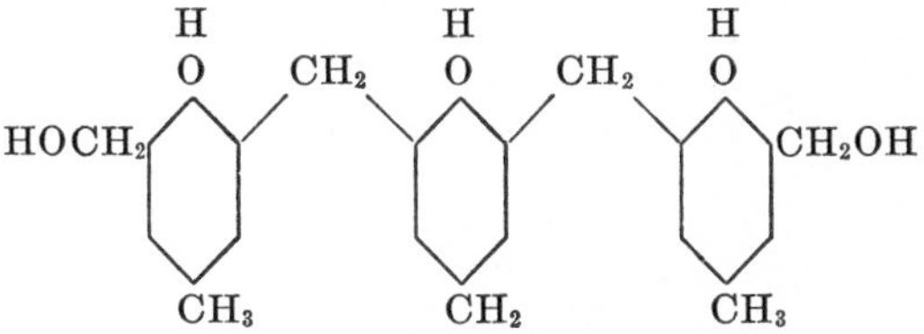

Trinuclear dialcohol
(m.p. 194°C)

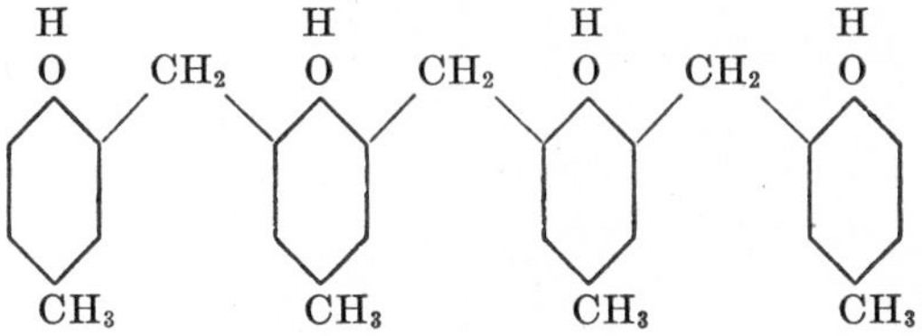

Tetranuclear phenol
(m.p. 173°C)

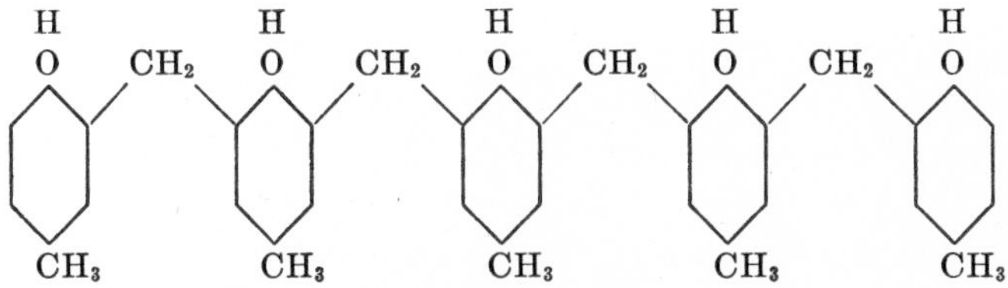

Pentanuclear phenol
(m.p. 202–206°C)

Koebner also obtained crude crystalline hexa- and heptanuclear phenols. As might be expected, the water-solubility of the polynuclear phenols decreases with increasing molecular weight. This is also true of the alkali phenates of these products. In the case of the heptanuclear product the alkali salt is stated to be completely insoluble. A homolog of the tri-nuclear methylene derivative of p-cresol was synthesized by Vanscheidt and co-workers[83] by condensing 2,6-dimethylol-4-benzylphenol with 4-benzyl-phenol in the presence of hydrochloric acid.

That methylene derivatives may also form as a result of the alkaline condensation of cresols and formaldehyde has been demonstrated by Granger[37]. He isolated a dinuclear methylol derivative as a by-product in the preparation of dimethylol-o-cresol and from o-cresol and two mols of aqueous formaldehyde by the method of Manasse. The product, which melted at 155°C, was shown to possess the empirical formula, $C_{17}H_{20}O_4$. It is probably one of several possible isomers indicated by the formula:

$$
\left.\begin{array}{l} CH_3 \\[1em] HO \\[1em] HOCH_2 \end{array}\right\} C_6H_2 \!-\! CH_2 \!-\! C_6H_2 \left\{\begin{array}{l} CH_3 \\[1em] OH \\[1em] CH_2OH \end{array}\right.
$$

Symmetrical tetramethylol dihydroxydiphenylmethanes were obtained by Martin[57] on neutralizing the trimethylol phenates obtained by reacting sodium phenate and sodium meta-cresylate with formaldehyde.

Xylenol. Xylenols yield methylene-bis derivatives in much the same manner as the isomeric cresols. In the case of 3,5-dimethylphenol, in which none of the active positions are blocked by methyl groups, Morgan[65] succeeded in preparing all three possible isomeric diphenylmethanes by reaction with formaldehyde under acidic conditions.

In the case of the xylenol, 2,6-dimethylphenol, a methylene bis-phenol is readily formed under alkaline conditions. That the 2,6-dimethyl-4-methylolphenol is first formed was demonstrated by Bamberger[9], who prepared it by the Manasse procedure. However, on long standing at room temperature, a reaction mixture containing equimolar proportions of xyle-nol, formaldehyde, and sodium hydroxide gave a 94.6 per cent yield of the crystalline methylenediphenol on neutralization with acetic acid[37]. Only half of the formaldehyde was consumed by this reaction. An almost quan-titative yield was obtained when a similar mixture was refluxed at 100°C. In this reaction the excess formaldehyde was converted to methanol and sodium formate by the Cannizzaro reaction. There is apparently little tendency for resin formation in this reaction, as might be predicted from the fact that the parent xylenol has only one active position which is not blocked by substituent groups.

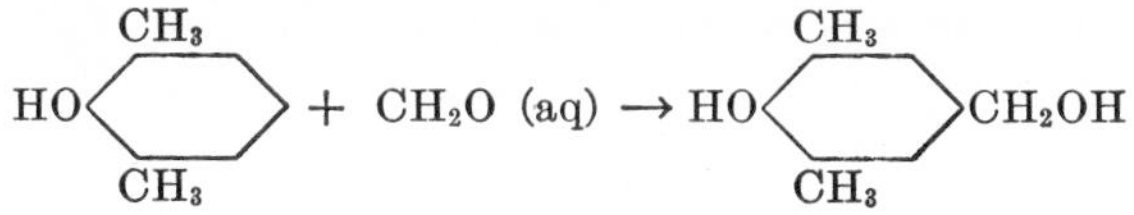

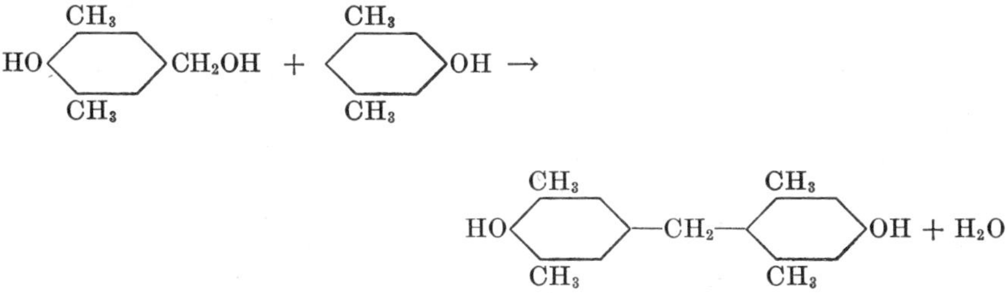

$$HO-\langle \rangle + CH_2O \text{ (aq)} \rightarrow HO-\langle \rangle CH_2OH$$

$$HO-\langle \rangle CH_2OH + \langle \rangle OH \rightarrow$$

$$HO-\langle \rangle -CH_2-\langle \rangle OH + H_2O$$

In view of the above findings, a similar result would be expected in the case of the ortho-para-xylenol (2,4-dimethylphenol); but according to Granger[37], this is not the case. An equimolar quantity of formaldehyde is stated to react with the phenol under alkaline conditions and a well-defined resin is reported to precipitate on neutralization of the reaction mixture after heating. The structure of this product has not been elucidated. One possibility is that it is an ether resin of the type:

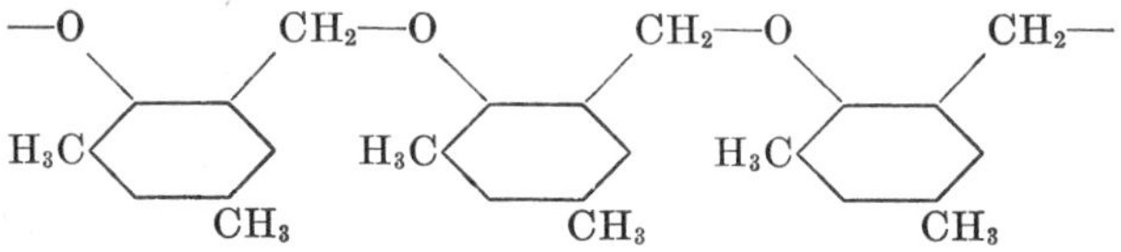

This is in agreement with Baekeland and Bender's theory[6] that the phenol ether arrangement is a resin structure. The product may be envisaged as forming when the sodium derivative of the theoretical methylolphenol is reacted with acid.

Mesitol. The work of Finn and Musty[35] indicates that a meta-linked trinuclear methylene derivative, as well as the dinuclear compound previously discussed, is produced by the acid-catalyzed condensation of mesitol and formaldehyde. This reaction is carried out by refluxing a solution made up of 7.2 g mesitol, 3.8 cc 40 volume per cent formaldehyde and 15 cc of concentrated hydrochloric acid on the water bath for 3 hours. Crude product precipitated from this reaction gives an alcohol insoluble fraction (1.8 g) containing the impure trinuclear derivative and a soluble fraction (2.3 g) yielding the pure dinuclear compound on recrystallization from dilute alcohol. The pure trinuclear compound was synthesized by reaction of the dinuclear derivative with meta-chloromethyl mesitol. The compounds and relationships involved are shown below:

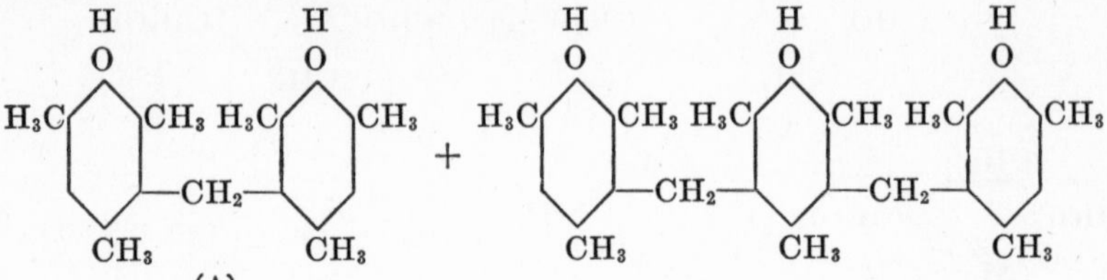

(A)

3,3'-Dihydroxy-2,2',4,4',6,6'-hexa-methyldiphenyl methane
(m.p. 188°C)

3,5-Bis(3-hydroxy-2,4,6-trimethyl-benzyl) 2,4,6-trimethylphenol
(m.p. 256 to 258°C)

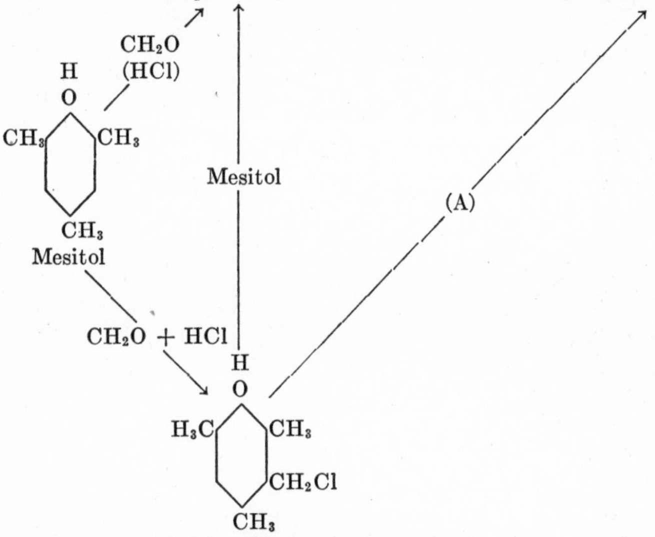

Chloromethyl mesitol
(2,4,6-Trimethyl-3-chloromethylphenol)
(m.p. 134°C)

Durenol. Buraway and Chamberlain[22] report the formation of bis(hydroxyduryl)methane in practically quantitative yield by the reaction of durenol with aqueous formaldehyde in approximately 10 per cent sodium hydroxide solution.

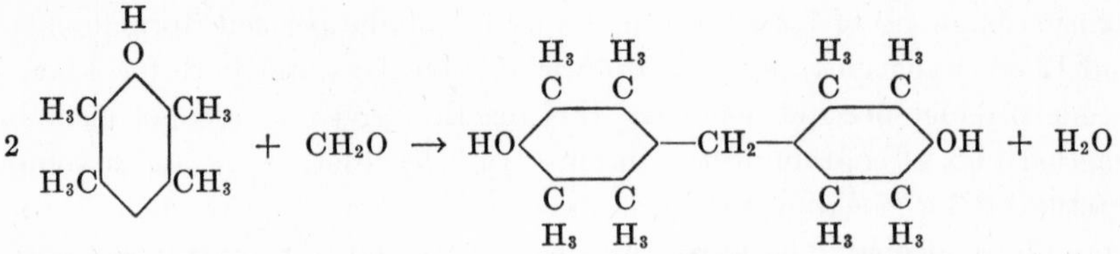

(m.p. 213 to 214°C)

The fully substituted diphenylmethane was found to split readily on reaction with concentrated hydrochloric acid. This indicates that a methylene bridge from a fully substituted benzene ring is easily ruptured, a characteristic confirmed by Finn and Musty's research[35] on mesitol derivatives.

Chlorophenols. Methylene derivatives of chlorophenols can be produced in approximately the same manner as those of the corresponding methyl phenols. Gump[40] obtains good yields of polychlorinated dihydroxydiphenyl methanes by adding phenols, such as 2,4,5-trichlorophenol, in admixture with formaldehyde to an excess of cold concentrated (approx. 93 per cent) sulfuric acid. After 4 to 20 hours reaction at 0 to 5°C, the reaction mixture is added to crushed ice and the crude product removed by filtration.

Naphthols. Beta-naphthol resembles 2,6-dimethylphenol in that it is readily converted to a methylene-bis derivative under alkaline conditions. The methylol derivative of this phenol has never been isolated. When Manasse[56] attempted the synthesis of the phenol alcohol by combining equimolar quantities of beta-naphthol, formaldehyde and caustic soda at room temperatures, the sodium salt of di-beta-naphtholmethane [methylene bis(beta-naphthol)] crystallized from the reaction mixture. On acidification, this methylene derivative was isolated as the sole reaction product. Hosaeus[46] obtained di-beta-naphtholmethane by heating beta-naphthol and formaldehyde with glacial acetic acid. According to Manasse[56], pure di-beta-naphtholmethane melts at 200°C. Alpha-naphthol differs from beta-naphthol in giving more complicated products both under alkaline and acidic conditions, bearing much the same relation to beta-naphthol that ortho-para-xylenol bears to the isomeric 2,4-xylenol. On reacting alpha-naphthol with formaldehyde in the presence of potassium carbonate, Breslauer and Pictet[20] obtained an amorphous product having the empirical formula $C_{23}H_{16}O_3$. Abel[1] obtained a brown, brittle, alkali-soluble resin on heating the same naphthol with formaldehyde in 50 per cent acetic acid, to which was added a small quantity of hydrochloric acid.

Nitrophenols. Good yields of simple dinuclear methylene derivatives may be obtained from a number of substituted phenols containing negative radicals. Methylene bis-nitrophenols are obtained from the isomeric ortho-, meta-, and para-nitrophenols by treatment with formaldehyde and concentrated sulfuric acid[61, 74]. Chattaway and Goepp[26] obtained a good yield of 2,2'-dihydroxy-5,5'-dinitrodiphenylmethane by adding a hot (75 to 80°C) solution containing 10 cc 37 per cent formaldehyde (one molar proportion), 6 cc water, and 24 cc concentrated sulfuric acid to 38 g p-nitrophenol (two molar proportions) plus 5 cc water held at 75°C. The temperature of this mixture rose to 120 to 130°C, and it was maintained at this temperature until the product separated as a greenish paste which was removed from the supernatant liquid, dissolved in 4 per cent caustic soda, filtered and acidified with hydrochloric acid to precipitate the product (37 g). The compound melted at 268°C after crystallization from hot acetic acid. Little or no alkali-insoluble nitro-saligenin formal (see page

244) was obtained in this preparation. When a slight excess over 1.5 mols of formaldehyde per mol of nitrophenol was employed a methylene bis-nitrosaligenin formal [di-8,8'-(6-nitrobenzdioxin)] was obtained in almost theoretical yield. The structure of this product is:

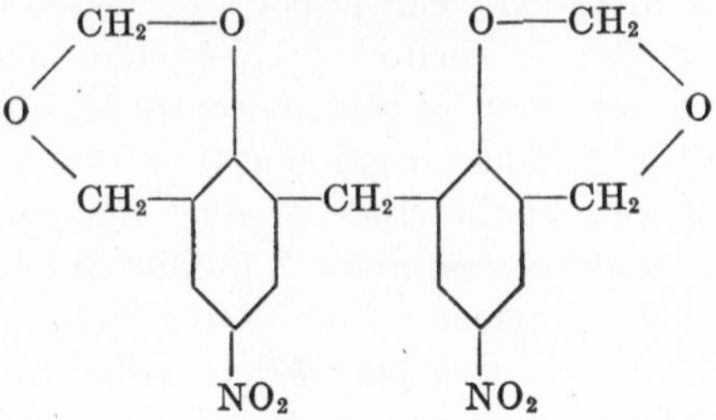

$$\begin{array}{c}\text{CH}_2\text{---O} \qquad\qquad \text{O---CH}_2\\[4pt]\text{O} \qquad\qquad\qquad\qquad\qquad \text{O}\\[4pt]\text{CH}_2\text{---}\ \bigcirc\ \text{---CH}_2\text{---}\ \bigcirc\ \text{---CH}_2\\[4pt]\text{NO}_2 \qquad\qquad \text{NO}_2\end{array}$$

On oxidation and hydrolysis this compound was converted to a benzophenone derivative, 2,2'-dihydroxy-3,3'-dicarboxy-5,5'-dinitrobenzophenone.

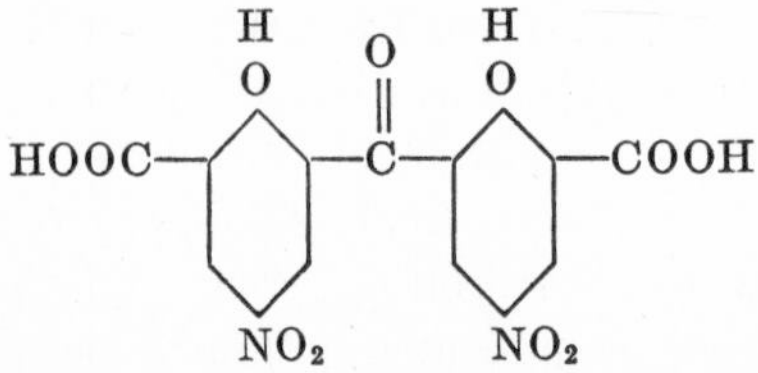

$$\text{HOOC---}\ \bigcirc\ \overset{\text{H}}{\underset{\text{NO}_2}{\text{O}}}\text{---}\overset{\text{O}}{\underset{}{\text{C}}}\text{---}\ \bigcirc\ \overset{\text{H}}{\underset{\text{NO}_2}{\text{O}}}\text{---COOH}$$

Salicylic Acid. Salicyclic acid (2-carboxyphenol) was reported by Caro[24] to give a methylene-bis derivative on heating with formaldehyde and concentrated hydrochloric acid. Clemmensen and Heitman[27] obtained a quantitative yield of this product by adding two mols salicylic acid and one mol formaldehyde to a 50 per cent solution of sulfuric acid and refluxing for 10 hours. The weight of 50 per cent acid was approximately 6 times the weight of salicylic acid in the charge. The product melts at 238°C. Madsen[55] reports a melting point of 243 to 244°C after recrystallization from hot water. Oxidation of a mixture of methylene bis-salicylic acid and salicylic acid leads to the formation of a triphenylmethane derivative:

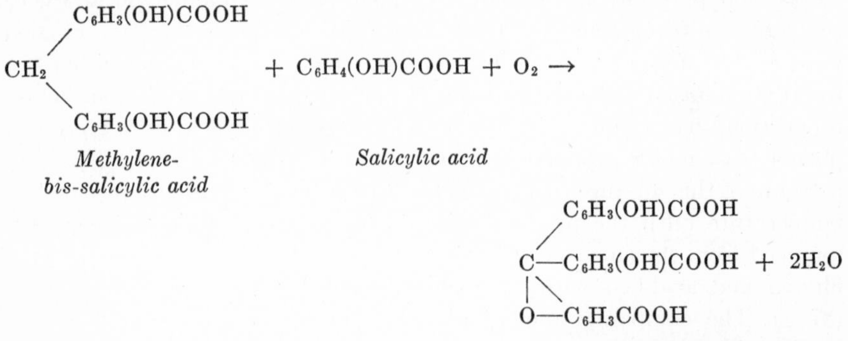

$$\underset{\substack{\textit{Methylene-}\\\textit{bis-salicylic acid}}}{\text{CH}_2\!\!\begin{array}{c}\diagup\ \text{C}_6\text{H}_3(\text{OH})\text{COOH}\\[6pt]\diagdown\ \text{C}_6\text{H}_3(\text{OH})\text{COOH}\end{array}} \quad + \underset{\textit{Salicylic acid}}{\text{C}_6\text{H}_4(\text{OH})\text{COOH}} + \text{O}_2 \rightarrow$$

$$\underset{\textit{Aurin tricarboxylic acid}}{\text{C}\!\!\begin{array}{c}\diagup\ \text{C}_6\text{H}_3(\text{OH})\text{COOH}\\[4pt]\text{---}\text{C}_6\text{H}_3(\text{OH})\text{COOH} \quad + \quad 2\text{H}_2\text{O}\\[4pt]\diagdown\ \text{O---C}_6\text{H}_3\text{COOH}\end{array}}$$

Caro[24] carried out this oxidation by adding sodium nitrite to the reagents in concentrated sulfuric acid. A related oxidation reaction may explain Traubenberg's[81] isolation of aurin from phenol-formaldehyde condensation products.

Polyhydric Phenols. As noted by Carleton Ellis[31], polyvalent phenols form resinous products so readily on reaction with formaldehyde that in many instances the isolation of intermediates is extremely difficult. Resorcinol, for example, is stated to react even in the absence of added catalysts. However, in some cases, simple products may be isolated and occasionally good yields can be obtained. As previously noted, Euler[32] has prepared methylol derivatives of hydroquinone. In the case of resorcinol, although there may be some doubt as to whether a true methylol derivative has even been isolated, the methylene-bis derivative has apparently been produced in almost quantitative yield by Möhlau and Koch[63]. This product was prepared by dissolving 10 g resorcinol in 100 g 35 per cent sulfuric acid and adding 3.5 g formaldehyde. Resin formation was prevented by keeping the mixture cold. On treatment with concentrated sulfuric acid, this product is converted to a fluorone derivative.

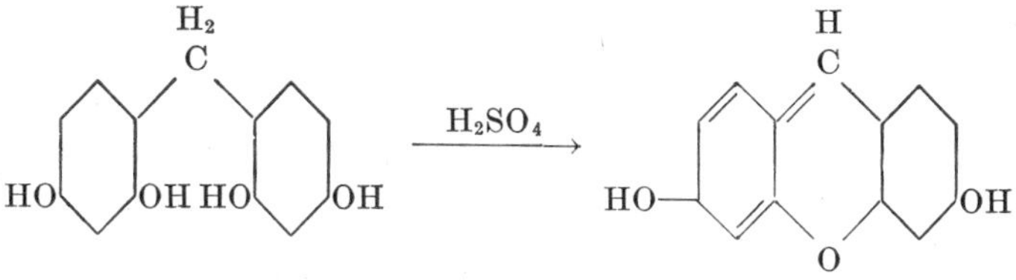

Möhlau and Koch also prepared methylene bis-orcinol, $CH_2[C_6H_2CH_3(OH)_2]_2$, from orcinol and formaldehyde by condensation with sulfuric acid in dilute aqueous solution.

Practical methods for controlling resorcinol-formaldehyde reactions have been worked out in recent years in connection with the commercial development of cold-set resorcinol adhesives (Cf. pp. 444–445).

The quantitative formation of the methylene-bis derivative of 5,5-dimethyldihydroresorcinol (dimedon) is employed for the detection and determination of formaldehyde (pages 390–393).

Methoxy derivatives of divalent phenols such as guaiacol react like simple monovalent phenols, forming methylol and methylene derivatives. Euler and co-workers[33] isolated vanillyl alcohol (methylolguaiacol) and a dimethylol derivative of a methylene bis-methoxyphenol from the products obtained by reacting guaiacol with formaldehyde under alkaline conditions. Under acidic conditions two isomeric methylenebis-methoxyphenols were obtained.

Phloroglucinol, according to Boehm[17], does not condense with formaldehyde at room temperature in the absence of added catalysts. When hydrochloric or sulfuric acid is added to the reaction mixture, a small quantity of symmetrical hexahydroxydiphenylmethane (m.p. 225°C) is formed, together with more complicated products. These products of further condensation are apparently nuclear polymethylene derivatives, since on reduction of the crude acid condensation product with zinc dust and caustic soda, mono-, di- and trimethylphloroglucinol as well as phloroglucinol itself are obtained. On heating with formaldehyde in the presence of relatively high concentrations of hydrochloric acid, phloroglucinol is converted to a dark reddish brown, water-insoluble resin whose composition is approximately that represented by the empirical formula $C_7H_6O_3$[28]. This end-product of the condensation reaction is apparently a polymethylene derivative in which the ratio of methylene radicals to aromatic nuclei is approximately 1:1.

Methylene bis-pyrogallol (m.p. 241°C with decompn.) was prepared by Caro[24] by treating 3 parts of pyrogallol in 20 parts of 10 per cent hydrochloric acid with one part commercial formaldehyde, and allowing the reaction to proceed for 12 hours at room temperature.

PHENOL-FORMALDEHYDE RESINS

Phenol-formaldehyde resins are the final products of most reactions involving formaldehyde and phenol. Although formaldehyde solution is the principle form of formaldehyde employed in their preparation, paraformaldehyde and hexamethylenetetramine (page 438) are also utilized to a considerable extent. Anhydrous gaseous formaldehyde may likewise be employed, as illustrated by a process devised by Aylsworth[3] in which formaldehyde gas is introduced into an autoclave containing hot phenol or cresol.

In general, these resins may be defined as mixtures of polymethylene compounds in which phenolic radicals are linked by methylene groups. They are broadly classified as belonging to two distinct types:

(1) Fusible resins in which the component molecules have a linear or chain-like structure.

(2) Infusible resins in which the component molecules are made up of a three-dimensional network consisting of cross-linked molecules.

However, even the above classification is by no means absolute. Megson[59] has concluded that the extremely irregular geometry of the phenol-methylene chains rules out all but a few cross-linkages and Pritchett[71] believes that cross-linkages may be relatively nonexistent. It is possible that the properties of the hardened resins may result entirely from entanglement of irregular linear chains.

The kinetics of the phenol-formaldehyde reaction have been the subject of several investigations. Nordlander[63] recognizes a primary reaction yielding water-soluble products and a secondary reaction involving condensation of the primary products to give water-insoluble derivatives. His findings indicate that both reactions are apparently of the first order but that the secondary reaction has a much greater temperature coefficient. When low concentrations of ammonia are employed as a catalyst, the primary reaction appears to be of the second order. Jones[49] reports that the acid-catalyzed reaction is of the second order at 80°C but approximates the first order at 30°C. His studies also indicate that the alkali-catalyzed reaction (pH = 8.5) was of the first order at 40°C until 45 per cent of the formaldehyde had been consumed. Sprung's studies[76] in which paraformaldehyde was employed with triethanolamine as a catalyst also indicate that the primary reaction follows the rate law prescribed by a first order reaction.

The type of resin obtained from a given condensation reaction is governed by the same factors that determine the nature of the primary reaction products, *viz.*, the structure of the phenol and the nature of the reaction catalyst. The molar ratio of formaldehyde to phenol is also of outstanding importance.

Fusible Resins: Novolacs. These resins are generally regarded as linear condensation products of simple monomethylol phenols. They are readily obtained from phenol, cresols, and many other phenols which possess two or more active nuclear positions unblocked by substituent groups. Baekeland gave the name of novolac to resins of this type. They are fusible, brittle solids which dissolve readily in some organic solvents and are unchanged by prolonged heating in the absence of air. When air is present, the hot resin is oxidized at the surface, forming a brown film which does melt and is not affected by organic solvents[38].

Baekeland[4] found that novolacs were obtained when phenol and formaldehyde were condensed with an acid catalyst when the molar ratio of formaldehyde to phenol was less than one. Under alkaline conditions, however, insoluble resins may be produced even when an excess of phenol is employed. This may be explained by the fact that since alkaline catalysis favors polymethylolation, subsequent condensation yields cross-linked resins which are infusible, whereas acid catalysts, as previously pointed out, favor condensation of the monomethylolated phenol as fast as formed.

Koebner's research[52] on the polymethylene derivatives of *p*-cresol, (pages 254–256) led him to the belief that fusible resins consist of mixtures of chain molecules of various lengths in which the nuclear phenolic carbon atoms are connected by methylene groups:

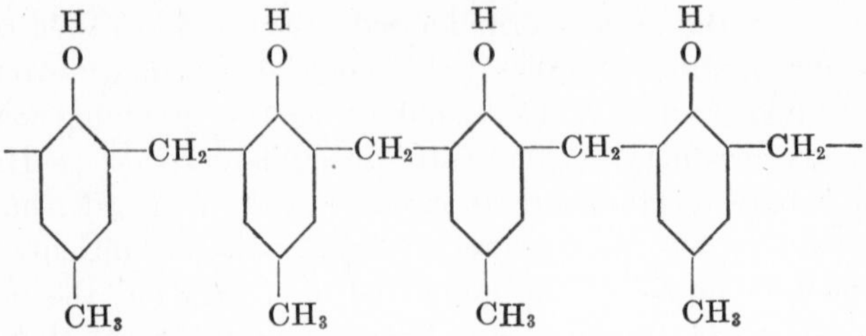

According to his work, the average molecular weights of these resins are low, ranging in the neighborhood of 300 to 700.

The presence of phenol hydroxy groups in novolacs is indicated by Koebner's finding that these products are soluble in alcoholic sodium hydroxide. The dielectric properties of thermoplastic phenolics are reported to be due to the rotation of these polar hydroxy groups[44]. Meyer's studies[62] on phenol-resin structures led him to the conclusion that the novolacs obtained by acid condensation of phenol and formaldehyde have terminal p-HOC$_6$H$_4$- groups.

As previously pointed out, these theories do not account for the resin from 2,4-dimethylphenol reported by Granger[37].

Since novolacs are permanently fusible and do not change on heating, they must be reacted with additional formaldehyde for most commercial uses. A number of the so-called "two-step" resins are prepared in this way. For example, they may be fused with hexamethylenetetramine which is equivalent to condensing with additional formaldehyde in the presence of an alkaline catalyst.

Infusible Resins. Infusible resins are most readily prepared from phenols in which the three active, 2,4 and 6 ring positions are unsubstituted. They are obtained by either alkaline or acid catalysis when the molar ratio of formaldehyde to phenol is one or greater. Granger[39] points out that, when acid catalysts are employed, considerable excess formaldehyde is required and much of it does not react and is wasted. Satisfactory results are obtained by acid condensation of polymethylolphenols produced under alkaline conditions. As indicated above, infusible resins are also obtained by reacting novolacs with formaldehyde. The water-soluble resins obtained by acid condensation of resorcinol with a low mol ratio of formaldehyde act as "cold set" adhesives because they become insoluble and infusible at room temperature when mixed with formaldehyde, paraformaldehyde or hexamethylenetetramine.

Baekeland[4] and Lebach[53] recognize three fairly distinct phases in the formation of infusible resins which, with minor modifications, have been generally accepted by resin chemists[64]. These are the "A" or resole stage, the "B" or resitol stage, and the "C" or resite stage. "A" resins are still soluble in alkali and probably consist mainly of phenol alcohols plus some phenol alcohol ethers and methylol derivatives of diphenyl methanes. The "B" resins have condensed to the point at which they are no longer soluble in alkalies but can be softened on heating and are, at least, partially soluble in organic solvents. The "C" resins are infusible and insoluble and represent the final cured resins whose free methylol groups have now undergone substantially complete condensation.

Structure. The way in which methylene groups are attached to the phenolic radicals in resins has been a subject of considerable controversy. However, the main body of evidence indicates that the majority of these linkages connect nuclear carbon atoms. Absorption spectra made on thin (0.03-mm) sections of hardened phenolics give definite indication that this is the case[78]. The results of destructive hydrogenation of both fusible and infusible resins prepared from phenol and formaldehyde also support this conclusion in that phenol and cresols are the principal products of this treatment[87]. The structure of the methylene derivatives, whose isolation as resin intermediates has been previously discussed, also constitutes good evidence. This theory of resin structure was apparently first advocated by Raschig[72] in 1912.

In contradistinction to Raschig's theory is Baekeland's hypothesis[4] advanced in 1909 that the resins should be regarded as saligenin or phenol-alcohol ethers having the structure:

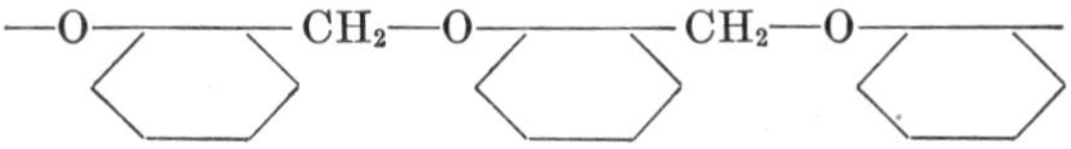

Although this hypothesis was modified by Baekeland and Bender[6] in 1925, the formation of an ether of this type was still assumed in the formation of thermoplastic or fusible resins. The fact that a crude product obtained by reaction of sodium phenate with formaldehyde and sodium bisulfite in methanol solution gave an analysis that indicated it to be methyl phenyl formal, $C_6H_5OCH_2OCH_3$, and resinified on all attempts at purification was presented as evidence for the ether structure. The properties of this formal checked with those of Reychler[73], who prepared it from chloromethyl ether and sodium phenate. It is water-soluble; and when the solution is acidified, resin precipitates. Breslauer and Pictet[20] isolated this same formal from products obtained when phenol and formaldehyde were reacted in the presence of sulfuric acid.

Although there is comparatively little direct evidence that phenoxy ether bridges are actually present in phenolic resins, the possibility of their formation in resin reactions cannot be lightly dismissed. The fact that fusible resins are obtained with alkaline catalysts from a phenol such as 2,4-dimethylphenol, which has only one active nuclear position, is not readily explained without recourse to the hypothesis that nuclear linkages may involve the phenolic hydroxy group, unless, of course, the possibility that the so-called inactive (meta) nuclear positions may be involved is considered likely. Finn and Musty[35] have now demonstrated that the formation of methylene linkages between meta positions on the phenolic nucleus is a definite reality and may account for resin formation in acid condensations but this reaction does not appear to take place in the presence of alkalies.

The formation of ether linkages between the methylol groups of phenol alcohols is clearly demonstrated in the primary phases of the alkaline catalyzed condensation and it is probable that some of these ether linkages survive in the final resin.

The possibility that linkages may be formed by condensation of formaldehyde with the methylene group of the diphenyl methane structure as indicated below has been proposed by Zinke and Hanus[90].

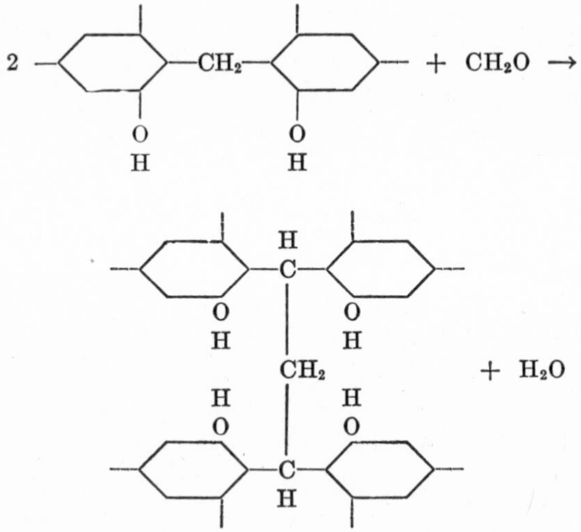

However, this hypothesis seems unlikely and cannot be supported by known reactions in the field of formaldehyde chemistry.

Another hypothesis of resin formation offered by Wohl and Mylo[88] envisages dehydration of a quinoid isomer of the methylolphenol to give a polymerizable product. This is illustrated in the case of the monomethylol derivative of 2,4-dimethylphenol.

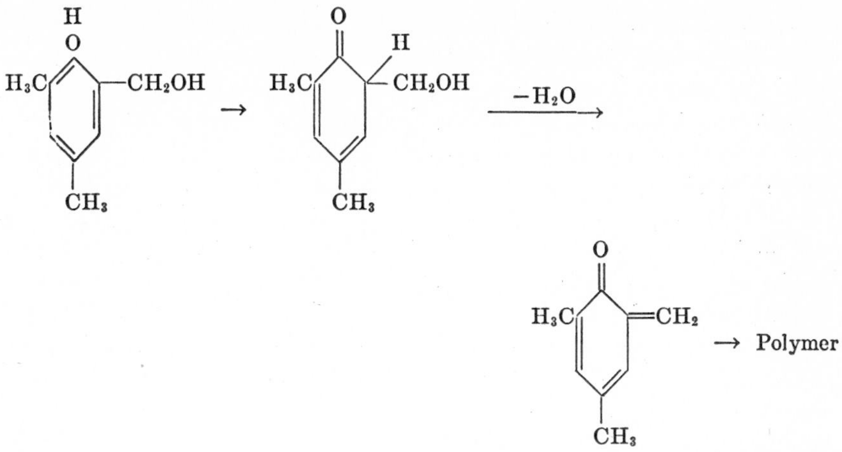

This reaction sequence bears a strong analogy to the formation of the polymerizable methyl vinyl ketone from formaldehyde and acetone. Paramethylene quinones analogous to the ortho derivative shown above are also envisaged. According to Hultzsch[47], the three fundamental phases in the phenolformaldehyde condensation are: (1) formation of methylol compounds by addition of formaldehyde, (2) intramolecular dehydration to form reactive methylene quinones, and (3) addition of the reactive

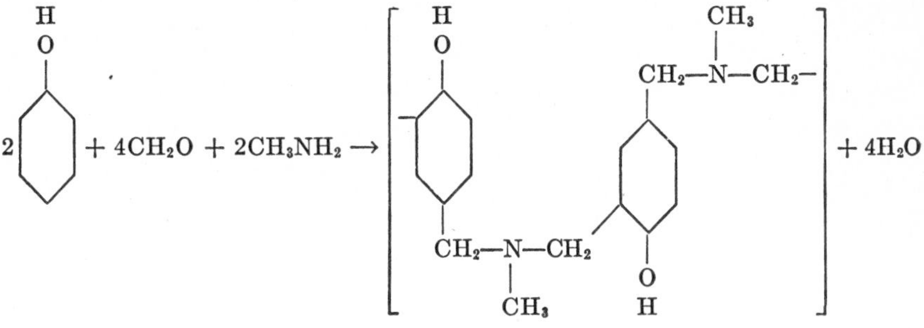

Amine resin (polymer unit)

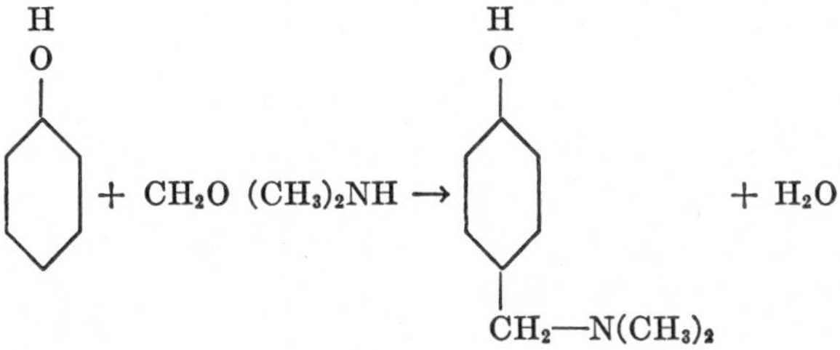

intermediates to the same or other materials with acid hydrogen atoms. Stilbene structures may result from polymerization of methylene quinones so that ethylenic resin linkages may occur. It should be noted that a methylene quinone structure may be formed from the methylol derivative of a di- or polynuclear methylene-linked phenol structure as well as a simple phenol alcohol.

Just as derivatives of methylolphenols can be obtained by joint reactions of phenols with formaldehyde and various polar compounds, complex resins can be obtained when the polar compounds are di- or polypolar. This may be illustrated by the formation of amine-resins from phenol, formaldehyde, and non-aromatic primary amines as described by Harmon and Meigs[43]. This reaction is contrasted on the preceding page with a similar reaction involving a secondary amine which is monopolar.

References

1. Abel, J., *Ber.*, **25,** 3477 (1892).
2. Auwers, K., and Daecke, S., *Ber.*, **32,** 3373–81 (1899).
3. Aylsworth, J. W., U. S. Patent 1,029,737 (1912); *see also* 1,033,044 (1912).
4. Baekeland, L. H., *Ind. Eng. Chem.*, **1,** 158 (1909).
5. *Ibid.*, **4,** 739 (1912).
6. Baekeland, L. H., and Bender, H. L., *Ind. Eng. Chem.*, **17,** 225–37 (1925).
7. Baeyer, A., *Ber.*, **5,** 25, 280, 1094 (1872).
8. *Ibid.*, **5,** 1095 (1872).
9. Bamberger, E., *Ber.*, **36,** 2036 (1903).
10. Bayer & Co., German Patent 85,588 (1894).
11. *Ibid.*, 87,335 (1896).
12. *Ibid.*, 89,979 (1896); 92,309 (1897).
13. *Ibid.*, 136,680 (1902); 132,475 (1902).
14. Bender, H. L., Farnham, A. G., and Guyer, J. W., (to Bakelite Corporation). U. S. Patent 2,464,207 (1949).
15. Bender, H. L., Farnham, A. G., Guyer, J. W., Apel, F. N., and Gibb, T. B., Jr., *Ind. Eng. Chem.*, **44,** 1619–23 (1952).
16. Biedermann, J., *Ber.*, **19,** 2374 (1886).
17. Boehm, R., *Ann.*, **329,** 270–2 (1903).
18. Boehm, T., and Parlasca, H., *Arch. Pharm.*, **270,** 168–82 (1932).
19. Borsche, W., and Berkhout, A. D., *Ann.*, **330,** 82–107 (1904).
20. Breslauer, J., and Pictet, A., *Ber.*, **40,** 3785 (1907).
21. Bruson, H. A., and MacMullen, C. W., *J. Am. Chem. Soc.*, **63,** 270–2 (1941).
22. Buraway, A., and Chamberlain, J. T., *J. Chem. Soc.*, **1949,** 624–6.
23. Caldwell, W. T., and Thompson, T. R., *J. Am. Chem. Soc.*, **61,** 765 (1939).
24. Caro, N., *Ber.*, **25,** 941 (1892).
25. Carswell, T. S., "Phenoplasts," New York, Interscience Publishers, Inc., (1947).
26. Chattaway, F. D., and Goepp, R. M., *J. Chem. Soc.*, **1933,** 699.
27. Clemmensen, E., and Heitman, A. H. C., *J. Am. Chem. Soc.*, **33,** 737 (1911).
28. Clowes, G. H. A., *Ber.*, **32,** 2841 (1899).
29. Décombe, J., *Compt. rend.*, **196,** 866–8 (1933); *Chem. Zentr.*, **1933,** I, 3188.
30. Ellis, Carleton, "The Chemistry of Synthetic Resins," pp. 277–395, New York, Reinhold Publishing Corp. (1935).

31. *Ibid.*, p. 371.
32. Euler, H. v., *Arkiv. Kemi. Mineral. Geol., Ser. B.*, **14**, No. 9, 1–7 (1940); *Chem. Zentr.*, **1941**, I, 128–9.
33. Euler, H. v., Adler, E., and Friedmann, O., *Arkiv. Kemi. Mineral. Geol., Ser. B*, **13**, No. 12, 1–7; *C. A.*, **34**, 1095 (1940).
34. Fabrique de produits de chimie organique De Laire, French Patent 361,539 (1905); *C. A.*, **1**, 2956 (1907).
34a. Finn, S. R., Lewis, G. J., and Megson, N. J. L., *J. Soc., Chem. Ind.*, **69**, S51–3 (1950).
35. Finn, A. R., and Musty, J. W. G., *Soc. Chem. Ind.*, **69**, S6–7 (1950).
36. Fishman, J. B., *J. Am. Chem. Soc.*, **42**, 2288–97 (1920).
37. Granger, F. S., *Ind. Eng. Chem.*, **24**, 442–8 (1932).
38. *Ibid.*, **29**, 860–6 (1937).
39. *Ibid.*, **29**, 1125–9 (1937).
40. Gump, W. S., (to Burton T. Bush, Inc.), U. S. Patents 2,250,480 (1941), 2,354,012 and 2,354,013 (1944).
41. Hanus, F., *J. prakt. Chem.*, **158**, 254–65 (1941); *C. A.*, **35**, 7949–50 (1941).
42. Hanus, F., and Fuchs, E., *J. prakt. Chem.*, **152**, 327 (1939).
43. Harmon, J., and Meigs, F. M., (to E. I. du Pont de Nemours & Co., Inc.), U. S. Patent 2,098,869 (1937).
44. Hartshorn, L., Megson, N. J. L., and Rushton, E., *J. Soc. Chem. Ind.*, **59**, 129–33T (1940).
45. Holmes, E. L., and Megson, N. J. L., *J. Soc. Chem. Ind.*, **52**, 415–8T (1933).
46. Hosaeus, H., *Ber.*, **25**, 3213–4 (1892).
47. Hultzsch, K., *Z. angew. Chem.*, **60A**, 253 (1948).
48. I. G. Farbenindustrie A. G., British Patent 347,887 (1931).
49. Jones, T. T., *J. Soc. Chem. Ind.*, **65**, 264–75 (1946).
50. Kennedy, T., *Chemistry & Industry*, **1940**, 297.
51. Kleeburg, W., *Ann.*, **263**, 283 (1891).
52. Koebner, M., *Z. angew. Chem.*, **46**, 251–6 (1933).
53. Lebach, H., *J. Soc. Chem. Ind.*, **32**, 559–64 (1913).
54. Lederer, L., *J. prakt. Chem. (2)*, **50**, 223 (1894); U. S. Patent 563,975 (1896).
55. Madsen, E. H., *Arch. Pharm.*, **245**, 42–8 (1907); *C. A.*, **1**, 1702 (1907).
56. Manasse, O., *Ber.*, **27**, 2409 (1894); U. S. Patent 526,786 (1894).
57. Martin, W. R., *J. Am. Chem. Soc.*, **73**, 3952 (1951); (to Gen. Electric Co.), U. S. Patent 2,579,329 (1951).
58. Megson, N. J. L., *J. Soc. Chem. Ind.*, **52**, 422T (1933).
59. *Ibid.*, **67**, 155 (1948).
60. Megson, N. J. L., and Drummond, A. A., *J. Soc. Chem. Ind.*, **49**, 251T (1930).
61. Meister, Lucius and Brunning, German Patents 72,490 (1893); 73,946 (1893); 73,951 (1893).
62. Meyer, G., *Materie plastiche*, **5**, 15–17 (1938); *C. A.*, **33**, 5539 (1939).
63. Möhlau, R., and Koch, P., *Ber.*, **27**, 2888 (1894).
64. Morgan, G. T., *J. Soc. Chem. Ind.*, **49**, 247–51T (1930).
65. Morgan, G. T., and Megson, N. J. L., *J. Soc. Chem. Ind.*, **52**, 418–420T (1933).
66. Muller, H. F., and Muller, I., *Kunstoffe*, **38**, 221–6 (1948).
67. Noelting, E., and Herzberg, W., *Chem. Ztg.*, **16**, 185 (1892).
68. Nordlander, B. W., *Oil, Paint Drug Reptr.*, **130**, 3, 27 (1936).
69. Piria, R., *Ann.*, **48**, 75 (1843).
70. *Ibid.*, **56**, 37 (1845).
71. Pritchett, E. G. K., *Chemistry & Industry*, **1949**, 295.

72. Raschig, F., *Z. angew, Chem.*, **25**, 1945 (1912); *C. A.*, **7**, 887 (1913).
73. Reychler, A., *Bull. soc. chim.* (*4*), **1**, 1189–95 (1907); *C. A.*, **2**, 1266 (1908).
74. Schöpff, M., *Ber.*, **27**, 2322–3 (1894).
75. Sen, R. N., and Sarker, N. N., *J. Am. Chem. Soc.*, **47**, 1084–5 (1925).
76. Sprung, M. M., *J. Am. Chem. Soc.*, **63**, 334–43 (1941).
77. Sprung, M. M., and Gladstone, M. T., *J. Am. Chem. Soc.*, **71**, 2907 (1949).
78. Stäger, I. H., Siegfried, W., and Sänger, R., *Schweiz. Arch. angew. Wiss. Tech.*, **7**, 129–139 (1941); *C. A.*, **35**, 7577 (1941).
79. Stoermer, R , and Behn, K., *Ber.*, **34**, 2459 (1901).
80. Suter, C. M., Bair, R. K., and Bordwell, F. G., *J. Org. Chem.*, **10**, 470 (1945).
81. Traubenberg, I. K., *Z. angew. Chem.*, **36**, 515 (1923).
82. Ullmann, F., and Brittner, K., *Ber.*, **42**, 2539 (1909).
83. Vanscheidt, A., *et al.*, *Plast. Massy*, **1935**, 11–17; *Chimie et Ind.*, **34**, 1404 (1936).
84. Wakeman, R. L., "Chemistry of Commercial Plastics," New York, Reinhold Publishing Corp., (1947).
85. Walker, J. F., (to E. I. du Pont de Nemours & Co., Inc.) U. S. Patent 2,304,431 (1942).
86. Walker, J. F., and Chadwick, A. F., *Ind. Eng. Chem.*, **39**, 974 (1947).
87. Waterman, H. I., and Veldman, A. R., *Brit. Plastics*, **8**, 125–8, 182–4 (1936).
88. Wohl, A., and Mylo, B., *Ber.*, **45**, 2046 (1912).
89. Ziegler, E., *et al.*, *Ber.*, **74**, 1871–9 (1941).
90. Zinke, A., and Hanus, F., *Ber.*, **74**, 205 (1941).

CHAPTER 13

REACTIONS OF FORMALDEHYDE WITH CARBOXYLIC ACIDS, ACID ANHYDRIDES, KETENE, ACYL CHLORIDES, AND ESTERS

Carboxylic Acids. Formaldehyde and paraformaldehyde are soluble in simple carboxylic acids, such as formic and acetic acid, but do not yield stable reaction products with these agents under ordinary circumstances. For this reason, acids of this type are often of value as solvent media for formaldehyde reactions. It is probable that solution is attended by the formation of loose addition compounds, just as is the case when formaldehyde dissolves in water or alcohols. These compounds are probably mono-esters of methylene and polyoxymethylene glycols, as indicated in the following equations involving formaldehyde and acetic acid:

$$CH_3COOH + CH_2O \ (g,p) \rightarrow CH_3COOCH_2OH$$

$$CH_3COOCH_2OH + CH_2O \ (g,p) \rightarrow CH_3COOCH_2OCH_2OH$$

Reactions with fatty acids at high temperatures involve pyrolysis and autocondensation of formaldehyde. Paquot and Perron[32] obtained a mixture of methyl and ethyl palmitates with a gaseous by-product, principally carbon dioxide, on heating 30 per cent aqueous formaldehyde with palmitic acid at 300°C in an autoclave for five hours. These investigators assume that ethanol is formed by the aldolization of formaldehyde to glycolaldehyde which is subsequently reduced by the hydrogen that is evolved together with carbon dioxide when formaldehyde is decomposed. Ralston and Vander Wal[36] obtained fatty aldehydes by reaction of commercial formaldehyde vapors with the calcium and sodium salts of fatty acids at 200 to 300°C as indicated in the type formula in which RCOONa represents the sodium salt of a fatty acid containing six or more carbon atoms.

$$2RCOONa + CH_2O \ (g) \rightarrow 2RCHO + Na_2CO_3$$

Unsaturated as well as saturated fatty acids are subject to this reaction.

Aliphatic hydroxy acids may be converted to cyclic methylene ether-esters by reaction with formaldehyde in the presence of strongly acidic catalysts. Derivatives of this type, whose preparation from tartaric and

271

citric acid is reported by Sternberg[42, 43], are shown below:

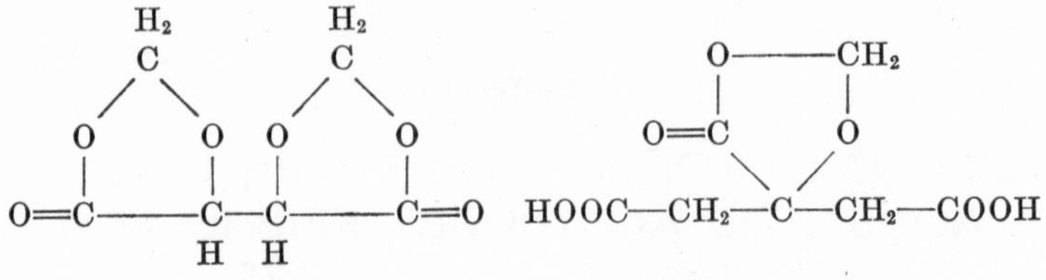

Methylene tartrate   *Methylene citrate*

Methylene tartrate (m.p. 120°C) was formed as a crystalline product by addition of a small quantity of concentrated sulfuric acid to the viscous oil obtained by heating molten tartaric acid with paraformaldehyde or alpha-polyoxymethylene ("trioxymethylene"). Methylene citrate was prepared by heating citric acid with paraformaldehyde at 140 to 160°C, or by evaporating a solution of citric acid in aqueous formaldehyde in the presence of hydrochloric acid.

Reactions of amino acids with formaldehyde are described on pages 311–312 in connection with the reactions of formaldehyde with organic nitrogen compounds.

Acids containing active alpha-hydrogen atoms, such as malonic acid, acetoacetic acid, cyanoacetic acid, etc., react readily with formaldehyde, producing derivatives in which the alpha-carbon is linked to a methylol or methylenic substituent. Florence[20], for example, obtained acrylic acid from the reaction mixture produced by treating an ether solution of malonic acid with gaseous formaldehyde in the presence of pyridine:

$$CH_2(COOH)_2 + CH_2O \text{ (g)} \rightarrow CH_2{:}CHCOOH + CO_2 + H_2O$$

Methylolmalonic acid, $HC(COOH)_2CH_2OH$, is probably formed as a reaction intermediate.

Mannich and Ganz[30] prepared amino-methylmalonic acid derivatives by reacting malonic acid or mono-substituted malonic acids with ammonia or amines and aqueous formaldehyde at ice-water temperatures. Reactions proceed as indicated in the following equation, in which R is equivalent to hydrogen or an aliphatic or aromatic radical, and R' to hydrogen or an aliphatic radical:

$$RHC(COOH)_2 + CH_2O \text{ (aq)} + HNR_2' \rightarrow RC(COOH)_2CH_2NR_2' + H_2O$$

Benzylmalonic acid reacts with ammonia and formaldehyde to give benzylaminomethylmalonic acid, $C_6H_5CH_2C(COOH)_2CH_2NH_2$, and under some conditions the bis-derivative, $(C_6H_5CH_2C(COOH)_2CH_2)_2NH$. These substituted malonic acids readily lose carbon dioxide and are thus converted to propionic acid derivatives. In some cases, the substituted malonic acid itself cannot be isolated. Phenylmalonic acid evolves carbon dioxide on reaction with formaldehyde and dimethylamine, giving phenyl-dimethylaminomethylpropionic acid.

$$C_6H_5CH(COOH)_2 + CH_2O \text{ (aq)} + HN(CH_3)_2 \rightarrow$$

$$C_6H_5CH(COOH)CH_2N(CH_3)_2 + CO_2 + H_2O$$

Piperazine, formaldehyde, and malonic acid react as shown below:

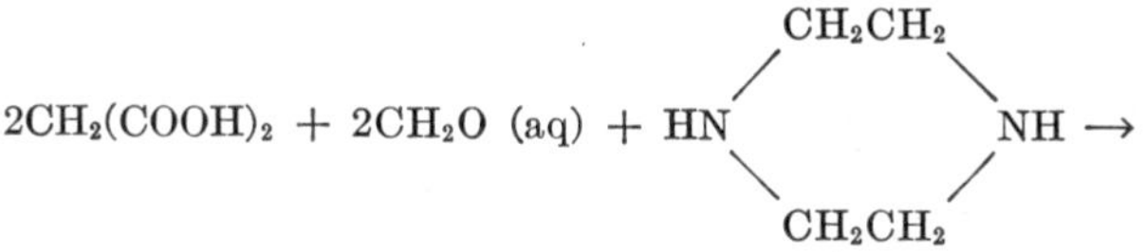

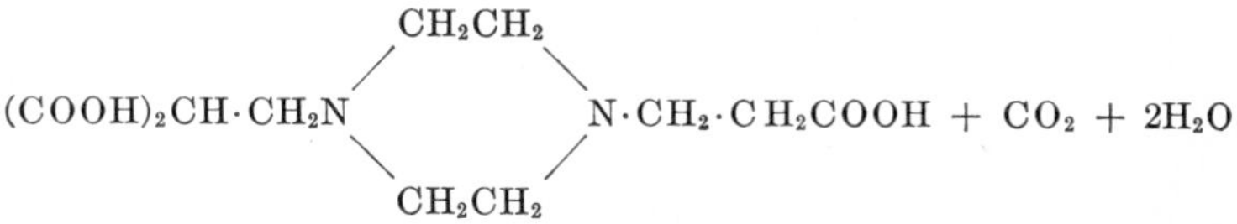

$$2CH_2(COOH)_2 + 2CH_2O \text{ (aq)} + HN\underset{\diagdown\ CH_2CH_2\ \diagup}{\overset{\diagup\ CH_2CH_2\ \diagdown}{}}NH \rightarrow$$

$$(COOH)_2CH\cdot CH_2N\underset{\diagdown\ CH_2CH_2\ \diagup}{\overset{\diagup\ CH_2CH_2\ \diagdown}{}}N\cdot CH_2\cdot CH_2COOH + CO_2 + 2H_2O$$

On warming, another molecule of carbon dioxide is evolved and pipera-zine-N,N'-bis(beta-propionic acid), $[COOH\cdot CH_2CH_2N(CH_2CH_2)_2NCH_2\text{-}CH_2COOH]$, is obtained.

Mannich and Bauroth[29] obtained a dimethylaminomethyl derivative from tartronic acid, dimethylamine, and formaldehyde:

$$HOCH(COOH)_2 + CH_2O \text{ (aq)} + HN(CH_3)_2 \rightarrow HOC(COOH)_2CH_2N(CH_3)_2 + H_2O$$

Cyanoacetic acid, dimethylamine, and formaldehyde react with evolu-tion of carbon dioxide to give a nitrile which, on hydrolysis, yields, acrylic acid. This nitrile is probably dimethylaminoethyl cyanide, $(CH_3)_2NCH_2\text{-}CH_2CN$[30].

Simple aromatic acids show little tendency to react with formaldehyde under neutral, alkaline, or mildly acidic conditions. However, according to Schopff[37], benzoic acid reacts slowly with formaldehyde in the presence of concentrated sulfuric acid, producing methylene-3,3'-dibenzoic acid:

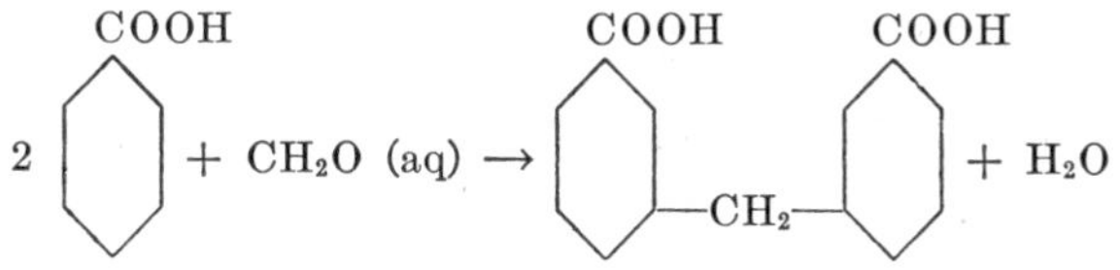

Nuclear derivatives of this type are readily obtained with aromatic hy-droxy acids, such as salicylic acid, which show the characteristic reactivity of phenolic compounds (pages 260–261). Smith, Sager and Siewers[38] report that pure, resin-free methylene bis-salicylic acid can be prepared by re-action of trioxane with salicylic acid in glacial acetic acid containing sulfuric acid as a reaction catalyst. The acid catalyzed reaction of meta-hydroxy-benzoic acid and formaldehyde yields 4-hydroxyphthalide and 6-methylol-1,3-benzodioxan-5-carboxylic acid lactone according to Harris, Shacklett and Block[24].

Methylene-bis derivatives of aryl sulfonic acids such as methylene-bis(beta-naphthalene sulfonic acid) can also be prepared by the acid condensation of these acids with formaldehyde as described by Tucker[44] in a patent covering the use of these materials as a dispersing agent for carbon black in water.

Acid Anhydrides. Acetic anhydride reacts with paraformaldehyde or alpha-polyoxymethylene ("trioxymethylene") in the presence of zinc chloride[17] or sulfuric acid[39] with formation of methylene and dioxymethylene diacetates:

$$(CH_3CO)_2O + CH_2O \text{ (p)} \rightarrow CH_3COOCH_2OOCCH_3$$

$$(CH_3CO)_2O + 2CH_2O \text{ (p)} \rightarrow CH_3COOCH_2OCH_2OOCCH_3$$

These products are colorless liquids boiling at 170 and 204 to 207°C, respectively. Methylene diacetate has an odor similar to that of other acetic esters, but has an irritating after-effect characteristic of formaldehyde itself, apparently due to hydrolysis on the mucous membranes. The preparation and properties of the higher polyoxymethylene diacetates are discussed in connection with formaldehyde polymers (page 133–137).

Methylene dipropionate and dibutyrate, which were prepared by Descudé[17] in the same manner from the acid anhydrides, boil at 190 to 192°C and 215 to 216°C, respectively.

Methylene dibenzoate[16], which may be obtained from benzoic anhydride, is a solid melting at 99°C and boiling at 225°C, with decomposition. Dry ammonia reacts with the molten ester to give methylene bis-benzamide.

Acyl Chlorides. When acyl chlorides are reacted with formaldehyde polymers, chloromethyl esters are obtained. According to Descudé[17], chloromethyl acetate is the principal product formed by the exothermic reaction which takes place when a mixture of acetyl chloride and paraformaldehyde is treated with a little anhydrous zinc chloride. Small quantities of dichloromethyl ether and methylene diacetate are formed as by-products. The following reactions are believed to take place:

$$CH_3COCl + CH_2O \text{ (p)} \rightarrow CH_3COOCH_2Cl$$
Chloromethyl acetate

$$2CH_3COOCH_2Cl \rightarrow (CH_3CO)_2O + (ClCH_2)_2O$$
Acetic   Dichloromethyl
anhydride   ether

$$(CH_3CO)_2O + CH_2O \text{ (p)} \rightarrow (CH_3COO)_2CH_2$$
Methylene diacetate

Dichloromethyl phthalate, phthalic anhydride, and dichloromethyl ether are produced in the same fashion from phthalyl chloride and paraformaldehyde under similar reaction conditions[19].

Chloromethyl acetate is a colorless liquid with an ester-like odor and an irritating after-odor; it boils at 110 to 112°C[15]. On heating for 3 hours at 160°C with alkali salts of organic acids, it gives methylene diesters[18]. Mixed esters such as methylene acetate benzoate (m.p. 38°C; b.p. 255 to 260°C) may be prepared thus:

$$CH_3COOCH_2Cl + C_6H_5COONa \rightarrow CH_3COOCH_2OOCC_6H_5 + NaCl$$

This method may be used with various chloromethyl esters and acid salts to synthesize a wide variety of methylene esters.

Methylene esters liberate formaldehyde on hydrolysis and may be readily detected by the odor of formaldehyde which is evolved on cautious addition of a little water to a mixture of ester and concentrated sulfuric acid.

Ketene. Ketene reacts with anhydrous formaldehyde in the presence of a suitable catalyst to give beta-propiolactone.

$$H_2C:O + H_2C:C:O \rightarrow \begin{array}{c} H_2C-C:O \\ | \quad | \\ H_2C-O \end{array}$$

Breyfogle and Steadman[40, 41] carry out the reaction by passing a mixture of paraformaldehyde vapors and ketene into liquid beta-propiolactone containing 0.5 per cent aluminum chloride and 0.05 per cent zinc chloride at 5 to 20°C. Numerous patents cover various reaction catalysts such as zinc perchlorate[10], zinc thiocyanate[11], zinc nitrate[12], esters of ortho- or meta-phosphoric acid[21], and metal fluoborates[22].

Esters. Simple esters of monocarboxylic acids apparently do not react with formaldehyde. Alpha halogenated esters undergo the Reformatsky reaction[3] with dry paraformaldehyde or polyoxymethylene and zinc, giving hydroxy esters as products. These reactions were studied by Blaise and co-workers[7, 8]. The reaction mechanism probably involves the formation of an organometallic intermediate as indicated below in the case of ethyl alpha-bromoisobutyrate:

$$(CH_3)_2CBrCOOC_2H_5 + Zn \rightarrow (CH_3)_2C(ZnBr)COOC_2H_5$$

$$(CH_3)_2C(ZnBr)COOC_2H_5 + CH_2O \text{ (p)} \rightarrow (CH_3)_2C(CH_2OZnBr)COOC_2H_5$$

$$(CH_3)_2C(CH_2OZnBr)COOC_2H_5 + H_2O \rightarrow (CH_3)_2C(CH_2OH)COOC_2H_5 + Zn(OH)Br$$

According to Blaise, the preparation may be carried out by suspending 14 g of the formaldehyde polymer, which has been desiccated at 120 to 130°C, in 120 g dry benzene and adding 35 g of fine zinc shavings and 40 g of the bromoester. The reaction is started by heating until it is self-sustaining, after which it may be moderated, if necessary, by cooling. An additional 40 g of ester is then added. The final reaction mixture is poured into water, to which small amounts of 20 per cent sulfuric acid are added to dissolve the zinc; the mixture is then filtered to remove unreacted formaldehyde polymer. The benzene layer is decanted and washed alternately with dilute ammonia and 20 per cent sulfuric acid. It is finally washed with water, dried with sodium sulfate, and distilled. The product, 2,2-dimethyl-beta-hydroxypropionic acid, distills at 84 to 86°C at 16 mm. Yields of 50 to 60 per cent are claimed. In order to start the organometallic reaction, it is often advisable to prime the reaction mixture by adding a little zinc which has been heated in a test tube with the bromoester until it has been attacked.

Vinyl acetate reacts with a solution of paraformaldehyde in acetic acid containing a strong mineral acid catalyst to give acrolein[45]. The reaction mechanism is probably similar to that of the Prins reaction (pages 327–335) between olefins and formaldehyde.

$$CH_2{:}CH{\cdot}OOC{\cdot}CH_3 + HOCH_2HSO_4(CH_2O + H_2SO_4) \rightarrow$$
$$HOCH_2{\cdot}CH_2{\cdot}CH(HSO_4)OOC{\cdot}CH_3$$

$$HOCH_2{\cdot}CH_2{\cdot}CH(HSO_4)OOC{\cdot}CH_3 \rightarrow CH_2{:}CH{\cdot}CHO + H_2SO_4 + CH_3COOH$$

Evidence for this mechanism is found in the fact that acetylation of the primary reaction mixture after neutralization of the mineral acid catalyst with sodium acetate yields hydracrylic aldehyde triacetate, CH_3COOCH_2-$CH_2CH(OOCCH_3)_2$ and allylidene diacetate $CH_2{:}CHCH(OOCCH_3)_2$.

Esters containing an activated alpha-hydrogen atom, such as malonic, acetoacetic, and cyanoacetic esters, react readily with formaldehyde. These reactions are similar in many respects to ketone-formaldehyde condensations.

According to Welch[46], the primary reaction product of formaldehyde and malonic ester is monomethylolmalonic ester:

$$CH_2O \text{ (aq)} + H_2C(COOC_2H_5)_2 \rightarrow HOCH_2CH(COOC_2H_5)_2$$

Unfortunately, this product is apparently too reactive to be isolated in the pure state. However, dimethylolmalonic ester, $(HOCH_2)_2C(COOC_2H_5)_2$, was isolated by Welch as a crystalline solid melting at 52 to 53°C when formaldehyde solution was reacted with malonic ester at a temperature below 50°C, using sodium hydroxide as a catalyst. The monomethylol derivative of ethyl malonic ester $[HOCH_2C(C_2H_5)(COOC_2H_5)_2]$ was obtained by a similar procedure. Ammonia and alkylamines were found to catalyze the formation of these simple derivatives at a rate inversely proportional to the hydrogen-ion concentration. Piperidine, however, appeared to possess a more specific catalytic action[47].

On heating the primary malonic ester-formaldehyde reaction mixtures, dehydration takes place with the formation of methylenemalonic ester, $CH_2:C(COOC_2H_5)_2$, and a series of products consisting of methylene-linked malonic ester radicals. Perkin and co-workers[25, 34] obtained these products by reacting formaldehyde and malonic ester in the presence of diethylamine and distilling the reaction mixture. In the primary exothermic stage of the reaction, the mixture was cooled with ice. It was then allowed to stand overnight at room temperature and finally heated in a water bath for several hours. When 96 g of malonic ester, 30 g of commercial formaldehyde, and 1.5 g of diethylamine were reacted in this way, the following products were isolated:

Methylenemalonic ester $CH_2:C(COOC_2H_5)_2$	3 g
Propane 1,1,3,3-tetracarboxylic ester (Methylene bis-malonic ester) $CH(COOC_2H_5)_2 \cdot CH_2 \cdot CH(COOC_2H_5)_2$	25 g
Pentane 1,1,3,3,5,5-hexacarboxylic ester (Methylene tris-malonic ester) $CH(COOC_2H_5)_2 \cdot CH_2 \cdot C(COOC_2H_5)_2 \cdot CH_2 \cdot CH(COOC_2H_5)_2$	48 g
High-boiling products (unidentified)	10 g

Welch[48] obtained methylene bis-malonic ester in better than 90 per cent yield by reacting paraformaldehyde and malonic ester in the presence of alcoholic potassium hydroxide, which was neutralized with alcoholic hydrogen chloride before isolating the product by distillation.

Pure methylenemalonic ester boils at 208 to 210°C. It is a colorless liquid with an irritating odor and polymerizes on standing to a clear organic glass, designated by Perkin as metamethylenemalonate. It adds bromine in a normal manner, giving a dibromo derivative. Crude methylenemalonic ester, as obtained by Perkin, polymerized to a low molecular weight substance, paramethylenemalonate, which melted at 154 to 156°C on recrystallization from hot alcohol. On heating, the para-polymer depolymerizes to give pure monomeric methylene malonate. Methylenemalonic ester was first prepared by Perkin[33] in 1886 by heating paraformaldehyde and malonic ester with acetic anhydride, employing the technique developed by Knoevenagel and applied by Komnenos[28] to acetaldehyde-malonic ester reactions. Yields of methylenemalonic ester obtained by Perkin and other early investigators were extremely low. A new procedure developed by Bachmann and Tanner[4, 5] is reported to give approximately 50 per cent yields. In this process glacial acetic acid is used as a reaction medium, formaldehyde is added as paraformaldehyde, and a mixture of copper and potassium acetate is employed as a catalyst. Methylenemalonic ester may also be prepared by the gas-phase reaction of malonic ester with formaldehyde in the presence of solid phosphates such as aluminum, copper, and di- or trisodium phosphates[6].

The methylene derivative of dimethylmalonic ester was prepared by Meerwein and Schürmann[31], who reacted dimethylmalonate with commercial formaldehyde solution, using piperidine as a catalyst. According to these investigators, it distills at 200 to 203°C and polymerizes on standing to a rubber-like mass.

Acetoacetic ester reacts readily with formaldehyde solution, giving methylene bis(ethyl acetoacetate) in good yield[35]:

$$2CH_3COCH_2COOC_2H_5 + CH_2O \text{ (aq)} \rightarrow \begin{matrix} CH_3COCHCOOC_2H_5 \\ | \\ CH_2 \\ | \\ CH_3COCHCOOC_2H_5 \end{matrix} + H_2O$$

The ester, which separates from the reaction mixture as an almost colorless, water-immiscible liquid, decomposes on heating and cannot be distilled. Klages and Knoevenagel[27], who first reported its preparation, carried out this reaction in the presence of pyridine or diethylamine. In the presence of these catalysts the condensation is highly exothermic and readily proceeds further with formation of cyclic products, as indicated in this sequence:

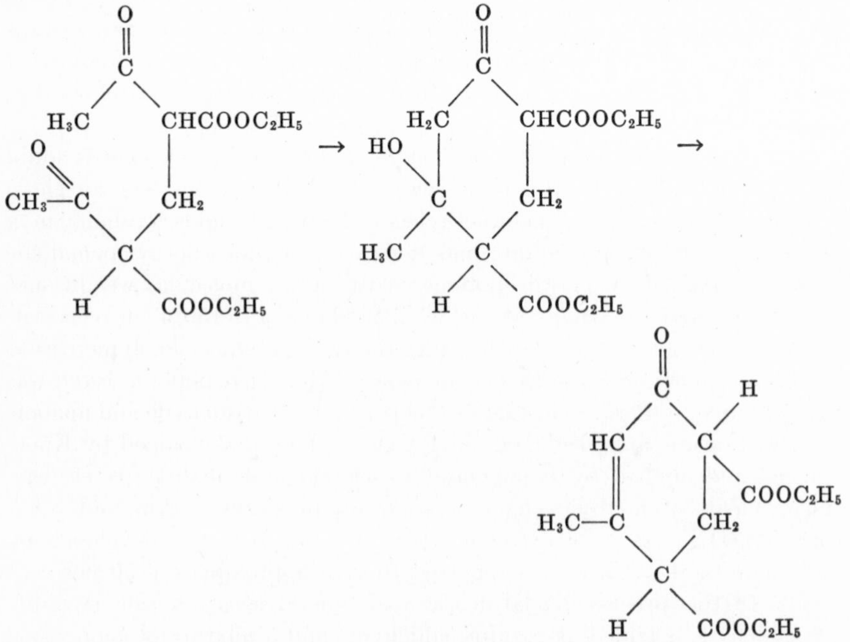

Methylene bis(ethyl acetoacetate), as isolated from the reaction of formaldehyde and acetoacetic ester, possesses a high degree of purity and reacts quantitatively with ammonia to give a dihydrolutidine dicarboxylic ester (m.p. 174 to 176°C)[27]:

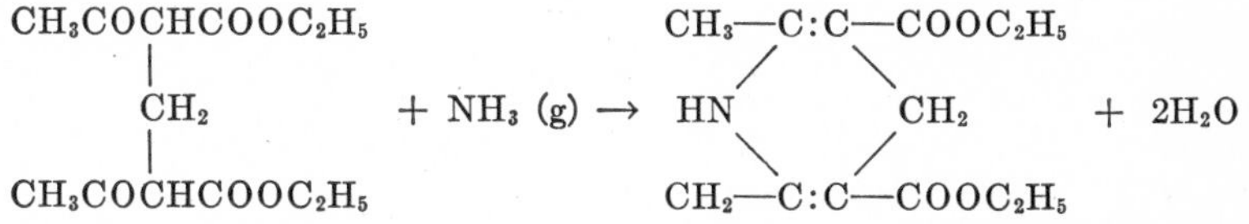

$$CH_3COCHCOOC_2H_5 \quad\quad CH_3-C:C-COOC_2H_5$$

(structure)

$$CH_2 \quad + NH_3\ (g) \rightarrow \quad HN \quad CH_2 \quad + 2H_2O$$

$$CH_3COCHCOOC_2H_5 \quad\quad CH_2-C:C-COOC_2H_5$$

By oxidation with nitric acid, followed by hydrolysis with alkali and decarboxylation with lime, this dihydrolutidine dicarboxylic ester may be converted to 2,6-lutidine in 63 to 65 per cent theoretical yield[9].

Harries[23] reports the preparation of methylol acetoacetic ester by the reaction of acetoacetic ester with cold, anhydrous liquid formaldehyde. A process for the preparation of resins by reaction of acetoacetic ester and formaldehyde in the presence of sodium hydroxide or morpholine has been patented by D'Alelio[14].

Alpha-cyanoacetic esters, such as ethyl cyanoacetate and ethyl alphacyanopropionate, react with formaldehyde in the presence of diethylamine, giving liquid methylene bis-derivatives. In each case, a solid amorphous product having the same empirical composition as the methylene derivative is also obtained. Auwers and Thorpe[2], who investigated these reactions, concluded that the solid product was a polymer of the methylene bis-cyanoester. A liquid resin is obtained by reacting formaldehyde solution with ethyl cyanoacetate in the presence of sodium carbonate[13]. Higson and Thorpe[26] prepared a cyanomethyl derivative of cyanoacetic ester by reacting it with formaldehyde cyanohydrin and elemental sodium:

$$HOCH_2CN + H_2C(CN)COOEt \rightarrow CNCH_2CH(CN)COOEt + H_2O$$

Ardis[1] obtains alkyl esters of alpha-cyanoacrylic acid by reaction of an alkyl cyanoacetate with formaldehyde in the presence of a basic catalyst at 50 to 90°C. This reaction gives a water-immiscible phase, reported to consist of partially polymerized cyanoacrylic ester, which depolymerizes on vacuum distillation.

References

1. Ardis, A. E., (to B. F. Goodrich Co.), U. S. Patent 2,467,927 (1949).
2. Auwers, K., and Thorpe, J. F., *Ann.*, **285**, 322 (1895).
3. Adams, R., and Schreiner, R. L., "Organic Reactions," pp. 2–36, New York, John Wiley and Sons (1942).
4. Bachmann, G. B., and Tanner, H. A., *J. Org. Chem.*, **4**, 493–501 (1939).
5. Bachmann, G. B., and Tanner, H. A., (to Eastman Kodak Co.), U. S. Patent 2,212,506 (1940).
6. *Ibid.*, U. S. Patent 2,313,501 (1943).

7. Blaise, E. E., *Compt. ren.d*, **134**, 551 (1902).
8. Blaise, E. E., and Marcilly, L., *Bull. soc. chim* (3), **31**, 111, 319 (1904); Blaise, E. E., and Lutringer, *Ibid.*, *(3)*, **33**, 635 (1905).
9. Blatt, A. H., "Organic Syntheses," Collective *Vol. II*, pp. 214–6, New York, John Wiley & Sons (1943).
10. Caldwell, J. R., (to Eastman Kodak Co.), U. S. Patent 2,450,116 (1948).
11. *Ibid.*, U. S. Patent 2,450,117 (1948).
12. *Ibid.*, U. S. Patent 2,450,118 (1948).
13. D'Alelio, G. F., (to General Electric Co.), U. S. Patent 2,276,828 (1942).
14. *Ibid.*, U. S. Patent 2,277,479 (1942).
15. Descudé, M., *Compt. rend.*, **132**, 1568 (1901).
16. *Ibid.*, **133**, 372 (1901).
17. Descudé, M., *Bull. soc. chim.* *(3)*, **27**, 867 (1902).
18. Descudé, M., *Compt. rend.*, **134**, 716 (1902).
19. *Ibid.*, **134**, 1065 (1902).
20. Florence, G., *Bull. soc. chim.* *(4)*, **41**, 444 (1927).
21. Hagemeyer, H. J., (to Eastman Kodak Co.), U. S. Patent 2,450,131 (1948).
22. *Ibid.*, U. S. Patent 2,450,133 (1948).
23. Harries, C., *Ber.*, **34**, 635 (1901).
24. Harris, J. O., Shacklett, C., and Block, B. P., *J. Am. Chem. Soc.*, **66**, 417–8 (1944); **68**, 574–7 (1946).
25. Haworth, E., and Perkin, W. H., *J. Chem. Soc.*, **73**, 339–345 (1898).
26. Higson, A., and Thorpe, J. F., *J. Chem. Soc.*, **89**, 1461 (1906).
27. Klages, A., and Knoevenagel, E., *Ann.*, **281**, 94–104 (1894).
28. Komnenos, T., *Ann.*, **218**, 145 (1883).
29. Mannich, C., and Bauroth, M., *Ber.*, **55**, 3504–10 (1922).
30. Mannich, C., and Ganz, E., *Ber.*, **55**, 3486–3504 (1922).
31. Meerwein, H., and Schürmann, W., *Ann.*, **398**, 214–218 (1913).
32. Paquot, C., and Perron, R., *Bull. Soc. Chim.* (*5*), **5**, 855–7 (1948).
33. Perkin, W. H., *Ber.*, **19**, 1053 (1886).
34. Perkin, W. H., and Bottomley, J. F., *J. Chem. Soc.*, **77**, 294–309 (1900).
35. Rabe, R., and Rahm, F., *Ann.*, **332**, 10–11 (1904).
36. Ralston, A. W., and Vander Wal, R. J., (to Armour and Co.), U. S. Patent 2,145, 801 (1939).
37. Schopff, M., *Ber.*, **27**, 2321–6 (1894).
38. Smith, W. H., Sager, E. E., and Siewers, I. J., *Anal. Chem.*, **21**, 1334 (1949).
39. Späth, E., *Monatsh.*, **36**, 30–34 (1915).
40 Steadman, T. R., (to B. F. Goodrich Co.), U. S. Patent 2,424,589 (1947).
41. Steadman, T. R. and Breyfogle, P. L., (to B. F. Goodrich Co.), U. S. Patent 2,424,590 (1947).
42. Sternberg, W., *Pharm. Ztg.*, **46**, 1003–4 (1901); *Chem. Zentr.*, **1902**, I, 299.
43. *Ibid.*, **46**, 1004 (1901); *Chem. Zentr.*, **1902**, I, 299–300.
44. Tucker, G. R., (to Dewey and Almy Chem. Corp.), U. S. Patent 2,046,757 (1936).
45. Walker, J. F., (to E. I. du Pont de Nemours & Co., Inc.), U. S. Patent 2,478,989 (1949).
46. Welch, K. N., *J. Chem. Soc.*, **1930**, 257.
47. *Ibid.*, **1931**, 653–657.
48. *Ibid.*, 673–4.

CHAPTER 14

REACTIONS OF FORMALDEHYDE WITH AMINES, AMIDES AND NITRILES

Organic nitrogen compounds, such as amines, amides, ureides, amino acids, and proteins containing hydrogen attached to nitrogen, react with formaldehyde to yield a wide variety of chemical products ranging from simple organic compounds of low molecular weight to complex resins. Nitriles are relatively unreactive but, under some circumstances, behave as amide precursors. Representative reactions of formaldehyde with these compounds will be reviewed in the following pages beginning with the simple aliphatic amines.

Aliphatic Amines

Formaldehyde reacts readily with primary and secondary amines which contain active hydrogen atoms, but does not react with tertiary amines.

Methylolamines (alkylaminomethanols) are apparently the primary reaction products of formaldehyde and mono- and dialkylamines. They are the ammono-analogs of the unstable hemiacetals whose formation from formaldehyde and alcohols has been previously discussed (pages 202–203). According to Henry[81, 82], methylolamines are readily prepared by the gradual addition of a mono- or dialkylamine to commercial formaldehyde containing a molecular equivalent of dissolved formaldehyde, and may be isolated from the reaction mixture as a separate phase by addition of potassium carbonate. Mono- and dimethylamine are postulated to react as indicated in the following equations, yielding methylaminomethanol and dimethylaminomethanol, respectively.

$$CH_3NH_2 + CH_2O \text{ (aq)} \rightarrow CH_3NHCH_2OH$$

$$(CH_3)_2NH + CH_2O \text{ (aq)} \rightarrow (CH_3)_2NCH_2OH$$

The above products, as well as the higher homologs obtained from mono- and diethylamine are described as colorless, viscous liquids which decompose on heating and cannot be distilled without decomposition. Since the commercial formaldehyde employed by Henry in these preparations probably contained methanol, it is possible that his products were contaminated with this impurity. However, in the case of benzylamine, he succeeded in isolating a pure crystalline benzylaminomethanol ($C_6H_5CH_2NHCH_2OH$) which melted at 43°C[81].

When simple primary amines are treated with a large excess of 30 per cent formaldehyde solution at low temperatures, cyclic triformals are produced. Bergmann and Miekeley[10] isolated these products by extracting the cold reaction mixture with ether and subjecting the extract to fractional distillation after drying with anhydrous sodium sulfate. Ethylamine triformal (b.p. 62 to 64°C at 42 mm) is obtained in this manner.

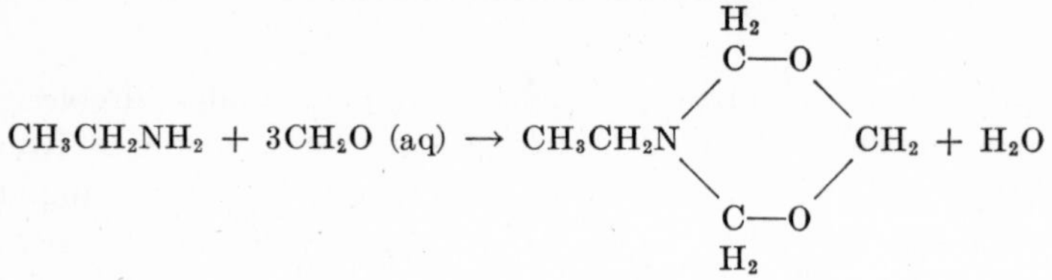

$$
CH_3CH_2NH_2 + 3CH_2O \text{ (aq)} \rightarrow CH_3CH_2N
\begin{array}{c}
H_2 \\
C-O \\
\diagup \qquad \diagdown \\
\qquad\qquad CH_2 + H_2O \\
\diagdown \qquad \diagup \\
C-O \\
H_2
\end{array}
$$

The mechanism of the above reaction may be envisaged as involving trioxymethylene glycol, whose presence in strong formaldehyde solution is indicated by the work of Auerbach and Barschall (pages 48–49).

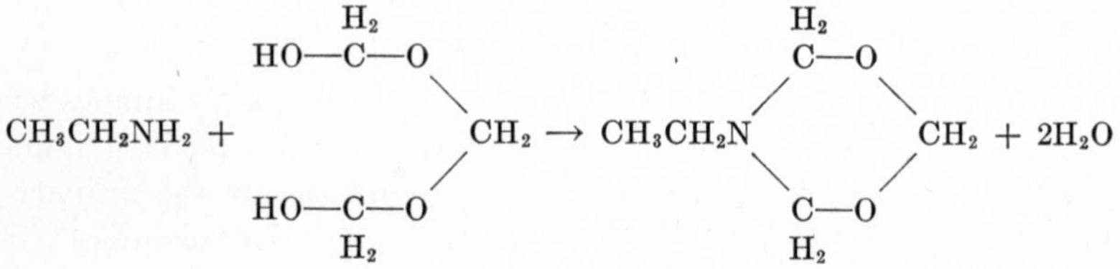

$$
CH_3CH_2NH_2 +
\begin{array}{c}
H_2 \\
HO-C-O \\
\diagdown \\
\quad CH_2 \\
\diagup \\
HO-C-O \\
H_2
\end{array}
\rightarrow CH_3CH_2N
\begin{array}{c}
H_2 \\
C-O \\
\diagup \quad \diagdown \\
\qquad CH_2 + 2H_2O \\
\diagdown \quad \diagup \\
C-O \\
H_2
\end{array}
$$

Methyleneamines are the final products of alkylamine-formaldehyde condensations. Primary and secondary alkylamines yield cyclic trialkyl trimethylenetriamines and tetraalkyl methylenediamines, respectively, as shown by the following equations, in which R and R′ stand for alkyl groups:

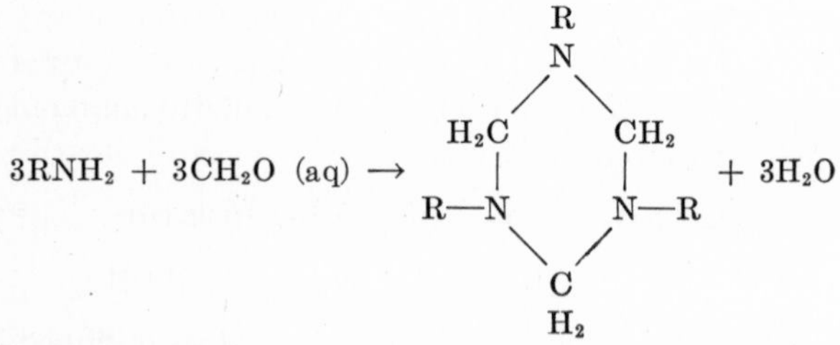

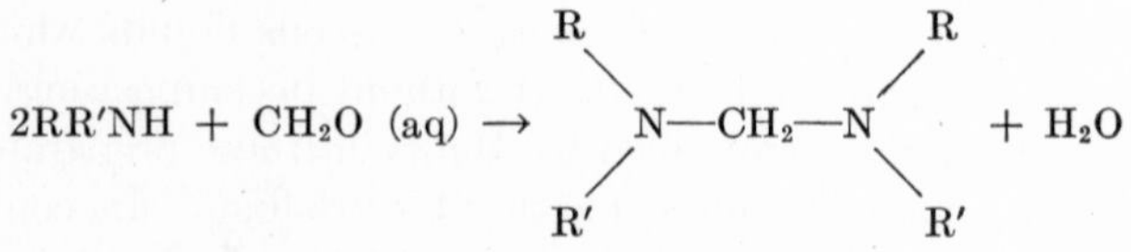

$$
3RNH_2 + 3CH_2O \text{ (aq)} \rightarrow
\begin{array}{c}
R \\
N \\
\diagup \quad \diagdown \\
H_2C \qquad CH_2 \\
| \qquad\quad | \\
R-N \qquad N-R \\
\diagdown \quad \diagup \\
C \\
H_2
\end{array}
+ 3H_2O
$$

$$
2RR'NH + CH_2O \text{ (aq)} \rightarrow
\begin{array}{c}
R \qquad\qquad R \\
\diagdown \qquad\qquad \diagup \\
N-CH_2-N \\
\diagup \qquad\qquad \diagdown \\
R' \qquad\qquad R'
\end{array}
+ H_2O
$$

These substituted methyleneamines are the ammono-analogs of formals. In general, they are colorless liquids possessing characteristic amine-like odors with an after-odor of formaldehyde. Like formals, they are stable in the dry state or in the presence of alkalies, but are readily hydrolyzed by acids.

They are best prepared by adding the amine gradually and with cooling to strong aqueous formaldehyde (approximately 37 per cent). The reaction mixture is then treated with caustic alkali, which causes the product to separate as an upper phase. The substituted methyleneamines are readily purified by distillation over solid caustic. If desired, the reaction mixture may be extracted with ether. This is of particular value in the case of the lower derivatives, which are the more soluble in water.

Trialkyl cyclotrimethylenetriamines are also obtained when monomethylolalkylamines are distilled over solid caustic[81]:

$$3RNHCH_2OH \rightarrow (RN \cdot CH_2)_3 + 3H_2O$$
$$| \quad |$$

In a similar manner, dialkyl methylolamines condense on warming with a mol of a dialkylamine to give tetraalkyl methylenediamines[81]:

$$R_2NCH_2OH + R_2NH \rightarrow R_2NCH_2NR_2 + H_2O$$

Trimethyl cyclotrimethylenetriamine boils in the range 163 to 166°C at ordinary pressures. It was apparently first prepared by Henry[80], who believed it to be methyl methyleneimine, $CH_3N:CH_2$. Cambier and Brochet later demonstrated that its molecular weight corresponded to that of the trimer, $(CH_3N:CH_2)_3$, whose cyclic structure was confirmed by the work of Duden and Scharff[43]. Other homologous trialkyl cyclotrimethylenetriamines include triethyl trimethylenetriamine (b.p. 196 to 198°C) prepared from ethylamine by Einhorn and Prettner[51], and triisobutyl trimethylenetriamine (b.p. 255°C) prepared from isobutylamine by Graymore[69].

On reduction with zinc dust and hydrochloric acid, trialkyl cyclotrimethylenetriamines yield methyl alkylamines as the principal product.

Unsubstituted cyclotrimethylenetriamine, which may be regarded as the parent compound of this series of products, is apparently an intermediate in the formation of hexamethylenetetramine from formaldehyde and ammonia (page 408).

Some monomeric, unsaturated methyleneimines are sufficiently stable to be isolated in a pure state. Hurwitz[85] has demonstrated the preparation of products of this type by reaction of formaldehyde with amines in which the amino group is attached to a tertiary carbon atom. Tertiary octyl methyleneimine[149] can be obtained in good yield by reacting a tertiary octylamine with an equimolar quantity of approximately 37 per cent aqueous formaldehyde while maintaining the temperature below 40°C, cooling to room temperature, adding 10 g potassium hydroxide, separating the organic layer and distilling at atmospheric pressure after drying with solid potassium hydroxide. Tertiary octyl methyleneimine distills at 147 to 151°C.

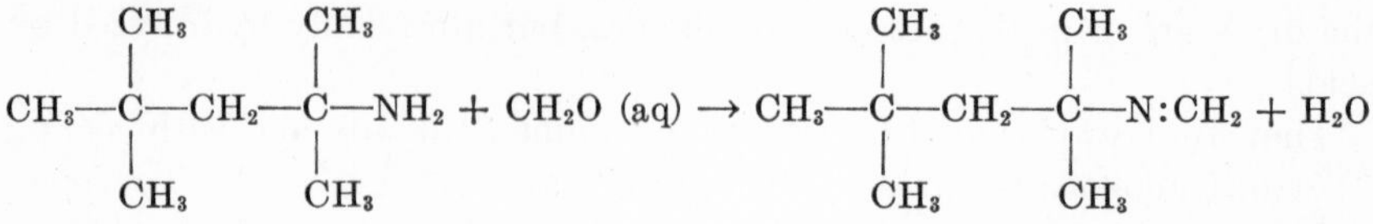

$$\underset{\overset{|}{CH_3}}{\overset{\overset{CH_3}{|}}{CH_3-C}}-CH_2-\underset{\overset{|}{CH_3}}{\overset{\overset{CH_3}{|}}{C}}-NH_2 + CH_2O \text{ (aq)} \rightarrow \underset{\overset{|}{CH_3}}{\overset{\overset{CH_3}{|}}{CH_3-C}}-CH_2-\underset{\overset{|}{CH_3}}{\overset{\overset{CH_3}{|}}{C}}-N:CH_2 + H_2O$$

t-Octyl methyleneimine

Luskin and co-workers[113] describe the preparation of tertiary butyl methyleneimine or azomethine by cracking the primary reaction product of *t*-butylamine and formaldehyde.

The simplest organic diamine, methylenediamine, $CH_2(NH_2)_2$, apparently reacts with formaldehyde and ammonia to give cyclotrimethylenetriamine and hexamethylene tetramine. Although this diamine cannot be isolated in a pure state, research on the mechanism of hexamethylene tetramine formation[146] (cf. Chapter 19) indicates that it may be present in formaldehyde-ammonia reaction mixtures under some circumstances and can be isolated as the sulfate whose stability was demonstrated by Knudsen[98] (page 292). Tetramethyl methylenediamine (b.p. 85°C) is the final reaction product of formaldehyde and dimethylamine. Homologous compounds of this type such as tetraethyl methylenediamine (b.p. 168°C) are obtained with the corresponding amines. The dinitro derivative of methylene diamine, prepared by the nitration and subsequent hydrolysis of methylenediacetamide[20], was shown by Woodcock[199] to yield a dimethylol derivative when treated with formaldehyde gas in ethyl acetate solution. The related dinitro derivative of ethylenediamine gives a monomethylol derivative with aqueous formaldehyde and dipiperodinomethyl derivative with aqueous formaldehyde and piperidine[199].

Diamines, such as ethylenediamine and trimethylenediamine, react with formaldehyde to give products whose exact structure has not been determined, but which are probably complex methyleneamines. The condensation usually results in the combination of one mol of diamine with two mols of formaldehyde[167].

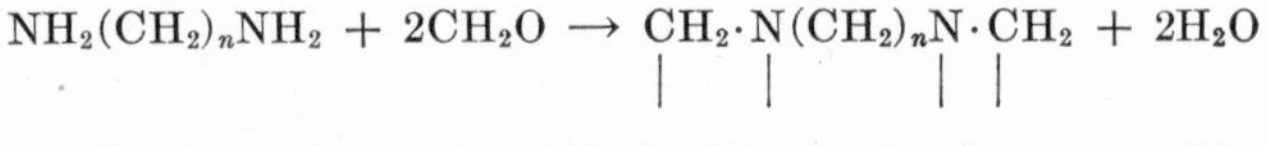

$$NH_2(CH_2)_nNH_2 + 2CH_2O \rightarrow \underset{|\quad\quad|}{CH_2 \cdot N(CH_2)_n N \cdot CH_2} + 2H_2O$$

In the case of ethylenediamine, Bischoff[14] obtained a crystalline product (m.p. 196°C) whose molecular weight indicated it to be a dimer of the unit shown in the type reaction above. It is believed to have the following structural formula:

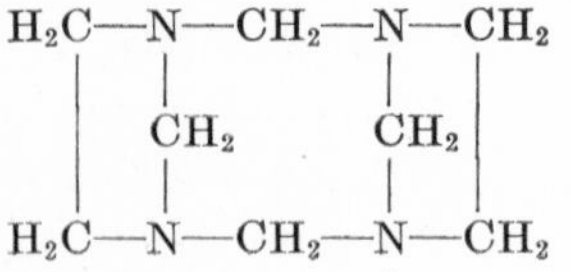

$$\begin{array}{c}H_2C-N-CH_2-N-CH_2 \\ |\qquad|\qquad| \\ CH_2\qquad CH_2 \\ |\qquad|\qquad| \\ H_2C-N-CH_2-N-CH_2\end{array}$$

The products from tri- and tetramethylenediamines are high-boiling liquids, whereas the product from pentamethylenediamine is a solid[15, 167]. Kern[95] describes the reaction of hexamethylenediamine with two molar equivalents of formaldehyde in dilute aqueous solution at room temperature as giving a colorless, voluminous, elastic product which is insoluble in water and organic solvents. However, on boiling with three times its weight of 30 per cent formaldehyde it goes into solution after four to six hours.

Under acidic conditions, amines react with formaldehyde to form compounds which, on neutralization, give those methyleneamines which would be normally expected from the reaction in alkaline media. However, according to Werner[196], there is evidence that in the case of primary amines methyleneamines are not formed in the acidic mixture, but that dissociated salts of methyleneimines are:

$$CH_3NH_2 \cdot HCl + CH_2O \text{ (aq)} \rightarrow CH_2{:}NCH_3(HCl) + H_2O$$

With secondary amines, salts of tetraalkyl methylenediamines are obtained, the reaction being practically unaffected by the presence of acid[196].

On heating amines, diamines, or their salts with excess formaldehyde, secondary or tertiary methyl alkylamines are formed; they may be isolated on treatment of the reaction mixture with caustic alkali[52, 53]. Reactions of this type are involved in the preparation of trimethylamine from formaldehyde and ammonium chloride (pages 182–183). As has been previously pointed out, a portion of the formaldehyde serves as a reducing agent and is converted to formic acid or carbon dioxide. Reactions take place slowly and incompletely in the neighborhood of 100°C, but are completed in a few hours when carried out under pressure at 120 at 160°C. Reactions of this type are also obtained when formaldehyde is employed in polymeric form, e.g., paraformaldehyde. The fact that hydrogenation catalysts are claimed to be of value in the preparation of tertiary methylamines by heating secondary amines with paraformaldehyde[36] may indicate that hydrogen is actually liberated in the course of the reaction, which is in some respects quite similar to the crossed Cannizzaro reaction.

Tertiary amines are obtained in yields of better than 80 per cent theoretical when simple aliphatic amines are warmed with formaldehyde and formic acid[32]. Under these conditions formic acid acts as the reducing agent.

The reduction of the cyclic trialkyl trimethylenetriamines with zinc and hydrochloric acid, which has already been noted, is indicative of the ease with which products of this type are hydrogenated. That methylolamines can also be reduced to methylamines is indicated by the fact that dimethylethanolamine can be prepared by a low-temperature, catalytic hydrogenation of the viscous liquid [probably dimethylolethanolamine, $HOCH_2$-

$CH_2N(CH_2OH)_2$] which is obtained by dissolving paraformaldehyde equivalent to two mols of formaldehyde in one mol of ethanolamine[25].

The Mannich reaction and other joint reactions[1] of formaldehyde with primary or secondary amines and various polar compounds lead to the formation of substituted aminomethyl derivatives.

Reactions of dimethylamine and formaldehyde with hydrogen cyanide[81], isobutanol[148], phenol[38], and acetone[54] are typical:

$$(CH_3)_2NH + CH_2O \text{ (aq)} + HCN \rightarrow (CH_3)_2NCH_2CN + H_2O$$

$$(CH_3)_2NH + CH_2O \text{ (aq)} + (CH_3)_2CHCH_2OH \rightarrow (CH_3)_2NCH_2OCH_2CH(CH_3)_2 + H_2O$$

$$(CH_3)_2NH + CH_2O \text{ (aq)} + C_6H_5OH \rightarrow$$

$$(CH_3)_2NCH_2C_6H_4OH \text{ (p)} + H_2O \quad \text{and} \quad C_6H_5OCH_2N(CH_3)_2 + H_2O$$

$$(CH_3)_2NH + CH_2O \text{ (aq)} + CH_3COCH_3 \rightarrow (CH_3)_2NCH_2CH_2COCH_3 + H_2O$$

Lieberman and Wagner[107] present the hypothesis that the Mannich reaction involves dual catalysis in an amphoteric system in which a cation, R_2NC^+, is formed from formaldehyde which then condenses with the anion of a reactive hydrogen compound. The formation of the cation is induced by acid and the formation of the anion by basic conditions. Excess acidity interferes with the primary condensation by preventing formation of the desired anion, whereas excess alkali suppresses formation of cation. Walker and Chadwick[190] have shown that trioxane can be employed as a special form of formaldehyde in the Mannich reaction of piperidine hydrochloride and acetophenone without use of a solvent.

Smith, Bullock, Bersworth and Martell[161] have demonstrated the synthetic utility of the joint reaction of sodium cyanide and formaldehyde with amines for the production of carboxymethyl derivatives such as ethylenediamine tetraacetic acid.

$$H_2NCH_2CH_2NH_2 + 4NaCN + 4CH_2O + 4H_2O \rightarrow$$

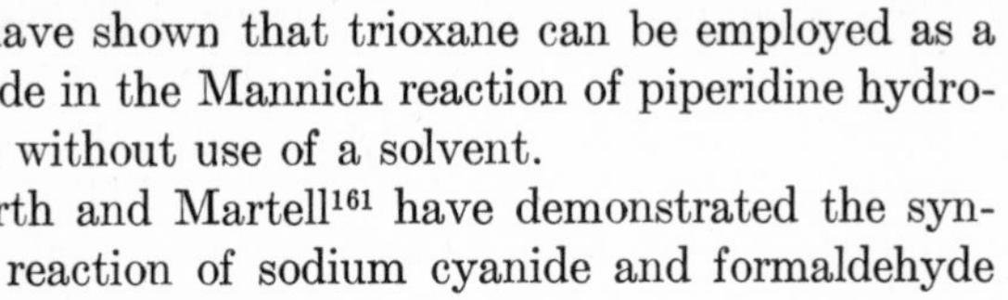

$$\begin{array}{ccc} NaOOCCH_2 & & CH_2COONa \\ & \diagdown \quad \diagup & \\ & NCH_2CH_2N & + 4NH_3 \\ & \diagup \quad \diagdown & \\ NaOOCCH_2 & & CH_2COONa \end{array}$$

The reaction gives high yields at about 50 to 80°C under strongly alkaline conditions. For best results the ammonia should be removed rapidly as formed by vacuum distillation since otherwise, it is carboxymethylated with formation of glycines (see page 186). It is believed that this reaction does not involve the formation of substituted alpha-aminonitriles as intermediates. The mechanism proposed is shown below:

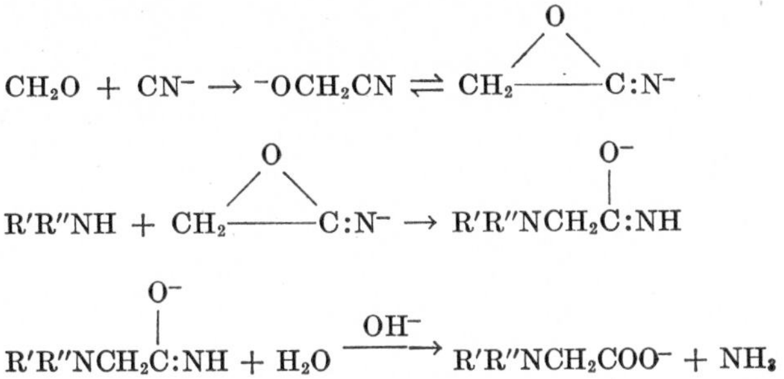

$$CH_2O + CN^- \rightarrow {}^-OCH_2CN \rightleftharpoons CH_2 \overset{O}{\diagup \diagdown} C:N^-$$

$$R'R''NH + CH_2 \overset{O}{\diagup \diagdown} C:N^- \rightarrow R'R''NCH_2C:NH \; \overset{O^-}{|}$$

$$R'R''NCH_2C:NH \; \overset{O^-}{|} + H_2O \xrightarrow{OH^-} R'R''NCH_2COO^- + NH_3$$

This reaction can be applied to a wide range of aliphatic amines and polyamines having replaceable hydrogen atoms. A commercial process involving this synthesis has been patented by Bersworth[11]. A related process by Ulrich and Ploetz[178] which takes place under neutral or acidic conditions yields the substituted alpha-amino nitriles as primary products.

Hallowell synthesizes dimethyl glycine by heating dimethylamine, formaldehyde and carbon monoxide under pressure in the presence of hydrochloric acid[78].

Production of alkyl propargylamines in high yields by the joint reaction of formaldehyde, acetylene, and dialkylamines is claimed[145]:

$$R_2NH + CH_2O \text{ (aq)} + HC:CH \rightarrow R_2NCH_2C:CH + H_2O$$

This reaction is carried out in acetic acid under pressure, using cuprous chloride as a catalyst. 1,4-Di-(alkylamine)-butines, $R_2NCH_2C:CCH_2NR_2$, are also formed in small amounts.

Methylamine, formaldehyde, and hydrogen sulfide react to form methylthioformaldine, $(CH_2)_3S_2NCH_3$[198].

A joint reaction of formaldehyde, secondary amines, and hydrogen selenide has also been described[13].

Aromatic Amines

Under ordinary conditions, the reaction of aniline with dilute formaldehyde gives a mixture of products. However, with proper control anhydroformaldehyde aniline (m.p. 140 to 141°C) may be obtained as the principal product[138, 176]. This compound, which is the trimer of the hypothetical Schiff's base, methylene aniline, $(C_6H_5N:CH_2)_3$, probably has a cyclic structure analogous to that of the trialkyl trimethylenetriamines[123, 141]:

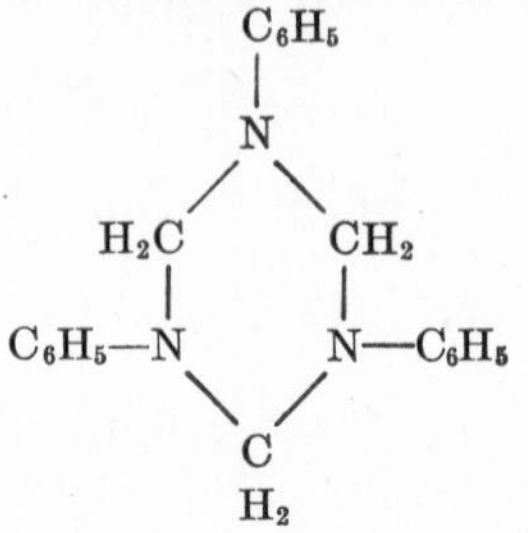

$$
\begin{array}{c}
C_6H_5 \\
| \\
N \\
\diagup \quad \diagdown \\
H_2C \qquad CH_2 \\
| \qquad\qquad | \\
C_6H_5{-}N \qquad N{-}C_6H_5 \\
\diagdown \quad \diagup \\
C \\
H_2
\end{array}
$$

Anhydro-formaldehyde Aniline

On heating in molten camphor to 176°C, anhydro-formaldehyde aniline dissociates, giving a molecular weight value corresponding approximately to the dimer of methyleneaniline[122].

When treated with anhydrous liquid formaldehyde at low temperatures, aniline gives a colorless microcrystalline product melting at 172 to 180°C, apparently an isomer, higher polymer, or metastable form of anhydro-formaldehyde aniline. Recrystallization of this material from toluene quantitatively converts it to the normal product[184].

Reaction rate studies by Ogata and co-workers[129a] indicate that the aniline-formaldehyde reaction is second order with respect to the initial aniline concentration and first order with respect to the sum of the concentrations of formaldehyde and methylene aniline. The reaction rate for substituents in the aniline nucleus decreases in the order: $mCH_3 > oCH_3 > $ None $ > mCl > pCH_3 > pCl$.

Other aryl monoamines, such as the toluidines, also give products analogous to anhydro-formaldehyde aniline when treated with formaldehyde[195].

The mechanism of the formation of cyclic triaryl trimethylenetriamines undoubtedly involves methylolamines as primary reaction products. According to Sutter[170], these methylol derivatives may be isolated by reacting two or more molecular proportions of aqueous formaldehyde with aryl monoamines in the presence of sodium carbonate. Avoidance of secondary reactions is facilitated by diluting with a water-immiscible solvent, such as benzene, ether, and the like. Dimethylolarylamines dissolve in these solvents and the extract thus obtained can be freed of formaldehyde by extraction with cold sodium sulfite solution. To isolate the monomethylolarylamines, the solvent is removed from the dried extract by vacuum evaporation. Dimethylolaniline [$C_6H_5N(CH_2OH)_2$] prepared in this way is a clear, almost colorless syrup. When its solution in alcohol is mixed with water, crystals of anhydro-formaldehyde aniline separate and formaldehyde is liberated. On long standing, or on heating above approximately 40°C, the product gradually decomposes. In strongly alkaline solutions, bis-

(phenylamino)-methane (m.p. 64 to 65°C) is obtained from aniline and formaldehyde[44].

$$2C_6N_5NH_2 + CH_2O \rightarrow C_6H_5NHCH_2NHC_6H_5 + H_2O$$

In acid solution aromatic amines form resins with formaldehyde. This reaction involves the formation of nuclear methylene linkages similar to those which occur in phenol-formaldehyde resins. The reaction type is characteristic of the aromatic nucleus rather than of the activating amine group, whose presence merely facilitates reaction. As a result, resins can be obtained even with tertiary amines such as dimethylaniline. Under these conditions the amine reacts as an ammono-phenol. Studies of the mechanism by which formaldehyde reacts with N-alkyl anilines in acid solution indicate that the primary reaction involves the nitrogen atom,

giving products containing the unit, $C_6H_5N\begin{smallmatrix}R\\ \diagup\\ \diagdown\\ CH_2-\end{smallmatrix}$, which then by rear-

rangement or condensation are converted to products containing the unit, R—$NC_6H_4CH_2$[182].

Diaminodiphenylmethanes are obtained by the action of one-half mol formaldehyde on aniline and toluidines in dilute acid solution[183].

The reactions of the various aromatic monoamines with formaldehyde in strong acid solution have yielded a number of complicated products, including heterocyclic hydroquinazolines such as Troeger's base, from *p*-toluidine and formaldehyde:

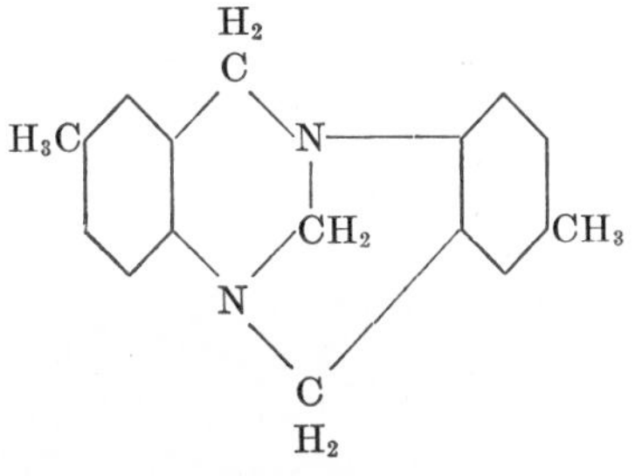

Troeger's Base

These reactions have been reviewed in considerable detail by Sprung[166].

On reaction with two mols formaldehyde in neutral solution aromatic orthodiamines, such as ortho-phenylenediamine and 1,2-diaminonaphthalene, give products whose empirical formulas correspond to the dimers of the respective Schiff bases. In weakly acid solution, diamines such as 1,3,4-toluylenediamine give imidazoles[56].

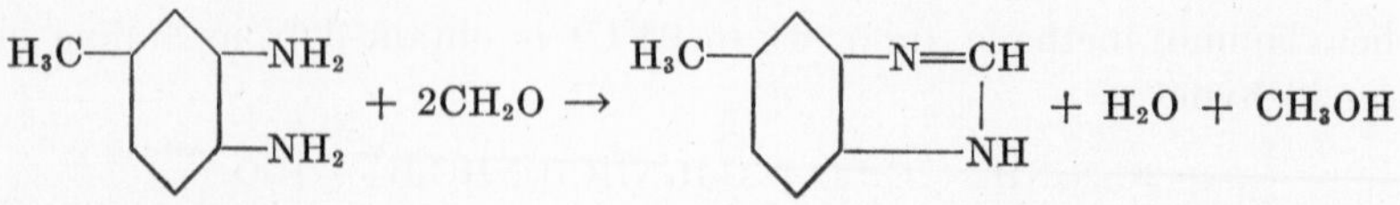

$$H_3C-\underset{NH_2}{\overset{NH_2}{\bigcirc}} + 2CH_2O \rightarrow H_3C-\underset{NH}{\overset{N=CH}{\bigcirc}} + H_2O + CH_3OH$$

Kondo and Ishida[99] report that symmetrical dimethylolbenzidine (m.p. 271 to 272°C) is formed by the reaction of benzidine and formaldehyde in alcohol solution. According to Schiff[153], dimethylenebenzidine (m.p. 140 to 141°C may be obtained from the same reagents. In the presence of acid a formaldehyde derivative of methylenebenzidine is obtained[154].

Secondary aromatic amines such as monomethyl- and monoethylaniline form alkyldiphenylaminomethanes in neutral or slightly alkaline solutions[19].

$$2C_6H_5NHCH_3 + CH_2O \rightarrow C_6H_5N(CH_3)\cdot CH_2\cdot N(CH_3)\cdot C_6H_5 + H_2O$$

In the presence of hot acids the benzidine rearrangement takes place and p-alkylaminodiphenylmethanes are produced[19, 66, 182, 183]. The behavior of diphenylamine is similar. With symmetrically diarylated 1,2- and 1,3-diamines, heterocyclic methylene derivatives are obtained[14, 159].

In general, secondary aromatic amines do not give resins with formaldehyde as readily as the primary amines. However, as previously stated, a brittle resin can be obtained even with a tertiary amine.

The reactions of hydrazobenzenes with formaldehyde were studied by Rassow[142] and Rassow and Lummerzheim[143], who obtained methylene derivatives as indicated:

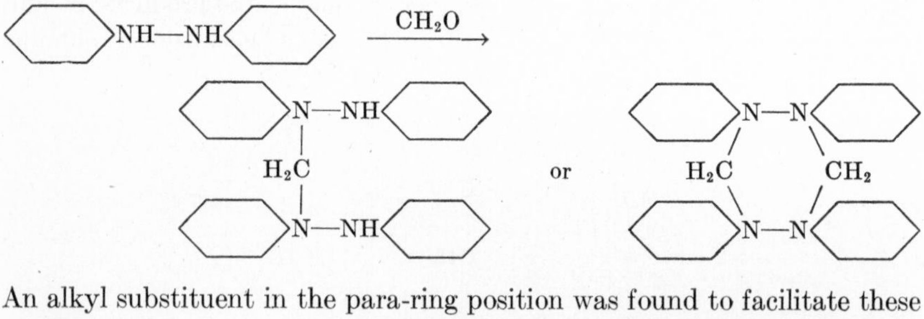

An alkyl substituent in the para-ring position was found to facilitate these reactions, whereas no reaction took place when a nitro group was thus substituted. Good yields of either the mono- or dimethylene derivative were reported. The mono-derivative is sensitive to alkalies or acids, whereas the cyclic dimethylene derivative is stable to alkalies and is sensitive only to acids.

Amides

Reactions of formaldehyde with amides involve the amino group and result in the formation of methylol and methylene derivatives. However, the nature of these reactions and the properties of the products formed are modified by the carbonyl group adjacent to the amine radical and accordingly differ in many respects from the corresponding reactions and reaction products of amines and amine derivatives.

The methylolamides, which are the primary reaction products, differ from methylolamines in possessing a higher order of chemical stability and in most cases can be readily isolated in a state of chemical purity. They are usually obtained by reacting amides with formaldehyde solution or paraformaldehyde under neutral (pH = 7) or alkaline conditions. Although they are generally produced at room temperature, heat is occasionally employed to accelerate the reaction. The use of a solvent, such as methyl or ethyl alcohol, plus a basic catalyst is claimed to be of particular advantage when polymers such as paraformaldehyde are employed[191, 193]. In some cases, methylolamides may also be obtained under acidic conditions[45].

Methylenediamides are generally stable, high-melting solids. They are formed when formaldehyde or paraformaldehyde is reacted with amides under acidic conditions, at high temperatures, or in the presence of dehydrating agents.

Monoamides. When formamide is heated for a short time at 120 to 150°C with a molecular equivalent of formaldehyde in the form of paraformaldehyde, monomethylolformamide is produced as a clear, colorless, syrup, compatible with water and alcohol in all proportions, but insoluble in ether[92]:

$$NH_2COH + CH_2O \text{ (p)} \rightarrow HOCH_2NHCOH$$

Methylolformamide is probably the least stable of the methylolamides. It has a slight odor of formaldehyde and cannot be distilled without decomposition.

On heating two mols of formamide with one of paraformaldehyde at 150 to 160°C, methylenediformamide is formed[93]. The same product is also obtained on refluxing a similar mixture for 4 to 5 hours at ordinary pressure. Methylenediformamide is a crystalline solid melting at 142 to 143°C[98].

$$2HCONH_2 + CH_2O \text{ (p)} \rightarrow HCONHCH_2NHOCH + H_2O$$

By reacting acetamide with formaldehyde solution in the presence of potassium carbonate, methylolacetamide ($CH_3CONHCH_2OH$) is produced, and may be isolated as a crystalline solid melting at 50 to 52°C[46]. Chwala[29], who prepared this product by melting equivalent amounts of acetamide and paraformaldehyde in the presence of 0.1 per cent of potassium hydroxide as as a 40 per cent aqueous solution and heating for 15 minutes at 60°C, reports a melting point of 56 to 57°C. Methylenediacetamide (m.p. 196°C), $CH_2(NHCOCH_3)_2$, is prepared by reacting acetamide and formaldehyde in the presence of hydrochloric acid[139]. The formation of these acetamide derivatives by heating acetamide and paraformaldehyde was reported by

Kalle & Co.[92, 93] in 1905. Methyl ethers of methylolformamide and methylol-acetamide were prepared by Chwala by reaction with excess methanol in the presence of hydrochloric acid[29].

The reaction of other aliphatic amides, both saturated and unsaturated, proceeds in a fashion similar to that of formamide and acetamide. Analogous reactions also take place with aromatic amides.

Benzamide and formaldehyde react to give methylolbenzamide in the presence of potassium carbonate. In the presence of sulfuric acid and aqueous formaldehyde, which represses dissociation of product, methylolbenzamide is converted to methylolmethylenedibenzamide[46].

$$2C_6H_5CONHCH_2OH \rightarrow C_6H_5CON(CH_2OH) \cdot CH_2 \cdot NH \cdot COC_6H_5 + H_2O$$

The latter compound loses formaldehyde readily to give methylenedibenzamide, which is also obtained by the action of hydrogen chloride on formaldehyde and benzamide in alcoholic solution[140].

According to Knudsen[98], methylenediformamide differs from methylenediacetamide and other methylenediamides in that it is converted to methylenedibenzamide on treatment with benzoyl chloride and aqueous alkali, whereas methylenediacetamide gives methylene benzoate under the same conditions. Also of interest in this connection is the preparation of methylenediamine salts* by the action of an excess of strong acid on methylenediformamide at low temperatures. When 77 g methylenediformamide were added to 500 g concentrated hydrochloric acid at approximately 12°C, Knudsen[98] obtained 57 g methylenediamine hydrochloride, $CH_2(NH_2)_2 \cdot 2HCl$, as a crystalline precipitate. Formic acid is obtained as a by-product of this reaction:

$$CH_2(NHCOH)_2 + 2HCl + 2H_2O \rightarrow CH_2(NH_2)_2 \cdot 2HCl + 2HCOOH$$

When treated in the same manner, methylenediacetamide is reported to undergo complete hydrolysis, with formation of ammonium chloride, formaldehyde, and acetic acid:

$$CH_2(NHCOCH_3)_2 + 2HCl + 3H_2O \rightarrow CH_2O \text{ (aq)} + 2CH_3COOH + 2NH_4Cl$$

Mixed methylene derivatives are obtained by the joint reaction of form-

---

* Methylenediamine is of particular interest since it is the ammono-analog of methylene glycol. Knudsen found that salts of this diamine are stable in the dry state and, in addition to the hydrochloride, prepared the nitrate $[CH_2(NH_2)_2 \cdot 2HNO_3]$ and the sulfate $[CH_2(NH_2)_2 \cdot H_2SO_4]$ by reacting methylenediformamide with an excess of concentrated nitric acid and 50 per cent sulfuric acid, respectively, at low temperatures. The nitrate of methylenediamine crystallizes in prisms, which melt and then explode when heated on platinum foil. Attempts to prepare methylenediamine itself by the action of alkali on the hydrochloride were unsuccessful, since the pure base is apparently unstable in the free state. However, when it is liberated in the presence of alcohol, solutions of methylenediamine are obtained and, according to Knudsen, are sufficiently stable for use in chemical reactions.

aldehyde, monoamides, and a number of polar compounds. On heating benzamide, piperidine, and formaldehyde, N-piperidyl methylbenzamide is produced[46]:

$$C_6H_5CONH_2 + CH_2O + HNC_5H_{10} \rightarrow C_6H_5CONH \cdot CH_2 \cdot NC_5H_{10} + H_2O$$

Benzamide, formaldehyde, and sodium bisulfite react in an analogous fashion[97]:

$$C_6H_5CONH_2 + CH_2O + NaHSO_3 \rightarrow C_6H_5CONH \cdot CH_2 \cdot NaSO_3 + H_2O$$

Stearylamide, paraformaldehyde, and anisole, when heated for 3 hours at 70°C in the presence of zinc chloride, give methoxybenzylstearamide[7]:

$$C_{18}H_{37}CONH_2 + CH_2O + C_6H_5OCH_3 \rightarrow C_{18}H_{37}CONHCH_2C_6H_4OCH_3 + H_2O$$

Mixed compounds of this type can also be obtained from the pre-formed methylolamides. Addition of methylolformamide to a cold solution of p-nitrophenol in concentrated sulfuric acid leads to the formation of N-formyl-nitro-hydroxybenzylamine (m.p. 236°C):

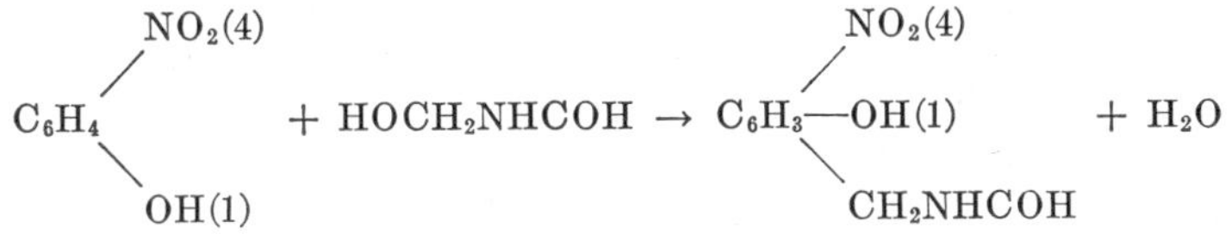

$$C_6H_4 \begin{matrix} NO_2(4) \\ \diagup \\ \diagdown \\ OH(1) \end{matrix} + HOCH_2NHCOH \rightarrow C_6H_3 \begin{matrix} NO_2(4) \\ \diagup \\ -OH(1) \\ \diagdown \\ CH_2NHCOH \end{matrix} + H_2O$$

Halomethyl derivatives of carboxylic acid amides are obtained by the joint reaction of formaldehyde and a hydrogen halide with amides having the type formula RCONHR', in which R and R' represent aliphatic and cycloaliphatic hydrocarbon radicals[6]. These products have the structure, RCONR'CH₂X, in which X represents a halogen atom. Unsubstituted amides yield chloromethylated derivatives of methylene linked amides, e.g., RCON(CH₂Cl)CH₂N(OCR)CH₂N(CH₂Cl)OCR[134]. Reactions may be carried out in an aromatic solvent, such as benzene, in which a mixture of the amide and paraformaldehyde is saturated with hydrogen chloride. Methylene bis-amides may also be chloromethylated in this way.

Sulfonamides, such as benzene sulfonamide, are converted to heterocyclic methylene derivatives by reaction with formaldehyde or its polymers in alcoholic hydrogen chloride[114, 118]. With benzene sulfonamide, benzene sulfotrimethylenetriimide is obtained (m.p. 217°C).

Urea

The primary reaction products of formaldehyde and urea are mono- and dimethylolurea. If the reaction is carried out under neutral or mildly alkaline conditions, these products may be isolated in a pure state. On heating, condensation reactions take place with the formation of methylene bridges between urea molecules, and resins are obtained. Under mildly acidic conditions, transparent resins may also be obtained, but in the presence of strong acids at low pH values insoluble poly-condensation products are precipitated directly from the reaction mixture. In recent years urea-formaldehyde resins have become of great commercial importance, and the chemical reactions of urea and formaldehyde have been the subject of intensive study.

Research on the mechanism of the formation of monomethylol urea from aqueous formaldehyde and urea demonstrates that it is a completely reversible bimolecular reaction. Studies by Crowe and Lynch[35] indicate that this reaction involves the urea anion and unhydrated formaldehyde so that the rates at which these reactants are formed in water solution are controlling factors. The anion is assumed to be a resonance structure.

$$HOCH_2OH \rightleftharpoons H_2C:O + H_2O$$

$$\underset{NH_2 \cdot C \cdot NH^-}{\overset{\overset{\textstyle O}{\|}}{\vphantom{x}}} \rightleftharpoons \underset{NH_2 \cdot C:NH}{\overset{\overset{\textstyle O^-}{|}}{\vphantom{x}}}$$

The urea anion concentration is controlled by pH, whereas the main effect of temperature is on the dehydration of methylene glycol. Smythe[162] reports that variations in reaction rate and pH on blending solutions of urea and formaldehyde are caused by small amounts of ammonium cyanate and carbonate which are formed in urea solutions on standing at room temperature. Although traces of these impurities do not change the general nature of the reaction, they retard its rate. Bettelheim and Cedwall[12] have found that the bimolecular reaction obtained with a 1:1 molal ratio of urea and formaldehyde gives the monomethylol derivative exclusively but that dimethylol urea also forms when the reactant ratio is 1:2.

The condensation of urea with paraformaldehyde and the autocondensation of dry solid methylol ureas has been studied by Kittel[94]. Reaction is slow under these conditions and catalysts have little effect in the absence of moisture.

Monomethylolurea, $H_2NCONHCH_2OH$, is a colorless solid, soluble in cold water and warm methanol. When pure, it is reported to melt at 111°C[47]. It was prepared by Einhorn and Hamburger[47, 49] in 1908 by gradually adding one molecular proportion of 37 per cent formaldehyde solution to a 50 per cent aqueous solution of urea cooled with ice and made alkaline with barium hydroxide. Following addition of formaldehyde, the reaction mixture was saturated with carbon dioxide, filtered to remove barium carbonate and evaporated in a vacuum desiccator.

Dimethylolurea, $HOCH_2NHCONHCH_2OH$, melts at 126°C, forming a clear liquid which solidifies on further heating. It is fairly soluble in cold water and may be purified by crystallization from 80 per cent alcohol. It was prepared by Einhorn and Hamburger[47, 49] by a procedure similar to that employed in making the monomethylol derivative. In this case, however, two equivalents of formaldehyde were employed and the reaction was allowed to proceed at 20 to 25°C until the solution gave a negative formaldehyde test with Tollen's ammoniacal silver reagent. A more practical method of preparation involves the addition of the calculated quantity of urea to 37 per cent formaldehyde which has been adjusted to a pH of 7 to 9 with sodium hydroxide buffered with sodium monophosphate. Cooling is required to keep the reaction mixture at 15 to 25°C. Dimethylolurea crystallizes and is filtered off after 15 to 24 hours, at which time the reaction is substantially complete. The crude product is then washed with water or alcohol and vacuum dried at a temperature of 50°C or below[147, 156, 191]. Dimethylolurea can also be prepared by the action of urea on an alkaline solution of paraformaldehyde in alcohol[192].

Dimethylolurea is a commercial chemical of considerable industrial importance. In general, its utility resides in the fact that it is a relatively stable, water-soluble resin intermediate. In this capacity, it has numerous and varied applications in the formulation of adhesives, resins, textile-modifying compositions etc. A number of these uses are described in connection with the industrial applications of formaldehyde treated in Chapters 20 and 21. Because of its commercial value, special methods for the preparation of dimethylolurea both in the pure state and in a variety of liquid, paste, and powder compositions have been the subject of industrial patents. Of special interest in this connection is Kvalnes'[103] finding that clear, room temperature stable formaldehyde-urea solutions can be obtained by adding one mol urea per 6 mols formaldehyde to a hot approximately 60 per cent formaldehyde solution at a pH of 7 to 9 and heating for a short period at 70 to 90°C. Since dimethylolurea undergoes condensation reactions giving products of high molecular weight at low pH values, in the presence of catalytic impurities, etc., the various product forms often contain buffer salts and other agents which have been found to improve stability.

On treatment with alcohols in the presence of an acidic catalyst, mono- and dimethylolureas may be converted to ethers[164]. The dimethyl ether of dimethylolurea, synthesized in this way, is a crystalline solid melting at 101°C:

$$2CH_3OH + HOCH_2NHCONHCH_2OH \rightarrow CH_3OCH_2NHCONHCH_2OCH_3 + 2H_2O$$

Catalytic hydrogenation of this ether of dimethylol urea in the presence of a nickel catalyst at 100°C yields 1,3-dimethylurea according to a process described by Farlow[55]. According to Kadowaki[91], when urea is heated with approximately four molecular equivalents of commercial strength formaldehyde in the presence of barium hydroxide and the reaction mixture evaporated to remove water, a sticky residue is obtained which, on treatment with acid methanol, gives the dimethyl ether of N,N'-dimethyloluron. The structural formula of this product, which can be vacuum-distilled at 82 to 83°C and 0.1 mm pressure without decomposition, is shown below:

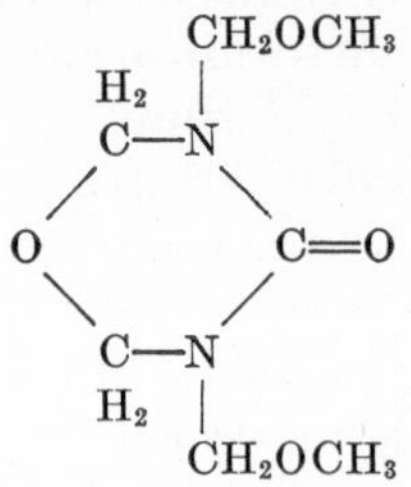

$$
\begin{array}{c}
CH_2OCH_3 \\
\underset{C-N}{\overset{H_2}{|}} \\
O \diagup \qquad \diagdown \\
\diagdown \qquad \diagup C=O \\
\underset{H_2}{\overset{C-N}{|}} \\
CH_2OCH_3
\end{array}
$$

By a controlled reaction of formaldehyde and urea under mildly acidic conditions[91], Kadowaki prepared methylene bis- or diurea (m.p. 218°C), $CH_2(NHCONH_2)_2$. By action of dilute acid on this product in aqueous solution, he also obtained trimethylenetetraurea and pentamethylenehexaurea.

As previously stated, insoluble condensation products are precipitated when formaldehyde and urea are reacted in the presence of strong acids. Substances of this type are described by Hölzer (cf. Tollens[177]), Goldschmidt[67], Lüdy[111], Litterscheid[108], Dixon[41], Van Laer[104], and other investigators. Although in some cases these products have the empirical composition of methylene urea, $NH_2CON:CH_2$, their infusible and insoluble nature makes it appear unlikely that this substance has been isolated in monomeric form. Depending on reaction conditions, such as the ratio of formaldehyde to urea, reagent concentrations, etc., products of varying composition are obtained. According to Dixon[41], a typical compound of this type, first described by Goldschmidt[67], has the structure:

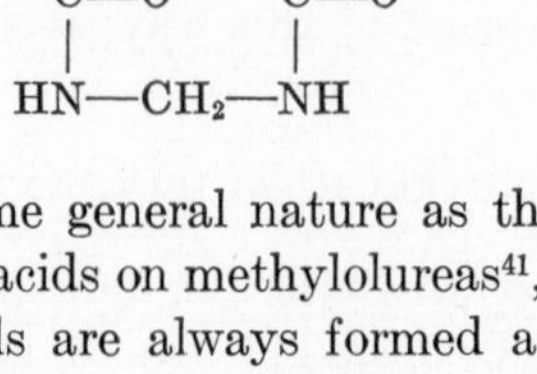

$$
\begin{array}{c}
HN-CH_2-N-CH_2OH \\
\ \ |\qquad\qquad | \\
C=O \qquad C=O \\
\ \ |\qquad\qquad | \\
HN-CH_2-NH
\end{array}
$$

Precipitates of the same general nature as those described above also result from the action of acids on methylolureas[41], and it is highly probable that methylol compounds are always formed as reaction intermediates.

According to Einhorn[47], monomethylolurea disproportionates on treatment with dilute hydrochloric acid, yielding urea and dimethylolurea prior to insoluble products.

Vaskevich and Reingach[179] report that polymethyleneureas are decomposed by heating with acid formaldehyde. Solution of polymer is favored by low pH and a high formaldehyde concentration. The scission process is believed to be a combination of a physical colloidal peptization and chemical destruction.

Urea-Formaldehyde Resins. It is beyond the scope of this book to give a detailed review of the voluminous literature dealing with urea-formaldehyde resins. For such information, the reader is referred to the many useful texts dealing specifically with resin chemistry.

As in the case of phenol-formaldehyde resins, the ratio of formaldehyde employed and the pH of the reaction medium are important factors in resin preparation. In general, the formaldehyde-urea ratio is about 1.5 and pH values are usually maintained in the range 4 to 8. In many cases the primary condensation is carried out under the neutral or mildly alkaline conditions, which favor the formation of methylol derivatives, whereas the resinification and final hardening of the finished product is usually conducted under acidic conditions. Precipitation of amorphous condensates in the early stages of the resin process is undesirable, and is prevented by avoiding the highly acid conditions which give rise to this effect. Water is removed by distillation at atmospheric or reduced pressure as the reaction takes place, and the solution of the product increases in viscosity. Paraformaldehyde is also used in the preparation of resins for scme purposes, and solvents other than water are also employed. Lacquer resins are sometimes obtained by carrying out the resin condensation in the presence of an alcohol, with the object of combining solvent and resin to obtain products having special solubility characteristics. Acidic catalysts are usually employed in preparations of this type.

Although the exact mechanism by which urea formaldehyde resins are formed is still incompletely understood, it is probable that a step-wise polycondensation is involved. The analogy between the formation of resins from methylolureas and polyoxymethylene glycols from formaldehyde has been pointed out by Staudinger[168] as a possible mechanism. For example, the condensation of monomethylolurea has been visualized as:

$$
\begin{array}{ccccccc}
\text{HN—CH}_2\text{OH} & & \text{HN—CH}_2\text{OH} & & \text{HN—CH}_2\text{—N—CH}_2\text{OH} & & \\
| & & | & & | \qquad\qquad | & & \\
\text{C=O} & + & \text{C=O} & \rightarrow & \text{C=O} \qquad \text{C=O} & + & \text{H}_2\text{O} \\
| & & | & & | \qquad\qquad | & & \\
\text{NH}_2 & & \text{NH}_2 & & \text{NH}_2 \qquad \text{NH}_2 & &
\end{array}
$$

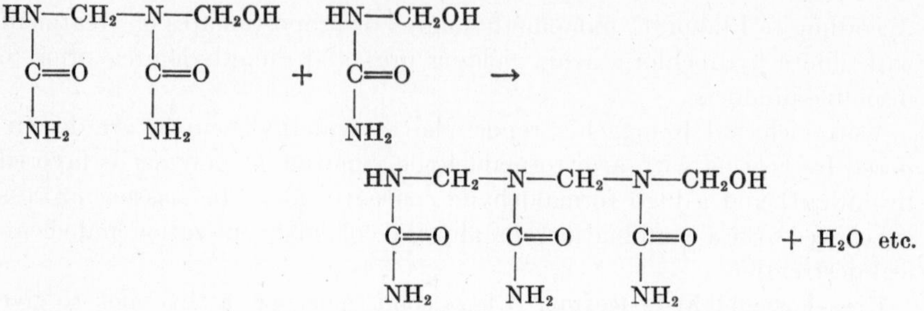

$$
\begin{array}{ccccc}
\text{HN—CH}_2\text{—N—CH}_2\text{OH} & & \text{HN—CH}_2\text{OH} & & \\
| \qquad\quad | & & | & & \\
\text{C=O} \quad\ \text{C=O} & + & \text{C=O} & \rightarrow & \\
| \qquad\quad | & & | & & \\
\text{NH}_2 \quad\ \text{NH}_2 & & \text{NH}_2 & &
\end{array}
$$

$$
\begin{array}{cccc}
\text{HN—CH}_2\text{—N—CH}_2\text{—N—CH}_2\text{OH} & & \\
| \qquad\quad | \qquad\quad | & & \\
\text{C=O} \qquad \text{C=O} \qquad \text{C=O} & + \text{H}_2\text{O etc.} \\
| \qquad\quad | \qquad\quad | & & \\
\text{NH}_2 \qquad \text{NH}_2 \qquad \text{NH}_2 & &
\end{array}
$$

or

$$\text{NH}_2\text{CONHCH}_2\text{OH} + \text{NH}_2\text{CONHCH}_2\text{OH} \rightarrow \text{H}_2\text{NCONHCH}_2\text{NHCONHCH}_2\text{OH} + \text{H}_2\text{O}$$

etc.

The polymeric acetate, $(\text{CH}_2\!:\!\text{NHCONH}_2)_{12}\cdot\text{CH}_3\text{COOH}$, and the corresponding hydrate, $(\text{CH}_2\!:\!\text{NHCONH}_2)_{12}\cdot\text{H}_2\text{O}$, which Scheibler, Trostler and Scholz[152] obtained by the action of glacial acetic acid on monomethylolurea, can be interpreted as compounds of the structural type indicated above.

In the case of dimethylolurea, complex cross-linked resins may be formed by a mechanism similar to that indicated in the case of monomethylolurea. Cyclic structures of the type whose formation is indicated below can also be considered primary resin building blocks:

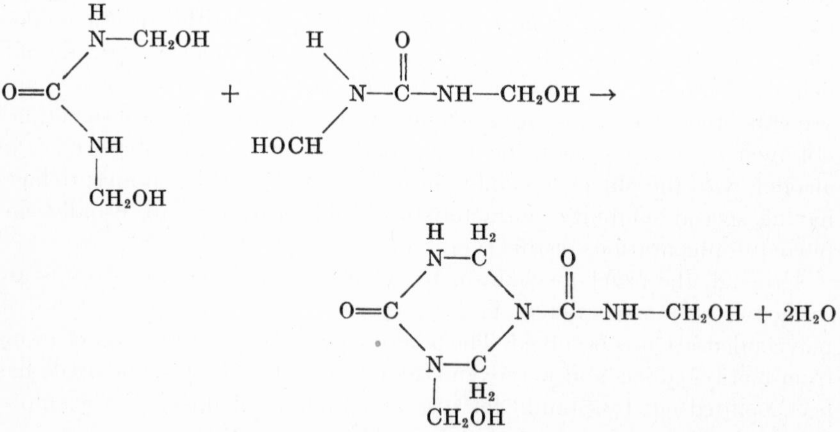

According to Hodgkins and Hovey[83], the cross-linking, which is probably responsible for the gelation and final set of urea-formaldehyde compositions, is possible only with dimethylolurea or mixtures of di- and monomethylolurea, and does not occur with methylolurea alone, or its equivalent, *viz.*, one-to-one molar ratios of urea and formaldehyde.

Some investigators favor the theory that urea-formaldehyde resins are formed by the polymerization of methyleneureas, which are first formed by the dehydration of mono- and dimethylolurea, as indicated below:

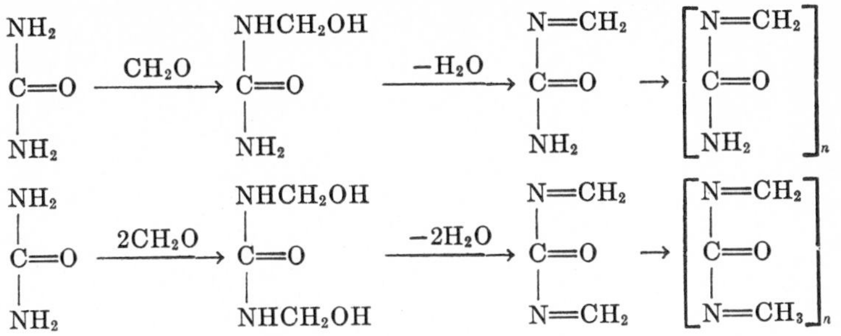

$$
\begin{array}{ccccc}
\mathrm{NH_2} & & \mathrm{NHCH_2OH} & & \mathrm{N{=}CH_2} & & \left[\begin{array}{c}\mathrm{N{=}CH_2}\end{array}\right. \\
| & \mathrm{CH_2O} & | & \mathrm{-H_2O} & | & \rightarrow & | \\
\mathrm{C{=}O} & \longrightarrow & \mathrm{C{=}O} & \longrightarrow & \mathrm{C{=}O} & & \mathrm{C{=}O} \\
| & & | & & | & & | \\
\mathrm{NH_2} & & \mathrm{NH_2} & & \mathrm{NH_2} & & \left.\mathrm{NH_2}\right]_n
\end{array}
$$

$$
\begin{array}{ccccc}
\mathrm{NH_2} & & \mathrm{NHCH_2OH} & & \mathrm{N{=}CH_2} & & \left[\begin{array}{c}\mathrm{N{=}CH_2}\end{array}\right. \\
| & \mathrm{2CH_2O} & | & \mathrm{-2H_2O} & | & \rightarrow & | \\
\mathrm{C{=}O} & \longrightarrow & \mathrm{C{=}O} & \longrightarrow & \mathrm{C{=}O} & & \mathrm{C{=}O} \\
| & & | & & | & & | \\
\mathrm{NH_2} & & \mathrm{NHCH_2OH} & & \mathrm{N{=}CH_2} & & \left.\mathrm{N{=}CH_3}\right]_n
\end{array}
$$

Thurston[64] has recently suggested that methyleneurea may polymerize to form a triamide derivative of trimethylenetriamine:

$$
\begin{array}{ccc}
 & \mathrm{H_2} & \\
 & \mathrm{C} & \\
 & \diagup \quad \diagdown & \\
\mathrm{H_2NOC{-}N} & & \mathrm{N{-}CONH_2} \\
| & & | \\
\mathrm{H_2C} & & \mathrm{CH_2} \\
 & \diagdown \quad \diagup & \\
 & \mathrm{N} & \\
 & | & \\
 & \mathrm{CONH_2} &
\end{array}
$$

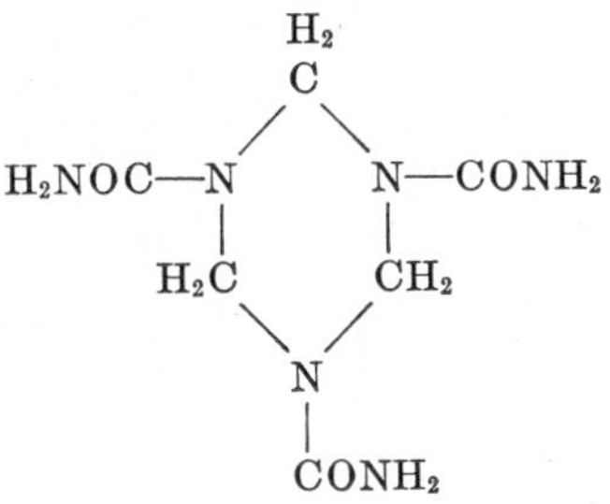

This trimer, which may also be produced by polycondensation, could then serve as a resin building block.

Marvel and co-workers[120] have recently suggested a mechanism for the formation of urea-formaldehyde resins based on the hypothesis that urea reacts as an amino acid amide. This involves the formation of cyclotrimethylene triamine rings of the type shown above followed by intercondensation of these rings by methylene bis-amide linkages. According to Marvel, rings are produced by trimerization of monomeric methyleneurea or monomeric methylol methyleneurea. Ring formation proceeds as in the case of a primary amine, whereas the polycondensation of these rings behaves like an amide reaction. Polymers produced would have the highly cross-linked structure shown at the top of the following page which has a theoretical methylene group content of 1.5 per urea residue. The amino-amide hypothesis accounts for the fact that cross-linked resins are not obtained by the condensation of alkyl ureas with formaldehyde as would be expected on the basis of the linear condensation mechanism. It is also supported by Marvel's demonstration that resins of this type are obtained by the reaction of formaldehyde with amino-amides such as glycinamide and epsilon-amino caproamide[119], whereas glycine methyl-

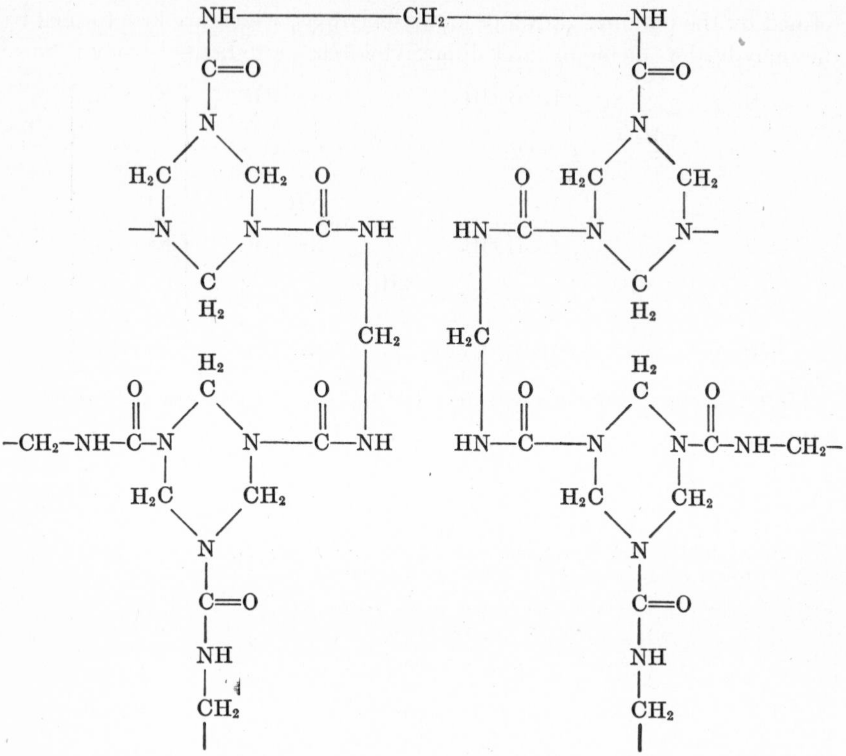

$$
\begin{array}{c}
\text{NH} \longrightarrow \text{CH}_2 \longrightarrow \text{NH} \\
| \qquad\qquad\qquad\qquad | \\
\text{C}{=}\text{O} \qquad\qquad\qquad \text{C}{=}\text{O} \\
| \qquad\qquad\qquad\qquad | \\
\text{N} \qquad\qquad\qquad\qquad \text{N}
\end{array}
$$

The NH₂ group in this structure would be expected to react as an amine, whereas the :NH group would react as an imide forming only simple monomethylol derivatives or methylene-bis derivatives under ordinary reaction conditions as would also be the case for an amide. This is in agreement with Crowe and Lynch's finding[35] that monomethylol urea is produced by the reaction of formaldehyde with the urea anion. This compound may accordingly possess the structure, $NH_2 \cdot C(OH){:}N \cdot CH_2OH$. If this is

amide yields only the trimeric methyleneamine derivative and sarcosin amide, $CH_3NHCH_2CONH_2$, gives a low molecular weight linear resin.

The fact that urea behaves like an amino acid amide may be due to the fact that it reacts in the tautomeric isourea form in which the two nitrogen atom groups are plainly differentiated as shown below:

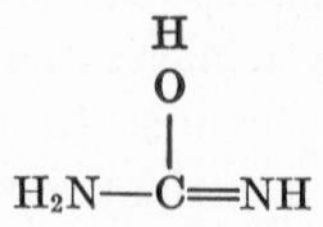

$$
\begin{array}{c}
\text{H} \\
\text{O} \\
| \\
\text{H}_2\text{N} \longrightarrow \text{C}{=}\text{NH}
\end{array}
$$

The NH₂ group in this structure would be expected to react as an amine, whereas the :NH group would react as an imide forming only simple monomethylol derivatives or methylene-bis derivatives under ordinary reaction conditions as would also be the case for an amide. This is in agreement with Crowe and Lynch's finding[35] that monomethylol urea is produced by the reaction of formaldehyde with the urea anion. This compound may accordingly possess the structure, $NH_2 \cdot C(OH){:}N \cdot CH_2OH$. If this is the case, the following structures may be representative of the primary urea-formaldehyde resin precursors.

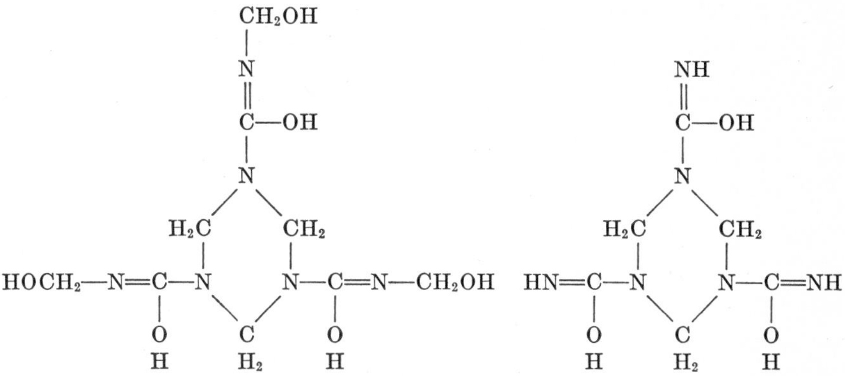

$$
\begin{array}{ccc}
\text{CH}_2\text{OH} & & \text{NH} \\
| & & \| \\
\text{N} & & \text{C—OH} \\
\| & & | \\
\text{C—OH} & & \text{N} \\
| & & \\
\text{N} & &
\end{array}
$$

Modified resins of the urea-formaldehyde type are also obtainable from the ethers of mono- and dimethylolureas. These intermediates are probably involved when urea and formaldehyde are condensed in alcohol solutions[83].

Substituted Ureas. Mono- and disubstituted ureas form methylol and methylene derivatives with formaldehyde. In general, the property of forming useful resins disappears when two or more of the urea hydrogen atoms have been replaced by alkyl or other non-reactive groups.

Einhorn[47] was unable to obtain the monomethylol derivative of ethylurea by reacting it with alkaline formaldehyde, succeeding only in isolating monomethylolmethylene-bis-ethylurea (m.p. 168–170°C), $\text{C}_2\text{H}_5\text{NHCON}$-$(\text{CH}_2\text{OH})\cdot\text{CH}_2\cdot\text{NH}\cdot\text{CONHC}_2\text{H}_5$. Under acidic conditions, he obtained methylene bis-monoethylurea (m.p. 115 to 116°C). Monomethylol derivatives of both symmetrical and asymmetrical dimethylurea were prepared by the same investigator, but dimethylol derivatives were not obtained. Complex methylenetricarbimides, derived from dimethylureas and formaldehyde, having the formulas $\text{C}_{12}\text{H}_{20}\text{O}_5\text{N}_{10}$, $(\text{C}_{10}\text{H}_{16}\text{O}_5\text{N}_6)_x$, and $(\text{C}_{14}\text{H}_{20}\text{O}_7\text{N}_6)_x$, have also been reported[104]. Einhorn[47] was unable to obtain a formaldehyde derivative of triethylurea under either acidic or alkaline conditions.

The fact that dimethylureas will not give dimethylol derivatives under alkaline conditions and that triethylurea will not react with formaldehyde at all, is possible evidence that the tautomeric isourea structure plays a part in determining the reactivity of these derivatives The iso-structures of the monomethylol derivatives of the dimethylureas, as based on this hypothesis, are:

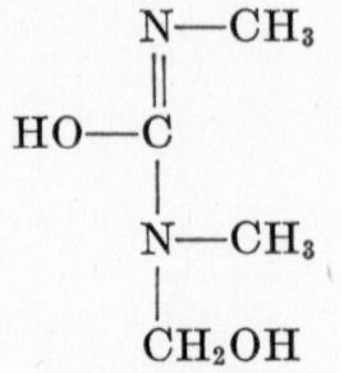

$$
\begin{array}{cc}
\underset{\substack{\text{N---CH}_3 \\ \| \\ \text{HO---C} \\ | \\ \text{N---CH}_3 \\ | \\ \text{CH}_2\text{OH}}}{} &
\underset{\substack{\text{H}_3\text{C---N---CH}_3 \\ | \\ \text{HO---C} \\ \| \\ \text{N} \\ | \\ \text{CH}_2\text{OH}}}{}
\end{array}
$$

Monomethylol-s-   *Monomethylol-as-*
dimethylurea   *dimethylurea*

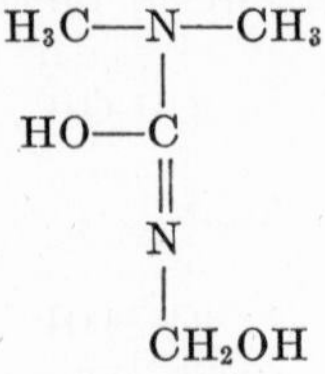

By employing an acid-catalyzed reaction similar to that used in preparing the dimethyl ether of dimethyloluron from urea (page 296), Kadowaki[91] prepared the methyl ether of N-methyl-N′-methyloluron from monomethylurea and N,N′-dimethyluron from symmetrical dimethylurea:

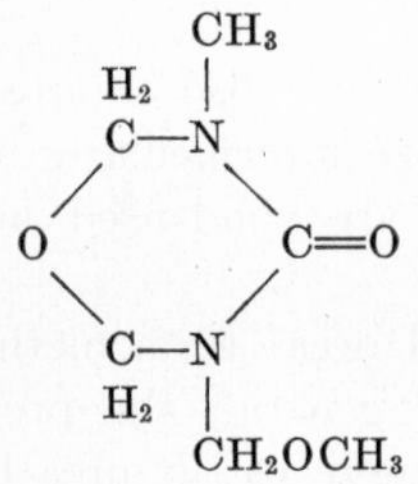

$$
\begin{array}{cc}
\underset{\text{CH}_2\text{OCH}_3}{\overset{\text{CH}_3}{
\begin{array}{c}
\text{H}_2 \mid \\
\text{C---N} \\
\diagup \quad \diagdown \\
\text{O} \qquad \text{C=O} \\
\diagdown \quad \diagup \\
\text{C---N} \\
\text{H}_2 \mid
\end{array}}} &
\underset{\text{CH}_3}{\overset{\text{CH}_3}{
\begin{array}{c}
\text{H}_2 \mid \\
\text{C---N} \\
\diagup \quad \diagdown \\
\text{O} \qquad \text{C=O} \\
\diagdown \quad \diagup \\
\text{C---N} \\
\text{H}_2 \mid
\end{array}}}
\end{array}
$$

Methyl ether of N-methyl-   *N,N′-Dimethyluron*
N′-methyloldimethyluron

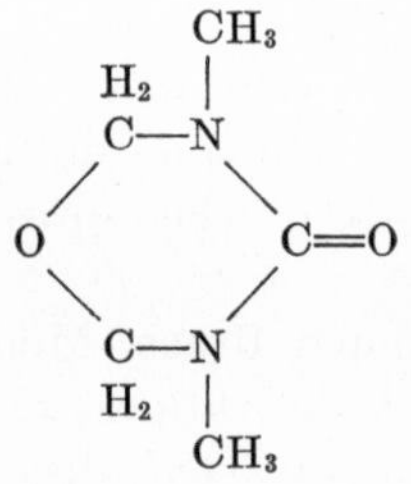

These compounds are derivatives of dimethyloldimethylureas.

Methylene bis-urea gives methylol and dimethylol derivatives with formaldehyde[91], reacting in much the same manner as urea itself.

Acetylurea gives methylene bis-acetylurea on treatment with formaldehyde in cold concentrated sulfuric acid[40]. On hydrolysis with caustic, this compound yields methylene bis-urea, which loses ammonia and cyclizes in the presence of hydrochloric acid:

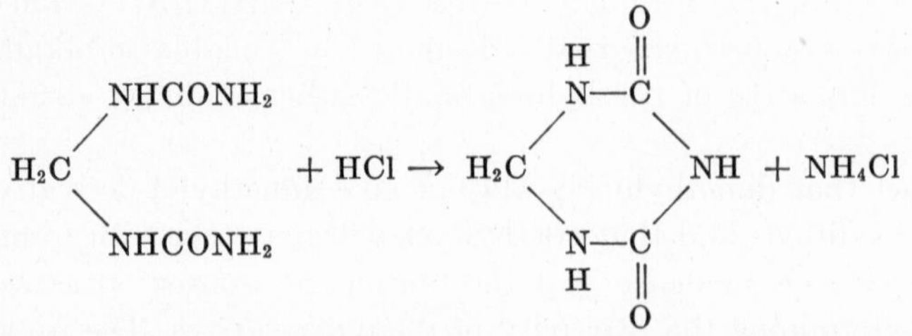

$$
\underset{\text{NHCONH}_2}{\overset{\text{NHCONH}_2}{\text{H}_2\text{C}\diagup \diagdown}} + \text{HCl} \rightarrow \text{H}_2\text{C}\underset{\substack{\text{N---C} \\ \text{H} \; \|}}{\overset{\substack{\text{H} \; \| \\ \text{N---C}}}{\diagup \diagdown}}\text{NH} + \text{NH}_4\text{Cl}
$$

Joint reactions of urea, formaldehyde, and secondary amines yield substituted bis(dialkylaminomethyl)-ureas[47]. Derivatives of this type can be made by adding the calculated quantity of diethylamine or piperidine and formaldehyde to a solution of urea in hot water:

$NH_2CONH_2 + 2HN(C_2H_5)_2 + 2CH_2O$ (aq) →

$$(C_2H_5)_2N \cdot CH_2 \cdot NHCONH \cdot CH_2 \cdot N(C_2H_5)_2 + 2H_2O$$

$NH_2CONH_2 + 2HNC_5H_{10} + 2CH_2O$ (aq) →

$$C_5H_{10}N \cdot CH_2 \cdot NHCONH \cdot CH_2 \cdot NC_5H_{10} + 2H_2O$$

The diethylamine derivative is an oil which decomposes on heating. The piperidine derivative is a crystalline product melting at approximately 136°C. Substituted mono-(dialkylaminomethyl) derivatives can be obtained from acylureas, formaldehyde, and secondary amines[48].

$RCONHCONH_2 + CH_2O$ (aq) $+ R'R''NH \rightarrow RCONHCONHCH_2NR'R'' + H_2O$

A solid organic peroxide is formed by the reaction of formaldehyde, urea and hydrogen peroxide in the presence of acid[65].

Polyureas, such as propylene diurea, $NH_2CONH(CH_2)_3NHCONH_2$, are reported to form both mono- and dimethylol derivatives on reaction with formaldehyde solution at pH values of 7.0 to 7.4[33].

Reactions of the cyclic ureide, hydantoin, with formaldehyde have been studied by Niemeyer and Behrend[126]. Monomethylolhydantoin is obtained by heating hydantoin with formaldehyde solution for a short time. Its exact structural formula has not been determined; two isomeric possibilities exist:

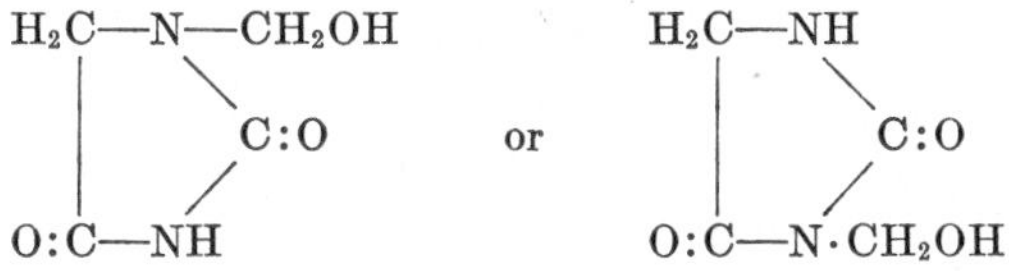

Monomethylolhydantoin

In the presence of hydrochloric acid, methylene bis-hydantoin, CH_2-$(C_3H_3N_2O_2)_2$, is precipitated from solutions containing hydantoin and formaldehyde. More complex products, such as methylolmethylene bis-hydantoin, tris-hydantoin derivatives, etc., may also be obtained.

Studies of reactions of formaldehyde with hydantoins containing one or two alkyl groups in the 5 position in Du Pont laboratories indicate that monomethylol 5,5-dimethylhydantoin functions as an odorless formaldehyde donor. The methylene-bis derivative of this dimethylhydantoin may be obtained in almost quantitative yield by reaction at ordinary temperature on addition of concentrated hydrochloric acid to a mixture of one mol of aqueous formaldehyde or paraformaldehyde and two mols of dimethylhydantoin provided that the water content of the reaction mixture is in the range 18 to 27 per cent[187].

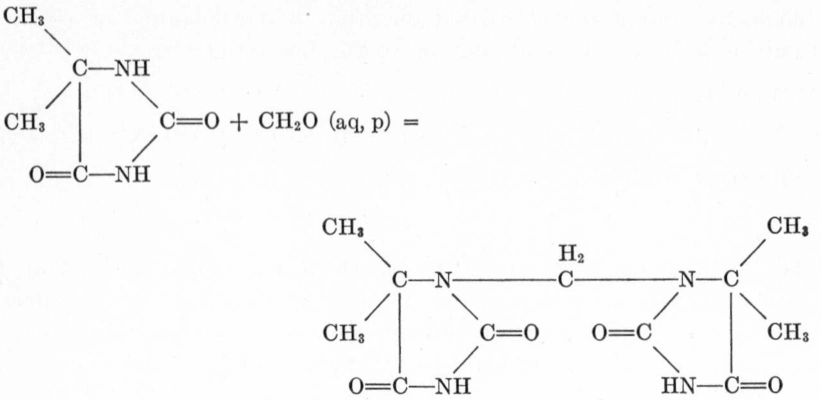

$$
\begin{array}{c}
\text{CH}_3 \\
\diagdown \\
\quad\text{C—NH} \\
\diagup\quad\quad\diagdown \\
\text{CH}_3\quad\Big|\quad\quad\text{C}{=}\text{O} + \text{CH}_2\text{O (aq, p)} = \\
\quad\quad\quad\diagup \\
\text{O}{=}\text{C—NH}
\end{array}
$$

$$
\begin{array}{c}
\text{CH}_3 \quad\quad\quad\quad\quad\quad\quad\quad\quad\quad\quad\quad\quad\text{CH}_3 \\
\diagdown\quad\quad\quad\quad\quad\quad\text{H}_2\quad\quad\quad\quad\quad\quad\diagup \\
\quad\text{C—N———————C———————N—C} \\
\diagup\quad\quad\diagdown\quad\quad\quad\quad\quad\quad\quad\quad\diagup\quad\quad\diagdown \\
\text{CH}_3\quad\Big|\quad\quad\text{C}{=}\text{O}\quad\quad\text{O}{=}\text{C}\quad\quad\text{CH}_3 \\
\quad\quad\quad\diagup\quad\quad\quad\quad\quad\quad\diagdown \\
\text{O}{=}\text{C—NH}\quad\quad\quad\quad\quad\text{HN—C}{=}\text{O}
\end{array}
$$

Methylene bis-dimethylhydantoin produced in the above reaction melts at 295° to 6°C and is only slightly soluble in water (0.5 per cent, 25°C; 2.0 per cent, 100°C). However, due to the two acidic imide groups indicated in its structure, it is readily soluble in aqueous sodium hydroxide forming a disodium salt. Similar methylene derivatives of dimethyl hydantoin containing 5-alkyl substituents containing more than one carbon atom are best prepared when the acid reaction catalyst is supplemented by addition of zinc or cadmium chlorides[188]. Resinous products of low molecular weight ranging from viscous liquids to glassy solids are formed by heating hydantoin and 5,5-dialkyl hydantoins with formaldehyde in the presence of alkaline catalysts (Jacobson[87] and Chadwick[26]). Resins of this type containing methyl or no substituents are readily soluble in water, whereas those containing higher alkyl substituents, although water sensitive, dissolve readily only in the presence of caustic alkalies.

Diamides

The reactions of formaldehyde with diamides of dibasic organic acids such as oxamide, succinamide and adipamide give methylol derivatives in much the same manner as urea. However, under mild reaction conditions, they show little capacity for polycondensation and cross-linkage unless they contain substituents, which are capable of reacting with formaldehyde, on the carbon chain.

Dimethyloloxamide was prepared by Bougault and Leboucq[16] in connection with an analytical study of methylol derivatives, and is apparently readily formed by the action of formaldehyde on oxamide.

Malonamide, since it possesses a methylene group activated by two adjacent carbonyl groupings, is able to react with three mols of formaldehyde. D'Alelio[36] claims that it yields compounds which are polymers of the trimethylene derivative indicated below when mixtures of malonamide and formaldehyde solution are boiled down in the presence of either alkaline or acidic catalysts.

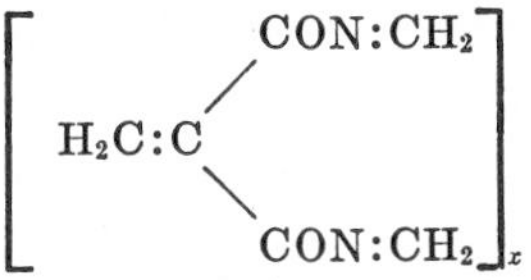

$$\left[H_2C:C \begin{array}{c} \diagup CON:CH_2 \\ \diagdown CON:CH_2 \end{array} \right]_x$$

By reacting succinamide with formaldehyde solution in the presence of potassium carbonate, Einhorn and Ladisch[50] obtained a dimethylolsuccinamide melting at 167°C, with decomposition. Arbuzov and Livshits[4] report the preparation of dimethylol adipamide (m.p. 148 to 150°C), dimethylol sebacamide (m.p. 138 to 140°C) and dimethylol suberamide (m.p. 130 to 131°C) by a similar reaction.

Resinous products have been obtained by reaction of a wide variety of di- and polyamides and sulfamides with formaldehyde[180]. Polysulfonamides also give resins on heating with formaldehyde and acids[21].

Polyamides

Investigations of Cairns and co-workers[23, 106] have shown that high molecular weight, linear polyamides react with formaldehyde to form methylol derivatives and cross-linked methylene or polyoxymethylene derivatives. Ternary reactions with formaldehyde and alcohols or mercaptans give N-alkoxymethyl and N-alkyl thiomethyl derivatives. These products are represented below with the polyamide chain indicated simply by the amide group.

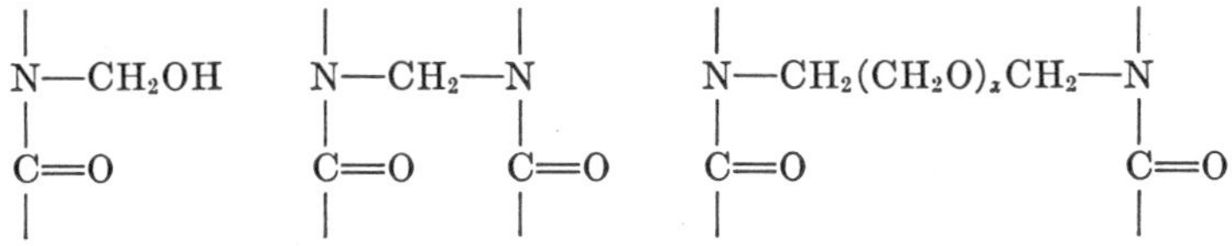

$$\begin{array}{cccc}
\mid & \mid \quad \mid & \mid \quad \quad \quad \mid \\
N-CH_2OH & N-CH_2-N & N-CH_2(CH_2O)_xCH_2-N \\
\mid & \mid \quad \mid & \mid \quad \quad \quad \mid \\
C=O & C=O \quad C=O & C=O \quad \quad \quad C=O \\
\mid & \mid \quad \mid & \mid \quad \quad \quad \mid \\
\text{I} & \text{II} & \text{III}
\end{array}$$

$$\begin{array}{cc}
\mid & \mid \\
N-CH_2OR & N-CH_2SR \\
\mid & \mid \\
C=O & C=O \\
\mid & \mid \\
\text{IV} & \text{V}
\end{array}$$

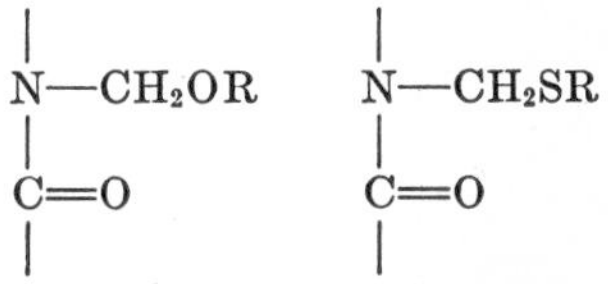

The N-methylol derivative (I) of polyhexamethylene adipamide (nylon 66) is obtained by the action of aqueous formaldehyde on solutions of the polymer in formic acid at approximately 60°C. Conditions of reaction must be carefully controlled to prevent the formation of cross-linkages

(II and III). Cross-linking results in a rapid increase of viscosity followed by formation of a tough, rubbery gel. A less critical method of preparation consists of treating the polymer with formaldehyde at 120 to 150°C in pyridine or an inert medium plus a basic catalyst, such as potassium carbonate. Methylol groups are readily removed on treatment with acids or alkalies.

The alkoxymethyl (IV) and alkylthiomethyl (V) derivatives are obtained by the action of a solution of formaldehyde in an alcohol or mercaptan on a polyamide dissolved in formic acid at 60°C. An alternate procedure is to treat the solid polyamide with an alcohol solution of formaldehyde at 150°C in the presence of a small concentration of phosphoric acid.

The substitution of alkoxymethyl groups in the polyamide chain reduces the hydrogen bonding, characteristic of these fiber-forming polymers. As a result, the originally hard, crystalline polymer becomes progressively

TABLE 50. EFFECT OF ALKOXYMETHYLATION ON THE PHYSICAL
PROPERTIES OF POLYAMIDES

	Percent Substitution of Methoxymethyl Radicals on Amide Groups in Polyhexamethylene Adipamide		
	0% (Control)	35%	50%
Solubility in hot 80% ethanol.....	None	About 40%	60%
Melting point (°C)*..............	264	150–160	100–110
Modulus of stiffness (100 psi).....	290	14.2	3.0
Recovery from 100 per cent stretch	30–35	70–80	90–95
Tensile strength (psi).............	8000–10,000	4500–6000	2000–3000

* Note: Melting points drop to room temperature and below at 70 to 80 per cent substitution.

softer and more elastic as the degree of substitution is increased. Solubility in organic solvents is also manifested in the alkoxymethyl derivatives. These effects of alkoxymethylation on the properties of polyhexamethylene adipamide are indicated in Table 50. On heating the alkoxymethylated polymers in the presence of an acid catalyst, methylene cross-linkages are produced and the polymer becomes insoluble and is no longer thermoplastic[24]. Further information on this subject will be found under the use of formaldehyde for the modification of nylon polyamide fibers (cf. pages 528–529).

Imides

Imides give both methylol and methylene derivatives with formaldehyde. N-methylolsuccinimide (m.p. 66°C) is obtained by heating formaldehyde solution and succinimide in the presence of potassium carbonate[28]. N-methylolphthalimide (m.p. 139 to 140°C) was prepared by Sachs in 1898[150].

Sakellarios[151] recently purified this product by means of its molecular compound with pyridine and reports a melting point of 149.5°C.

Passerini[132] reports that methylenedisuccinimide (m.p. 290 to 295°C) may be obtained by heating "trioxymethylene" (paraformaldehyde or alpha-polyoxymethylene) with succinimide in acetic acid containing a little sulfuric acid. The same investigator obtained methylenediphthalimide (m.p. 226°C) by heating phthalic acid and hexamethylenetetramine.

Urethanes (Carbamates)

The reactions of urethanes (carbamic esters) with formaldehyde follow the pattern characteristic of the related amides. Under alkaline conditions at room temperature ethylurethane reacts with formaldehyde solution to give methylolurethane, melting at 53°C[47].

$$C_2H_5O \cdot CONH_2 + CH_2O \text{ (aq)} \rightarrow C_2H_5O \cdot CONH \cdot CH_2OH$$

This compound is stable at ordinary temperatures but readily liberates formaldehyde on heating. In the presence of acid, it loses formaldehyde, with the formation of methylenediurethane[47]:

$$2C_2H_5OCONH \cdot CH_2OH \rightarrow (C_2H_5O \cdot CONH)_2CH_2 + CH_2O \text{ (aq)} + H_2O$$

On warming ethylurethane with formaldehyde and caustic, the reaction proceeds further and methylolmethylenediurethane is formed[47].

$$2C_2H_5OCONH_2 + 2CH_2O \text{ (aq)} \rightarrow C_2H_5O \cdot CONH \cdot CH_2 \cdot N(CH_2OH) \cdot CO \cdot OC_2H_5 + H_2O$$

Methyleneurethane ($C_2H_5O \cdot CON:CH_2$), also obtainable by reacting formaldehyde and urethane in the presence of hydrochloric acid, gives tri- and tetrameric products on treatment with acetic acid.

Thiourea

There is comparatively little information concerning the reactions of the simple thioamides with formaldehyde, with the exception of thiourea, which has received considerable attention because of the importance of thiourea-formaldehyde resins.

Although the formation of simple thiourea-formaldehyde addition products was undoubtedly recognized by early investigators in the field of formaldehyde-amide chemistry and has been mentioned in the patent literature[133, 191, 192], a detailed study of the preparation, structure, and chemical properties of these compounds was not published until 1939[136].

Since these addition products are highly soluble in water and alcohol and are easily decomposed at only slightly elevated temperatures, they are less easily isolated than the corresponding methylolureas.

Pollak[136] reports that crystalline mono- and dimethylolthioureas are produced in good yield by reaction of formaldehyde and thiourea in aqueous solution at temperatures not exceeding 50°C under mildly acidic or mildly alkaline conditions. The products obtained in acid media differ from those obtained under alkaline conditions, indicating the existence of isomeric forms whose structural formulas as postulated by Pollak are indicated below:

$$
\begin{array}{ll}
\text{NH·CH}_2\text{OH} & \text{NH} \\
\quad| & \quad\| \\
\text{A}\quad\text{C:S} & \text{C—S·CH}_2\text{OH}\quad\text{B} \\
\quad| & \quad| \\
\text{NH}_2 & \text{NH}_2
\end{array}
$$

Monomethylolthiourea isomers

$$
\begin{array}{ll}
\text{NH·CH}_2\text{OH} & \text{N·CH}_2\text{OH} \\
\quad| & \quad\| \\
\text{A}\quad\text{C:S} & \text{C·S·CH}_2\text{OH}\quad\text{B} \\
\quad| & \quad| \\
\text{NH·CH}_2\text{OH} & \text{NH}_2
\end{array}
$$

Dimethylolthiourea isomers

The A isomers react very slowly with silver nitrate as compared with thiourea itself, whereas the B isomers (derived from isothiourea) react rapidly. A preponderance of the A isomers is obtained when thiourea is reacted with formaldehyde under mildly acidic conditions, whereas the B isomers are the principal product under alkaline conditions. Since these isomeric forms are readily converted to one another, Pollak does not believe that the materials obtained in his work were 100 per cent isomerically pure. Their melting points are:

	A Isomers Prepared under mildly acidic conditions	B Isomers Prepared under mildly alkaline conditions
Monomethylolthiourea	97–98°C	92–94°C
Dimethylolthiourea	86–88°C	83–85°C

On exposure to dilute acetic or mineral acids, methylolthioureas are converted to high molecular weight insoluble products and resins.

When thiourea is added to dilute, acidified solutions of aqueous formaldehyde, a white powder is precipitated whose composition indicates that it may be a polymer of methylenethiourea, $(\text{NH}_2\text{CSN}:\text{CH}_2)_x$[42]. This substance melts at 202 to 203°C with decomposition, and is probably similar to the products formed by the action of acids on methylolthioureas[136]. Hot mineral acids were found to hydrolyze this product to formaldehyde and thiourea[42]. Excess formaldehyde hinders the precipitation of the methylene polymer, stopping it completely at high formaldehyde concentrations.

Hydrogen chloride, formaldehyde, and thiourea react to give a clear syrup whose empirical formula indicates that it is composed of equimolar proportions of these raw materials. On neutralizing a solution of this liquid, the methylene polymer is precipitated[42].

Alkyl and aralkyl ammonium dithiocarbamates, which are closely related to thioureas, react readily with formaldehyde in aqueous solution, precipitating insoluble methylene derivatives[105].

Aminonitriles, Cyanamide and Cyanamide Polymers

Cyanamide is the simplest aminonitrile. According to Griffith[73], it reacts with formaldehyde in aqueous solution to give a white precipitate whose structure has not been determined but whose empirical formula is $C_4N_4H_6O$. When a solution of this product in dilute hydrochloric acid is subjected to prolonged heating at 60°C and then neutralized, a precipitate is obtained which becomes crystalline on drying and has the formula, $C_4N_4H_8O_2$. This crystalline material is apparently formed by the addition of one molecule of water to the product originally obtained. On reaction with formaldehyde, it is converted to a resin[73]. Formaldehyde-cyanamide condensation products are also obtained by reacting cyanamide with formaldehyde in the presence of strong acids[15] and alkalies[158].

The writer[186] obtained a partially hydrated polymer of methylene cyanamide by reaction of equimolar proportions of sodium cyanamide and formaldehyde in aqueous solution at 15 to 20°C. This product is soluble in dilute acids and strong alkalies but is practically insoluble in water at a pH of 8.0 to 9.0.

The dimer of cyanamide, dicyandiamide, forms a monomethylol derivative, melting at 118°C when heated with aqueous formaldehyde[135]:

$$NH_2C(:NH)NH \cdot CN + CH_2O \text{ (aq)} \rightarrow HOCH_2NHC(:NH)NH \cdot CN$$

Dicyandiamide also gives resins on reaction with formaldehyde, preferably in alkaline solutions having a pH of 8 to 10[154].

Melamine, the cyclic trimer of cyanamide, reacts with almost neutral or mildly alkaline formaldehyde solution, producing methylolmelamines[163, 197]. From one to six mols of formaldehyde react readily with one mol of melamine, apparently yielding all the methylol derivatives theoretically possible. Hexamethylolmelamine, for example, is obtained by dissolving 0.1 mol of melamine in 75 parts of neutral 32 per cent formaldehyde (0.8 mol CH_2O) while heating with a bath of boiling water for 10 minutes. On cooling, the product gradually crystallizes in the course of two days and may then be filtered from solution, washed with alcohol, and dried for 5 hours at 60°C. The product as isolated apparently contains water of crystallization and is a monohydrate of hexamethylolmelamine [$C_3N_6(CH_2$

OH)$_6 \cdot$H$_2$O]. The formation of this derivative is indicated thus:

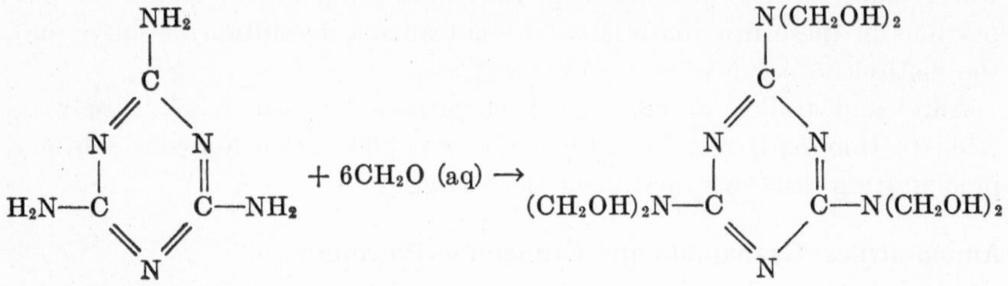

$$
\begin{array}{ccc}
\text{NH}_2 & & \text{N(CH}_2\text{OH)}_2 \\
| & & | \\
\text{C} & & \text{C} \\
\diagup\!\!\diagdown & & \diagup\!\!\diagdown \\
\text{N}\quad\text{N} & + 6\text{CH}_2\text{O (aq)} \rightarrow & \text{N}\quad\text{N} \\
|\qquad\| & & |\qquad\| \\
\text{H}_2\text{N}\!-\!\text{C}\quad\text{C}\!-\!\text{NH}_2 & (\text{CH}_2\text{OH})_2\text{N}\!-\!\text{C}\quad\text{C}\!-\!\text{N(CH}_2\text{OH})_2 \\
\diagdown\!\!\diagup & & \diagdown\!\!\diagup \\
\text{N} & & \text{N}
\end{array}
$$

On heating, hexamethylolmelamine hydrate melts at approximately 150°C to give a clear liquid which is converted on continued heating to a clear, colorless, water-insoluble resin.

All the methylolmelamines are water-soluble products which resinify on heating. The products thus obtained are the well-known melamine-formaldehyde resins. These resins are also made directly by heating melamine and formaldehyde[172]. In general, the reactions of melamine with formaldehyde resemble formaldehyde-amide reactions and the resin chemistry involved is undoubtedly similar in many respects to the chemistry of urea-formaldehyde resins (pages 297–301).

Aminoacetonitrile reacts with formaldehyde under alkaline conditions in the same manner as other primary amines, giving trimeric methylene-aminoacetonitrile[96]:

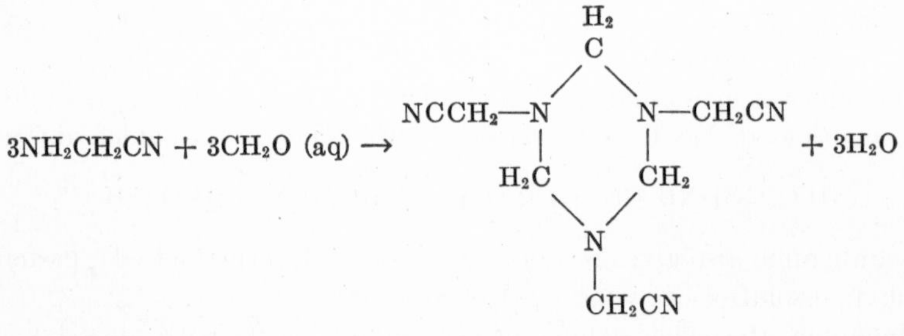

$$
\begin{array}{c}
\text{H}_2 \\
\text{C} \\
\diagup\!\!\diagdown \\
3\text{NH}_2\text{CH}_2\text{CN} + 3\text{CH}_2\text{O (aq)} \rightarrow \quad \text{NCCH}_2\!-\!\text{N}\qquad\text{N}\!-\!\text{CH}_2\text{CN} \quad + 3\text{H}_2\text{O} \\
|\qquad\qquad| \\
\text{H}_2\text{C}\qquad\text{CH}_2 \\
\diagdown\!\!\diagup \\
\text{N} \\
| \\
\text{CH}_2\text{CN}
\end{array}
$$

This product is also obtained by the joint reaction of ammonium chloride and sodium cyanide with formaldehyde[90]. Loder[110] prepares nitrilotriacetonitrile, N(CH$_2$CN)$_3$, by reaction of amino acetonitrile with formaldehyde and hydrogen cyanide. Gresham and Schweitzer describe the preparation of hydantoin by reaction of aminoacetonitrile (ex formaldehyde, hydrogen cyanide and ammonia) with ammonium carbonate[70].

When polyaminonitriles, such as 2,2,11,11-tetramethyl-3,10-diazododecane dicarboxylonitrile, which is readily prepared by the reaction of hexamethylenediamine with two mols of acetone cyanohydrin, are reacted with 4 or more mols of formaldehyde, a white rubber-like polymeric product results[88]. The course of this reaction is believed to be quite different from the ordinary amine aldehyde reactions. In the example cited, one mol of acetone is liberated for each mol of polyaminonitrile taking part in the resin-forming reaction.

Amino Acids and Esters

The reactions of formaldehyde with amino acids and esters are essentially reactions of the amino group, similar in most respects to those involving the amines and amine derivatives which have already been discussed. Of specific interest in the case of amino acids is the fact that the basicity of the amino group is reduced by combination with formaldehyde. Schiff's finding[155] that the lower amino acids, such as glycine, alanine and asparagine, although practically neutral in ordinary aqueous solution, react as strong acids in the presence of formaldehyde is the basis for the well known Sorenson method[165] for titrating amino acids.

The reactions of glycine (amino-acetic acid) and its derivatives may be taken as representative of this group of compounds. In neutral solution formaldehyde and glycine combine readily, giving methyleneaminoacetic acid[109].

$$NH_2CH_2COOH + CH_2O \text{ (aq)} \rightarrow CH_2{:}NCH_2COOH + H_2O$$

This compound behaves as a normal monobasic acid and may be readily titrated[89,155], whereas glycine itself is neutral. In strong acid solutions glycine and formaldehyde react to give methylenediglycine[109].

$$2NH_2CH_2COOH + CH_2O \rightarrow H_2C(NHCH_2COOH)_2 + H_2O$$

In the presence of hydrochloric acid and tin, the primary reaction products of formaldehyde and glycine are reduced to methyl derivatives; sarcosine (methylaminoacetic acid) and dimethylaminoacetic acid are obtained[109]:

$$NH_2CH_2COOH + CH_2O \text{ (aq)} + H_2 \rightarrow CH_3NHCH_2COOH + H_2O$$

$$CH_3NHCH_2COOH + CH_2O \text{ (aq)} + H_2 \rightarrow (CH_3)_2NCH_2COOH + H_2O$$

Although not readily prepared under ordinary conditions, salts of methylolglycine, $HOCH_2NHCH_2COOH$, are formed when solutions of glycinates are reacted with formaldehyde at temperatures in the neighborhood of 0 to 5°C[102]. When an excess of commercial formaldehyde solution is employed and the reaction mixture is not chilled, a compound having the empirical formula $C_7H_{14}O_5N_2$ is obtained as the principal product. According to Krause[100], it is probably hydroxytrimethyleneglycine, $HOCH(CH_2NHCH_2COOH)_2$. This formula is supported by the following facts: (a) it gives a monoacetyl derivative, as would be expected, and (b) acetone has been identified among its oxidation products. Although the reaction mechanism has not been clearly established, the methanol present in commercial formaldehyde apparently plays an important part in ths process. If methanol-free formaldehyde is employed, very little product is obtained and considerable formic acid results. According to Krause, the reaction involving commercial formaldehyde takes place as follows:

$$2NH_2CH_2COOH + 2CH_2O \ (aq) + CH_3OH \rightarrow HOCH(CH_2NHCH_2COOH)_2 + 2H_2O$$

whereas, methanol-free formaldehyde reacts according to the equation:

$$2NH_2CH_2COOH + 4CH_2O \ (aq) \rightarrow HOCH(CH_2NHCH_2COOH)_2 + HCOOH + H_2O$$

More complex derivatives, such as $C_{12}H_{21}N_3O_5$, which are also obtained by reactions of formaldehyde and glycine, suggest the possibility that formaldehyde may take part in the formation of polypeptides and other natural products in living organisms[62]. Noncrystalline hygroscopic masses similar to the polypeptides in their chemical behavior, were obtained by Galeotti[62] on digesting formaldehyde with glycine, alanine, leucine, and other amino acids for 8 to 10 hours.

Two reaction products of glycine ethyl ester hydrochloride and formaldehyde are reported: triformaldehyde glycine ester (or glycine ester triformal[9] and diethylhydroxytrimethyleneglycine[101]:

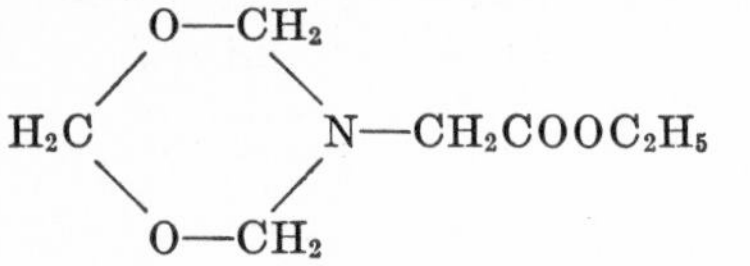

Triformaldehyde glycine ester 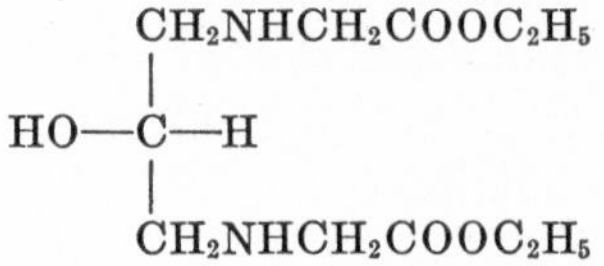   *Diethylhydroxytrimethyleneglycine*

Cyclic glycine anhydride forms a dimethylol derivative with formaldehyde[27].

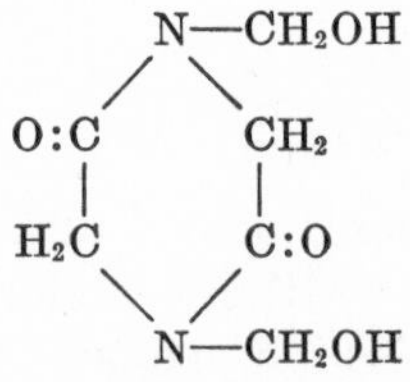

Dimethylolglycine anhydride

Production of resins from glycinamide and related amino acid amides by reaction with formaldehyde has been described by Marvel[119] (cf. page 299).

Proteins

In general, reaction with formaldehyde hardens proteins, decreases their water-sensitivity, and increases their resistance to the action of chemical reagents and enzymes. These tanning effects have been the subject of considerable study from a practical standpoint (pages 447, 483) but only in recent years has the mechanism of the reactions by which they are produced received serious attention. It has now been demonstrated that the tanning effect is due, in large part, to the cross-linkage of protein chains and micellar units by methylene bonds connecting reactive groups[3]. The reactions involved are apparently similar to those observed with simple amines, amides and other compounds containing polar structural units of the type known to occur in proteins. The formation of simple methylol derivatives takes place as would be expected but does not produce the increase in hydrothermal stability characteristic of tannage.

Comparison of the tanning action of various aldehydes by Gustavson[74, 76] shows that formaldehyde is, by far, the most effective agent. This is demonstrated by the elevation of the shrinkage temperature in water for hides treated with $\frac{2}{3}$ molar solutions of the aldehydes for 24 hours. Under these conditions, formaldehyde raises the shrinkage temperature approximately 20°C. Glyoxal, acrolein and crotonaldehyde, which also show tanning activity, raise the shrinkage temperature around 10°C. Acetaldehyde is less active, whereas other aldehydes such as propionaldehyde, octyl aldehyde, benzaldehyde and furfural do not raise the shrinkage temperature at all under the conditions of this test. Of the other aldehydes tested to date, glyoxal is the most efficient tanner.

Research on the mechanism of the reaction of formaldehyde with proteins by Nitzchmann and Hadorn[128] established that water was split off in the tanning process. Fraenkel-Conrat and Olcott[59] working with the protamine, salmine, proved that an increase in molecular weight was also definitely involved since soluble reaction product could no longer be dialyzed through cellulose films. These findings, coupled with the known chemistry of amides, amines, amino acids, and synthetic polypeptides confirm the formation of cross-linkages in protein-formaldehyde reactions.

Protein groupings taking part in formaldehyde reactions include primary amino and amido groups, guanidyl groups, secondary amide groups in peptide chains, indole groups, mercaptan radicals, imidazole groups and phenolic nuclei. These reactions are governed by the relative reactivity of the groups concerned and the conditions of pH, temperature, etc., favoring reaction. It is generally accepted that formaldehyde reacts most readily with primary amino groups even at low temperatures. Studies of the action of formaldehyde on collagen and deaminated collagen by Gustavson[75, 74, 77] support the hypothesis that the formation of methylene cross-linkages in this protein is dependent upon the presence of the epsilon amino groups of the lysine residue. Fraenkel-Conrat and Olcott[60, 61] present evidence that methylol-amino groups react with amide groups to form cross-linkages under mildly acidic conditions (pH approx. 3.5) where neither grouping binds formaldehyde appreciably alone. This reaction is analogous to the Mannich synthesis. In this connection, these investigators have also shown that proteins which are rich in amide groupings but contain few amino radicals can be cross-linked by a joint reaction with formaldehyde and ammonia or a primary amine. The reaction apparently makes use of a methylene amine linkage such as —CH$_2$—N—CH$_2$—. Methylolamino rad-

$$\begin{array}{c} | \\ R \end{array}$$

icals can also cross-link with guanidyl, phenol, imidazole and indole groups.

Reaction of formaldehyde with free amino groups is not limited to the formation of methylol derivatives and methylene linkages, but may also involve the production of triformals[175]. This is indicated in the following equation, in which the letter P designates a protein radical to which is attached an amino group:

$$[P]\!-\!NH_2 + 3CH_2O \text{ (aq)} \rightarrow [P]\!-\!N \overset{\displaystyle CH_2\!-\!O}{\underset{\displaystyle CH_2\!-\!O}{\diagup \diagdown}} CH_2 + H_2O$$

It is probable that simple methylol derivatives are the primary products formed in all protein-formaldehyde reactions.

Formaldehyde reacts most rapidly with proteins in the dissolved state. Gustavson[74, 76] has shown that it is equally effective when applied in water, ethanol, acetone, dioxane, chloroform and benzene. In aqueous solution, rapidity of reaction increases with formaldehyde concentration up to 10 per cent, according to Lumiere and co-workers[112], and is then stated to remain practically constant. In most cases, the rate of reaction increases with increasing temperature. Hrubesky and Browne[84] report that para-formaldehyde gives a delayed effect with glue, since depolymerization and solution must apparently precede reaction. Paraformaldehyde, which has been polymerized further by heating at 100°C, is less reactive than the untreated polymer. This modified polymer is accordingly especially suit-able for use in waterproof glues, since glue baths must not set before the glue can be applied. Acids, such as oxalic acid, have a marked retarding effect on the paraformaldehyde-glue reaction and are of special value in the preparation of water-resistant glues. It is probable that they produce a mild acidity in the glue solution which reduces the rate at which the poly-mer depolymerizes before it can enter the reaction medium. Gaseous for-maldehyde reacts with gelatin producing the same effect as the aqueous solution but at a much slower rate[112]. Water-soluble fatty acid amides, such as formamide[37], alcohol[42, 169], and acetone[169] have a retarding effect on some protein-formaldehyde reactions.

The pH of the reaction media is a controlling factor in the action of formaldehyde on proteins in aqueous media. However, its effect is dependent on the type of reactive groups available in the protein involved and the reaction temperature. Primary and secondary amino groups form methylol derivatives readily over a wide range of pH values even at low temperatures, whereas amide, guanidyl and indole groups require elevated temperatures or acid or alkaline conditions[58, 61]. Secondary amide groups are less reactive than primary ones and require strong acid catalysts as shown by studies with synthetic polyamides (pages 305–306). Cross-linking of methylolamino groups with amide or guanidyl radicals takes place over the pH range 3 to 8[61]. Results encountered in commercial tanning of leather indicate, in general, that the amount of formaldehyde which reacts is less on the acid side of the isoelectric point. This amount increases as the pH advances from 4 to 9[131]. Theis[174] reports that collagen amino groups combine with two mols of formaldehyde at pH 8 and with three at higher pH values. Gustavson's data[75] indicate that formaldehyde is fixed on the epsilon nitrogen atoms of the lysine radicals of collagen at pH 6–8, and at pH 12 on both lysine and arginine nitrogen. The presence of salts such as sodium and calcium chloride increases the amounts of formaldehyde fixed by collagen in tanning[174]. Anderson[2] claims that formaldehyde does not combine with proteins on the acid side of the isoelectric point. However, some proteins apparently differ considerably in this respect, since Oku[130] reports that maximum fixation of formaldehyde by sericin is obtained in the presence of 2 per cent hydrochloric or sulfuric acid. Croston[34] has shown that zein, which is practically free of amino groups, forms strong cross-linked fibers when cured by reaction with substantially anhydrous formaldehyde in an inert solvent using a strong acid catalyst, such as hydrogen chloride.

The combining capacity of various proteins for formaldehyde is dependent on the nature and abundance of reactive groups. The maximum amount of formaldehyde fixed by gelatin is reported to be approximately 4.0 to 4.8 grams per 100 grams of protein[112], whereas casein fixes a maximum of only 0.6 to 2.5 grams per 100 grams[131]. The best grades of glue, which possess high jelly values and set quickly to give strong bonds, require less formaldehyde to effect insolubilization than do the lower grades[17]. Reiner[144] reports that partially tanned gelatin is more reactive than the original protein.

Degradation or denaturing of proteins may take place under some conditions when powerful reagents or high temperatures are employed. On heating casein with an excess of acid formaldehyde, degradation of the protein molecule takes place with the formation of primary, secondary, and tertiary methylamines equivalent to from 12 to 40 per cent of the total protein nitrogen. When this reaction is carried out under pressure at 180°C with formaldehyde solution containing 5 per cent acetic acid, trimethylamine is obtained[200]. These reactions are similar to those which take place when simple amine salts are reacted with formaldehyde.

Since the formation of methylene bridges between protein molecules results in a product of increased molecular weight, increased hardness and reduced water-sensitivity[173] would naturally be expected in the reaction product. Furthermore, since such reactions result in the removal of reactive hydrogen atoms from the protein molecule, the stability of formaldehyde-treated proteins to heat, enzymes, and chemical reagents is increased. As illustrated in the case of gelatin[63], the isoelectric point of proteins is lowered by reaction with formaldehyde. X-ray diffraction patterns are also modified[31].

Proteins containing formaldehyde give characteristic colors both with sulfuric acid alone and in the presence of nitrous acid[79, 181]. These tests are extremely sensitive.

Formaldehyde-treated proteins are usually stable to the action of alkalies, as illustrated by the fact that formaldehyde-treated wool is highly resistant to the action of caustic soda[30]. This is in agreement with the known stability of methyleneamines and formals to alkaline agents. As would be expected, however, these formaldehyde-protein products are not as resistant under acidic conditions. Formaldehyde-gelatin condensation products liberate formaldehyde gradually on treatment with warm water and decompose rapidly on exposure to cold hydrochloric acid[112]. Liberation of formaldehyde by distillation with dilute acids is substantially quantitative in some cases and is often employed for the determination of combined formaldehyde[127]. However, this method of analysis is not reliable since some proteins combine irreversibly with a portion of the formaldehyde with which they are reacted.

The existence of non-hydrolyzable formaldehyde in modified proteins was first suggested by Baudouy[8] who pointed out that this was the case with cyclized derivatives of histidine and tryptophane. Nitschmann and Lauener[129] have also pointed out this property of these amino acids estimating that casein cured by formaldehyde gas at 70°C might contain up to 5 per cent of irreversibly combined formaldehyde. Fraenkel-Conrat[57] and co-workers have since shown that the tryptophane units in gramicidin fix formaldehyde irreversibly forming a methylol indole with formation of a carbon-nitrogen or a carbon-carbon bond.

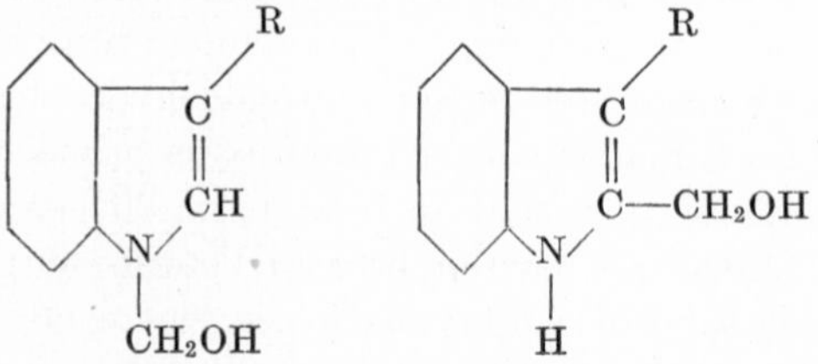

Swain and co-workers[171] have shown that when formaldehyde is reacted with casein at 100 to 120°C alone or in the presence of acids, about 10 per cent of the combined formaldehyde cannot be recovered on acid hydrolysis. It is suggested that this formaldehyde is bound by the carbon-carbon linkage. If not already present, this linkage may be formed by migration of an N-methylol group on treatment with acid. The nonaqueous curing of zein, described by Croston[34], also yields irreversible methylene cross-linkages and it is suggested by this investigator that phenolic nuclei of tyrosine residues in zein may be involved.

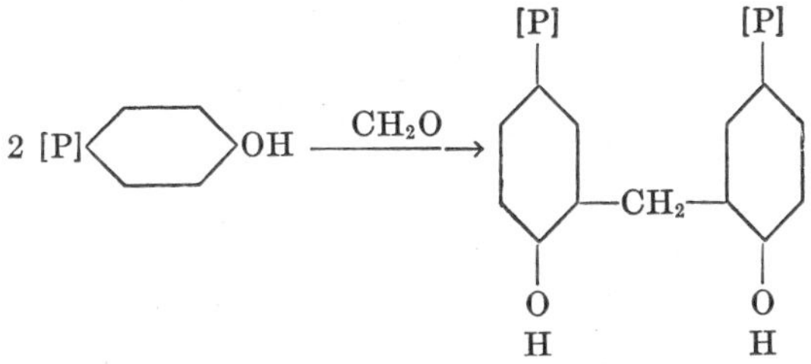

Middlebrook[121] suggests that irreversible combination of formaldehyde with asparagine amide groups results in formation of 6-hydroxytetrahydropyrimidine-4-carboxylic acid which is stable to dilute boiling phosphoric acid.

Nitriles

In general, nitriles do not react with formaldehyde at ordinary temperatures in the absence of catalytic agents unless they contain substituent groups which are normally reactive under these conditions. However, at temperatures in the neighborhood of 200 to 400°C, Brant and Hasche[18] have demonstrated that aliphatic nitriles and arylacetonitriles undergo a vapor phase condensation with formaldehyde in the presence of a solid dehydration catalyst to give vinyl derivatives. This reaction is illustrated by the conversion of propionitrile to methacrylonitrile.

$$CH_3CH_2CN + CH_2O \text{ (g)} \rightarrow CH_2{:}C(CH_3){\cdot}CN + H_2O$$

Shand[160] prepared alpha-methoxypropionitrile by adding a hot slurry of paraformaldehyde and propionitrile to a catalyst bed of alumina maintained at about 200°C.

Aliphatic nitriles do not give alpha-methylol or methylene-bis derivatives with formaldehyde unless the alpha-hydrogen atoms are activated by adjacent substituent groups. This may be effected by an additional nitrile group as in malononitrile, an aryl radical, a carboxyl group (see page 272), etc.

Dimethylol malononitrile was prepared by Ardis and co-workers[5] by reacting malononitrile with formaldehyde gas dissolved in glacial acetic acid to which potassium acetate had been added as a reaction catalyst. The process was carried out at 100°C and the product, which precipitated on adding benzene to the reaction mixture, melted and resinified at 93 to 98°C. The dimethylol derivative reacts with additional malononitrile on heating in dioxane at 100°C for two hours to give a methylene derivative as shown below:

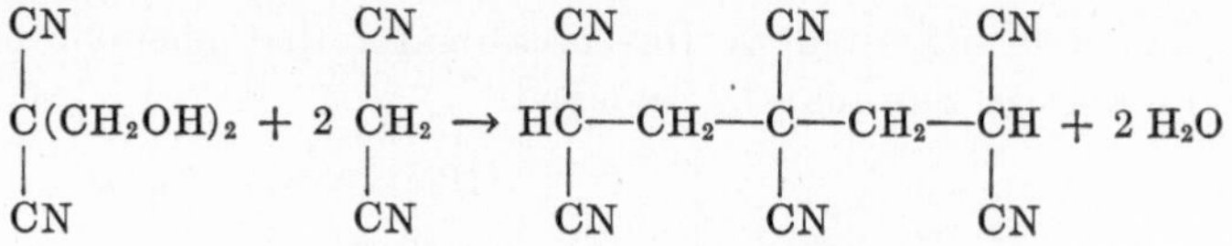

$$
\underset{\overset{|}{CN}}{\overset{\overset{CN}{|}}{C}}(CH_2OH)_2 + 2\ \underset{\overset{|}{CN}}{\overset{\overset{CN}{|}}{CH_2}} \rightarrow \underset{\overset{|}{CN}}{\overset{\overset{CN}{|}}{HC}}-CH_2-\underset{\overset{|}{CN}}{\overset{\overset{CN}{|}}{C}}-CH_2-\underset{\overset{|}{CN}}{\overset{\overset{CN}{|}}{CH}} + 2\ H_2O
$$

(m.p. 206° to 7°C)

An alcoholic solution of malononitrile reacts directly with aqueous formaldehyde at 0°C to give methylene bis-malononitrile (melting point 137°C)[39]. In this case the methylol derivative cannot be isolated. On pyrolysis, methylene bis-malononitrile gives vinylidene cyanide, $CH_2:C(CN)_2$, (m.p. 9.7°C, b.p. 154°C), which polymerizes instantly in the cold on contact with water, alcohol and amines.

Benzyl cyanide reacts with formaldehyde in methanol containing sodium methylate to give a viscous liquid product, probably containing the methylol derivative, which pyrolyzes to give atroponitrile[189].

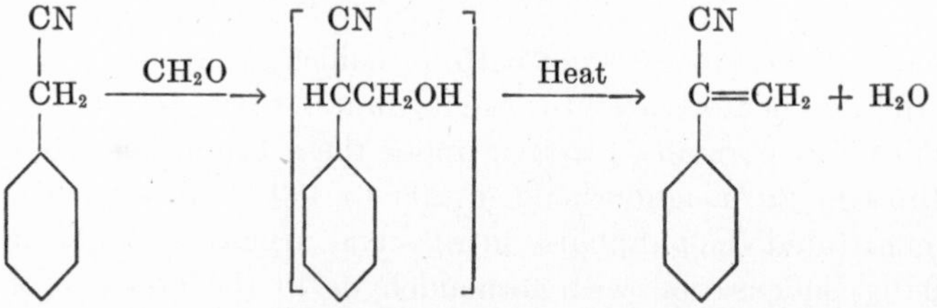

Monomeric atroponitrile, produced in this way, polymerizes on standing to give a crystalline dimer melting at 123°C.

Strong sulfuric acid causes nitriles to react with formaldehyde producing methylene derivatives of the corresponding amides. At low temperatures (under 50°C) in the presence of excess sulfuric acid, methylene bis-amides are obtained; whereas, at high temperatures (70°C and higher), with small amounts of acid, the principal products are amides of cyclotrimethylene-triamine. Products are precipitated in fair to excellent yields by pouring the crude reaction mixture into cold water or a mixture of water and ice. These reactions are indicated below for a nitrile, RCN, in which R signifies an aliphatic, olefinic or aromatic radical.

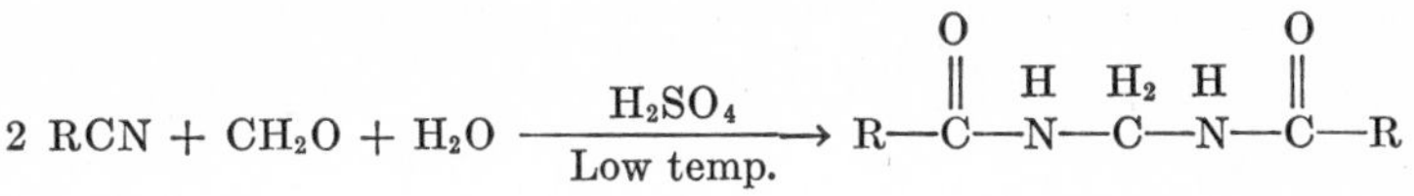

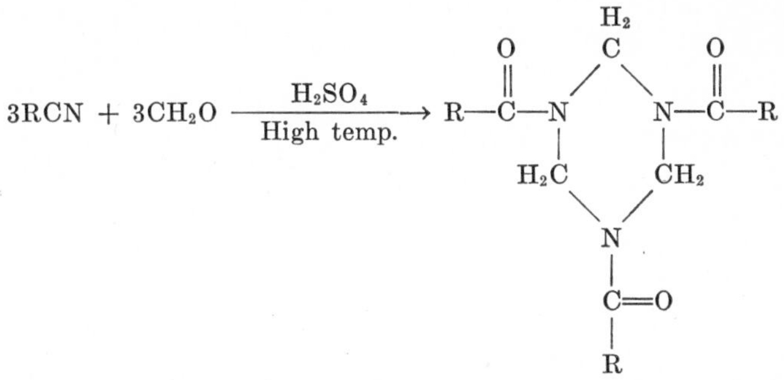

$$2\ \text{RCN} + \text{CH}_2\text{O} + \text{H}_2\text{O} \xrightarrow[\text{Low temp.}]{\text{H}_2\text{SO}_4} \underset{\text{O}}{\overset{\text{O}}{\text{R—C—N—C—N—C—R}}}$$

$$3\text{RCN} + 3\text{CH}_2\text{O} \xrightarrow[\text{High temp.}]{\text{H}_2\text{SO}_4}$$

Acrylamides prepared from acrylonitrile and formaldehyde have proved of special interest to chemical investigators. Dinitriles, such as adiponitrile and azelaonitrile, when subjected to this reaction, give high molecular weight polyamides.

A reaction of this type yielding a methylene bis-amide appears to have been first observed by Hepp and Spiess in 1876. As a result of increasing industrial interest in the reactions of acrylonitrile and the synthesis of polyamides, these formaldehyde-nitrile syntheses have recently received intensive study by Gradsten[63], Gresham[71, 72], Magat[115, 116], Mowry[124, 125], Pollock[137], Wegler[194] and numerous co-workers associated with these investigators.

Although these reactions may be carried out by use of aqueous formaldehyde, polymers or anhydrous monomer are preferred since best results are obtained when the amount of water in the reaction mixture is limited. Trioxane, which can be dissolved in the nitrile, is a highly satisfactory form of formaldehyde for these syntheses. Chlorsulfonic acid can be employed in place of sulfuric acid[137] and combinations of sulfuric acid with formic and acetic are of special value in some instances[115,116].

The reaction mechanism may involve hydration of nitrile to amide followed by reaction with formaldehyde or a carbonium ion mechanism as suggested by Magat[116].

Magat and Salisbury[117] have shown that methylene amides can also be prepared by the reaction of mono- and dimethylol amides (e.g., dimethylol adipamide) with nitriles in strong sulfuric acid at room temperature.

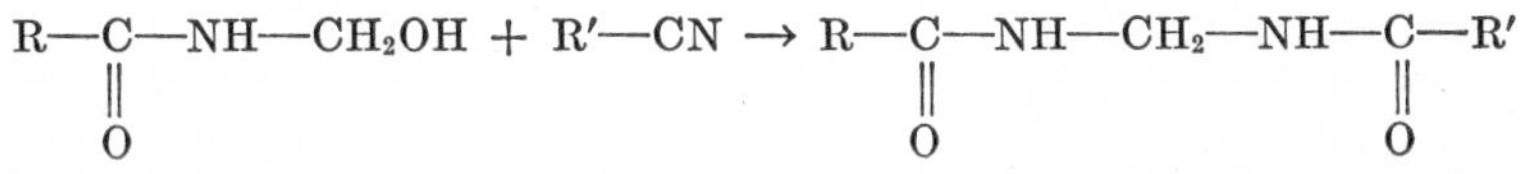

$$\underset{\text{O}}{\text{R—C—NH—CH}_2\text{OH}} + \text{R}'\text{—CN} \rightarrow \underset{\text{O}}{\text{R—C—NH—CH}_2\text{—NH—C—R}'}$$

Acrylonitrile also reacts with aqueous formaldehyde solutions and solutions of formaldehyde in alcohol to give methylene ethers of ethylene cyanohydrin in the presence of strong alkalies. Under these circumstances, the dissolved aqueous formaldehyde reacts as methylene glycol.

$$HOCH_2OH + CH_2:CHCN \rightarrow HOCH_2OCH_2CH_2CN$$

$$HOCH_2OH + 2CH_2:CHCN \rightarrow CH_2(OCH_2CH_2CN)_2$$

Solutions of formaldehyde in an alcohol, such as capryl alcohol ($C_8H_{17}OH$), react as the hemiacetal.

$$C_8H_{17}OCH_2OH + CH_2:CHCN \rightarrow C_8H_{17}OCH_2OCH_2CH_2CN$$

The aqueous reaction was demonstrated by the writer[185]; the alcoholic reaction by Bruson and Riener[22].

References

1. Adams, R., and Blicke, F. F., "Organic Reactions," pp. 304–330, New York, John Wiley and Sons (1942).
2. Anderson, H., *J. Intern. Soc. Leather Trades' Chemists*, **18**, 197–200 (1934).
3. Anson, M. L., and Edsall, J. T., "Advances in Protein Chemistry," II, pp. 278–336, New York, Academic Press (1945).
4. Arbuzov, B. A., and Livshits, D. A., *C. A.*, **42**, 6335 (1948).
5. Ardis, A. E., Averill, S. J., Gilbert, H., Miller, F. F., Schmidt, R. F., Stewart F. D., and Trumbull, H. L., *J. Am. Chem. Soc.*, **72**, 1305–7 (1950).
6. Baldwin, A. W., and Piggott, H. A., (to Imperial Chemical Industries, Ltd.), U. S. Patent 2,131,362 (1938).
7. Baldwin, A. W., Piggott, H. A., and Statham, F. S., (to Imperial Chemistry Industries, Ltd.), U. S. Patent 2,237,296 (1941).
8. Baudouy, C., *Compt. rend.*, **214**, 692–5 (1942).
9. Bergmann, M., Jacobsohn, M., and Schotte, H., *Z. Physiol. Chem.*, **131**, 18–28 (1924); *C. A.*, **18**, 2129.
10. Bergmann, M., and Miekeley, A., *Ber.*, **57**, 662 (1924).
11. Bersworth, F. C., (to The Martin Dennis Co.), U. S. Patents 2,387,735 (1945), 2,407,645 (1946); 2,461,519 (1949).
12. Bettelheim, L., and Cedwall, J., *Svensk. Kem. Tid.*, **60**, 208–14 (1948); *C. A.*, **43**, 1247 (1949).
13. Binz, A. H., Reinhart, F. E., and Winter, H. C., *J. Am. Chem. Soc.*, **62**, 7–8 (1940).
14. Bischoff, C. A., *Ber.*, **31**, 3248 (1898).
15. Bischoff, C. A., and Reinfeld, F., *Ber.*, **36**, 35 (1903).
16. Bougault, M. J., and Leboucq, J., *J. Pharm. Chim.*, **17**, 193 (1933).
17. Bogue, R. H., *Chem. Met. Eng.*, **23**, 5 (1920).
18. Brant, J. H., and Hasche, R. L., (to Eastman Kodak Co.), U. S. Patent 2,386,-586 (1945).
19. Braun, J. v., *Ber.*, **41**, 2145 (1908).
20. Brian, R. C., and Lambertson, A. H., *J. Chem. Soc.*, **1949**, 1633–5.
21. Bruson, H. A., and Eastes, J. W., (to Resinous Products & Chemical Co.), U. S. Patent 2,160,196 (1939).

22. Bruson, H. A., and Riener, T. W., (to Resinous Products & Chemical Co.), U. S. Patent 2,435,869 (1944).
23. Cairns, T. L., Foster, H. D., Larchar, A. W., Schneider, A. K., and Schreiber, R. S., *J. Am. Chem. Soc.*, **71**, 651–5 (1949).
24. Cairns, T. L., Gray, H. W., Schneider, A. K., and Schreiber, R. S., *J. Am. Chem. Soc.*, **71**, 655–7 (1949).
25. Cass, O. W., and K'Burg, R. T., (to E. I. du Pont de Nemours & Co., Inc.), U. S. Patent 2,194,294 (1940).
26. Chadwick, A. F., (to E. I. du Pont de Nemours & Co., Inc.), U. S. Patent 2,532,-278 (1950).
27. Cherbuliez, E., and Feer, E., *Helv. Chim. Acta*, **5**, 678 (1923).
28. Cherbuliez, E., and Sulzer, G., *Helv. Chim. Acta*, **8**, 567–71 (1925); *C. A.*, **20**, 365.
29. Chwala, A., *Monatsh.*, **78**, 172–3 (1948).
30. Clark, A. M., British Patent 3,492 (1903).
31. Clark, G. L., and Shenk, J. H., *Radiology*, **28**, 357–61 (1937).
32. Clarke, H. T., Gillespie, H. B., and Weisshaus, S. Z., *J. Am. Chem. Soc.*, **55**, 4571–87 (1933).
33. Cordier, D. E., (to Plaskon Co., Inc.), U. S. Patent 2,213,578 (1940).
34. Croston, C. B., *Ind. Eng. Chem.*, **42**, 482–4 (1950).
35. Crowe, G. A., and Lynch, C. C., *J. Am. Chem. Soc.*, **70**, 3795 (1948); **71**, 3731–3 (1949).
36. D'Alelio, G. F., (to General Electric Co.), U. S. Patents 2,239,440–1 (1941).
37. Dangelmajer, C., and Perkins, E. C., (to E. I. du Pont de Nemours & Co., Inc.), U. S. Patent 2,061,063 (1936).
38. Décombe, J., *Compt. rend.*, **196**, 866–8 (1933).
39. Diels, O., Gärtner, H., and Kaack, R., *Ber.*, **55**, 3439–3448 (1922).
40. Diels, O., and Lichte, R., *Ber.*, **59**, 2778–84 (1926).
41. Dixon, A. E., *J. Chem. Soc.*, **113**, 238–9 (1918).
42. Dixon, A. E., and Taylor, J., *J. Chem. Soc.*, **109**, 1253 (1916).
43. Duden, P., and Scharff, M., *Ann.*, **288**, 228, 252 (1895).
44. Eberhardt, C., and Welter, A., *Ber.*, **27**, 1804 (1894).
45. Einhorn, A., German Patent 158,088 (1905).
46. Einhorn, A., *Ann.*, **343**, 207–310 (1905).
47. *Ibid.*, **361**, 113–165 (1908).
48. Einhorn, A., German Patent 284,440 (1915).
49. Einhorn, A., and Hamburger, A., *Ber.*, **41**, 24–8 (1908).
50. Einhorn, A., and Ladisch, C., *Ann.*, **343**, 277 (1905).
51. Einhorn, A., and Prettner, A., *Ann.*, **334**, 210–233 (1904).
52. Eschweiler, W., German Patent 80,520 (1893).
53. Eschweiler, W., *Ber.*, **38**, 8801 (1905).
54. Farbenfabriken vorm. F. Bayer & Co., British Patent 11,360 (1912); *C. A.*, **7**, 3673.
55. Farlow, M. W., (to E. I. du Pont de Nemours & Co., Inc.), U. S. Patent 2,422,-400 (1947).
56. Fischer, O., and Wreszinski, H., *Ber.*, **25**, 2711 (1892).
57. Fraenkel-Conrat, H., Brandon, B. A., and Olcott, H. S., *J. Biol. Chem.*, **168**, 99–118 (1947).
58. Fraenkel-Conrat, H., and Olcott, H. S., *J. Am. Chem. Soc.*, **67**, 950–4 (1945).
59. Fraenkel-Conrat, H., and Olcott, H. S., *J. Am. Chem. Soc.*, **68**, 34–7 (1946).
60. Fraenkel-Conrat, H., and Olcott, H. S., *J. Am. Chem. Soc.*, **70**, 2673 (1948).

61. Fraenkel-Conrat, H., and Olcott, H. S., *J. Biol. Chem.*, **174**, 827–42 (1948).
62. Galeotti, G., *Biochem. Z.*, **53**, 474, 477 (1913); *C. A.*, **8**, 1792.
63. Gerngross, O., and Bach, S., *Collegium*, **1922**, 350–1; *C. A.*, **17**, 2071; *Collegium*, **1923**, 377–85; *C. A.*, **18**, 1411.
64. Gilman, H., "Organic Chemistry," Vol. I, p. 729, New York, John Wiley & Sons (1943).
65. Girsewald, C. F. v., German Patent 281,045 (1914).
66. Gnehm, R., and Blumer, E., *Ann.*, **304**, 115 (1899).
67. Goldschmidt, C., *Chem. Ztg.*, **46**, 460 (1897).
68. Gradsten, M. A., and Pollock, M. W., *J. Am. Chem. Soc.*, **70**, 3079–81 (1948).
69. Graymore, J., *J. Chem. Soc.*, **1932**, I, 1353.
70. Gresham, W. F., and Schweitzer, C. E., (to E. I. du Pont de Nemours & Co., Inc.), U. S. Patent 2,402,134 (1946).
71. Gresham, T. L., and Steadman, T. R., *J. Am. Chem. Soc.*, **71**, 1872 (1949).
72. Gresham, T. L., and Steadman, T. R., (to The B. F. Goodrich Co.), U. S. Patent 2,568,620 (1951).
73. Griffith, P. W., (to American Cyanamid Co.), U. S. Patent 2,019,490 (1935).
74. Gustavson, K. H., *J. Intern. Soc. Leather Trades' Chemists*, **24**, 377 (1940).
75. Gustavson, K. H., *Svensk. Kem. Tid.*, **52**, 261–77 (1940).
76. Gustavson, K. H., *Särtryk ur Svensk. Kem. Tid.*, **59**, 98 (1947).
77. Gustavson, K. H., *J. Am. Leather Chemists Assoc.*, **43**, 741, 744 (1948).
78. Hallowell, A. T., (to E. I. du Pont de Nemours & Co., Inc.), U. S. Patent 2,413,-968 (1947).
79. Hehner, O., *Analyst*, **21**, 94–7 (1896); *Chem. Zentr.*, **1896**, I, 1146.
80. Henry, L., *Ber.*, **26**, 934, Ref. (1893).
81. Henry, L., *Bull. acad. roy. Belg.* (*3*), **28**, 359, 366 (1894).
82. *Ibid.*, **29**, 355 (1895).
83. Hodgkins, T. S., and Hovey, A. G., *Ind. Eng. Chem.*, **31**, 673–7 (1939).
84. Hrubesky, C. E., and Browne, F. L., *Ind. Eng. Chem.*, **19**, 217 (1927).
85. Hurwitz, M. D., (to Rohm and Haas Co.), U. S. Patent 2,582,128 (1952).
86. I. G. Farbenindustrie Akt.-Ges., British Patent 436,414 (1934).
87. Jacobson, R. A., (to E. I. du Pont de Nemours & Co., Inc.), U. S. Patent 2,155,-863 (1939).
88. Jacobson, R. A., and Mighton, C. J., (to E. I. du Pont de Nemours & Co., Inc.), U. S. Patent 2,228,271 (1941).
89. Jodidi, S. L., *J. Am. Chem. Soc.*, **40**, 1031–5 (1918).
90. Johnson, T. B., and Rinehart, H. W., *J. Am. Chem. Soc.*, **46**, 768, 1653 (1924).
91. Kadowaki, H., *Bull. Chem. Soc. Japan*, **11**, 248 (1936); *C. A.*, **30**, 5944.
92. Kalle & Co., Akt.-Ges., German Patent 164,610 (1905).
93. Kalle & Co., Akt.-Ges., German Patent 164,611 (1905).
94. Kittel, H., *C. A.*, **42**, 8521–2 (1948).
95. Kern, R., (to Alien Property Custodian), U. S. Patent 2,390,153 (1945)—Patent vested in A.P.C.
96. Klages, A., *J. prakt. Chem.*, **65**, 188 (1902).
97. Knoevenagel, E., and Mercklin, E., *Ber.*, **37**, 4087–4104 (1904).
98. Knudsen, P., *Ber.*, **47**, 2698–2701 (1914).
99. Kondo, H., and Ishida, S., *J. Pharm. Soc. Japan*, **489**, 979 (1922); *C. A.*, **17**, 1456.
100. Krause, H., *Ber.*, **51**, 136–150 (1918).
101. *Ibid.*, **51**, 1556–7 (1918).
102. *Ibid.*, **52**, 1211–1222 (1919).

103. Kvalnes, H. M., (to E. I. du Pont de Nemours & Co., Inc.), U. S. Patent 2,467,212 (1949); reissue 23,174 (1949).
104. Laer, M. H. van, *Bull soc. chim. Belg.*, **28**, 381–92 (1919); *C. A.*, **16**, 2113.
105. Levi, T. G., *Gazz. chim. ital.*, **61**, 803–814 (1931).
106. Lewis, J. R., and Reynolds, R. J. W., *Chem. & Ind.*, **1951**, 958–61.
107. Lieberman, S. V., and Wagner, E. C., *J. Org. Chem.*, **14**, 1001–12 (1949).
108. Litterscheid, F. M., *Ann.*, **316**, 180 (1901).
109. Loeb, W., *Biochem. Z.*, **51**, 116–27 (1913); *C. A.*, **8**, 679.
110. Loder, D. J., (to E. I. du Pont de Nemours & Co., Inc.), U. S. Patent 2,405,966 (1946).
111. Lüdy, E., *Monatsh.*, **10**, 295–316 (1889); *J. Chem. Soc.*, **56**, 1059 (1889).
112. Lumiere, A., Lumiere, L., and Seyewetz, A., *Bull. soc. chim.* (*3*), **35**, 872 (1906); *C. A.*, **1**, 67; *Rev. gén. chim.*, **11**, 295–9 (1908); *C. A.*, **2**, 3201.
113. Luskin, L. S., *et al.*, Abstracts of Papers, 119th Meeting of the Am. Chem. Soc., page 82M, April 1951.
114. McMaster, L., *J. Am. Chem. Soc.*, **56**, 204–6 (1934).
115. Magat, E. E., Chandler, L. B., Faris, B. F., Reith, J. E., and Salisbury, L. F., *J. Am. Chem. Soc.*, **73**, 1028 (1951).
116. Magat, E. E., Faris, B. F., Reith, J. E., and Salisbury, L. F., *J. Am. Chem. Soc.*, **73**, 1028 (1951).
117. Magat, E. E., and Salisbury, L. F., *J. Am. Chem. Soc.*, **73**, 1035 (1951).
118. Magnus-Levy, A., *Ber.*, **26**, 2148–50 (1893).
119. Marvel, C. S., U. S. Patent 2,436,363 (1948).
120. Marvel, C. S., Elliott, J. R., Boettner, F. E., and Yuska, H., *J. Am. Chem. Soc.*, **68**, 1681–6 (1946).
121. Middlebrook, W. R., *Biochem. J.*, **44**, 17–23 (1949).
122. Miller, J. G., and Wagner, E. C., *J. Am. Chem. Soc.*, **54**, 3698 (1932).
123. Miller, W. v., and Plöchl, J., *Ber.*, **25**, 2028 (1892).
124. Mowry, D. T., (to Monsanto Chemical Co.), U. S. Patent 2,534,204 (1950).
125. Mowry, D. T., and Ringwald, E. L., *J. Am. Chem. Soc.*, **72**, 4439 (1950).
126. Niemeyer, R., and Behrend, R., *Ann.*, **365**, 38–49 (1909).
127. Nitschmann, H., and Hadorn, H., *Helv. Chim. Acta*, **24**, 237–42 (1941).
128. *Ibid.*, **27**, 299–312 (1944).
129. Nitschmann, H., and Lauener, H., *Helv. Chim. Acta*, **29**, 174–9 (1949).
129a. Ogata, Y., Okano, M., Sugowara, M., *J. Am. Chem. Soc.*, **73**, 1715–1717 (1951).
130. Oku, M., *J. Agr. Chem. Soc. Japan*, **16**, 895–7 (1940); *Bull. Agr. Chem. Soc. Japan*, **16**, 141–2 (1940) (in English).
131. Ozawa, T., *Bull. Sericulture Silk Ind.* (*Japan*), **12**, 56–60 (1940); *C. A.*, **35**, 3272–3.
132. Passerini, M., *Gazz. chim. ital.*, **53**, 333–8 (1923); *C. A.*, **18**, 69.
133. Petroff, G., British Patent 283,002 (1928).
134. Pikl, J., (to E. I. du Pont de Nemours & Co., Inc.), U. S. Patent 2,426,790 (1947).
135. Pohl, F., *J. prakt. Chem.* (*2*), **77**, 537 (1908).
136. Pollak, F., *Modern Plastics*, **16**, No. 10, 45, 74, 76 (1939).
137. Pollock, M., and Zerner, E., (to Sun Chemical Corp.), U. S. Patent 2,493,360 (1950).
138. Pratesi, L., *Gazz. chim. ital.*, **14**, 352 (1884).
139. Pulvermacher, G., *Ber.*, **25**, 307–310 (1892).
140. *Ibid.*, p. 311.
141. *Ibid.*, p. 2765.
142. Rassow, B., *J. prakt. Chem.*, **64**, 129–35 (1901).
143. Rassow, B., and Lummerzheim, M., *J. prakt. Chem.*, **64**, 136–65 (1901).

144. Reiner, L., *Kolloid-Z.*, 195–7 (1920); *C. A.*, **15,** 610.
145. Reppe, W., and Hecht, O., (to General Aniline and Film Corp.), U. S. Patent 2,273,141 (1942).
146. Richmond, H. H., Myers, G. S., and Wright, G. F., J. *Am. Chem. Soc.*, **70,** 3659–64 (1948).
147. Ripper, K., (to Pollak, F.) U. S. Patent 1,460,606 (1923).
148. Robinson, G. M., and Robinson, R., *J. Chem. Soc.*, **123,** 532–3 (1923); *C. A.*, **17,** 1949.
149. Rohm and Haas Chem. Co., Bulletin SP-33, Dec. 12, 1950.
150. Sachs, B., *Ber.*, **31,** 1225, 3230 (1898).
151. Sakellarios, E. J., *J. Am. Chem. Soc.*, **70,** 2822 (1948).
152. Scheibler, H., Trostler, F., and Scholz, E., *Z. angew. Chem.*, **41,** 1305–9 (1928); *C. A.*, **23,** 2425.
153. Schiff, H., *Ber.*, **24,** 2130 (1891).
154. *Ibid.*, **25,** 1936 (1892).
155. Schiff, H., *Ann.*, **319,** 59–76 (1901).
156. Schmihing, M., (to Unyte Corp.), U. S. Patent 1,989,628 (1935).
157. Schmidt, H., (to I. G. Farbenindustrie, Akt.-Ges.), U. S. Patent 1,791,433 (1931).
158. *Ibid.*, U. S. Patent 1,791,434 (1931).
159. Schultz, M., and Jaross, K., *Ber.*, **34,** 1504 (1901).
160. Shand, E. W., (to Sinclair Refining Co.), U. S. Patent 2,511,653 (1950).
161. Smith, R., Bullock, J. L., Bersworth, F. C., and Martell, A. E., *J. Org. Chem.*, **14,** 355–61 (1949).
162. Smythe, L. E., *J. Am. Chem. Soc.*, **73,** 2735–8 (1951).
163. Société pour l'Industrie Chimique à Bâle, French Patent 811,804 (1937).
164. Sorenson, B. E., (to E. I. du Pont de Nemours & Co., Inc.), U. S. Patent 2,201,-927 (1940).
165. Sorensen, S. P. L., *Biochem. Z.*, **7,** 45–101 (1908); *C. A.*, **2,** 1288.
166. Sprung, M. M., *Chem. Revs.*, **26,** 297–338 (1940).
167. *Ibid.*, p. 305.
168. Staudinger, H., *Ber.*, **59,** 3019 (1926).
169. Supf, F., (to The Arobol Manufacturing Co.), U. S. Patent 974,448 (1910).
170. Sutter, T., (to Society of Chemical Industry in Basel), U. S. Patent 2,088,143 (1937).
171. Swain, A. P., Kokes, E. L., Hipp, N. J., Wood, J. L., and Jackson, R. W., *Ind. Eng. Chem.*, **40,** 465–9 (1948).
172. Talbot, W. F., (to Monsanto Chemical Co.), U. S. Patent 2,260,239 (1941).
173. Theis, E. R., and Ottens, E. F., *J. Am. Leather Chemists' Assoc.*, **35,** 330–47 (1940).
174. *Ibid.*, **36,** 22–37 (1941).
175. Thomas, A. W., Kelly, M. W., and Foster, S. B., *J. Am. Leather Chemists' Assoc.* **21,** 57–76 (1926).
176. Tollens, B., *Ber.*, **17,** 657 (1884).
177. *Ibid.*, p. 659.
178. Ulrich, H., and Ploetz, E., (to I. G. Farbenindustrie Akt.), U. S. Patent 2,205,-995 (1940).
179. Vaskevich, D. N., and Reingach, B. Ya., *Bull. acad. sci. U.S.S.R., Clases sci. chim.*, **1946,** 71–5; *C. A.*, **42,** 6160 (1948).
180. Vereinigte Chemische Fabriken Kreidl, Heller & Co., French Patent 852,079 (1939).
181. Voisenet, E., *Bull. soc. chim.* (*3*), **33,** 1198–1214 (1905); *Chem. Zentr.*, **1906,** I, 90.

182. Wagner, E. C., *J. Am. Chem. Soc.*, **55,** 724 (1933).
183. *Ibid.*, **56,** 1944 (1934).
184. Walker, J. F., *J. Am. Chem. Soc.*, **55,** 2821 (1933).
185. Walker, J. F., (to E. I. du Pont de Nemours & Co., Inc.), U. S. Patent 2,352,671 (1944).
186. *Ibid.*, U. S. Patent 2,411,396 (1946).
187. *Ibid.*, U. S. Patent 2,417,999 (1947).
188. *Ibid.*, U. S. Patent 2,418,000 (1947).
189. *Ibid.*, U. S. Patent 2,478,990 (1949).
190. Walker, J. F., and Chadwick, A. F., *Ind. Eng. Chem.*, **39,** 977 1947).
191. Walter, G., British Patent 262,148 (1927).
192. *Ibid.*, British Patent 284,272 (1928).
193. *Ibid.*, British Patent 291,712 (1928).
194. Wegler, R., and Ballauf, A., *Ber.*, **81,** 527 (1948).
195. Wellington, C., and Tollens, B., *Ber.*, **18,** 3298 (1885).
196. Werner, E. A., *J. Chem. Soc.*, **111,** 844–853 (1917).
197. Widmer, G., and Fisch, W., (to Ciba Products Corp.), U. S. Patent 2,197,357 (1940).
198. Wohl, A., *Ber.*, **19,** 2345–6 (1886).
199. Woodcock, D., *J. Chem. Soc.*, **1949,** 1635–8.
200. Zeleny, L., and Gortner, R. A., *J. Biol. Chem.*, **90,** 427–41 (1931).

CHAPTER 15

REACTIONS OF FORMALDEHYDE WITH HYDROCARBONS AND HYDROCARBON DERIVATIVES

Under appropriate conditions of catalysis, temperature, etc., formaldehyde reacts with a wide variety of unsaturated aliphatic and aromatic hydrocarbons. These reactions, together with those encountered with hydrocarbon derivatives such as halogenated hydrocarbons, nitrohydrocarbons, and organometallic compounds, will be reviewed in the following pages. To date, we know of no clear-cut instances in which paraffins have been successfully reacted with formaldehyde.

Although nitroparaffins and organometallic compounds react readily with formaldehyde, unsubstituted olefins, aromatics and nitro-aromatics as well as their halogen derivatives react only in the presence of strong acids. The simplicity of the formaldehyde molecule gives it a good degree of chemical stability to acids. This property is not generally shared by the more complicated aldehydes and carbonyl compounds, which tend to decompose or undergo auto-condensation reactions when treated with acidic compounds. As a result of its stability, formaldehyde can be successfully employed for many organic syntheses in cases where extremely stringent conditions are required to initiate reactions. In some cases acids apparently unite with formaldehyde to give simple addition products which function as reaction intermediates without destroying the integrity of the formaldehyde molecule. This is similar to the manner in which methylene glycol acts as an intermediate in all reactions involving aqueous formaldehyde.

Reactions with Olefins and Cyclo-olefins

Reactions of formaldehyde with hydrocarbons containing the ethylenic linkage usually result in the formation of 1,3-glycols, unsaturated alcohols, and functional derivatives of these compounds. Strong acid catalysts are normally required but uncatalyzed reactions are known in some instances.

Copolymerization of formaldehyde and olefins has also been reported. The structure of these products is not clear. The ethylene copolymer apparently contains hydroxyl groups and is formed under pressure in the presence of an organic peroxide catalyst at around 150°C[58, 86]. Peterson[86] employs formaldehyde as trioxane with a small concentration of sulfuric acid to liberate the monomeric aldehyde.

Fundamentally, the acid catalyzed, formaldehyde-olefin reaction or Prins reaction may be envisaged as involving the addition of methylene glycol to the carbon-carbon double bond as shown below:

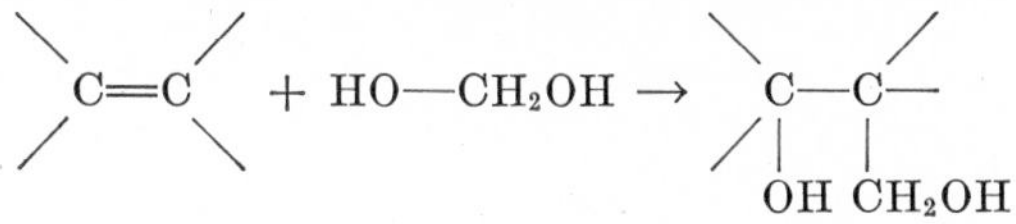

$$\diagdown_{\diagup} C = C _{\diagdown}^{\diagup} \quad + \text{ HO—CH}_2\text{OH} \rightarrow \quad \diagdown_{\diagup} \underset{\text{OH}}{C} — \underset{\text{CH}_2\text{OH}}{C} —$$

Since strongly acidic catalysts are required, it is also possible that a methylol derivative is primarily formed by reaction of formaldehyde with acid and that it is this intermediate which adds to the unsaturated linkage,

$$\text{CH}_2\text{O} + \text{H}_2\text{SO}_4 \rightarrow \text{HOCH}_2\cdot\text{HSO}_4$$

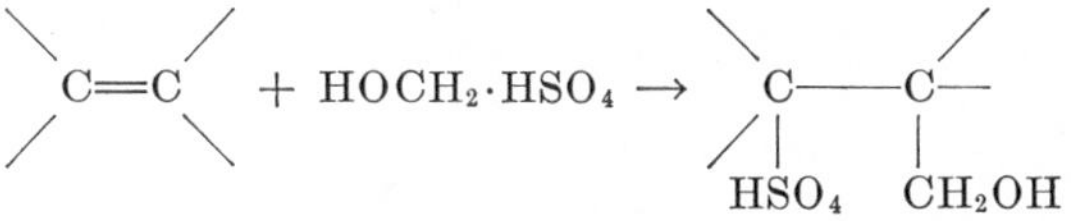

$$\diagdown_{\diagup} C = C _{\diagdown}^{\diagup} \quad + \text{ HOCH}_2\cdot\text{HSO}_4 \rightarrow \quad \diagdown_{\diagup} \underset{\text{HSO}_4}{C} — \underset{\text{CH}_2\text{OH}}{C} —$$

Hydrolysis of the above product would give a 1,3-glycol, whereas cleavage of acid would lead to the formation of an unsaturated alcohol:

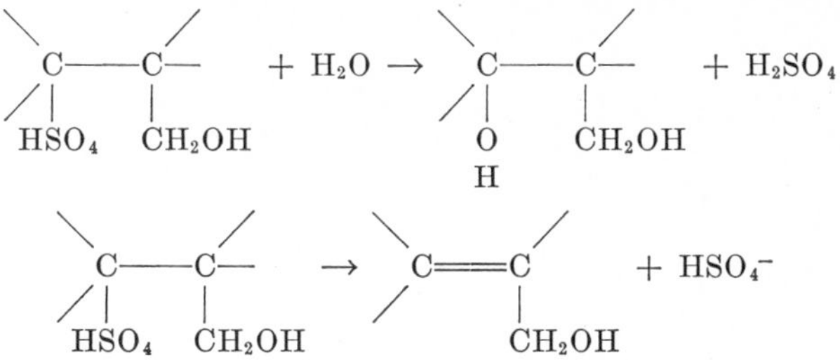

$$\diagdown_{\diagup} \underset{\text{HSO}_4}{C} — \underset{\text{CH}_2\text{OH}}{C} — \quad + \text{ H}_2\text{O} \rightarrow \quad \diagdown_{\diagup} \underset{\underset{\text{H}}{\text{O}}}{C} — \underset{\text{CH}_2\text{OH}}{C} — \quad + \text{ H}_2\text{SO}_4$$

$$\diagdown_{\diagup} \underset{\text{HSO}_4}{C} — \underset{\text{CH}_2\text{OH}}{C} — \quad \rightarrow \quad \diagdown_{\diagup} C = C _{\underset{\text{CH}_2\text{OH}}{}}^{\diagup} \quad + \text{ HSO}_4^{-}$$

Reaction of the glycol with additional formaldehyde would give a cyclic formal, and reaction with a fatty acid reaction medium would give a diester. All these products are obtained.

Recent investigators tend to favor a reaction mechanism involving an intermediate carbonium ion.

$$\text{CH}_2\text{O (aq)} + \text{H}_2\text{SO}_4 \rightarrow \overset{+}{\text{C}}\text{H}_2\text{OH} + \text{HSO}_4^{-} \qquad (1)$$

$$\text{R}\cdot\text{CH}_2\cdot\text{CH:CH}_2 + \overset{+}{\text{C}}\text{H}_2\text{OH} \rightarrow [\text{R}\cdot\text{CH}_2\cdot\overset{+}{\text{C}}\text{H}\cdot\text{CH}_2\cdot\text{CH}_2\text{OH}] \qquad (2)$$

The intermediate ion shown in reaction (2) can then yield unsaturated alcohols, glycols or an ester.

$$[\text{R}\cdot\text{CH}_2\cdot\overset{+}{\text{C}}\text{H}\cdot\text{CH}_2\cdot\text{CH}_2\text{OH}] \rightarrow \text{R}\cdot\text{CH:CH}\cdot\text{CH}_2\cdot\text{CH}_2\text{OH} + \text{H}^{+} \qquad (3a)$$

$$[\text{R}\cdot\text{CH}_2\cdot\overset{+}{\text{C}}\text{H}\cdot\text{CH}_2\cdot\text{CH}_2\text{OH}] + \text{H}_2\text{O} \rightarrow \text{R}\cdot\text{CH}_2\cdot\text{CHOH}\cdot\text{CH}_2\cdot\text{CH}_2\text{OH} + \text{H}^{+} \qquad (3b)$$

$$[\text{R}\cdot\text{CH}_2\cdot\overset{+}{\text{C}}\text{H}\cdot\text{CH}_2\cdot\text{CH}_2\text{OH}] + \text{CH}_3\text{COOH} \rightarrow$$

$$\text{R}\cdot\text{CH}_2\cdot\text{CH(OOCCH}_3)\cdot\text{CH}_2\cdot\text{CH}_2\text{OH} + \text{H}^{+} \qquad (3c)$$

In some cases, an H—C bond conjugated with an olefinic bond may be involved. This was demonstrated by Baker[12] who demonstrated the formation of 4-acetoxytetrahydro-gamma-pyran as a product of the sulfuric acid catalyzed reaction of propylene with formaldehyde in acetic acid.

$$
\begin{array}{c}
H\cdot CH_2\cdot CH:CH_2 \\
\\
HO\overset{+}{C}H_2 \cdot O:CH_2
\end{array}
\;\rightarrow\;
\begin{array}{c}
H\cdot CH_2:\overset{+}{CH}\cdot CH_2 \\
| \\
HO\cdot CH_2\cdot O\cdot CH_2
\end{array}
\;\xrightarrow[\text{Acid cat.}]{\text{AcOH}}\;
$$

$$
\left[\begin{array}{c}
AcO\!-\!H \\
CH_2:\overset{}{CH}\cdot CH_2 \\
\overset{+}{H_2O}\cdot CH_2\cdot O\cdot CH_2
\end{array}\right]
\rightarrow
\begin{array}{c}
OAc \\
| \\
CH_2\!-\!CH\!-\!CH_2 \\
| \qquad\quad / \\
CH_2\!-\!O\!-\!CH_2
\end{array}
+ H_2O + H^+
$$

Arnold[4] has postulated a six-membered ring as a transitory intermediate in the uncatalyzed reaction which takes place when formaldehyde and tertiary olefins are heated together at 150 to 200°C.

$$
RR'CH\cdot CR'':CH_2 + CH_2O \rightarrow
\left[\begin{array}{c}
R'' \\
| \\
R'\quad C=CH_2 \\
\diagdown\diagup \quad (\\
C \;\rangle \quad CH_2 \\
\diagup\diagdown \quad \| \\
R \quad H \;\; O
\end{array}\right]
\rightarrow
\begin{array}{c}
R'' \\
| \\
R'\quad C\!-\!CH_2 \\
\diagdown\diagup \quad \diagdown \\
C \qquad CH_2 \\
\diagup \qquad \diagup \\
R \qquad HO
\end{array}
$$

Although the production of unsaturated alcohols by the action of formaldehyde on limonene and pinene was clearly demonstrated by Kriewitz[64] in 1899, it was not until 1917 that Prins[87, 88] made the first really comprehensive study of the reactions of formaldehyde with ethylenic hydrocarbons. Prins studied reactions of formaldehyde with styrene, anethole, pinene, d-limonene, camphene and cedrene in the presence of sulfuric acid, using water, glacial acetic, or formic acid as solvent. In general, the most satisfactory results were obtained in the acid solvents. In aqueous media, the products were generally isolated as formals of 1,3-glycols or unsaturated alcohols, whereas acetic or formic esters were usually obtained when the corresponding acids were employed as solvents.

By reaction of formaldehyde with styrene in glacial acetic acid containing approximately 13 per cent by weight of concentrated sulfuric acid at 40 to 50°C, Prins obtained the diacetate of a phenyl propylene glycol. Prins did not determine the exact structure of this glycol but proposed the alternative formulas $C_6H_5CHOHCH_2CH_2OH$ and $C_6H_5CH(CH_2OH)_2$. Fourneau and co-workers[31] have since demonstrated that Prins' glycol was

$C_6H_5CHOHCH_2CH_2OH$. They also obtained cinnamyl alcohol, $C_6H_5CH:$
$CH \cdot CH_2OH$, as a by-product of the hydrolysis of the diacetate.

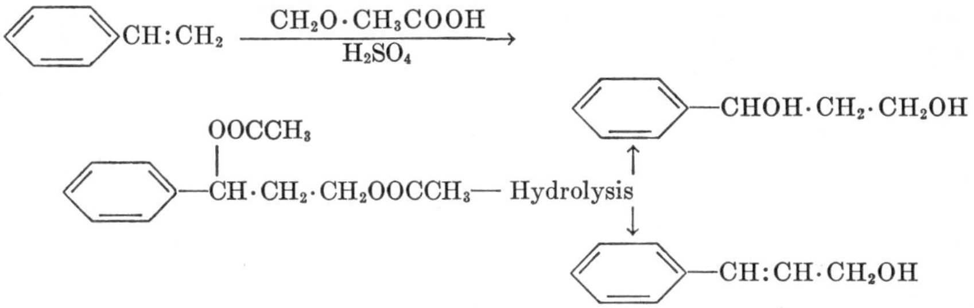

A process for the preparation of aliphatic or hydroaromatic 1,3-glycols
by reaction of formaldehyde with ethylenic hydrocarbons in mixtures of
acetic and sulfuric acids followed by hydrolysis of the primary products
thus obtained was patented in France in 1932[102]. Since then, numerous
patents covering novel process variations of these reactions and special
products have been granted.

Olsen[82] has pointed out that the structural principals of the olefin-formaldehyde reaction can frequently lead to compounds which may be considered as forerunners of natural materials that occur in growing plants.

The reaction of ethylene with formaldehyde in glacial acetic acid containing sulfuric acid as a catalyst gives the diacetate of trimethylene glycol.
Olsen[81], who carried out the reaction under pressure in a steel autoclave
at 130 to 140°C isolated a by-product acetate, which he identified as pentaglycerol triacetate, $CH_3C(CH_2OOCCH_3)_3$. Baker[13], who passed ethylene
into the reaction mixture at 65 to 70°C at atmospheric pressure, also obtained a by-product acetate, $(C_7H_{12}O_4)_x$, which he later recognized as
identical with Olsen's triacetate derivative. The mechanism for the formation of pentaglycerol triacetate in this process has not been explained.
Olsen[83] synthesized glycerol from ethylene and formaldehyde by cracking
the trimethylene glycol diacetate to allyl acetate, reacting the latter with
hypochlorous acid and saponifying the resultant chlorohydrin.

Gresham[46] produced 1,3-dichloropropane by reacting ethylene with formaldehyde and hydrogen chloride at 900 atmospheres and 150°C. At lower
pressures, 450 atmospheres and below, the chlorohydrin was obtained.
Paraformaldehyde and concentrated hydrochloric acid were employed
in this reaction.

$$CH_2:CH_2 + CH_2O \text{ (p)} + HCl \rightarrow ClCH_2 \cdot CH_2 \cdot CH_2OH$$

$$CH_2:CH_2 + CH_2O \text{ (p)} + 2HCl \rightarrow ClCH_2 \cdot CH_2 \cdot CH_2Cl + H_2O$$

The reaction of propylene and formaldehyde is illustrated by Hamblet
and McAlevy's process[49] for making 1,3-butylene glycol by heating aque-

ous formaldehyde with propylene under pressure at 100 to 200°C in the presence of sulfuric acid. An equilibrium concentration of butylene glycol formal is produced and is maintained in the reaction media by recycling.

$$CH_2{:}CH{\cdot}CH_3 + CH_2O \text{ (aq)} \rightarrow CH_2OH{\cdot}CH_2{\cdot}CHOH{\cdot}CH_3$$

By reaction of formaldehyde solution or paraformaldehyde with hydrogen chloride and an olefin under pressure, Fitzky[29] obtained chloro-alcohols of the type which would be expected by the addition of chloromethanol ($ClCH_2OH$) to the unsaturated linkage. In the case of propylene, the following reaction takes place:

$$CH_2O \text{ (aq, p)} + HCl + CH_2{:}CH{\cdot}CH_3 \rightarrow CH_2OH{\cdot}CH_2{\cdot}CHCl + CH_3$$

Production of 1,3-butylene glycol by reaction of aqueous formaldehyde and propylene using hydrogen chloride as a catalyst is also claimed by Fitzky[30].

Processes involving reaction of propylene with formaldehyde and dehydrating to give butadiene have been developed by Workman[114] and Mottern[75]. The latter also describes the preparation of 2-methyl-1,3-butadiene from formaldehyde and isobutylene. A vapor phase diene process by Marsh[70] involves passing isobutylene, formaldehyde and steam over an alumina catalyst at 650°F. Arundale and Mikeska[7a] prepared 2,3-dimethyl-1,3-butadiene by heating trimethylethylene with paraformaldehyde at about 100°C in the presence of 5 per cent sulfuric acid.

Mikeska and Arundale[73] have developed a process for the preparation of isobutenylcarbinol (b.p. 130°C) by the reaction of "trioxymethylene" (paraformaldehyde) with isobutylene in chloroform, using tin tetrachloride as a catalyst. The reaction is carried out by shaking the reagents in a bomb at room temperature for $7\frac{1}{2}$ hours. A small quantity of 4,4-dimethyl-*m*-dioxane is also obtained. The crude product mixture distills in the range 125 to 130°C and is reported to contain 90 per cent isobutenylcarbinol. The *m*-dioxane derivative is removed by extraction with water. Zinc chloride, silicon tetrachloride, and zinc dichloroacetate are also stated to act as catalysts. Similar reactions are claimed for tertiary olefins containing up to 16 carbon atoms.

Cyclic formals of 1,3-glycols are prepared by patented processes involving the reactions of olefins with formaldehyde in the presence of acidic catalysts, such as boron fluoride[24, 96], mineral acids, zinc chloride, etc.[56, 57]. According to Arundale[6], a cyanometadioxane of this type is prepared by heating methallyl cyanide and paraformaldehyde with 50 per cent sulfuric acid at 50 to 65°C.

Following the earlier studies of Matti[72], Olsen[80] and Olsen and Padberg[84] have made a detailed study of the reaction of cyclohexene and formaldehyde using the general technique developed by Prins. Cyclohexyl acetate was detected in the reaction mixture and later shown to yield the same products when employed in place of cyclohexene. Experiments with formic, propionic, butyric and isovaleric acid in place of the acetic medium also yielded similar results except that esterified products were derived from the acid used as a medium. The results obtained are indicated in the following scheme of reaction.

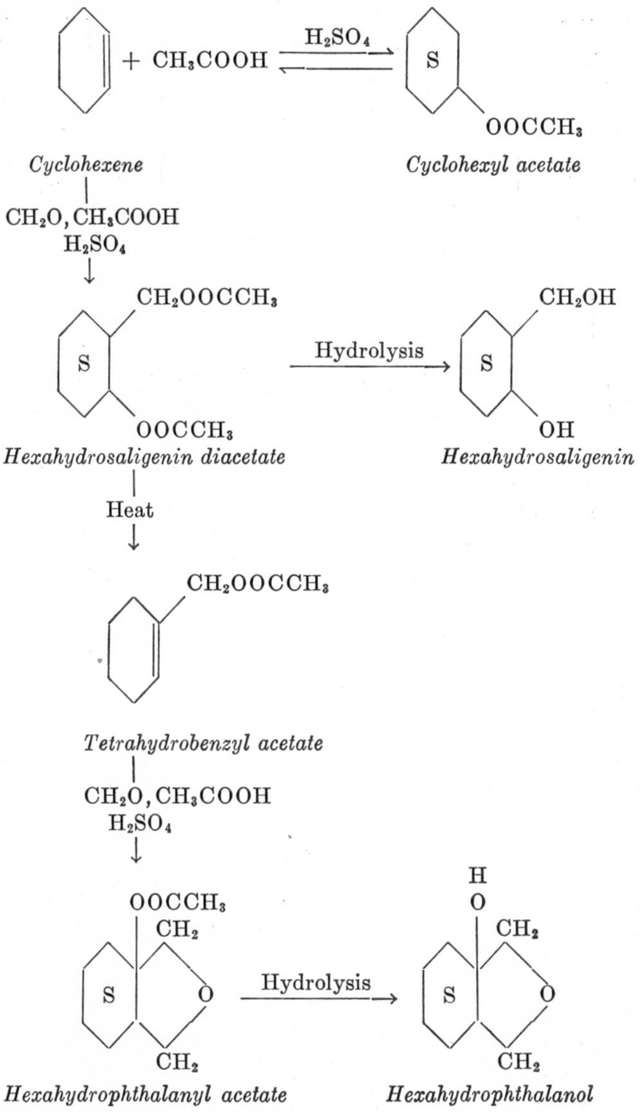

Cyclohexene   *Cyclohexyl acetate*

CH_2O, CH_3COOH
H_2SO_4

Hexahydrosaligenin diacetate   *Hexahydrosaligenin*

Heat

Tetrahydrobenzyl acetate

CH_2O, CH_3COOH
H_2SO_4

Hexahydrophthalanyl acetate   *Hexahydrophthalanol*

Hexahydrophthalanol (m.p. 205 to 208°C) was identified as the product believed by Matti[72] to be the cyclic dimethylene bis-ether of hexahydrosaligenin.

The uncatalyzed reaction of formaldehyde is illustrated by Bain's synthesis[11] of nopol in which beta-pinene is heated with paraformaldehyde at 180°C for 4 hours giving a high yield (approx. 95 per cent) of the unsaturated alcohol.

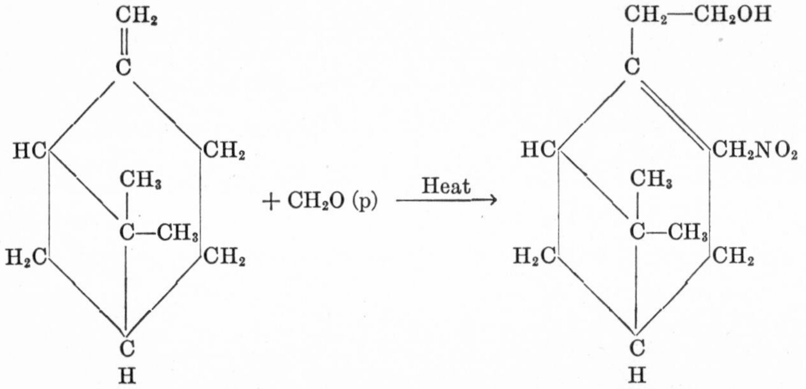

β-Pinene   *Nopol*

Kriewitz[64] carried out this reaction in ethanol obtaining a 51 per cent yield after 12 hours at 180 to 200°C with alpha-pinene, the uncatalyzed process gives only traces of alcohol products. Ritter[94] also uses the uncatalyzed reaction in obtaining derivatives of diisobutylene and related tertiary olefins.

Isolable derivatives obtained by reactions involving formaldehyde and acids also possess the ability to add to the ethylenic linkage in some cases. Monochloromethyl ether ($ClCH_2OCH_3$), which may be regarded as the methyl ether of the theoretical addition product of formaldehyde and hydrogen chloride (*viz.*, chloromethanol, $ClCH_2OH$), also adds to olefins in the presence of bismuth chloride, tin tetrachloride, zinc chloride, and related catalysts, as demonstrated in a process devised by Scott[100]. Propylene reacts thus with monochloromethyl ether:

$$CH_3CH:CH_2 + CH_3OCH_2Cl \rightarrow CH_3\underset{\underset{\displaystyle Cl}{|}}{C}HCH_2CH_2OCH_3$$

Halogenated Olefins

Halogenated olefins apparently react in much the same manner as the unsubstituted hydrocarbons. Vinyl chloride reacts with formaldehyde and hydrogen chloride giving 3,3-dichloropropyl alcohol and 2,3-dichloropropyl alcohol (dichlorohydrin), the former being the principal product[29].

Methallyl chloride reacts with "trioxymethylene" (paraformaldehyde) in the presence of 50 per cent sulfuric acid to give a chloro- derivative of a meta-dioxane[7]:

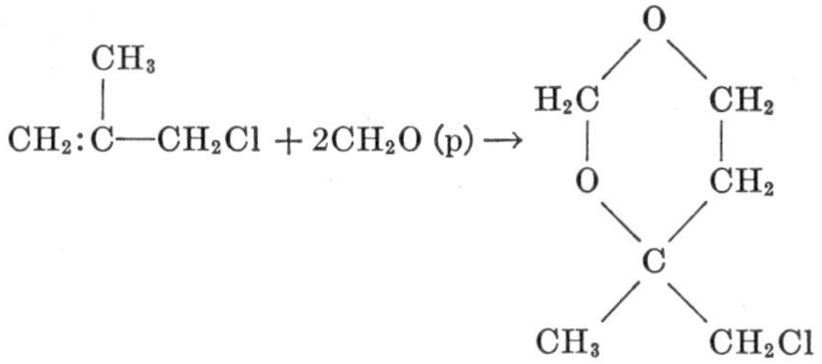

$$
\begin{array}{c}
CH_3 \\
| \\
CH_2:C\!-\!CH_2Cl + 2CH_2O\ (p) \rightarrow
\end{array}
\qquad
\begin{array}{c}
O \\
\diagup\ \diagdown \\
H_2C \qquad CH_2 \\
|\qquad\quad | \\
O \qquad CH_2 \\
\diagdown\ \diagup \\
C \\
\diagup\ \diagdown \\
CH_3 \qquad CH_2Cl
\end{array}
$$

When polychloroethylenes are reacted with paraformaldehyde in the presence of concentrated sulfuric acid and the reaction mixture is then treated with water, chlorocarboxylic acids are obtained. From a procedure of this sort involving paraformaldehyde and trichloroethylene, Prins[89] succeeded in isolating the ether of monochlorohydracrylic acid, $O(CH_2CHClCOOH)_2$, (m.p. 124 to 126°C). With tetrachloroethylene, 2,2-dichlorohydracrylic acid, $CH_2OH\cdot CCl_2\cdot COOH$, (m.p. 88 to 89°C), was obtained. (Dichloroethylene gave only a resinous mass.) The following reaction mechanisms are suggested for the formation of these acids:

Trichloroethylene Reaction:

$$CHCl:CCl_2 + CH_2O\ (p) + H_2SO_4 \rightarrow HOCH_2\cdot CHCl\cdot CCl_2\cdot HSO_4$$

$$2HOCH_2\cdot CHCl\cdot CCl_2\cdot HSO_4 + 3H_2O \rightarrow O(CH_2\cdot CHCl\cdot COOH)_2 + 4HCl + 2H_2SO_4$$

Tetrachloroethylene Reaction:

$$CCl_2:CCl_2 + CH_2O\ (p) + H_2SO_4 \rightarrow HOCH_2\cdot CCl_2\cdot CCl_2\cdot HSO_4$$

$$HOCH_2\cdot CCl_2\cdot CCl_2\cdot HSO_4 + 2H_2O \rightarrow HOCH_2\cdot CCl_2\cdot COOH + 2HCl + H_2SO_4$$

Short[101] controls the reaction of vinylidene chloride (asymmetrical dichloroethylene) by gradual addition of an approximately 50 per cent paraformaldehyde-water slurry to a mixture of strong sulfuric acid and the chloro-olefin at 55 to 60°C to obtain acrylic acid.

Londergan[66] prepares tetrachloropropanol by reacting paraformaldehyde and hydrogen chloride with trichlorethylene in the presence of phosphoric acid and zinc or aluminum chloride.

$$Cl_2C:CHCl + CH_2O + HCl \rightarrow Cl_3C\cdot CHCl\cdot CH_2OH$$

On heating trichloroethylene with formaldehyde solution or paraformaldehyde in the presence of approximately 80 per cent sulfuric acid, chloroacrylic acid ($CH_2:CHCl\cdot COOH$), the dehydration product of monochlorohydracrylic acid, is obtained. Esters of this acid can be obtained by adding alcohols to the reaction mixture and distilling. These processes have been patented by Crawford and McLeish[19,20,69]. A one-step process for making these esters by adding a mixture of formaldehyde solution and methanol to a mixture of trichlorethylene and sulfuric acid is described by Roberts[95].

Diolefins

The uncatalyzed reaction of formaldehyde with 2-methyl butadiene takes the form of a Diels-Alder reaction as demonstrated by T. L. Gresham and Steadman[45] but could not be reproduced with butadiene. The reaction was carried out by heating 2-methyl butadiene with formaldehyde as alpha-polyoxymethylene in an autoclave for 6.5 hours at 185°C.

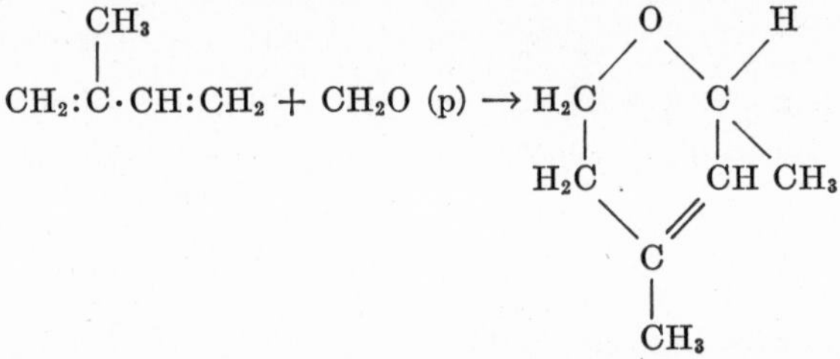

$$CH_2{:}C{\cdot}CH{:}CH_2 + CH_2O \text{ (p)} \rightarrow$$

(with CH_3 substituent shown above the C)

W. F. Gresham and Grigsby[47] carried out the acid catalyzed reaction with butadiene at 100 to 200°C under autogenous pressure in the presence of a polymerization inhibitor and reported the following products: 4-vinyl-1,3-dioxane (I), 5,6-dihydro-2H-pyran (II), 3-methylol-4-hydroxy tetrahydropyran (III) and the cyclic formal of the dihydroxy compound (IV).

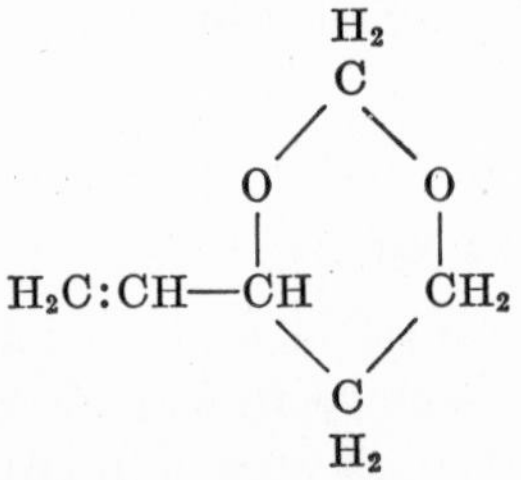

I 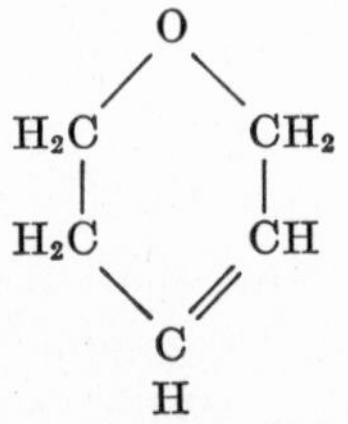   II

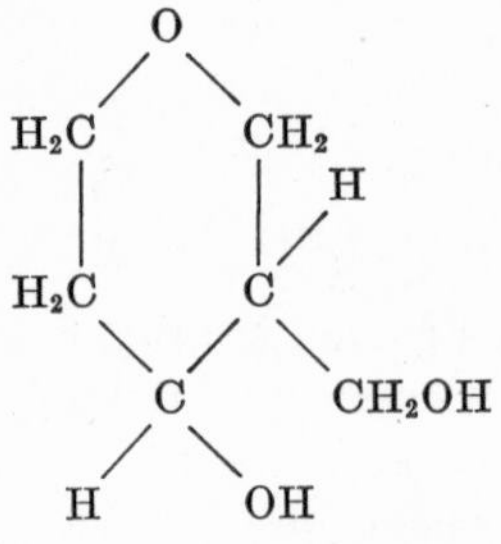

III 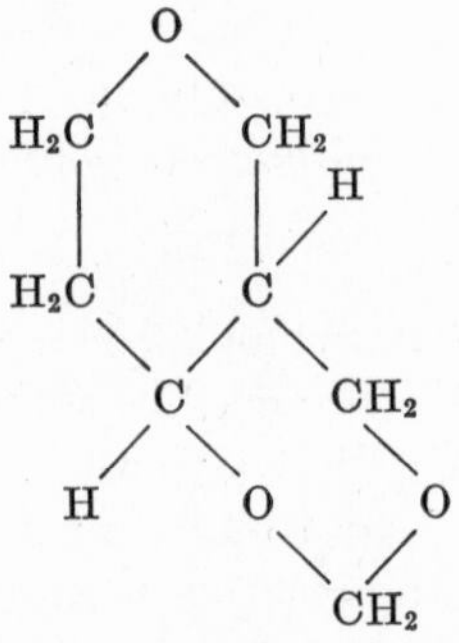   IV

The formation of the above products has recently been confirmed by Dermer, Kohn and Nelson[23], who employed diethyl ether or glacial acetic acid as reaction media with 15 to 35 per cent by volume of concentrated sulfuric acid as the reaction catalyst. Experiments carried out at 20 to 30°C gave yields, based on butadiene, of 29 per cent 4-vinyl-1,3-dioxane (I), 13 per cent of the cyclic formal of the dihydroxy pyran derivative (III) and about 3 per cent of the free 5,6-dihydro-2H-pyran (II). These investigators discount claims that the 5-vinyl-1,3-dioxane has been obtained and have conclusively demonstrated the structure of the 4-vinyl derivative.

Dermer and Hawkins[22] have reacted methylal with butadiene at 0 to 10°C in the presence of concentrated sulfuric acid to obtain the methyl ethers that would be expected by Prins-type additions of this compound. These products include 3,5-dimethoxy-1-pentene, 1,5-dimethoxy-2-pentene, 3-methoxy-4-methoxymethyl-tetrahydropyran and 1,2,5-trimethoxy-3-methoxymethyl pentane.

Addition of the chlormethyl ethers, chloromethyl ethyl ether and chloromethyl butyl ether, to butadiene in the presence of zinc chloride has been investigated by Pudovik and co-workers[90]. Addition is 28 to 35 per cent in the 1,4-position and 65 to 72 per cent in the 1,2-position.

$$CH_2:CH\cdot CH:CH_2 \xrightarrow{\text{ROCH}_2\text{Cl}} \begin{array}{l} ROCH_2CH_2CH:CH\cdot CH_2Cl \\ \quad\quad\quad\quad Cl \\ \quad\quad\quad\quad | \\ ROCH_2\cdot CH_2\cdot CH\cdot CH:CH_2 \end{array}$$

Arundale and Mikeska[7b] have presented a detailed review of the Prins reaction.

Acetylenic Hydrocarbons

The reaction of acetylene with concentrated aqueous formaldehyde and hydrogen chloride in the presence of a high concentration of zinc chloride gives 5-dichloromethyl-1,3-dioxane and a mixture of chlorinated formals having the empirical formula, $C_9H_{15}O_3Cl_5$[111].

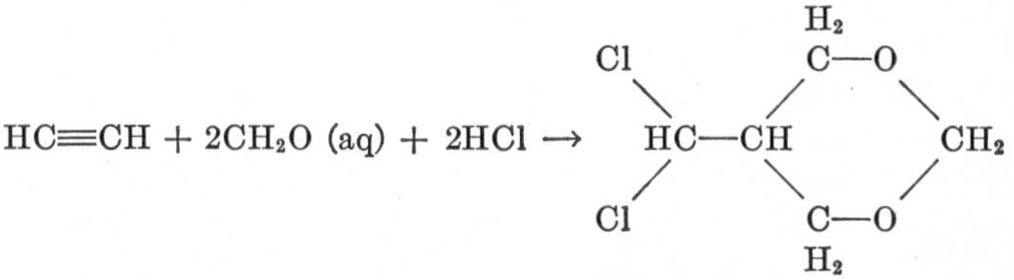

$$HC\equiv CH + 2CH_2O \text{ (aq)} + 2HCl \rightarrow \begin{array}{c} Cl \quad\quad\quad H_2 \\ \diagdown \quad\quad\quad C-O \\ \quad \diagup \quad \diagdown \\ HC-CH \quad\quad CH_2 \\ \diagup \quad\quad \diagdown \quad \diagup \\ Cl \quad\quad\quad C-O \\ \quad\quad\quad\quad H_2 \end{array}$$

5-Dichloromethyl-1,3-dioxane

Keyssner and Reppe[62] have developed a method for the preparation of acetylenic alcohols (methylolacetylenes) by the direct reaction of formaldehyde and acetylenes. This process may be carried out by heating formaldehyde solution at 40 to 150°C with the acetylenic hydrocarbon in the presence of a catalyst consisting of an acetylide of a metal of the Ib group of the Periodic System, or mercury. Since these catalysts are highly explosive in the dry state, *they must be kept wet at all times to minimize the hazards* involved in their use. In general, a mixture of copper and silver acetylides plus an inert diluent is the preferred catalyst; for example, a mixture of 1 part copper acetylide, 0.15 part silver acetylide and 2 parts fuller's earth is said to give satisfactory results.

With monosubstituted acetylenes, the reaction is:

$$R—C\vdots CH + CH_2O \text{ (aq)} \rightarrow R—C\vdots C—CH_2OH$$

When acetylene itself is employed, propargyl alcohol and butynediol-1,4 are obtained:

$$HC\vdots CH + CH_2O \text{ (aq)} \rightarrow HC\vdots C \cdot CH_2OH$$

$$HC\vdots CH + 2CH_2O \text{ (aq)} \rightarrow HOCH_2C\vdots C \cdot CH_2OH$$

Butynediol-1,4 is a crystalline compound melting at 58°C and boiling at 125 to 127°C at 2 mm. It can also be produced by heating propargyl alcohol with formaldehyde in the presence of the acetylide catalysts described above[61]·

$$HC\vdots C \cdot CH_2OH + CH_2O \text{ (aq)} \rightarrow HOCH_2 \cdot C\vdots C \cdot CH_2OH$$

According to Reppe and Keyssner's patents, the reaction of acetylene and formaldehyde is best carried out under pressure. For this purpose the acetylene is mixed with nitrogen in a 2:1 ratio and charged into an autoclave containing the hot 30 to 40 per cent formaldehyde solution and catalyst until a total pressure of 15 to as high as 30 atmospheres is obtained. Butynediol yields of 90 per cent or better are reported when the reaction is carried out at 80 to 100°C.

A related process[63] for the production of dialkylaminomethyl derivatives of acetylene from reactions of dialkylamines and formaldehyde with acetylene has been previously discussed (page 287).

Londergan and the writer[67, 112] have prepared butynediol at atmospheric pressure by passing acetylene into a suspension of paraformaldehyde in a special solvent, such as acetonyl acetone, dimethyl formamide *et al.*, at approximately 120°C in the presence of copper acetylide supported on active carbon.

Aromatic Hydrocarbons

Reactions of formaldehyde with aromatic hydrocarbons are similar in some respects to those involving olefins and may involve a somewhat similar mechanism. However, reactions apparently proceed further than in the case of olefins, and the simple addition products of methylene glycol or substituted methylene glycol have not been isolated. With aromatic hydrocarbons, formaldehyde and hydrogen halides, the primary reaction products isolated are compounds in which one or two halomethyl groups are substituted for hydrogen on the aromatic nucleus. On further reaction, compounds are obtained in which two or more aromatic nuclei are linked together by methylene groups. When sulfuric acid is employed as a reaction catalyst, methylene derivatives of this latter type are apparently the principal products obtained.

Halomethylation Reactions. The reaction of aromatic hydrocarbons with formaldehyde or its polymers and hydrogen chloride leads to the introduction of chloromethyl groups into the aromatic nucleus, and is known as the chloromethylation reaction[1]. Related halomethylations involving hydrogen bromide and hydrogen iodide also occur in some instances, but have received less attention by chemists since they offer less promise as methods of organic synthesis. Both chloro- and bromomethylation reactions are also obtained when aromatic hydrocarbons are reacted with the chloro- and bromomethyl ethers produced from formaldehyde and the corresponding hydrogen halides.

The exact mechanism of the chloromethylation of benzene and other aromatics with formaldehyde and hydrogen chloride has not been demonstrated. The following reaction sequence based on a probable analogy with olefin-formaldehyde syntheses seems to offer a rational explanation. Reactions with chloromethyl ethers may be explained in a similar fashion.

$$CH_2O \text{ (aq, p)} + HCl \rightarrow HOCH_2Cl$$

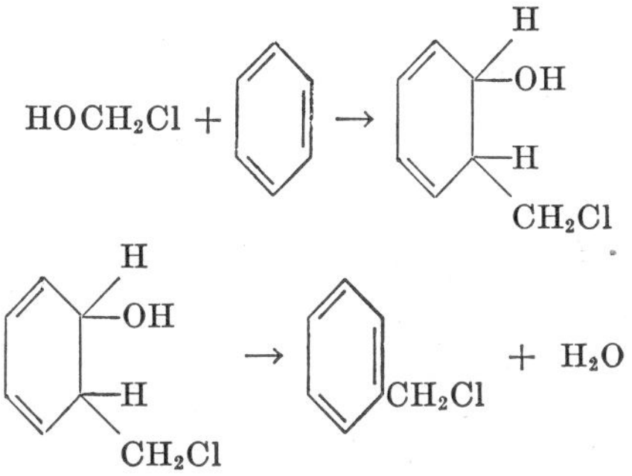

It is probable that an ionic mechanism is involved.

The chloromethylation reaction appears to have been first reported by Grassi and Maselli[44] in 1898. These investigators synthesized benzyl chloride by treating benzene with "trioxymethylene" (paraformaldehyde or alpha-polyoxymethylene) and hydrogen chloride in the presence of zinc chloride, which is an excellent catalyst for the reaction. In 1923, Blanc[14] demonstrated that benzene could be converted to benzyl chloride in 80 per cent yield by this procedure and showed that analogous products could also be obtained from toluene, xylene, and naphthalene. On further reaction with formaldehyde polymer and hydrogen chloride, benzyl chloride was found to give a mixture of ortho- and para-xylylene dichlorides.

$$
\underset{\text{CH}_2\text{Cl}}{\bigcirc} \xrightarrow{\text{CH}_2\text{O (p)} + \text{HCl}} \underset{}{\overset{\text{CH}_2\text{Cl}}{\bigcirc}\text{CH}_2\text{Cl}} \quad \text{and} \quad \underset{\text{CH}_2\text{Cl}}{\overset{\text{CH}_2\text{Cl}}{\bigcirc}} + \text{H}_2\text{O}
$$

Studies of the chloromethylation of benzene by reaction with paraformaldehyde and hydrogen chloride in the presence of zinc chloride are also reported by Sabetay[97] and Quelet[91]. In 1931, Vorozhtzov and Yuruigina[110] carried out reactions with aqueous formaldehyde and obtained a 46 per cent yield of benzyl chloride plus a small amount of xylylene dichlorides by heating a saturated solution of hydrogen chloride in 36 per cent formaldehyde for 6 hours at 60°C with equal weights of benzene and zinc chloride. Detailed studies by Lock[65] and Mraz[77] indicate that optimum yields of benzyl chloride (around 70 per cent) are obtainable by reaction of paraformaldehyde with excess benzene in the presence of zinc chloride. Pilot plant studies of this preparation are described by Ginsberg and co-workers[40].

On reacting paraformaldehyde ("trioxymethylene") with excess toluene and hydrogen chloride in the presence of zinc chloride (1 gram for every 2 grams of formaldehyde polymer), Blanc[14] obtained a monochloromethyl derivative in 82 per cent yield (based on formaldehyde) plus a small quantity of dichloromethyl derivatives. Blanc believed his monochloromethyltoluene was *p*-methyl benzyl chloride. However, recent studies by Hill and Short[52] have demonstrated that this product was an approximately 50-50 mixture of both ortho and para derivatives. Darzens[21] obtained monochloromethyltoluene in 85 per cent yield by adding one molar equivalent (approximately 30 grams) of formaldehyde polymer to 300 grams glacial acetic acid, saturating with hydrogen chloride, and heating the resulting solution with one gram mol of toluene (92 grams) for 90 to 100 hours at 60°C in a closed vessel.

Von Braun and Nelles[15] made an exhaustive study of the chloromethylation of *o*-, *m*-, and *p*-xylene obtaining both mono- and dichloromethyl derivatives. In this work, they secured good results by agitating the hydrocarbon with 37 per cent formaldehyde at 60 to 70°C while passing hydrogen chloride into the reaction mixture. Products isolated from the various isomeric xylenes are shown below:

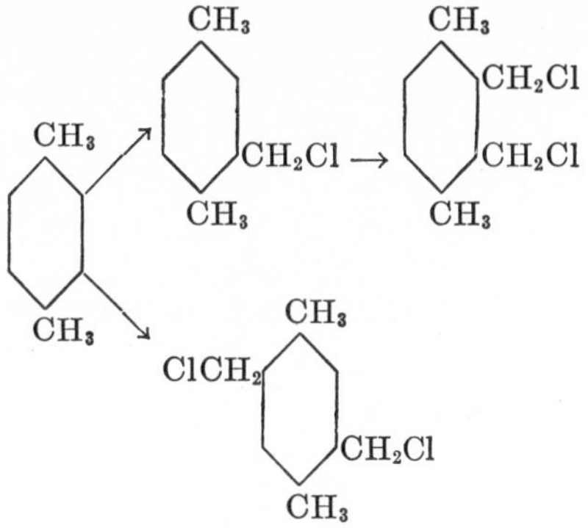

Para-xylene

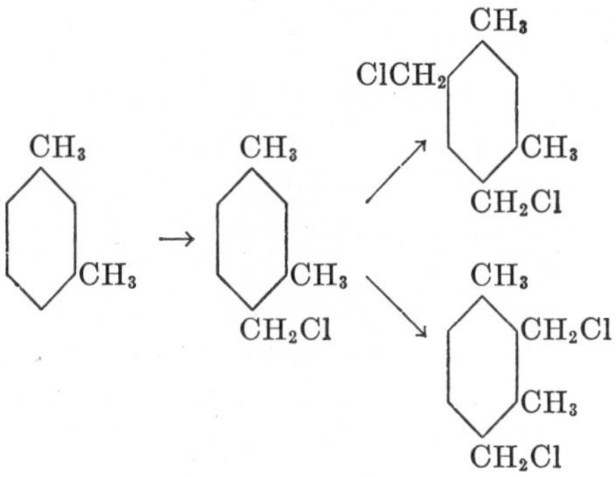

Meta-xylene

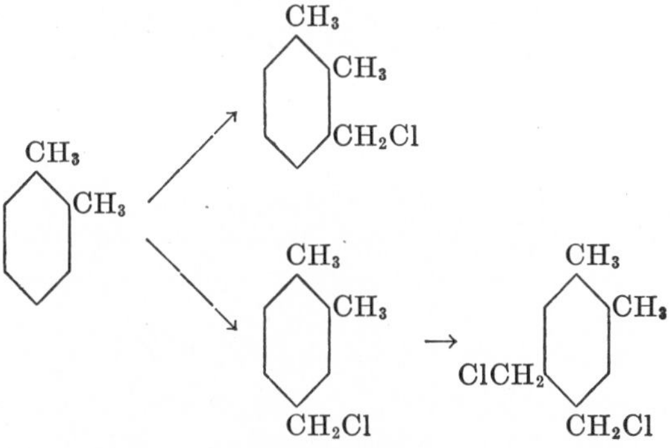

Ortho-xylene

In the case of naphthalene and other polynuclear aromatics, chloromethylations may be carried out in the absence of zinc chloride. In 1929, Reddelien and Lange[93] obtained 1-chloromethyl naphthalene by heating an agitated mixture of naphthalene, formaldehyde solution and concentrated hydrochloric acid at 60 to 70°C while introducing hydrogen chloride gas. A net yield of 83 per cent theoretical, based on naphthalene consumed, was thus obtained. The same investigators[92] also reported the preparation of 6-chloromethyl-1,2,3,4-tetrahydronaphthalene from tetralin by a similar reaction. Chloromethylation of naphthalene can also be accomplished by passing hydrogen chloride into an agitated mixture of molten naphthalene and paraformaldehyde as illustrated by Funk[36]. Arnold and Barnes[5] report that 1-chloromethyl and 2-chloromethyl naphthalene are produced in a 7:3 ratio by chloromethylation. Production of bis-chloromethyl aromatics was patented by Brunner and Greune[17] in 1933. Badger, Cook and Crosbie[8] have shown that this bis-chloromethylation gives a mixture of the 1,4 and 1,5-dichloromethyl naphthalenes and have isolated 5,8-dichloromethyl-1,2,3,4-tetrahydronaphthalene from the chloromethylation of tetralin.

Chloromethyl derivatives of phenolic and thiophenolic ethers and their chloro- and nitro-substitution products have been obtained by Brunner[16].

Nitrobenzene is not readily chloromethylated but Matsukawa and Shirakawa[71] have prepared meta-nitro benzyl chloride by reacting nitrobenzene with paraformaldehyde and hydrogen chloride in concentrated sulfuric acid.

In 1923, Stephen, Short and Gladdings[103] succeeded in chloromethylating benzene and other mononuclear aromatics by reaction with symmetrical-dichloromethyl ether in the presence of zinc chloride. This chloroether, $ClCH_2OCH_2Cl$, is prepared by the reaction of formaldehyde and hydrogen chloride (page 196).

Numerous processes for the chloromethylation of a wide variety of aromatic compounds are covered by a constantly growing volume of technical articles and industrial patents.

Comparative chloromethylation rates for a number of aromatic hydrocarbons were measured by Vavon and co-workers[109] in 1939 and by Szamant and Dudek[104] in 1949. The former investigators followed reactions with monochloromethyl ether in glacial acetic acid determining the extent of the reaction by hydrolysis followed by determination of the hydrogen chloride liberated by the unreacted chloroether. Szamant and Dudek followed hydrogen chloride consumption in a glacial acetic acid solution to which measured quantities of hydrocarbon, paraformaldehyde and concentrated hydrochloric acid had been added. All reactions were carried out at 85°C, whereas Vavon carried out some experiments at 100°C and others at 65°C assuming that comparative rates would be independent of temperature. Results obtained by these investigators are shown in Table 51. The Vavon figures were all adjusted to 65°C before calculation of comparative rates.

However, the parenthetic temperature values in the table, give the actual temperatures at which the individual reactions were carried out. Although the data disagree in several instances, the results agree qualitatively to the effect that alkyl and alkoxy substituents accelerate reaction. Vavon's findings also showed that Cl, Br, I, CH_2Cl and COOH groups retarded

TABLE 51. RELATIVE RATES OF CHLOROMETHYLATION FOR VARIOUS AROMATIC COMPOUNDS

	Relative Reaction Rate	
Compound	Vavon et al.[109] (65°C)	Szamant and Dudek[104] (85°C)
Benzene..........................	1.0 (100°C)	1.0
Toluene..........................	3.0 (100°C)	3.1
n-Butyl benzene..................	—	2.9
t-Butyl benzene..................	—	2.8
o-Xylene.........................	6.0 (100°C)	—
m-Xylene.........................	24 (65°C)	—
p-Xylene.........................	2.0 (100°C)	1.6
Mesitylene.......................	600 (65°C)	13
Diphenyl.........................	2.0 (100°C)	—
Diphenylmethane..................	—	0.77
Naphthalene......................	6.7 (65°C)	—
α-Methyl naphthalene.............	60 (65°C)	—
Tetralin.........................	7.5 (100°C)	—
Bromobenzene.....................	—	0.48
Nitrobenzene.....................	0	—
Anisole..........................	1333 (65°C)	23
p-Methyl cresyl ether............	1200 (65°C)	7.4
Chloromesitylene.................	18 (100°C)	—
Bromomesitylene..................	4.0 (100°C)	—
Iodomesitylene...................	8.6 (65°C)	—
Nitromesitylene..................	0 (100°C)	—
Chloromethyl mesitylene..........	13 (65°C)	—
Methoxymesitylene................	100 (65°C)	—
Diphenyl oxide...................	100 (65°C)	1.5
Diphenyl sulfide.................	—	0.88

reaction. With nitrobenzene and nitromesitylene, Vavon found the reaction too slow to measure.

Bromomethylation of aromatic hydrocarbons may be accomplished by the same general technique as that employed in chloromethylations. However, according to Darzens[21], lower yields are obtained in bromomethylations than in chloromethylations. Vavon[109] states that bromomethyl ether reacts with aromatics in glacial acetic ten times as rapidly as chloromethyl ether. However, although he states that better yields of bromomethyl derivatives are obtained in special cases, secondary reactions involving diarylmethane formation, etc., are more rapid and cause poor results in many instances. In a process patented by Jones[60] it is claimed that bromomethyl as well as chloromethyl derivatives may be readily prepared by reacting aromatic hydrocarbons with a solution of paraformaldehyde or 30 to 40 per cent formaldehyde containing glacial acetic acid and hydrogen bromide or chloride.

That iodomethylations may also occur is illustrated by Fieser's preparation[28] of 9-methyl-10-iodomethyl-1,2-benzanthrene by treating 9-methyl-1,2-benzanthrene with paraform and hydrogen iodide in glacial acetic acid.

The technique of halomethylation is adaptable for use with a wide range of aromatic compounds and is a valuable procedure of synthetic chemistry.

Formation of Diarylmethanes and Hydrocarbon Resins. The formation of diarylmethanes and compounds in which several aromatic nuclei are linked by methylene groups represents a further stage of formaldehyde condensation than that involved in halomethylations. These substances are obtained as by-products of halomethylations and predominate when the reaction is run for an excessive length of time or otherwise subjected to extremely stringent conditions. With catalytic quantities of sulfuric and phosphoric acids as well as hydrochloric acid, the principal products are methylene-linked aromatics and aryl methyl ethers. High concentrations of acid and high temperatures lead almost exclusively to the polyaryl methanes and their sulfonates, chloromethyl derivatives, etc. The formation of oxygen containing hydrocarbon resins has recently been clarified by the work of Wegler[113] and his associates.

In 1872, Baeyer[9] studied reactions of formaldehyde with aromatic compounds by treating them with methylene diacetate $[CH_2(OOCCH_3)_2]$ in the presence of concentrated sulfuric acid which liberated formaldehyde from the methylene ester and caused it to react with the aromatic employed. In reactions with benzene, a hydrocarbon resin smelling of benzyl alcohol was the principal product. Since benzyl alcohol (methylolbenzene, C_6H_5-CH_2OH) is resinified by acids, Baeyer suggested its formation as the primary reaction product. In the case of mesitylene, dimesitylmethane was obtained. In later experiments in which methylal was employed as the formaldehyde donor, Baeyer[10] isolated diphenylmethane as a reaction product of formaldehyde and benzene. Related investigations carried out by other workers yielded similar results. Grabowski[43], for example, obtained dinaphthylmethane from naphthalene and methylal.

The reaction of aqueous formaldehyde and benzene in the presence of sulfuric acid was apparently first studied by Nastyukov[78] in 1903. A solution of 2 volumes of benzene in an equal volume of glacial acetic acid was added to a cooled solution of one volume of 40 per cent formaldehyde in two volumes of concentrated sulfuric acid. This reaction mixture was heated on a steam bath for about 20 minutes and then poured into water. The product, which separated as a water-immiscible oil, was freed of acid by washing with a soda solution, dried, filtered, and subjected to fractionation. A yield of high-boiling hydrocarbons was thus obtained from which diphenylmethane was isolated. By subjecting a similar reaction mixture to a longer heating period at a higher temperature, a light yellow, infusible solid was obtained. This product, which Nastyukov called phenyl formol, was extremely resistant to oxidizing agents and insoluble in all common solvents. On destructive distillation, it yielded liquid products containing diphenylmethane.

In 1914, Frankforter and Kokatnur[32, 33] reacted paraformaldehyde ("trioxymethylene") with aromatic hydrocarbons, using aluminum chloride as a catalyst. Good yields of diphenylmethane and anthracene in approximately equimolar amount were claimed as reaction products when benzene was the hydrocarbon employed. Mixtures of diarylmethanes and methylanthracenes were reported in the case of toluene, o-xylene, and mesitylene. Durene (1,2,4,5-tetramethyl benzene) did not react and was recovered unchanged from the reaction mixture. Huston and Ewing[54] reacted p-xylene with paraformaldehyde and aluminum chloride, but did not obtain anthracene derivatives. Dixylylmethane was the principal product, but was accompanied by compounds in which three and four p-xylene molecules were linked by methylene groups.

A study of diphenylmethane formation by Moshchinskaya and Globus[74] indicates that methanol and traces of iron salts have a strong catalytic effect on the sulfuric acid catalyzed reaction of benzene with formaldehyde. With a large excess of benzene and commercial methanol-containing formaldehyde, these investigators obtained a 70 to 76 per cent yield of diphenylmethane plus 13 to 15 per cent dibenzyl benzene when using 78 per cent sulfuric acid as a catalyst. According to their findings, the minimal effective concentration of the sulfuric acid used is 69 to 70 per cent.

The studies of Wegler[113] and his co-workers demonstrate that although diarylmethanes and higher polynuclear methylene-linked hydrocarbons resins are the major products when highly reactive hydrocarbons such as anthracene or naphthalene are condensed with formaldehyde in the presence of strongly acid catalysts, aryl methyl ethers may be the major products with benzene, alkyl benzenes, chlorinated benzene, alkyl naphthalenes, etc. However, even in these cases, high acid concentrations, high reaction temperatures, long reaction periods, and the presence of solvents for the acids or aromatics, such as acetic acid, favor the formation of oxygen-free resins. Excessive formaldehyde when used under conditions favoring oxygen-rich resins also results in the formation of aryl methyl polyoxymethylene ethers. The general reaction scheme as suggested by Wegler is illustrated below for the condensation of meta-xylene with formaldehyde.

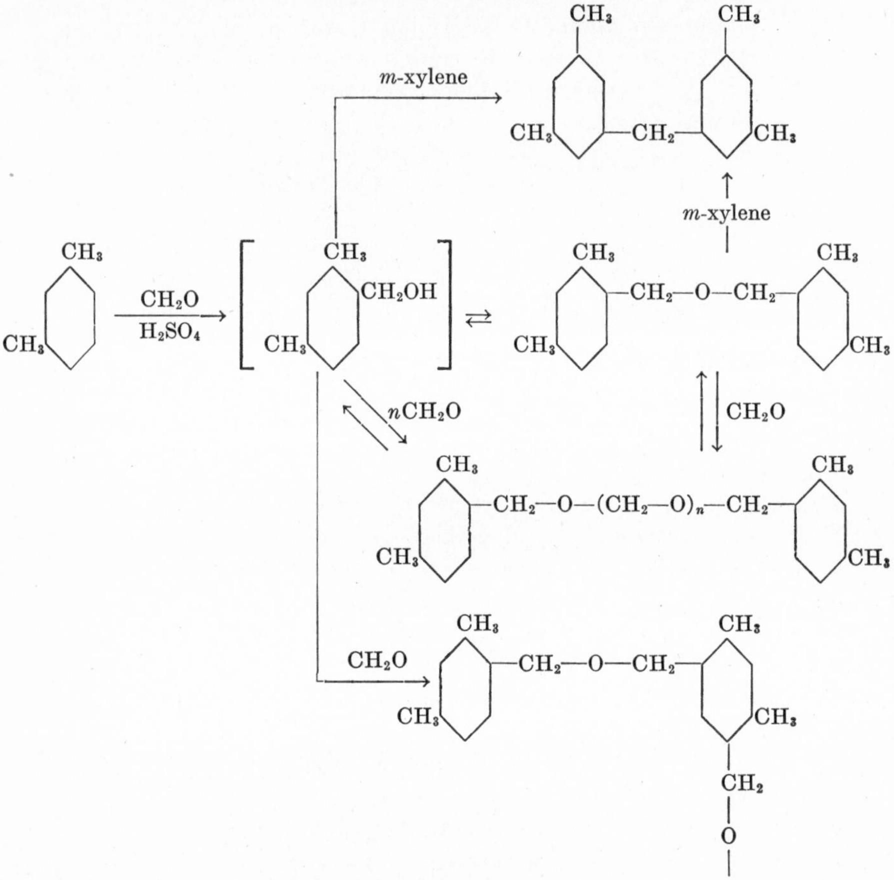

m-xylene

m-xylene

CH_2O / H_2SO_4

CH_2OH

$\rightleftarrows$

nCH_2O

$-CH_2-O-CH_2-$

CH_2O

$-CH_2-O-(CH_2-O)_n-CH_2-$

CH_2O

$-CH_2-O-CH_2-$

CH_2

O

Wegler reports the preparation of xylene-formaldehyde ether resins which can be further modified by reaction with active resin formers such as anthracene, phenols, etc. He reports that m-xylene reacts much faster than o-xylene and a little faster than p-xylene. He also states that 1,2,4,5-tetramethyl benzene reacts readily indicating that six-fold substitution of the benzene nucleus is possible. Cyclic formaldehyde ethers have also been isolated in the reaction products, e.g.,

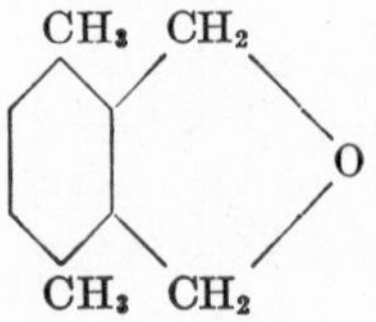

CH_3 CH_2

O

CH_3 CH_2

Production of hydrocarbon resins by treating crude petroleum distillates with formaldehyde and sulfuric acid was demonstrated by Nastyukov[79], who found that in general the best yields of resins were obtained from the high-boiling naphthas. This reaction is frequently referred to as the "formolite reaction" and is employed both for extracting aromatics from hydrocarbon mixtures[34] and preparing resins. Recent studies by Fulton and Gleason[35, 41] have demonstrated that light-colored, brittle resins can be prepared from light petroleum aromatic distillates by reaction with paraformaldehyde in the presence of zinc chloride and acetic acid. Catalysts, such as sulfuric acid, phosphoric acid, ferric chloride, etc., are unsatisfactory, since they yield dark-colored products. The aromatic fractions giving the best results contained ethylbenzene and xylenes. The products had molecular weights of the order of 700 or greater. Patented processes for preparing hydrocarbon resins with formaldehyde also include the use of mixtures of sulfuric acid with alkyl ethers[2] and sulfuric acid adsorbed on activated clay[3, 26] as condensation catalysts. These resins are evidently made up of linear molecules of methylene-linked aromatic nuclei. Ready fusibility contraindicates the existence of more than minimal cross-linkages between chains. Infusible, insoluble resins such as Nastyukov's phenyl formol are probably cross-linked.

Organometallic Hydrocarbon Derivatives

Anhydrous formaldehyde reacts readily with organometallic hydrocarbon derivatives, yielding substituted carbinols having the type formula R-CH_2OH. Syntheses of this type were first carried out by Grignard[48], who employed desiccated paraformaldehyde or polyoxymethylene. Unfortunately this polymer is insoluble in the ether solvents employed for the Grignard reagent and depolymerizes slowly in the reaction mixture. Refluxing for 2 to 3 days is necessary to complete the reaction, and even then only poor yields are obtained. Better results were obtained in 10 to 15 minutes by Ziegler[119], who employed monomeric formaldehyde gas produced by heating the polymer. By using this technique Ziegler obtained a 70 per cent yield of benzyl alcohol from the action of phenyl magnesium bromide on formaldehyde gas:

$$C_6H_5MgBr + CH_2O \text{ (g)} \rightarrow C_6H_5CH_2OMgBr$$

$$C_6H_5CH_2OMgBr + H_2O \rightarrow C_6H_5CH_2OH + Mg(OH)Br$$

With this procedure, alpha-naphthylcarbinol was also produced from naphthyl magnesium bromide. Less satisfactory results were obtained with aliphatic Grignard compounds. Ethyl magnesium bromide gave a small yield of n-propyl alcohol plus diethyl formal, which was formed as a reaction by-product.

When benzyl magnesium chloride is reacted with formaldehyde, ortho-tolycarbinol is obtained instead of the expected phenyl ethyl alcohol:

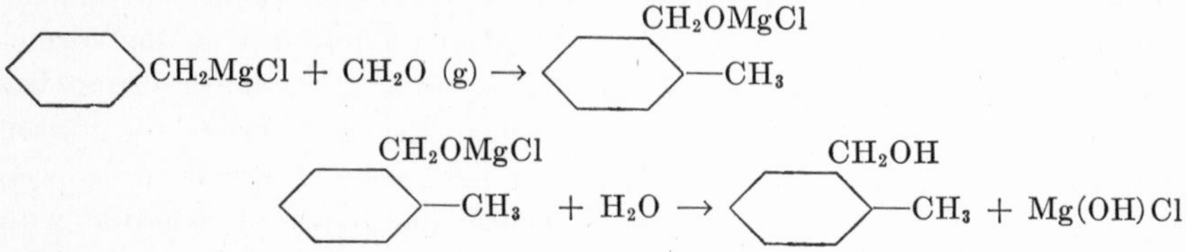

$$\langle \rangle CH_2MgCl + CH_2O \ (g) \rightarrow \langle \rangle -CH_3 \quad [CH_2OMgCl]$$

$$\langle \rangle -CH_3 \ [CH_2OMgCl] + H_2O \rightarrow \langle \rangle -CH_3 \ [CH_2OH] + Mg(OH)Cl$$

This reaction was first reported by Tiffeneau and Delange[106] in 1903. Recent studies of the reaction mechanism by Gilman and Kirby[39] indicate that a rearrangement takes place which correlates with other allylic or 3-carbon rearrangements for compounds having the type formula

$$-\overset{|}{C}=\overset{|}{C}-\overset{|}{\underset{|}{C}}-CH_2MgBr.$$

Prior to the development of methods for carrying out the direct reaction of formaldehyde with acetylene or monosubstituted acetylenes (page 336), the only satisfactory procedure for preparing acetylenic alcohols from formaldehyde involved the intermediate preparation of organometallic compounds. Substituted propargyl alcohols are obtained in yields of 70 per cent or better by reaction of monosubstituted acetylenic Grignard compounds with gaseous formaldehyde[105]:

$$RC \vdots CMgBr + CH_2O \ (g) \rightarrow RC \vdots C \cdot CH_2OMgBr$$

$$RC \vdots C \cdot CH_2OMgBr + H_2O \rightarrow RC \vdots C \cdot CH_2OH + Mg(OH)Br$$

Yocich[118] prepared butynediol-1,4 from the acetylenic Grignard compound:

$$BrMgC \vdots CMgBr + 2CH_2O \rightarrow BrMgOCH_2C \vdots CCH_2OMgBr$$

$$BrMgOCH_2C \vdots CCH_2MgBr + 2H_2O \rightarrow HOCH_2C \vdots CCH_2OH + 2Mg(OH)Br$$

Alkali-metal hydrocarbon derivatives react with formaldehyde in the same manner as the Grignard compounds. Schlenk and Ochs[98] report the formation of triphenylmethylcarbinol by reaction of formaldehyde and sodium triphenylmethane. The allylic rearrangement also takes place when lithium benzyl is reacted with formaldehyde gas, ortho-tolylcarbinol being obtained[38]. Sodium acetylides react normally giving acetylenic alcohols. Moureu and Demots[76] obtained hexyl propargyl alcohol from sodium amyl acetylide and formaldehyde.

Nitro- Hydrocarbons

Nitro- derivatives of aliphatic hydrocarbons react readily with formaldehyde in the presence of alkaline catalysts showing a striking difference in this respect from other compounds commonly classified as hydrocarbon derivatives. This reactivity is not manifested by nitro-aromatics, which behave in much the same manner as the unsubstituted hydrocarbons.

Reactions of nitro-paraffins with formaldehyde were apparently first studied in 1895 by Henry[51], who demonstrated that the hydrogen atoms on the carbon adjacent to the nitro radical were replaced by methylol groups in the presence of alkaline catalysts. The following reactions take place when nitromethane, nitroethane, and 2-nitropropane are treated with commercial formaldehyde solution to which potassium carbonate has been added:

$$CH_3NO_2 + 3CH_2O \text{ (aq)} \rightarrow NO_2 \cdot C(CH_2OH)_3$$
Nitromethane   *Trimethylolnitromethane*

$$CH_3CH_2NO_2 + 2CH_2O \text{ (aq)} \rightarrow NO_2 \cdot C(CH_2OH)_2CH_3$$
Nitroethane   *2-Nitro-2-methylpropane-*
diol-1,3 (Dimethylolnitroethane)

$$(CH_3)_2CHNO_2 + CH_2O \text{ (aq)} \rightarrow NO_2 \cdot C(CH_3)_2CH_2OH$$
2-Nitropropane   *2-Nitro-2-methylpropanol*
(Methyl-2-nitropropane)

All these substances are crystalline, and dissolve readily in cold water. Trimethylolnitromethane melts at 165 to 170°C, 2-nitro-2-methylpropane-diol-1,3 at 147 to 149°C, and 2-nitro-2-methylpropanol-1 at 90 to 95.5°C[37]. The trinitrate of trimethylolnitromethane was prepared by Hofwimmer[33], who found it to be a viscous liquid explosive resembling nitroglycerol. However, unlike nitroglycerol, it did not freeze even when cooled to a temperature of −35°C.

Trowell[107] has shown that trimethylolnitromethane can be employed as a hardening or curing agent for phenolic resins and proteins.

Special processes for the preparation of trimethylolnitromethane have been covered in patents by Wyler[115, 116, 117] and Cox[118]. McLean[68] has disclosed that methylol nitroparaffins can be prepared in good yield by reacting nitroparaffins with paraformaldehyde in alcohols containing 4 to 6 carbon atoms.

Studies of the mechanism of the formaldehyde-nitromethane reaction by Gorski and Makarov[42] have shown that methylolation is reversible and proceeds in a step-wise manner:

$$CH_3NO_2 + CH_2O \text{ (aq)} \rightleftharpoons NO_2 \cdot CH_2CH_2OH \qquad (1)$$
2-Nitroethanol

$$NO_2 \cdot CH_2CH_2OH + CH_2O \text{ (aq)} \rightleftharpoons NO_2CH(CH_2OH)_2 \qquad (2)$$
2-Nitropropanediol-1,3

$$NO_2CH(CH_2OH)_2 + CH_2O \text{ (aq)} \rightleftharpoons NO_2C(CH_2OH)_3 \qquad (3)$$
Trimethylolnitromethane

All three reaction products can be isolated and may be obtained in fair yield by varying the ratio of nitromethane to formaldehyde in the reaction mixture. This may be accomplished by heating nitromethane and paraformaldehyde with potassium carbonate at the boiling point of nitromethane, cooling to room temperature, neutralizing with sodium bisulfite or sulfuric acid, and separating the products by vacuum distillation. With 0.01 mol of formaldehyde per mol nitromethane, 2-nitroethanol is obtained as the sole product. With 0.1 mol formaldehyde, it is obtained in 61 per cent yield. The maximum yield of 2-nitropropanediol-1,3 (42 per cent) is obtained with 0.2 formaldehyde. With one mol formaldehyde, a 78 per cent yield of trimethylolnitromethane is produced. The reversibility of these reactions is demonstrated by Gorski and Makarov's preparation[42] of 2-nitropropanediol-1,3 and 2-nitroethanol by heating trimethylolnitromethane with nitromethane and potassium carbonate.

Schmidt and Wilkendorf[99] isolated a colorless, solid sodium derivative of 2-nitropropanediol-1,3 from the mixture obtained by reacting trimethylolnitromethane with alkali:

$$NO_2 \cdot C(CH_2OH)_3 + NaOH \rightarrow Na \cdot C(NO_2)(CH_2OH)_2 + H_2O + CH_2O \text{ (aq)}$$

Studies of nitroparaffin-formaldehyde reactions reported by Vanderbilt and Hass[108] indicate that conditions for securing optimum yields involve:
(1) Use of minimum alkali concentration to obtain a reasonable reaction velocity.
(2) Maintaining a homogeneous reaction mixture.
(3) Keeping temperatures to a practical minimum.
(4) Use of excess nitroparaffin.
(5) Removal of alkali before isolation of products.
A detailed review of nitroparaffin-formaldehyde chemistry by Hass and Riley[50] was published in 1943.

The aminomethylation of nitropropane by a joint reaction with formaldehyde and ammonia is reported by Johnson[59]. Fieser and Gates[27] report the reaction of formaldehyde with phenyl nitromethane and nitrophenyl nitromethanes. These reactions give phenyldimethylolnitromethanes.

Nitroaromatics are relatively inert to formaldehyde. However, when nitrobenzene, o- and p-nitrotoluene, respectively, are heated with formaldehyde in the presence of concentrated sulfuric acid at 40 to 50°C for 24 to 36 hours, dinitrodiphenylmethane and two isomeric dinitrotoluenes are obtained[25]. According to Parkes and Morley[85], *sym*-3,3-dinitrodiphenylmethane is produced from nitrobenzene and formaldehyde by this reaction. Chloromethylations of aromatic nitro compounds may also be brought about with formaldehyde and hydrogen chloride, as exemplified by an I. G.

Farbenindustrie patent[55] which describes a process of this type for the preparation of chloromethyl derivatives of nitroxylenes.

References

1. Adams, R., Fuson, R. C., and McKeever, C. H., "Organic Reactions," pp. 64–89, New York, John Wiley and Sons (1942).
2. Anderson, G. K., Tayler, E. A., and Fishel, J. B., (to The Neville Company), U. S. Patent 2,200,762 (1940).
3. *Ibid.*, U. S. Patent 2,200,763 (1940).
4. Arnold, R. T., National Org. Chem. Symposium, Boston, Mass. (June, 1947).
5. Arnold, R. T., and Barnes, R., *J. Am. Chem. Soc.*, **65**, 2393–2395 (1943).
6. Arundale, E., (to Standard Oil Development Co.), U. S. Patent 2,384,268 (1945).
7. Arundale, E., and Mikeska, L. A., (to Standard Oil Development Co.), U. S. Patent 2,296,375 (1942).
7a. Arundale, E., and Mikeska, L. A., (to Jasco Inc.), U. S. Patent 2,350,485 (1944).
7b. Arundale, E., and Mikeska, L. A., *Chem. Revs.*, **51**, 505 (1952).
8. Badger, G. M., Cook, J. W., and Crosbie, G. W., *J. Chem. Soc.*, **1937**, 1432–4.
9. Baeyer, A., *Ber.*, **5**, 1096–1100 (1872).
10. *Ibid.*, **6**, 221 (1873).
11. Bain, J. P., *J. Am. Chem. Soc.*, **68**, 638–41 (1946).
12. Baker, J. W., *J. Chem. Soc.*, **1944**, 296–301.
13. *Ibid.*, **1948**, 89–93; **1949**, 770.
14. Blanc, G., *Bull. soc. chim.* (*4*), **33**, 313 (1923).
15. Braun, J. v., and Nelles, J., *Ber.*, **67**, 1094 (1934).
16. Brunner, A., (to General Aniline Works, Inc.), U. S. Patent 1,887,396 (1932).
17. Brunner, A., and Greune, H., (to General Aniline Works, Inc.), U. S. Patent 1,910,462 (1933).
18. Cox, R. F. B., (to Hercules Powder Co.), U. S. Patent 2,250,256 (1941).
19. Crawford, J. W. C., McLeish, N., and Imperial Chemical Industries, Ltd., British Patent 528,761 (1940).
20. Crawford, J. W. C., and McLeish, N., (to Imperial Chemical Industries, Ltd.), U. S. Patent 2,233,835 (1941).
21. Darzens, G., *Compt. rend.*, **208**, 818 (1939).
22. Dermer, O. C., and Hawkins, J. J., (to Cities Service Research and Development Co.), U. S. Patents 2,524,777–2,524,778 (1950).
23. Dermer, O. C., Kohn, L., and Nelson, W. J., *J. Am. Chem. Soc.*, **73**, 5869–5870 (1951).
24. Du Pont de Nemours, E. I., & Co., Inc., British Patent 483,828 (1938).
25. Farbenfabriken vorm. F. Bayer & Co., German Patent 67,001 (1892).
26. Feasley, C. F., (to Socony-Vacuum Oil Co.), U. S. Patent 2,501,600 (1950).
27. Fieser, L. F., and Gates, M., *J. Am. Chem. Soc.*, **68**, 2249 (1946).
28. Fieser, L. F., and Sandin, R. B., *J. Am. Chem. Soc.*, **62**, 3098 (1940).
29. Fitsky, W., (to I. G. Farbenindustrie, A. G.), U. S. Patent 2,124,851 (1938).
30. *Ibid.*, U. S. Patent 2,143,370 (1939).
31. Fourneau, E., Benoit, G., and Firmenich, R., *Bull. soc. chim.* (*4*), **47**, 860 (1930).
32. Frankforter, G. B., and Kotaknur, V., *J. Am. Chem. Soc.*, **36**, 1529–37 (1914).
33. *Ibid.*, **37**, 2399 (1915).
34. Fulton, S. C., (to Standard Oil Development Co.), U. S. Patent 2,018,715 (1935).
35. Fulton, S. C., and Gleason, A. H., *Ind. Eng. Chem.*, **32**, 304–9 (1940).

36. Funk, C. E., Jr., (to American Cyanamid Co.), U. S. Patent 2,387,702 (1945).
37. Gabriel, C. L., *Chem. Ind.*, **45,** 666–8 (1939).
38. Gilman, H., and Breuer, F., *J. Am. Chem. Soc.*, **56,** 1127–8 (1934).
39. Gilman, H., and Kirby, J. E., *J. Am. Chem. Soc.*, **54,** 345–55 (1932).
40. Ginsburg, A., and co-workers, *Ind. Eng. Chem.*, **38,** 478–85 (1946).
41. Gleason, A. H., (to Standard Oil Development Co.), U. S. Patent 2,216,941 (1940).
42. Gorski, I. M., and Makarov, S. P., *Ber.*, **67,** 996–1000 (1934).
43. Grabowski, J., *Ber.*, **7,** 1605 (1874).
44. Grassi, G., and Maselli, C., *Gazz. chim. ital.*, **28,** II, 477 (1898).
45. Gresham, T. L., and Steadman, T. R., *J. Am. Chem. Soc.*, **71,** 737 (1949).
46. Gresham, W. F., (to E. I. du Pont de Nemours & Co., Inc.), U. S. Patent 2,405,-948 (1946).
47. Gresham, W. F., and Grigsby, W. E., (to E. I. du Pont de Nemours & Co., Inc.), U. S. Patent 2,493,964 (1950), reissue 23,248 (1950).
48. Grignard, V., *Compt. rend.*, **134,** 107 (1902).
49. Hamblet, C. H., and McAlevy, A., (to E. I. du Pont de Nemours & Co., Inc.), U. S. Patent 2,426,017 (1947).
50. Hass, H. B., and Riley, E. F., *Chem. Revs.*, **32,** 406 (1943).
51. Henry, L., *Compt. rend.*, **120,** 1265 (1895); **121,** 210 (1895).
52. Hill, P., and Short, W. F., *J. Chem. Soc.*, **1935,** 1124–6.
53. Hofwimmer, F., *Z. ges. Schiess-Sprengstoffw.*, I, 43 (1912); *Chem. Zentr.*, **1912,** I, 1265.
54. Huston, R. C., and Ewing, D. T., *J. Am. Chem. Soc.*, **37,** 2394 (1915).
55. I. G. Farbenindustrie, A. G., French Patent 802,365 (1936).
56. *Ibid.*, British Patent 507,571 (1939).
57. *Ibid.*, French Patent 847,255 (1939).
58. Imperial Chemical Industries, Ltd., British Patent 583,173 (1946).
59. Johnson, H. G., (to Commercial Solvents Corp.), U. S. Patent 2,408,172 (1946).
60. Jones, F. D., (to American Chemical Paint Co.), U. S. Patent 2,212,099 (1940).
61. Keyssner, E., and Eichler, E., (to General Aniline & Film Corp.), U. S. Patent 2,238,471 (1941).
62. Keyssner, E., and Reppe, W., (to General Aniline & Film Corp.), U. S. Patent 2,232,867 (1941).
63. Keyssner, E., Reppe, W., and Hecht, O., (to General Aniline & Film Corp.), U. S. Patent 2,273,141 (1942).
64. Kriewitz, O., *Ber.*, **32,** 57–60 (1899).
65. Lock, G., *Ber.*, **74B,** 1568–74 (1941).
66. Londergan, T. E., (to E. I. du Pont de Nemours & Co., Inc.), U. S. Patent 2,461,906 (1949).
67. *Ibid.*, U. S. Patents 2,487,008 and 2,487,009 (1949).
68. McLean, A., (to Imperial Chemical Industries, Ltd.), U. S. Patent 2,399,686 (1946).
69. McLeish, N., Crawford, J. W. C., and Imperial Chemical Industries, Ltd., British Patent 514,619 (1939).
70. Marsh, N. H., (to Standard Oil Development Co.), U. S. Patent 2,389,205 (1945).
71. Matsukawa, T., and Shirakawa, K., *J. Pharm. Soc. Japan*, **70,** 25–8 (1950); *C. A.*, **44,** 4435 (1950).
72. Matti, J., *Bull. soc. chim.* (*4*), **51,** 974–9 (1932).
73. Mikeska, L. A., and Arundale, E., (to Standard Oil Development Co.), U. S. Patent 2,308,192 (1943).

74. Moshchinskaya, N. K., and Globus, R. L., *J. Applied Chem.*, (*U. S. S. R.*), **17**, 76–82, 137–143 (1944).
75. Mottern, H. O., (to Jasco, Inc.), U. S. Patent 2,335,691 (1943).
76. Moureu, C., and Desmots, H., *Bull. soc. chim.* (*3*), **27**, 360–6 (1902)
77. Mraz, R. G., Abstrs. Doctoral Dissertations (Penn. State College), **11**, 75–6 (1948).
78. Nastyukov, A. M., *J. Russ. Phys. Chem. Soc.*, **35**, 824 (1903); *Chem. Zentr.*, **1903**, II, 1425.
79. *Ibid.*, **36**, 881 (1904); *J. Chem. Soc.*, **86**, 801 (1904).
80. Olsen, S., *Z. Naturforschg.*, **1**, 671–6 (1946).
81. *Ibid.*, **1**, 676–82 (1946).
82. *Ibid.*, **3b**, 314–320 (1948).
83. Olsen, S., Norw. Patent 74,890 (1949).
84. Olsen, S., and Padberg, H., *Z. Naturforschg.*, **1**, 448–458 (1946).
85. Parkes, G. D., and Morley, R. H. H., *J. Chem. Soc.*, **1936**, 1478–9.
86. Peterson, M. D., (to E. I. du Pont de Nemours & Co., Inc.), Example 6, U. S. Patent 2,425,638 (1947).
87. Prins, H. J., *Chem. Weekblad.*, **14**, 932–9 (1917); **16**, 1072–3 (1919); **16**, 1510–26 (1919); *Chem. Zentr.*, **1918**, I, 168–9; *C. A.*, **13**, 3155 (1919); *C. A.*, **14**, 1662 (1920).
88. Prins, H. J., *Proc. Acad. Sci. Amsterdam*, **22**, 51–6 (1919).
89. Prins, H. J., *Rec. trav. chim.*, **51**, 469–74 (1932).
90. Pudovik, A. N., Nikitina, V. I., and Aigistova, S. Kh., *J. Gen. Chem.* (*U.S.S.R.*), **19**, 279–89 (1949).
91. Quelet, R., *Bull. soc. chim.* (*4*), **53**, 222–34, 510 (1933).
92. Reddelien, G., and Lange, H., (to Winthrop Chemical Co., Inc.), U. S. Patent 1,853,083 (1932).
93. Reddelien, G., and Lange, H., (to General Aniline Works, Inc.), U. S. Patent 1,910,476 (1933).
94. Ritter, J. J., (to Standard Oil Development Co.), U. S. Patent 2,335,027 (1943).
95. Roberts, A. M., (to Imperial Chemical Industries, Ltd.), U. S. Patent 2,379,104 (1945).
96. Rosen, R., and Arundale, E., (to Standard Oil Development Co.), U. S. Patent 2,504,732 (1950).
97. Sabetay, S., *Compt. rend.*, **192**, 1109–10 (1931).
98. Schlenk, W., and Ochs, R., *Ber.*, **49**, 608 (1916).
99. Schmidt, E., and Wilkendorf, R., *Ber.*, **52**, 392 (1919).
100. Scott, N. D., (to E. I. du Pont de Nemours & Co., Inc.), U. S. Patent 2,024,749 (1935).
101. Short, N., (to Imperial Chemical Industries, Ltd.), U. S. Patent 2,408,889 (1946).
102. Société des Usines Chimiques Rhône-Poulenc, French Patent 717,712 (1932).
103. Stephen, H., Short, W. F., and Gladding, G., *J. Chem. Soc.*, **117**, 510 (1920).
104. Szamant, H. H., and Dudek, J., *J. Am. Chem. Soc.*, **71**, 3763–5 (1949).
105. Tchao, Y. L., *Bull. soc. chim.* (*4*), **53**, 682–7 (1933).
106. Tiffeneau, M., and Delange, R., *Compt. rend.*, **137**, 573 (1903).
107. Trowell, W. W., (to Hercules Powder Co.), U. S. Patent 2,426,128 (1947).
108. Vanderbilt, B. M., and Hass, H. B., *Ind. Eng. Chem.*, **32**, 34–8 (1940).
109. Vavon, G., Bolle, J., and Calin, J., *Bull. soc. chim.* (*3*), **6**, 1025 (1939).
110. Vorozhtzov, N. N., and Yuruigina, E. N., *Zhur. Obschchei Khim., Khim. Ser. I*, 49–64 (1931).

111. Walker, J. F., (to E. I. du Pont de Nemours & Co., Inc.), U. S. Patent 2,463,227 (1949).
112. Walker, J. F., and Londergan, T. E., (to E. I. du Pont de Nemours & Co., Inc.), U. S. Patents 2,439,765 (1948), 2,487,006 (1949) and 2,487,007 (1949).
113. Wegler, R., *Z. angew Chem.*, **A60,** 88–96 (1948).
114. Workman, A. R., (to Cities Service Oil Co.), U. S. Patent 2,412,762 (1946).
115. Wyler, J. A., (to Trojan Powder Co.), U. S. Patent 2,164,440 (1939).
116. *Ibid.*, U. S. Patent 2,231,403 (1941).
117. *Ibid.*, U. S. Patent 2,261,788 (1941).
118. Yocich, I., *J. Russ. Phys. Chem. Soc.*, **38,** 252 (1906).
119. Ziegler, K., *Ber.*, **54,** 737 (1921).

CHAPTER 16

REACTIONS OF FORMALDEHYDE WITH HETEROCYCLIC COMPOUNDS

The purpose of this chapter is to review formaldehyde chemistry involving typical representative heterocyclic nuclei. In this connection, reactions with furans, thiophenes, pyrroles, pyridines and quinolines are covered, as well as those of some of their partially or completely hydrogenated derivatives. Reactions in which hydroxyl, mercaptan, amine and amide groupings behave in an essentially normal manner are not generally included since the chapters dealing with the reactions of alcohols, aldehydes, ketones, amines, etc., describe these processes.

Furans

Zerweck, Heinrich and Pinten[67] have shown that furan, as well as furfuryl alcohol, condenses with aqueous formaldehyde in the presence of hydrochloric acid to give infusible, insoluble opaque resins which have an excellent degree of resistance to boiling water, acids and alkalies. In the case of furfuryl alcohol, Harvey[20] describes the preparation of fusible resins by carrying out the reaction at a controlled pH in the range 1.5 to 3.5. Furfural also condenses with formaldehyde to give resins in the presence of acid catalyst[32].

Although methylol derivatives have apparently not been prepared by reactions of formaldehyde with furan or alkyl furans, simple 2-aminomethyl derivatives have been obtained by the Mannich synthesis in some instances. Holdren and Hixon[30] isolated a number of furfurylamines from the reaction products of sylvan (2-methyl furan) with a mixture of aqueous formaldehyde and primary and secondary amine hydrochlorides at 30 to 35°C. Reactions involving ethylamine hydrochloride are examples.

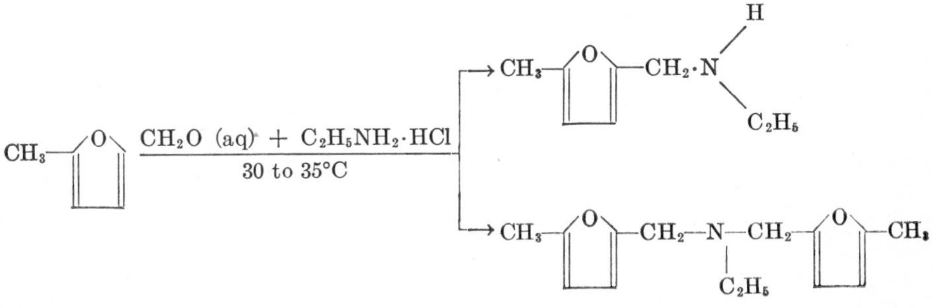

Holdren[31] also reports that when a negative group is attached to the furan ring, as in 2-furoic acid or its ethyl ester, no definite product can be isolated from the amine hydrochloride-formaldehyde reaction. Although, furfuryl alcohol tends to give resins in these reactions, 5-dimethylamino furfuryl alcohol can be obtained by reaction with formaldehyde and dimethylamine hydrochloride.

Furan does not give simple aminomethyl derivatives with ammonium chloride and formaldehyde. Holdren[30] reports resins as the sole product. Mooney[56] obtains a complex aminomethyl furan derivative possessing the empirical formula $C_{37}H_{73}N_5O_8$ when the reaction mixture obtained at 32 to 35°C is neutralized with ammonia. The product is a light tan powder which dissolves in dilute acids and is reprecipitated on neutralization. This product can also be obtained by adding furan to a solution of hexamethyl-enetetramine containing hydrochloric acid.

The chloromethylation of methyl furoate by reaction with formaldehyde and hydrogen chloride in the presence of aluminum chloride is reported by Mackay and Jones[51]. A methylene bis-derivative can also be obtained by a reaction of this type[6].

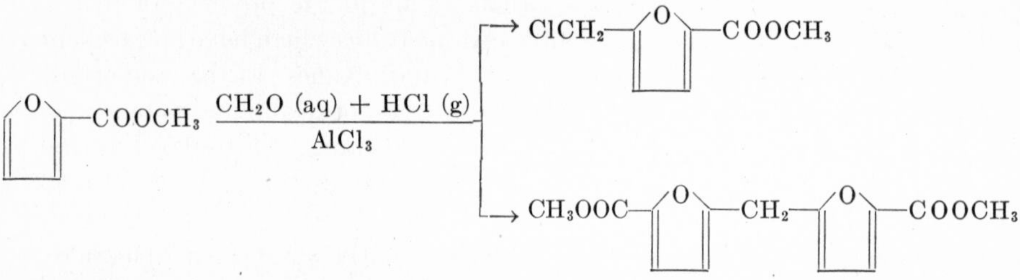

Berkenheim and Dankova[3] state that the crossed Cannizzaro reaction of furfural with formaldehyde and alkali gives furfuryl alcohol in high yield.

Tetrahydrofuran. Mooney and the writer[66] obtained delta-chlorobutyl chloromethyl ether and delta-chlorobutyl formal by reaction of paraform-aldehyde and hydrogen chloride with tetrahydrofuran at 15 to 30°C.

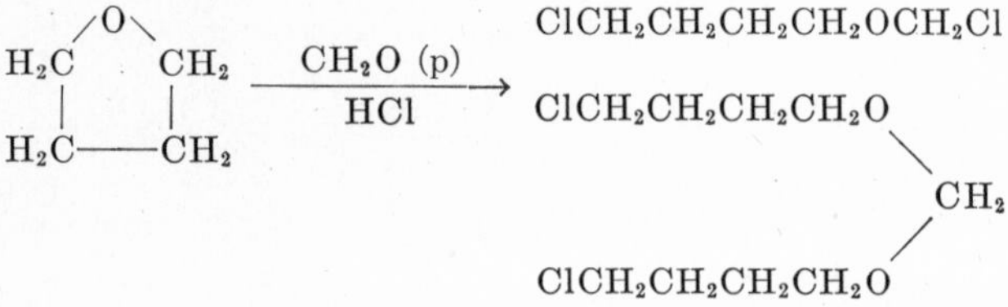

Thiophenes

Thiophene condenses with formaldehyde in the presence of acids to produce methylene-linked resins. Since this reaction proceeds much more readily than the formaldehyde-benzene condensation, it was utilized in an early process for benzene purification (page 479). Recent studies by Caesar and Sachanen[7, 8] cover the preparation of viscous liquid and solid plastics by acid condensation of thiophene and some alkyl thiophenes with formaldehyde in equimolar proportions. These resins may be cured by prolonged heating with acid in the absence of water, but preferably with excess formaldehyde. Resins are not formed on reaction with alkali. However, a thiophene thiol resembles phenol in that it does form resins on reacting with aqueous formaldehyde and a caustic alkali[9]. Modified resins are obtained by reactions involving thiophene and phenol[7, 8].

An approximately 50 per cent yield of a methylene bis-thiophene (di-2-thienyl methane) was obtained by Cairns, McKusick and Weinmayr[10] by reaction of trioxane with thiophene using 48 per cent hydrofluoric acid as a catalyst. This compound is also obtained as a by-product by Blicke and co-workers[4, 5] in chloromethylating thiophene with aqueous formaldehyde at 0°C. The chloromethylation product is 2-thienyl chloride.

Mannich reactions of thiophene with ammonium chloride and formaldehyde (or hexamethylenetetramine and hydrochloric acid) have been studied by Hartough, Lukasiewicz and various co-workers[50, 24, 25, 26, 21, 22]. Products include 2-thenylamine, di-2-thenylamine and polymeric amines. A reaction of 2 mols thiophene with 1 mol aqueous formaldehyde and 3 mols ammonium chloride gives 40 per cent of the primary amine, 20 per cent secondary amine and 40 per cent polymer. With hexamethylenetetramine, the yield of the polymeric product is 68 per cent[50].

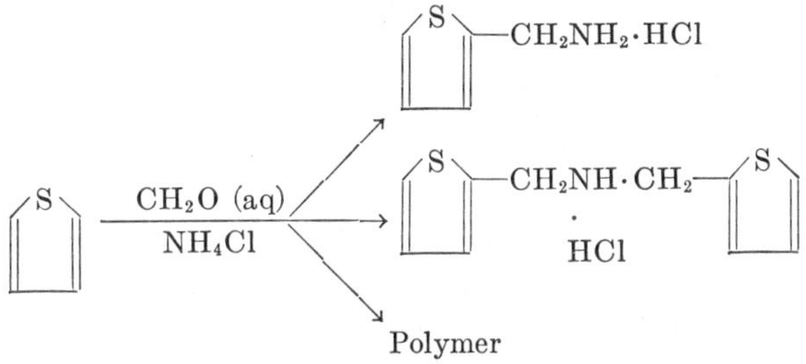

Isolation of N-(2-thenyl)-formaldimines from the Mannich reaction indicates that the mechanism is probably as shown in the formula at the top of page 356[25]. As noted, a methylene bis-thiophene is obtained as a by-product.

Related Mannich reactions involving hydroxylamine hydrochloride, formaldehyde and thiophene are also reported by Hartough[23]. An approximately 80 per cent yield of alpha-thenylhydroxylamine is obtained if the reaction of the hydroxylamine hydrochloride and formaldehyde is completed before the thiophene is added.

A reaction of 2-methyl thiophene and formaldehyde is described by

$$CH_2O + NH_4Cl \rightleftharpoons CH_2{:}NH{\cdot}HCl + H_2O$$

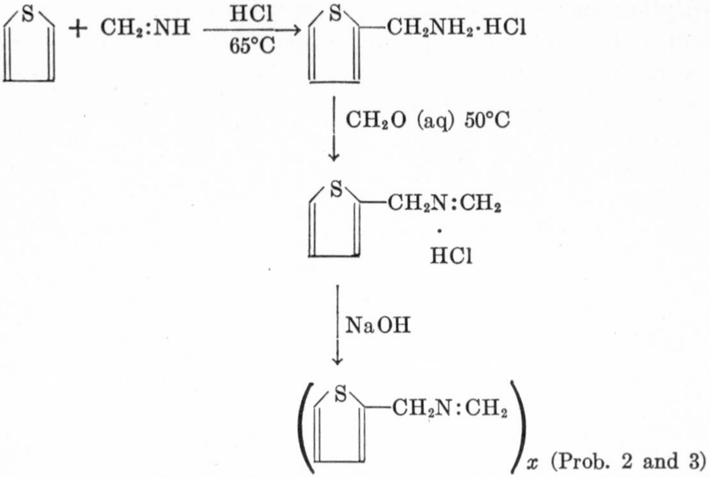

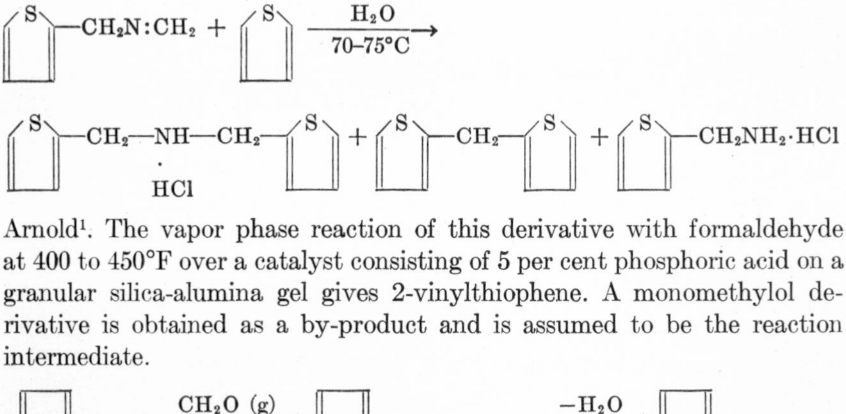

$$\text{[thiophene]} + CH_2{:}NH \xrightarrow[65°C]{HCl} \text{[thiophene]}{-}CH_2NH_2{\cdot}HCl$$

$$\Big\downarrow CH_2O \text{ (aq) } 50°C$$

$$\text{[thiophene]}{-}CH_2N{:}CH_2 \\ \cdot \\ HCl$$

$$\Big\downarrow NaOH$$

$$\left(\text{[thiophene]}{-}CH_2N{:}CH_2 \right)_x \quad \text{(Prob. 2 and 3)}$$

$$\text{[thiophene]}{-}CH_2N{:}CH_2 + \text{[thiophene]} \xrightarrow[70-75°C]{H_2O}$$

$$\text{[thiophene]}{-}CH_2{-}NH{-}CH_2{-}\text{[thiophene]} + \text{[thiophene]}{-}CH_2{-}\text{[thiophene]} + \text{[thiophene]}{-}CH_2NH_2{\cdot}HCl \\ \cdot \\ HCl$$

Arnold[1]. The vapor phase reaction of this derivative with formaldehyde at 400 to 450°F over a catalyst consisting of 5 per cent phosphoric acid on a granular silica-alumina gel gives 2-vinylthiophene. A monomethylol derivative is obtained as a by-product and is assumed to be the reaction intermediate.

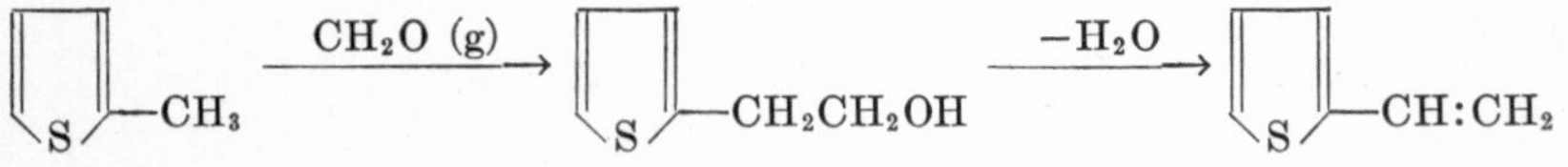

$$\text{[thiophene]}{-}CH_3 \xrightarrow{CH_2O \text{ (g)}} \text{[thiophene]}{-}CH_2CH_2OH \xrightarrow{-H_2O} \text{[thiophene]}{-}CH{:}CH_2$$

Formaldehyde solution, paraformaldehyde or trioxane vapors may be employed in this reaction. Substituted vinyl thiophenes may also be obtained by using other 2-alkyl thiophenes as well as nuclear substituted derivatives of these compounds.

Pyrroles

Pyrroles yield alpha-methylol derivatives by reaction with alkaline formaldehyde solution. This was apparently first demonstrated by Chelintzev and Maksorov[13] in 1916, who prepared the 2,5-dimethylol derivatives of pyrrole and N-methyl pyrrole. Taggart and Richter[63] obtained N-methylol pyrrole by reacting pyrrole with formaldehyde in ethanol using calcium hydroxide as a catalyst. Chadwick[12] has shown that N-methylol pyrrole is formed when paraformaldehyde is heated with pyrrole at 40 to 55°C in the presence of approximately 0.2 per cent of anhydrous potassium carbonate. On heating to 75 to 90°C, the N-methylol derivative rearranges and disproportionates to give the 2,5-dimethylol derivative.

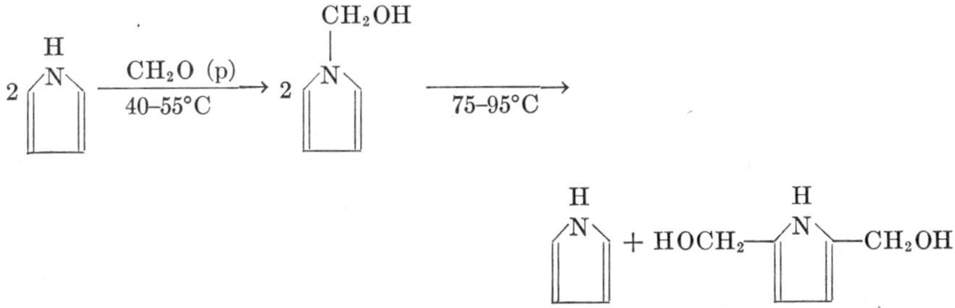

With two molecular equivalents of formaldehyde as paraformaldehyde, a good yield of almost colorless 2,5-dimethylol pyrrole (m.p. 112 to 115°C) can be obtained. Uncontrolled reactions of pyrrole with formaldehyde give dark-colored resins.

Reaction of formaldehyde with potassium pyrrole in the presence of zinc chloride gives α,α'-dipyrryl methane according to Colacicchi[14]. Rothemund[59, 60] has shown that prolonged reactions of formaldehyde and pyrrole in methanol yield products related to porphyrins. Calvin, Ball and Aronoff[11] report the formation of porphyrin and beta-dihydroporphyrins as products of this reaction. Although the uncatalyzed condensation of substituted pyrroles containing at least one free alpha position commonly leads to symmetrical dipyrryl methanes, Corwin and Brunings[15] obtained a **75** per cent yield of unsymmetrical dipyrryl methane by carrying out the following reaction in methanol using hydrochloric acid as a catalyst.

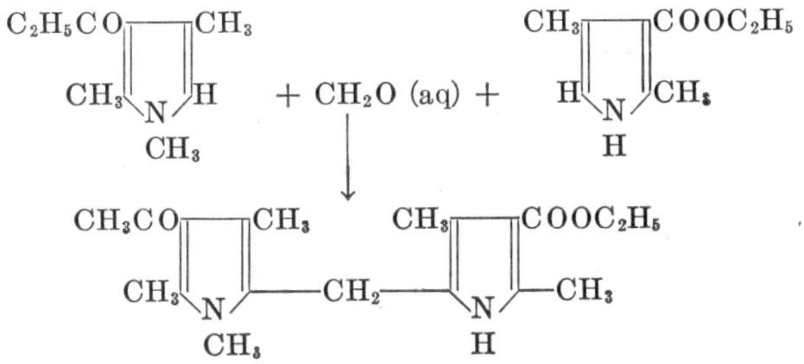

Fischer and Nenitzescu[17] obtained beta-methylol pyrroles by condensation of derivatives in which the alpha positions were blocked by substituents. In general, these substituted methylol pyrroles are less stable than the simple alpha derivatives. They lose formaldehyde readily on heating and are easily converted to dipyrryl methanes by treatment with alcoholic potash, dilute acids or merely by boiling with water in the absence of added catalysts.

Mannich reactions of pyrrole with formaldehyde and primary or secondary amines and their hydrochlorides have been carried out by Bachman and Heisey[2] and Herz, Dittmer and Cristol[27]. Mono- and diaminoalkyl derivatives were prepared. In these reactions, only the alpha positions of the pyrrole nucleus were involved. Good yields are apparently obtained only with secondary amines. Bachman and Heisey[2] obtained their best results with morpholine and piperidine.

Pyrrolidine

Potokhin[57] has reported the preparation of N-methylol pyrrolidine, alpha-methylol pyrrolidine and N-methylene bis-pyrrolidine, $C_4H_8N \cdot CH_2 \cdot NC_4H_8$, by heating pyrrolidine with a polyoxymethylene for 6 hours at 140 to 150°C in a sealed tube. In general, pyrrolidine reacts as would be expected for a secondary amine.

Hess and co-workers[28, 29] have demonstrated the N-methylation of alpha-substituted pyrrolidines containing a secondary alcohol group by heating with formaldehyde in the presence of hydrochloric acid. This reaction probably involves formation of an N-methylol intermediate followed by an intramolecular oxidation-reduction reaction, yielding a ketonic product.

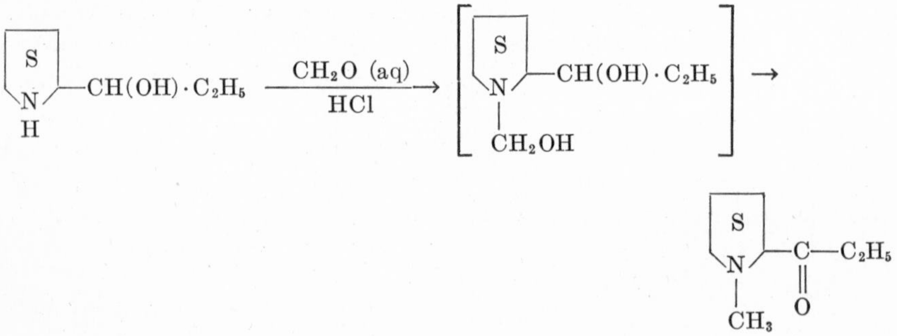

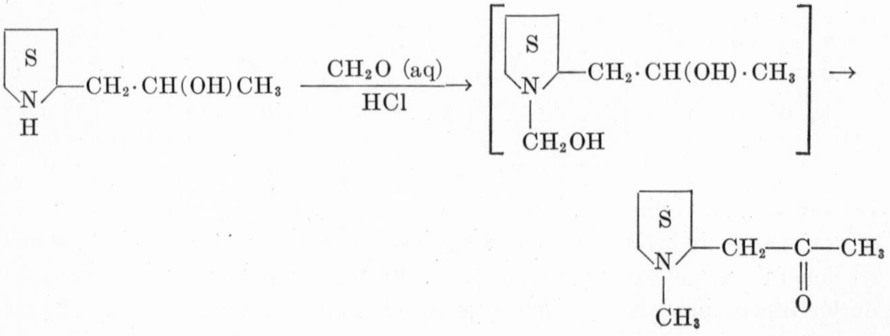

Pyridines

Pyridine is reported by Formanek[18] to form a crystalline compound on heating with excess 40 volume per cent formaldehyde on a water bath for several hours. The product which crystallizes from the reaction mixture on cooling is soluble in water and is purified by recrystallization from hot alcohol. Its empirical formula corresponds to $C_5H_5N \cdot CH_2O$. On heating, it decomposes rapidly to pyridine and formaldehyde, a change which takes place slowly even at room temperature. The structure of this product has not been determined. Shmidt and Petrov[61, 62] describe the formation of resinous products by condensation of pyridine bases with formaldehyde in the presence of acid or alkaline catalysts.

Beta-hydroxy pyridine is reported on Urbanski[65] to form an alpha-methylol derivative with aqueous formaldehyde in the presence of caustic soda. A yield of approximately 50 per cent is obtained and the reaction is stated to be readily controllable with respect to the formation of resinous condensates indicating that its reactivity is much less than that of phenol. The reaction of alpha-aminopyridine with formaldehyde in the presence of formic acid is reported by Tschitschibabin and Knunyants[64] to give a methylene bis-dimethylaminopyridine.

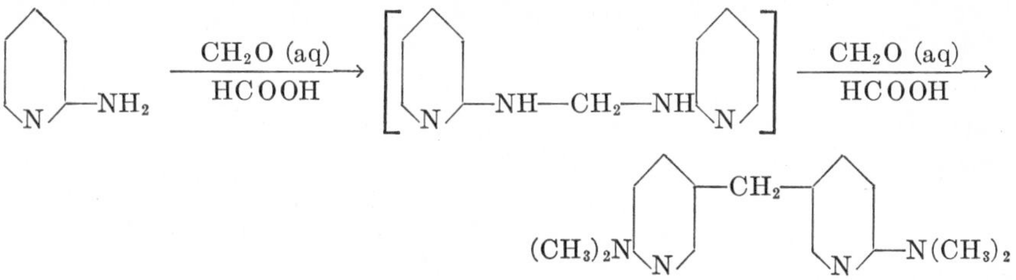

The intermediate formation of the methylene bis-amino derivative is suggested by the fact that alpha-dimethylaminopyridine does not react under the same conditions. Further, 2-amino-5-nitropyridine gives only the methylene bis-amine. The latter cannot isomerize in the manner indicated above.

Alkyl Pyridines. Compounds of the pyridine (or quinoline) series with an alkyl group in the alpha or gamma position to the nitrogen ring react with formaldehyde in a manner similar to the alkyl ketones as indicated in the formula at the top of the following page in the case of alpha-picoline. Because of the basicity of the nitrogen atom, addition of a basic catalyst is unnecessary and the reactions may be carried out simply by heating with aqueous formaldehyde or paraformaldehyde. The vinyl derivatives are produced by pyrolysis of the corresponding methylol derivatives. Because of the importance of butadiene-vinylpyridine copolymers as synthetic

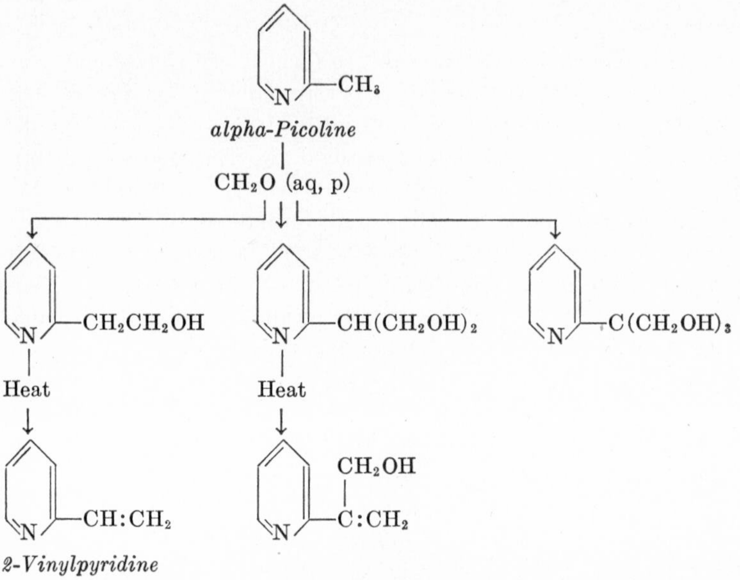

alpha-Picoline
|
CH₂O (aq, p)

2-Vinylpyridine

rubbers, the preparation of this vinyl derivative from alpha-picoline has been the subject of considerable industrial research in recent years.

The formation of the monomethylol derivative of alpha-picoline was apparently first observed by Ladenburg[40] in 1889. He named the product alpha-picolyl alkine and showed that it could be converted to vinyl pyridine by treatment with concentrated hydrochloric acid, distillation with caustic potash or simple distillation. Further studies were reported in 1898[42]. Initial experiments in which picoline was reacted with methylal at high temperatures (280 to 290°C) under acidic conditions gave a methylene bis-picoline as the only isolable product[41].

Following Ladenburg's discovery, the chemistry of the reactions of alkyl pyridines and quinolines with formaldehyde was the subject of detailed studies by Koenigs[35], Lipp[44, 46] and other investigators. It was shown that the reaction was characteristic of alpha and gamma alkyl substituents containing one or more hydrogen atoms on the carbon attached to the pyridine nucleus and that beta substituents did not react. Koenigs and Happe[36] demonstrated the formation of dimethylol picoline and mono-methylol ethyl picoline, $C_5H_4N \cdot CH(CH_2OH)CH_3$. Lipp and co-workers[44, 46] studied the formation of mono-, di- and trimethylol derivatives and the dehydration reactions yielding vinylpyridine and methylol vinylpyridine. Lipp and Zirngibl[46] found that when alpha-picoline is heated with two equivalents of aqueous formaldehyde, nearly equal amounts of mono- and dimethylol picoline are obtained together with a small amount of the tri-methylol derivative. The trimethylol derivative can also be obtained by heating the mono- and di- derivatives with excess formaldehyde. Recent industrial patents by Mahan[52, 53] cover special processes for the production of vinylpyridine.

Reactions to gamma-methyl pyridine with formaldehyde, similar to those of the alpha derivative, were investigated by Löffler and Stietzel[49], Meisenheimer[54] and Koenigs and Happe[38].

Higher alkyl pyridines such as alpha-ethyl pyridine yield mono- and dimethylol derivatives. On pyrolysis, the monomethylol compound from ethyl pyridine yields methyl vinylpyridine, $CH_2:C(CH_3)\cdot C_5H_4N$[47].

In the case of the dimethyl pyridines, Löffler and Remmuler[48] have shown that with 2,6-lutidine, both monomethylol and symmetrical dimethylol derivatives can be obtained. Both methyl groups are equally reactive as would be expected because of ring resonance. With 2,4-lutidine, Engels[16] found that only the 2-methyl substituent reacted with formaldehyde on heating in a sealed tube at 135 to 140°C.

On reaction with aqueous formaldehyde at zero degrees to room temperature, N-methyl tetrahydropicoline does not form a methylol derivative like picoline as was first believed, but yields 1-methyl-3-acetyl piperidine. The mechanism for this reaction as reported by Lipp and Widnmann[45] is indicated below:

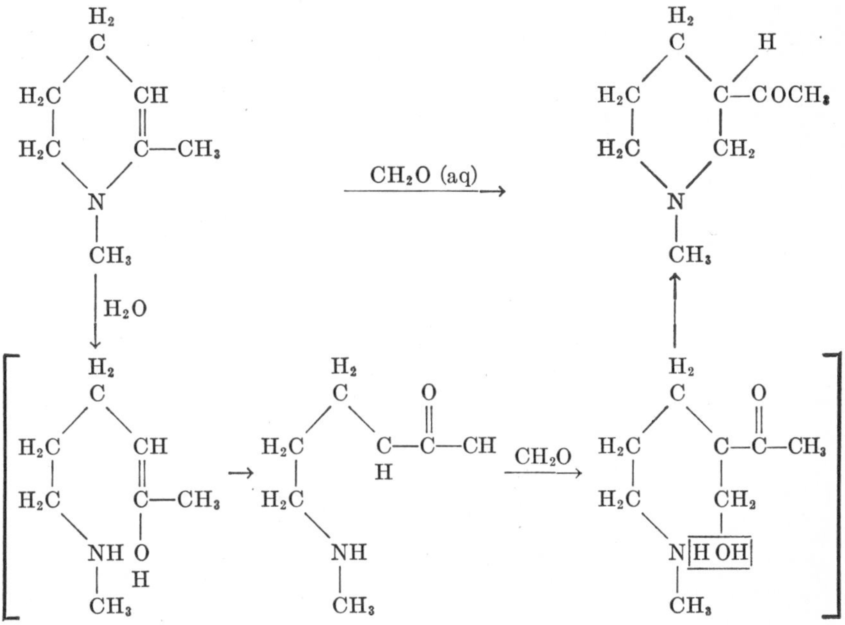

Piperidine

The reactions of piperidine with formaldehyde are those of a normal secondary amine. (See Chapter 14.)

Quinolines

Quinoline does not react with formaldehyde under ordinary conditions but will form methylene-linked resinous products in the presence of strong acids. It is soluble in strong formaldehyde but insoluble in dilute solution. Alkyl substituted quinolines react with formaldehyde in essentially the same manner as the corresponding pyridine derivatives when the alkyl groups are in the alpha and gamma positions. Koenigs[34] showed that one, two or all three hydrogen atoms on alpha-methyl quinoline (quinaldine) could be substituted by methylol groups on reaction with formaldehyde, whereas gamma-methyl quinoline yields only the mono- and dimethylol substituents. Dehydration of the monomethylol derivative of quinaldine gives vinyl quinoline as would be expected[55]. Koenigs and Mengel[39] found that the methylol group enters 2,4-dimethyl quinoline on the 2-methyl group and that a second methylol group also enters at this position.

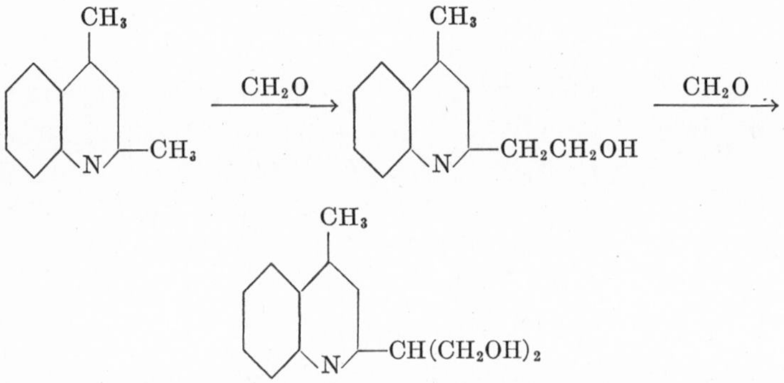

The presence of other substituent groups naturally affects this reaction. Koenigs[34, 37] has shown that in the case of 2-ethyl-3-methyl quinoline, only one methylol group can be introduced at the alpha position.

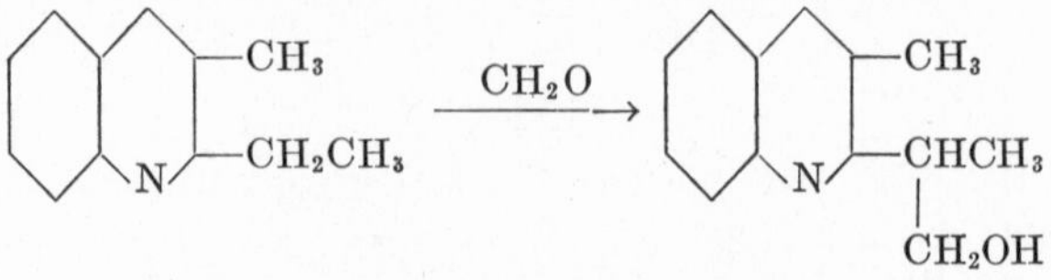

However, if the beta-methyl group is replaced by a carboxyl group, complete substitution of the available hydrogen atoms on the carbon attached to the alpha position on the quinoline nucleus is possible.

Quinaldine can also be involved in Mannich syntheses involving formaldehyde and a secondary amine hydrochloride as illustrated by its reaction with diethylamine reported by Kermach and Muir[33].

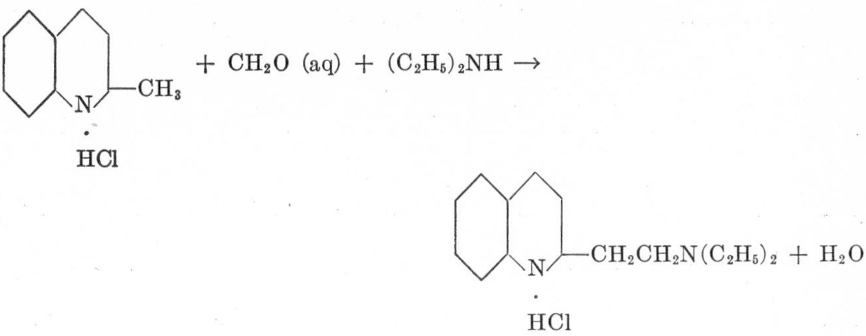
N
CH3
HCl
+ CH2O (aq) + (C2H5)2NH →
N
CH2CH2N(C2H5)2 + H2O
HCl

The carbocyanine dye, pinacyanol, can be synthesized by heating quinaldine ethiodide with formaldehyde and alkali[43, 58].

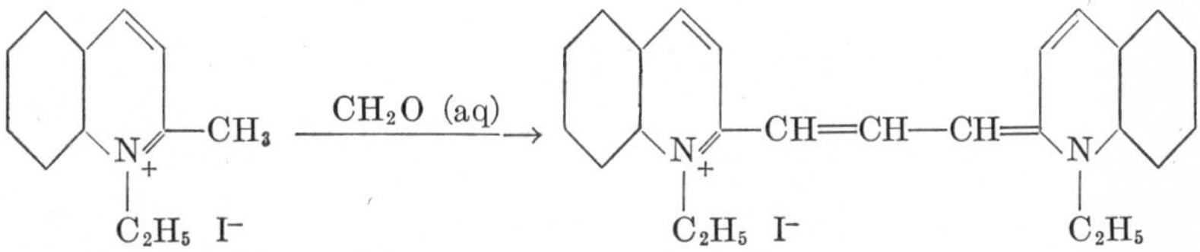
N+
CH3
C2H5 I−
CH2O (aq)
N+
CH=CH—CH=
N
C2H5 I−
C2H5

A similar reaction with lepidine (gamma-methyl quinoline) ethiodide yields the analogous cryptocyanine. Hamer[19] obtained methylene diquinaldine methiodide by heating quinaldine methiodide with aqueous formaldehyde in absolute alcohol in the presence of piperidine, and obtained pinacyanol by heating this product with alkali in the presence of quinoline methiodide which acted as a catalyst for the reaction. The formation of methylene diquinaldine methiodide is shown below:

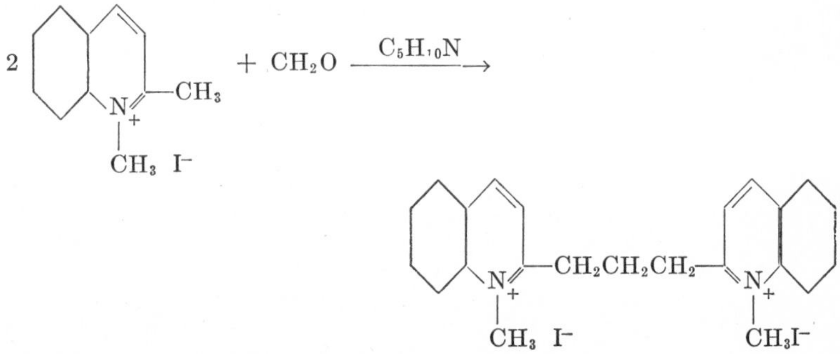
2
N+
CH3
CH3 I−
+ CH2O
C5H10N
N+
CH2CH2CH2
N+
CH3 I−
CH3I−

The conversion of the methylene derivative to pinacyanol apparently involves loss of one proton and oxidation in the alkaline medium.

References

1. Arnold, P. M., (to Phillips Petroleum Co.), U. S. Patent 2,512,596 (1950).
2. Bachman, G. B., and Heisey, L. V., *J. Am. Chem. Soc.*, **68**, 2496 (1946).

3. Berkenheim, A. M., and Dankova, T. F., *J. Gen. Chem. (U.S.S.R.)*, **9**, 924–31 (1939); *Brit. C. A.*, **1939 AII**, 529; *C. A.*, **34**, 368 (1940).
4. Blicke, F. F., and Burckhalter, J. H., *J. Am. Chem. Soc.*, **64**, 477–80 (1942).
5. Blicke, F. F., and Leonard, F., *J. Am. Chem. Soc.*, **68**, 1934–6 (1946).
6. Bremner, J. G. M., and Jones, D. G., (to Imperial Chemical Industries, Ltd.), British Patent 588,377 (1944).
7. Caesar, P. D., and Sachanen, A. N., *Ind. Eng. Chem.*, **40**, 922–8 (1948).
8. Caesar, P. D., and Sachanen, A. N., (to Socony-Vacuum Oil Co.), U. S. Patents 2,453,085 (1948); 2,453,086 (1948).
9. Caesar, P. D., and Sachanen, A. N., (to Socony-Vacuum Oil Co.), U. S. Patent 2,533,764 (1950).
10. Cairns, T. L., McKusick, B. C., and Weinmayr, V., *J. Am. Chem. Soc.*, **73**, 1270–3 (1951).
11. Calvin, M., Ball, R. H., and Aronoff, S., *J. Am. Chem. Soc.*, **65**, 2259 (1943).
12. Chadwick, A. F., (to E. I. du Pont de Nemours & Co., Inc.), U. S. Patent 2,492,414 (1949).
13. Chelintzev, V. V., and Maksorov, B. V., *J. Russ. Phys. Chem. Soc.*, **48**, 748–79 (1916); *C. A.*, **11**, 782; *Chem. Zentr.*, **1923**, I, 1505.
14. Colacicchi, U., *Gazz. chim. ital.*, **42**, I, 10–24 (1912); *Atti. accad. Lincei*, **20**, II, 312–7 (1911); *C. A.*, **6**, 230 (1912).
15. Corwin, A. H., and Brunings, K. J., *J. Am. Chem. Soc.*, **64**, 2106–15 (1942).
16. Engels, O., *Ber.*, **33**, 1087–1090 (1900).
17. Fischer, H., and Nenitzescu, C., *Ann.*, **443**, 113–129 (1925).
18. Formanek, E., *Ber.*, **38**, 944–5 (1905).
19. Hamer, Frances Mary, *J. Chem. Soc.*, **123**, 246–59 (1923).
20. Harvey, M. T., (to Harvel Research Corp.), U. S. Patents 2,343,972 (1944); 2,343,973 (1944).
21. Hartough, H. D., and Lukasiewicz, S. J., (to Socony-Vacuum Oil Co.), U. S. Patent 2,497,067 (1950).
22. *Ibid.*, U. S. Patent 2,559,567 (1951).
23. Hartough, H. D., *J. Am. Chem. Soc.*, **69**, 1355–8 (1947).
24. Hartough, H. D., Lukasiewicz, S. J., and Murray, E. H., Jr., *J. Am. Chem. Soc.*, **70**, 1146–9 (1948).
25. Hartough, H. D., Meisel, S. L., Koft, E., and Schick, J. W., *J. Am. Chem. Soc.*, **70**, 4013–17 (1948).
26. Hartough, H. D., and Meisel, S. L., *J. Am. Chem. Soc.*, **70**, 4018–19 (1948).
27. Herz, W., Dittmer, K., and Cristol, S. J., *J. Am. Chem. Soc.*, **69**, 1698–1700 (1947).
28. Hess, K., *Ber.*, **46**, 4104–4115 (1913).
29. Hess, K., Merck, F., and Uibrig, C., *Ber.*, **48**, 1886–1906 (1915).
30. Holdren, R. F., and Hixon, R. M., *J. Am. Chem. Soc.*, **68**, 1198–1200 (1946).
31. Holdren, R. F., *J. Am. Chem. Soc.*, **69**, 464–5 (1947).
32. Kappeler, H., (to Société anon. pour l'ind. chim. à Bâle), U. S. Patent 1,873,599 (1932). British Patent 345,891, June 12, 1929, *C. A.*, **26**, 2073 (1932).
33. Kermack, W. O., and Muir, W., *J. Chem. Soc.*, **1931**, 3089–96.
34. Koenigs, W., *Ber.*, **31**, 2364; **32**, 223–31, 3599–3613 (1899); **34**, 4322–26 (1901).
35. Koenigs, W., *Ber.*, **34**, 4322–6 (1902).
36. Koenigs, W., and Happe, G., *Ber.*, **35**, 1343–49 (1902).
37. Koenigs, W., and Bischkopf, E., *Ber.*, **34**, 4327–4330 (1901).
38. Koenigs, W., and Happe, G., *Ber.*, **36**, 2904–2912 (1903).
39. Koenigs, W., and Mengel, A., *Ber.*, **37**, 1322–37 (1904).

40. Ladenburg, A., *Ber.*, **22**, 2583–90 (1889).
41. *Ibid.*, **21**, 3099–3104 (1888).
42. Ladenburg, A., *Ann.*, **301**, 117–153 (1898).
43. Lauer, K., and Horio, M., *J. prakt. Chem.*, **143**, 305–24 (1935); *C. A.*, **29**, 7983; *Chem. Zentr.*, **1935**, II.
44. Lipp, A., and Richard, J., *Ber.*, **37**, 737–46 (1904).
45. Lipp, A., and Widnmann, E., *Ann.*, **409**, 79–147 (1915).
46. Lipp, A., and Zirngibl, E., *Ber.*, **39**, 1045–1054 (1906).
47. Löffler, K., and Grosse, A., *Ber.*, **40**, 1325–36 (1907).
48. Löffler, K., and Remmuler, H., *Ber.*, **43**, 2048–60 (1910).
49. Löffler, K., and Stietzel, F., *Ber.*, **42**, 124–32 (1909).
50. Lukasiewicz, S. J., and Murray, E. H., Jr., *J. Am. Chem. Soc.*, **68**, 1389–90 (1946).
51. Mackay, J. G., and Jones, D. G., (to Imperial Chemical Industries, Ltd.), U. S. Patent 2,450,108 (1948).
52. Mahan, J. E., (to Phillips Petroleum Co.), U. S. Patent 2,512,660 (1950).
53. *Ibid.*, U. S. Patent 2,534,285 (1950).
54. Meisenheimer, J., *Ann.*, **420**, 129–239 (1920).
55. Methner, T., *Ber.*, **27**, 2689–93 (1894).
56. Mooney, T. J., (to E. I. du Pont de Nemours & Co., Inc.), U. S. Patent 2,572,371 (1951).
57. Potokhin, N., *J. Russ. Phys. Chem. Soc.*, **59**, 761–817 (1927); *Chem. Zentr.*, **1928**, I, 2942.
58. Rosenhauer, E., Schmidt, A., and Unger, H., *Ber.*, **59B**, 2356–60 (1926).
59. Rothemund, P., *J. Am. Chem. Soc.*, **57**, 2010–11 (1935); **58**, 625–7 (1936).
60. *Ibid.*, **61**, 2912–15 (1939).
61. Shmidt, Ya. A., *Org. Chem. Ind. (U.S.S.R.)*, **5**, 339–43 (1938); *C. A.*, **33**, 266. Russ. Patent 45,602.
62. Shmidt, Ya. A., and Petrov, G. S., Russ. Patent 45,602 (1936); *C. A.*, **32**, 5535 (1938).
63. Taggart, M. S., and Richter, G. H., *J. Am. Chem. Soc.*, **56**, 1385–6 (1934).
64. Tschitschibabin, A. E., and Knunyants, I. L., *Ber.*, **62**, 3048–53 (1929).
65. Urbanski, T., *J. Chem. Soc.*, **1946**, 1104–5.
66. Walker, J. F., and Mooney, T. J., (to E. I. du Pont de Nemours & Co., Inc.), U. S. Patent 2,532,044 (1950).
67. Zerweck, W., Heinrich, E., and Pinten, P., (to General Aniline & Film Corp.), U. S. Patent 2,306,924 (1942).

CHAPTER 17

DETECTION AND ESTIMATION OF SMALL QUANTITIES OF FORMALDEHYDE

The chemical reactivity of formaldehyde, its many characteristic derivatives, and its pronounced reducing action in alkaline solutions provide a wide variety of methods for its detection. However, many of these methods are not specific for formaldehyde and some are not even specific for aldehydes in general. A complete survey of all these procedures would be pointless. It is our object to review in detail only those methods which are most generally applicable for detecting formaldehyde alone or in the presence of other compounds with which it is commonly associated. In this connection, special emphasis will be placed on methods which make it possible to obtain some estimate of the quantity of formaldehyde involved. Methods for the detection of free or combined formaldehyde in materials which have been subjected to formaldehyde treatment will also be included. Procedures for the quantitative determination of formaldehyde, etc. will be discussed in the next chapter. In dealing with the estimation of formaldehyde and other aldehydes in the primary products of hydrocarbon oxidation, the reader is also referred to Reynolds and Irwin's excellent study[42] of this subject.

Since paraformaldehyde and alpha-polyoxymethylene depolymerize readily to give formaldehyde solution on heating with water and react as formaldehyde with many chemical reagents, they are readily detected by methods applicable to formaldehyde itself. Ethers and esters of polyoxymethylene glycols depolymerize less readily and must be heated with dilute solutions of strong acids to convert them to formaldehyde solutions. Trioxane is particularly difficult to hydrolyze, requiring vigorous treatment with strongly acidic catalysts.

COLORIMETRIC PROCEDURES

Denigès' Method

Denigès' method for the detection and approximate determination of small quantities of formaldehyde has the advantages of being both extremely simple and highly sensitive. It is based on Denigès' discovery[15, 16] that Schiff's fuchsin-bisulfite reagent, developed in 1867 as a general reagent for aldehyde detection, gives a specific coloration with formaldehyde when employed in the presence of strong acids. The test is specific for formaldehyde in the presence of acetaldehyde and the higher aliphatic aldehydes, but does not differentiate acrolein or glyoxalic acid[45]. Recent improvements in the modified Schiff's reagent employed and in the testing technique have made it an accurate and useful tool for the chemical investigator.

Procedure. *Preparation of Modified Schiff's Reagent:* Dissolve 0.2 g of pure rosaniline hydrochloride* in 120 ml of hot water. Cool and dissolve 2 g of anhydrous sodium bisulfite followed by 2 ml of concentrated hydrochloric acid. Dilute solution to 200 ml with water and store in well-filled amber bottles. The reagent is ready for use after standing at room temperature for about one hour. According to Georgia and Morales[23], this modified reagent, which was developed by Elvove[18], may be kept for as long as two years without loss of sensitivity, if properly stored at temperatures of 15°C or below.

Test: Add 1 ml of concentrated sulfuric acid to 5 ml of the solution to be tested, cool to room temperature, and add 5 ml of the modified Schiff's reagent. The presence of formaldehyde is indicated by the development of a blue-violet color after 10 to 15 minutes. This solution has a definite absorption band in the orange. If much formaldehyde is present the color will be too dark for proper recognition and another test should be made with a more dilute sample.

To obtain quantitative results, solutions containing known concentrations of formaldehyde (*e.g.*, 0.0001, 0.0005, 0.001, 0.0015%, etc.) should be tested at the same time so that the colors may be compared in similar test tubes or Nessler tubes. The sensitivity of the test is better than 1 ppm. Use of a colorimeter is recommended for the highest speed and accuracy if many tests are to be made.

Acetaldehyde does not interfere with this method of formaldehyde detection even if present in enormously greater amounts. Thus, one part of formaldehyde is readily detected in the presence of 10,000 parts of acetaldehyde. Acetaldehyde sometimes gives an immediate purple-red coloration if present in border-line concentration in the neighborhood of 2 per cent, but this color fades rapidly in 10 minutes, whereas the characteristic color produced by formaldehyde itself does not fade in 6 to 12 hours[26]. According to Georgia and Morales[23], 2 per cent solutions of acetone, furfural, camphor, dextrose, sucrose, salicylic acid, citric acid, formic acid, and oils of almond, cinnamon, cloves, spearmint, and wintergreen give negative results when subjected to this test.

Veksler[50] reports that this procedure can be used for the spectrophotometric determination of formaldehyde. The dilute sample is placed in a graduate flask and diluted with water to 40 ml. After this, 5 ml each of concentrated hydrochloric acid and the modified Schiff's reagent are added and the flask is set in the dark at 20°C for 6 hours to allow the maximum color to develop. The color intensity is then measured for a wave length of 570 to 590 millimicrons. Solutions are stated to follow the Lambert-Beer law up to a concentration 0.002 mg/ml. The color intensity remains constant for 2 to 4 hours.

* Georgia and Morales[23] specify Kahlbaum's product.

Denigès' method has found special use in connection with the detection of methanol in alcoholic beverages[18, 23]. For this purpose, the reagent is added to a sample of beverage distillate which has been oxidized by treatment with acid permanganate and bleached with a test solution of oxalic acid.

Eegriwe's Method

Formaldehyde can be detected and estimated colorimetrically by the violet color which develops on heating with 1,8-dihydroxynaphthalene-3,6-disulfonic acid (chromotropic acid) or Eegriwe's reagent[17] in the presence of strong sulfuric acid. This reagent is highly specific when properly used and available information on its specificity is more complete than is the case for Denigès' use of the modified Schiff's reagent. An improved technique for employing this reagent, which has been recently reported by Boyd and Logan[10], is described below.

Procedure. *Preparation of Reagent:* Dissolve 0.9 g pure chromotropic acid crystals in 25 ml water and add 50 mg stannous chloride. Shake well in a stoppered flask and centrifuge until clear. This solution is approximately 0.1 molal in chromotropic acid.

Purified acid for use in making this reagent is obtained by dissolving 25 g of impure, colored chromotropic acid in 100 ml of water in a steam bath. To this solution 2 g of lead carbonate is added. When the carbonate has dissolved, hydrogen sulfide is passed into the mixture until all the lead sulfide has precipitated. If the supernatant liquid is not pale yellow when the sulfide settles more carbonate and hydrogen sulfide should be employed. The hot solution should now be centrifuged or filtered without exposure to air and then cooled to 4°C. Crystals obtained from the cold solution should be filtered off, washed with alcohol and ether, and vacuum-dried. The purified acid should be nearly colorless.

Test: Add 5 ml of solution to be tested to a 1 x 8-in test tube graduated at 50 ml. Add 0.5 ml of the 0.1 molal test solution and enough water to make up to 17 ml. Cool solution in an ice bath and add 10 ml concentrated sulfuric acid with agitation in the course of 40 to 45 seconds. When the resulting solution has reached the temperature of the ice bath, add sufficient concentrated sulfuric acid to make a total of 50 ml, pouring acid down the center of the tube. This causes the temperature to rise to approximately 80°C. Heat the mixture in a boiling water bath for ten minutes and cool to room temperature.

For quantitative estimation, similar tests should be made with a standard formaldehyde solution and compared with sample in a colorimeter equipped with paraffin-protected or all-glass cups. A solution containing 0.015 mg is satisfactory for comparison. Best results are obtained with solutions containing 0.015 to 0.1 mg. The sensitivity of the test is approximately 3 ppm.

According to Eegriwe[17], the chromotropic acid reagent gives negative results with acetaldehyde, propionaldehyde, butyraldehyde, isobutyraldehyde, isovaleraldehyde, crotonaldehyde, chloral hydrate, glyoxal, benzaldehyde, salicylaldehyde, phthalaldehyde, vanillin, and many other aromatic aldehydes. Glyceraldehyde gives a yellow color with a green fluorescence when present in quantity, and furfural gives a yellow brown color. Acetone, glucose, glycerol, formic acid, glycolic acid, gallic acid, and levulinic acid also give negative results. One part of formaldehyde can be determined in the presence of 94 parts fructose, 380 parts furfural, or 10,100 parts glucose. Even acrolein gives a negative test with this reagent.

Bricker and Johnson[11] have adapted this procedure for use with the spectrophotometer and report that as little as one microgram of formaldehyde in one ml of solution can be determined rapidly with an accuracy of better than 5 per cent. Their method is as follows:

Procedure. *Preparation of Reagent:* Dissolve 2.5 g of dry powdered chromotropic acid in 25 ml of water and filter if necessary to obtain a clear solution. This reagent changes color gradually on storage, but is satisfactory for two weeks. Because of this change, a reagent blank should be run each day and all readings for the day should be made against this blank.

Test: Place sample in a glass-stoppered, approximately 18- by 150-mm test tube. Samples should not contain over 100 micrograms formaldehyde, and if liquid, should have a volume of 0.4 to 0.9 ml. (In the case of dry samples, add 0.5 ml water after placing in test tube.) Add 0.5 ml of chromotropic acid reagent and follow by gradual addition of 5 ml concentrated sulfuric acid with continuous shaking. Stopper test tube and place in boiling water for 30 minutes. Cool and dilute with water to a volume of under 50 ml and cool again. Place in a 50-ml volumetric flask and adjust to volume at room temperature. Read transmission, T, against reagent blank with a spectrophotometer for light with a wave length of 570 millimicrons. Calculate optical density ($\log 1/T$) and determine formaldehyde equivalent for this figure from a calibration curve prepared by plotting optical densities, obtained with known formaldehyde samples, against the corresponding weights of formaldehyde. Determine formaldehyde concentration by dividing the formaldehyde reading by the weight of the original sample and multiplying by 100. (Calibration curves are linear up to an optical density of 0.750 but beyond this value absorption starts to deviate slightly from Beer's law.)

Methanol and ethanol do not interfere with this procedure, but higher alcohols hinder color formation. Acetaldehyde, acrolein and beta-hydroxypropionaldehyde interfere, but formaldehyde can be determined in the presence of 100 times as much acetaldehyde if 300 mg of reagent is employed in place of the customary 50 mg. Benzaldehyde does not interfere. Acetone causes the color to fade when the solution is diluted with water, but this can be avoided by diluting to 50 ml with 18N sulfuric acid. Methyl ethyl ketone and diacetone alcohol interfere.

Miscellaneous Color Tests

A wide range of color tests of varying degrees of sensitivity and specificity have been reported for the detection of formaldehyde in addition to the two authenticated methods already described. These tests involve the use of phenols, aromatic amines, alkaloids, and other miscellaneous reagents.

Illustrative of tests involving phenolic compounds is Lebbin's resorcinol test[34]. A few ml of the liquid to be tested is boiled with an equal volume of 40 to 50 per cent sodium hydroxide containing 0.05 g resorcinol. A yellow coloration changing to red is obtained if formaldehyde is present. A sensitivity of 0.1 ppm is claimed. Jorissen[33] obtained a scmewhat similar reaction with phloroglucinol.

Pittarelli[40] reports a characteristic test involving the addition of 5 to 6 drops of a saturated phenylhydrazine hydrochloride solution to 25 to 30 ml of the solution to be tested. The solution is then heated to 100°C and 5 to 6 drops of 1 per cent metol (methyl-*p*-aminophenol sulfate) and 3 drops of 25 per cent sodium hydroxide are added. A scarlet color is produced when formaldehyde is present. This solution is stated to turn blue when a saturated solution of magnesium sulfate is added. A sensitivity of 1 ppm is claimed.

Schryver's test[48] involves the addition of 1 ml of a freshly prepared 5 per cent potassium ferricyanide solution, 2 ml of a freshly prepared 1 per cent phenylhydrazine hydrochloride, and 5 ml of concentrated hydrochloric acid to 10 ml of the solution to be tested. If formaldehyde is present a fuchsin-red color develops. Chlorophyll obtained from green grass is reported to give a distinct test by this method when it is exposed to sunlight in the presence of moist carbon dioxide.

Alkaloids have long been used for formaldehyde detection and are reported to have a high degree of sensitivity. Formaldehyde solutions give a purple coloration with morphine hydrochloride in the presence of sulfuric acid[8, 28, 33, 39]. A convenient procedure is to add a few drops of $\frac{1}{2}$ per cent morphine hydrochloride to a test tube containing the dilute sample, then form an underlayer of sulfuric acid. A characteristic violet blue ring forms when the sample contains formaldehyde[28]. Codeine[41] and apomorphine[24, 54] give similar reactions.

Fulton[22] found that the addition of an oxidizing agent, such as a 10 per cent solution of ferric sulfate or 5 drops of concentrated nitric acid in 50 ml water, increases the sensitivity of the alkaloid tests for formaldehyde. Thus, apomorphine, codeine, and pseudomorphine become sensitive to 2 ppm, whereas papaverine, claimed to be the most nearly specific, detects 10 ppm formaldehyde.

Voisenet[51] reports that an extremely sensitive test capable of detecting 0.1 ppm formaldhyde can be produced with albumin and pctassium nitrite. One drop of the unknown solution is added to 2 to 3 ml of a solution containing 0.05 g albumin per ml, followed by 6 to 9 ml of a solution of 0.5 ml of 3.6 per cent potassium nitrite in 1 liter of concentrated hydrochloric acid. A rose color forms immediately and changes to violet on standing for 5 minutes when formaldehyde is present.

Tollen's ammoniacal silver nitrate, although not specific, is a sensitive formaldehyde reagent possessing a sensitivity of about 2 ppm under optimum conditions.

According to Aloy and Valdiguie[4], it is possible to identify formaldehyde in the presence of methyl alcohol, ethyl alcohol, acetic acid, acetone, sucrose, dextrose, or levulose by adding one or two drops of the mixture to a reagent composed of 2 drops of very dilute ferric sulfate, 1 per cent alcoholic codeine and 3 to 4 ml pure sulfuric acid. Formaldehyde causes an intense blue coloration within 1 to 2 minutes[4].

POLAROGRAPHIC METHOD

Jahoda[31, 32] reported in 1935 that formaldehyde could be determined in small concentrations with a fair degree of accuracy by measuring the current-voltage curves obtained with a dropping mercury electrode. The height of the current wave in the polarogram is directly proportional to the formaldehyde concentration. Since acetaldehyde and higher aldehydes in the aliphatic series are reduced at higher voltages than formaldehyde, they do not interfere in the detection process. Methanol, ethanol, acetone, and benzaldehyde are also known to be without effect.

Research studies of this procedure by Winkel and Proske[53], Barnes and Speicher[7], and Boyd and Bambach[9] are in essential agreement with Jahoda's findings and have resulted in the development of improved techniques. The method appears to be one of growing importance. Its use is naturally restricted to solutions which are free from materials whose half-wave potentials are in the region -1.4 to -1.8 volts.

Procedure: According to Boyd and Bambach[9], best results are obtained when 3 ml of the solution to be tested are mixed with 1 ml of a solution which is $0.4N$ in potassium chloride and $0.2N$ in potassium hydroxide. Nitrogen is bubbled through the resulting solution for 15 minutes, after which a polarogram is taken from -1.4 to -1.8 volts. The Leeds and Northrup Electro-chemograph is recommended. The half-wave of the formaldehyde step is reported[9] to occur at -1.63 volts (normal calomel electrode). The solution should be kept in a constant-temperature bath controlled to $0.1°C$ during the determination, since the wave height is stated to change by about 6.5 per cent for each $1°C$ change in temperature. Boyd and Bambach state that 3 ppm of formaldehyde can be determined. Jahoda[32] claimed that 0.07 ppm may be estimated with an accuracy of 10 per cent.

In general the method appears applicable to any formaldehyde solution which can be isolated by distillation or is otherwise free of interfering substances.

Boyd and Bambach state that acetaldehyde is reduced at a potential of -1.8 volts (normal calomel electrode). This aldehyde and formaldehyde can be estimated simultaneously if the potassium electrolyte is replaced by lithium hydroxide. The potassium wave interferes with acetaldehyde detection. A polarogram obtained by the above investigators showing both formaldehyde and acetaldehyde is shown in Figure 29. Whithack and Moshier[52] have also employed the polarographic method using a lithium electrolyte to determine formaldehyde in the presence of acrolein, acetaldehyde and propionaldehyde. (See also Warshowsky and Elving[51a].)

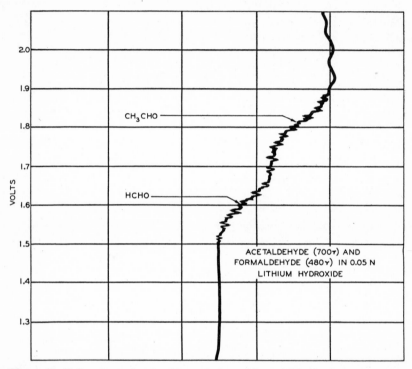

Figure 29. Polarogram showing the presence of formaldehyde and acetaldehyde in dilute aqueous solution. (From Boyd, M. J., and Bambach, K., *Ind. Eng. Chem., Anal. Ed.*, **15**, 314 (1943).)

FORMALDEHYDE DERIVATIVES

For purposes of identification, it is often desirable to prepare characteristic formaldehyde derivatives whose identity can be certified by melting points, analyses, etc. The following reagents are of value for this purpose in cases involving small quantities of formaldehyde.

Methone or Dimedon (dimethyldihydroresorcinol, 5,5-dimethylcyclohexanedione). A dinuclear methylene derivative of this agent, methylene bis-methone, is precipitated when a few drops of a 5 to 10 per cent solution of the reagent in alcohol is added to the solution to be tested, which has been made faintly acid with acetic acid. The precipitate obtained with formaldehyde melts at 189°C after recrystallization from hot alcohol. The presence of salt in the reaction mixture is stated to increase sensitivity, which is claimed to be approximately 4 ppm. Methylene bis-methone is soluble in dilute alkali and does not become alkali-insoluble after heating in a boiling water bath with a small amount of glacial acetic acid for 6 to 7 hours. Methone derivatives of other aliphatic aldehydes become alkali-insoluble when treated in this way (pages 391–393).

p-**Nitrophenylhydrazine Hydrochloride.** This agent precipitates a *p*-nitrophenylhydrazone from mildly acid formaldehyde solutions. On purification from alcohol, the product melts at 181 to 182°C[6, 55].

2,4-Dinitrophenylhydrazine Hydrochloride. This agent is used in the same manner as the one mentioned above. Formaldehyde-2,4-dinitrophenylhydrazone melts at 166 to 167°C[12, 13].

Beta-Naphthol. Beta-naphthol reacts with formaldehyde in acid solution to give a precipitate of methylene dinaphthol[20]. Application of this test involves the treatment of a few drops of formaldehyde solution with dilute alcohol (33 per cent), beta-naphthol, and a little hydrochloric acid. After gently boiling the mixture to precipitate small white needles and purifying these, a melting point determination is made on the product, which turns brown at 180°C and melts with decomposition to a brownish-red liquid at 189 to 192°C[29].

Detection of Higher Aldehydes in the Presence of Formaldehyde

Although, as we have previously pointed out, formaldehyde may be readily detected in the presence of acetaldehyde, there is comparatively little available information concerning adequate chemical methods for detecting acetaldehyde in the presence of formaldehyde. A test of this sort is often desirable in connection with problems of formaldehyde chemistry.

The Leys reagent affords an extremely sensitive test for acetaldehyde in the presence of large quantities of formaldehyde. This reagent is prepared by adding one gram of finely powdered mercuric oxide to 100 ml of 5 per cent aqueous sodium sulfite (Na_2SO_3). The mixture is warmed until substantially all the oxide dissolves and is then cooled and filtered. Our findings indicate that when 5 ml of formaldehyde solution (*e.g.*, 37 per cent formaldehyde), rendered alkaline by addition of a few drops of normal sodium hydroxide, are treated with 1 ml of the Leys reagent, a white precipitate will be obtained in less than 30 minutes, if as little as 0.05 to 0.1 per cent aldehyde is present. Small quantities of formic acid (1 per cent or less) do not interfere with this test. Solutions containing 1 per cent propionaldehyde, isobutyraldehyde, furfuraldehyde, and acetone give negative results. According to Leys[36], aldehydes containing the group —CH_2CHO give precipitates when an equal volume of 10 per cent potassium hydroxide is added to the reagent and a few drops of the solution to be tested are then added. Under these conditions, a dilute propionaldehyde solution gives a white precipitate.

Rosenthaler[46] states that diazobenzenesulfonic acid gives a red color with aqueous acetaldehyde even when it is present in small quantity, and that the presence of several drops of 37 per cent formaldehyde does not interfere with this test. Larger quantities of formaldehyde are reported to cause some coloration but this color is said to fade completely in at least two hours, whereas the color caused by acetaldehyde is stable.

According to Schibstead[47], higher aldehydes can be determined by making use of their solubility in petroleum ether and employing a specially modified Schiff's reagent for their estimation. (See also the silver oxide-chromatographic method of Bailey and Knox[5a].)

Detection and Estimation of Small Quantities of Formaldehyde in Air

Formaldehyde in air may be determined by the analysis of an aqueous scrubber solution obtained by bubbling air through a suitable water scrubber. The American Standards Association[5] recommends that the air should be sucked through water using a fritted glass bubbler with a plate diameter of 6 cm at a rate of 2 to 4 liters per minute. They also recommend the use of an impinger containing a 1.25 per cent solution of potassium hydroxide. However, in the writer's opinion, the latter device is questionable since formaldehyde may be lost by the Cannizzaro reaction.

The concentration of formaldehyde in the scrubber liquid may then be determined by Denigès' method, or any other sensitive and relatively specific procedure for formaldehyde detection.

Goldman and Yagoda[25] claim that highly accurate results can be obtained by a procedure employing a scrubber or impinger containing 1 per cent sodium bisulfite which reacts with the aldehyde forming the nonvolatile bisulfite compound. Their results indicate that a single impinger containing 100 ml of this solution picks up 95 per cent of the formaldehyde in air containing 7 or more ppm formaldehyde at scrubbing rates as high as 28 liters per minute, whereas water absorbs only about 70 to 75 per cent of the formaldehyde under the same conditions. According to these investigators, formaldehyde is strongly absorbed by rubber tubing, and exposure to rubber of the gas to be analyzed should be avoided.

Goldman and Yagoda[25] determine formaldehyde in the bisulfite scrubber liquid by destroying unreacted bisulfite with iodine, making the resulting solution mildly alkaline and titrating the bisulfite compound with iodine under these conditions. Their procedure is carried out as indicated below:

Procedure: Pass air to be tested through a midget impinger containing 10 ml of 1 per cent sodium bisulfite at a rate of 1 to 3 liters per minute, as described by Jacobs[30]. When the scrubbing process, is complete, transfer solution and washings to a 250- or 300-ml Erlenmeyer flask. Use a large impinger containing 100 ml of bisulfite[30] when scrubbing at a rate of 28 liters per minute and employ a 10-ml aliquot of the scrubber solution for analysis.

Titrate solution to be analyzed with approximately $0.1N$ iodine to a dark blue end point with starch as an indicator; destroy excess iodine with $0.05N$ thiosulfate and then titrate to a faint blue with standard $0.01N$ iodine. If acetone is present in the air tested add 2 ml of 5 per cent sodium bicarbonate and destroy sulfite liberated from combination with acetone by addition $0.01N$ iodine.

Add 25 ml of a sodium carbonate buffer prepared by dissolving 80 g sodium carbonate (Na_3CO_3) in 500 ml water and adding 20 ml glacial acetic acid.

When this addition is completed, titrate with $0.01N$ iodine to a faint blue end point. When the titer is less than 1 ml, carry out a blank analysis on 10 ml 1 per cent bisulfite and subtract result (normally less than 0.1 ml) from sample titer.

Results are calculated on the basis that 1.0 ml of $0.01N$ iodine is equivalent to 0.15 mg formaldehyde.

By the above procedure it is claimed that 1 mg formaldehyde in 10 ml of test solution can be measured with an accuracy of 1 per cent. With smaller quantities the error increases rapidly so that sufficient air should always be scrubbed to absorb at least 1 mg CH_2O per 10 ml of scrubber solution. The accuracy of the method is not impaired by the presence of methanol, acetic acid, and bromine in the air to be tested. Errors due to acetone may be eliminated as indicated in the procedure. Formic acid should be without effect, since it does not interfere with the iodimetric procedure for formaldehyde analysis (pages 385–386).

The bisulfite method cannot be employed in the manner described above in the presence of other aldehydes. However, it is possible that the bisulfite compounds obtained in such instances could be hydrolyzed and the formaldehyde determined by a more specific method.

Detection of Formaldehyde in Foodstuffs

Although the use of formaldehyde as a preservative for foods is prohibited, the methods developed for its identification in various foods and drinks are of sufficient versatility to warrant interest in their own right. There are, for example, a variety of special tests which have been developed for the detection of formaldehyde in milk. Of these the best known is Hehner's milk test[27, cf. 19, 35, 43, 44] in which 90 to 94 per cent sulfuric acid containing a trace of iron is carefully poured into a test tube containing milk so that an underlayer of acid is formed. The presence of formaldehyde is indicated by the appearance of a violet ring at the junction of the two liquids. Difficulties encountered in executing this test have been described by McLachlan[38] to impurities present in commercial sulfuric acid. He recommends that pure acid be employed with the addition of a ferric salt, *e.g.*, ferric chloride, and that a control test be established for the acid used.

It has also been recommended that a few crystals of potassium sulfate be added to the milk before underlying it with acid[2]. When no formaldehyde is present, the test shows a greenish tint; occasionally a brown ring will also form below the junction of the two liquids[3]. Fulton[21] reports that it is possible to attain a sensitivity of 1 ppm by using bromine as an oxidizing agent and carrying out the reaction in dilute sulfuric acid.

A similar milk test is that of Leach[49], in which milk (10 ml) is heated to 80 to 90°C with 10 ml hydrochloric acid (sp. gr. 1.2) containing about 1 ml of a 10 per cent ferric chloride solution in 500 ml of acid. The curd formed is broken by agitation; a violet-colored solution indicates the presence of formaldehyde. Lyons[37] found that when phenylhydrazine and potassium ferricyanide, or some other oxidizing agent such as ammonium persulfate, are used a red coloration develops quite promptly.

The characteristic reactions of formaldehyde with phenolic compounds and amines have also been adapted to the detection of the aldehyde in foodstuffs. In such instances the material is generally steam-distilled and a qualitative analysis is then carried out on the distillate obtained. Cohn[14] shakes 2 ml of the distillate with an equal volume of 0.1 per cent resorcinol solution, then carefully adds 2 ml of concentrated sulfuric acid to produce a dense precipitate at the zone of contact between acid and solution with a dark violet-red zone immediately below. Formic, oxalic, and tartaric acids are claimed not to interfere with this color reaction, which is sensitive to 0.00005 mg formaldehyde. Gallic acid reacts with like sensitivity to form an emerald green band[1]. The test devised by Pittarelli (page 370) is reported to be of value for the detection of formaldehyde in wine, milk, spirits, beer, and solid foods. In general, the test is best applied to 25 to 30 ml of distillate. In the case of solid foods, an aqueous suspension in water or dilute sulfuric acid may be steam-distilled to obtain a suitable solution for analysis.

Detection of Formaldehyde in Products Which Have Been Subjected to Formaldehyde Treatment

Formaldehyde that is combined with hydroxy compounds in the form of formals or methylene ethers can be liberated by heating with dilute solutions of strong acids. Methylene derivatives of amino and amido compounds may also be hydrolyzed in the same manner. As a result, formaldehyde solutions suitable for analytical tests can usually be readily obtained from such varied products as textiles and paper which have been treated with formaldehyde and urea-formaldehyde condensates, polyvinyl formals, protein products which have been hardened or insolubilized by formaldehyde treatment, formaldehyde-tanned leather, etc. In general, the material to be examined is heated with a solution of a strong mineral acid and then subjected to distillation or steam distillation. Sulfuric acid is usually preferable because of its nonvolatile nature and a 10 to 20 per cent solution is generally satisfactory. In carrying out such distillations the condenser outlet should dip below the level of a little distilled water which is first placed in the receiver so that any gaseous formaldehyde will be dissolved. A solution obtained in this way will naturally contain free formaldehyde originally present in the sample material. At low concentrations (less than approximately 10 per cent), formaldehyde can be quantitatively distilled from aqueous solutions usually coming over in the first 10 to 60 per cent of the distillate.

References

1. "Allen's Commercial Organic Analysis," 5th Ed., Vol. I., p. 302, Philadelphia, P. Blakiston's Son & Co. (1923).
2. *Ibid.*, p. 328.
3. *Ibid.*, Vol. IX, pp. 92–3.
4. Aloy, J., and Valdiguie, A., *J. Pharm. chim.* (8), **4**, 390–3 (1926); *C. A.*, **21**, 2358–9 (1927).
5. American Standards Association, "Allowable Concentration of Formaldehyde" (pamphlet), New York, American Standards Association (1944).
5a. Bailey, H. C., and Knox, J. H., *J. Chem. Soc.*, **1951**, 2741–2.
6. Bamberger, E., *Ber.*, **32**, 1807 (1899).
7. Barnes, E. C., and Speicher, H. W., *J. Ind. Hyg.*, **24**, 10 (1942).
8. Bonnet, F., Jr., *J. Am. Chem. Soc.*, **27**, 601 (1905).
9. Boyd, M. J., and Bambach, K., *Ind. Eng. Chem., Anal. Ed.*, **15**, 314–5 (1943).
10. Boyd, M. J., and Logan, M. A., *J. Biol. Chem.*, **146**, 279 (1942).
11. Bricker, C. E., and Johnson, H. R., *Ind. Eng. Chem.*, **17**, 400–402 (1945).
12. Bryant, W. M. D., *J. Am. Chem. Soc.*, **54**, 3760 (1932).
13. Campbell, N. R., *Analyst*, **61**, 392 (1936).
14. Cohn, R., *Chem. Ztg.*, **45**, 997–8 (1921); *C. A.*, **16**, 222 (1922).
15. Denigès, G., *J. pharm. chim.* (4), **6**, 193 (1896).
16. Denigès, G., *Compt. rend.*, **150**, 529 (1910).
17. Eegriwe, E., *Z. anal. Chem.*, **110**, 22 (1937).
18. Elvove, E., *Ind. Eng. Chem.*, **9**, 295 (1917).
19. Fillinger, F. v., *Z. Untersuch. Nahr. u. Genussm.*, **16**, 232–4 (1908).
20. Fosse, R., Graeve, P. de, and Thomas, P.-E., *Compt rend.*, **200**, 1450–4 (1935); *C. A.*, **29**, 7869–70 (1935).
21. Fulton, C. C., *Ind. Eng. Chem., Anal. Ed.*, **3**, 199–200 (1931); *C. A.*, **25**, 2665 (1931).
22. *Ibid.*, pp. 200–1.
23. Georgia, F. R., and Morales, R., *Ind. Eng. Chem.*, **18**, 305 (1926).
24. Gettler, A. O., *J. Biol. Chem.*, **42**, 311 (1920).
25. Goldman, F. H., and Yagoda, H., *Ind. Eng. Chem., Anal. Ed.*, **15**, 376 (1943).
26. Heerman, P., *Textilber.*, **3**, 101–21 (1922); *C. A.*, **16**, 2406 (1922).
27. Hehner, O., *Analyst*, **21**, 94–7 (1896); *Chem. Zentr.*, **1896**, I, 1145–6.
28. Hinkel, L. E., *Analyst*, **33**, 417–9 (1908); *Chem. Zentr.*, **1909**, I, 46.
29. Huntress, E. H., and Mulliken, S. P., "Identification of Pure Organic Compounds-Order I," p. 50, New York, John Wiley & Sons (1941).
30. Jacobs, M. B., "The Analytical Chemistry of Industrial Poisons, Hazards and Solvents," New York, Interscience Publishers, Inc. (1941).

31. Jahoda, F. G., *Casopis Ceskoslov. Lekarnictva,* **14,** 225–34 (1935); *Chem. Zentr.,* **1935,** I, 1091.
32. Jahoda, F. G., Collection Czechoslov. Chem. Commun., **7,** 415 (1935).
33. Jorissen, A., *J. pharm. chim.,* **6,** 167 (1897).
34. Lebbin, *Pharm. Ztg.,* **42,** 18–9 (1897); *Chem. Zentr.,* **1897,** I, 270.
35. Leffemann, H., *Am. J. Pharm.,* **96,** 366–7 (1924); *C. A.,* **18,** 2056 (1924).
36. Leys, A., *J. pharm. chim. (6),* **22,** 107–112 (1897).
37. Lyons, A. B., *J. Am. Pharm. Assoc.,* **13,** 7–9 (1924); *C. A.,* **18,** 1260 (1924).
38. McLachlan, T., *Analyst,* **60,** 752 (1935).
39. Mayer, J. L., *J. Am. Pharm. Assoc.,* **12,** 698–700 (1923); *C. A.,* **18,** 32 (1924).
40. Pittarelli, E., *Arch. farm. sper.,* **30,** 148–160 (1920).
41. Pollacci, G., *Boll. chim.-farm.,* **38,** 601–3 (1899); *Chem. Zentr.,* **1899,** II, 881.
42. Reynolds, J. G., and Irwin, M., *Chem. & Ind.,* **1948,** 419–24.
43. Richmond, H. D., and Boseley, L. K., *Analyst,* **20,** 154–6 (1895); *Chem. Zentr.,* **1895,** II, 463.
44. *Ibid.,* **21,** 92 (1896); *Chem. Zentr.,* **1896,** I, 1145.
45. Rosenthaler, L., "Der Nachweis organischer Verbindungen: Die chemische Analyze," 2nd Ed., Vol. XIX/XX, p. 128, Stuttgart, F. Enke (1923).
46. *Ibid.,* p. 134.
47. Schibstead, H., *Ind. Eng. Chem. (Anal. Ed.),* **4,** 204 (1932).
48. Schryver, S. B., *Proc. Roy. Soc. London (B),* **82,** 226 (1910).
49. Scott, W. W., "Standard Methods of Chemical Analysis," 4th Ed., Vol. II, p. 1754, New York, D. van Nostrand Co. (1925).
50. Veksler, R. I., *Zhur. Anal. Khim.,* **1,** 301–10 (1946); *C. A.,* **43,** 4976 (1949).
51. Voisenet, E., *Bull. soc. chim. (3),* **33,** 1198–1214 (1905); *Chem. Zentr.,* **1906,** I, 90.
51a. Warshowsky, B., and Elving, P. J., *Ind. Eng. Chem., Anal. Ed.,* **18,** 253–4 (1946).
52. Whithack, G. C., and Moshier, R. W., *Ind. Eng. Chem. (Anal. Ed.),* **16,** 496 (1944).
53. Winkel, A., and Proske, G., *Ber.,* **69,** 693 (1936).
54. Wolff, H., *Chem. Ztg.,* **43,** 555 (1919).
55. Zerner, E., *Monatsh.,* **34,** 957–61 (1913).

CHAPTER 18

QUANTITATIVE ANALYSIS OF FORMALDEHYDE SOLUTIONS AND POLYMERS

Considerable research has been devoted to developing satisfactory methods for the quantitative determination of formaldehyde, and numerous chemical and physical procedures involving volumetric, gravimetric, colorimetric, and other techniques have been developed. These methods possess various degrees of accuracy and selectivity. The optimum procedure for the solution of a given analytical problem is best determined by a study of the available technique. Particular attention must be given to the nature of the impurities in whose presence formaldehyde must be determined, the concentrations of formaldehyde involved, and the accuracy required. Generalizations concerning the relative utility of the various methods will be made in our discussion to serve as a rough guide to the analyst.

In the previous chapter, we have reviewed those methods which are best suited for the detection and estimation of formaldehyde in compositions in which it is a minor constituent. Such methods require a maximum of selectivity and sensitivity. We shall now consider methods of quantitative analysis which are best adapted to the accurate determination of formaldehyde in commercial solutions, polymers and other compositions in which formaldehyde is present in quantity and in which impurities such as ketones, higher aldehydes, etc., which interfere with the determination of formaldehyde, are either completely absent cr present in minor amounts. These methods possess a good degree of accuracy but do not require the extreme sensitivity demanded by the procedures previously discussed.

Physical Methods for Determining Formaldehyde

The formaldehyde content of pure aqueous formaldehyde solutions or solutions containing only very small percentages of impurities may be readily determined by specific gravity or refractivity (pages 380, 396). However, since methanol is normally present in most commercial formaldehyde solutions, neither of these measurements is sufficient when taken by itself to fix the formaldehyde content. By means of a ternary diagram constructed by Natta and Bacceraredda[36], (Figure 30), both formaldehyde and methanol may be estimated from a solution of the two when both the density (18°/4°) and refractive index ($n_D^{18°}$) are known.

Formaldehyde may also be determined by means of the polarograph. This procedure is discussed in detail in connection with the detection and estimation of extremely small quantities of formaldehyde (pages 371–372).

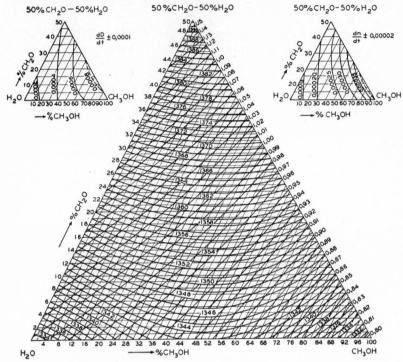

Figure 30. Ternary diagram showing densities and refractive indexes for the system formaldehyde-methanol-water. From Natta, G., and Baccaredda, M., *Giorn. chim. ind. applicata*, **15**, 273–81 (1933).

Chemical Methods

Because of their simplicity and accuracy, chemical procedures, particularly those of the volumetric type, are widely employed for formaldehyde analysis. The following methods, and in some cases special variants of these methods, have been selected on the basis of general utility and reliability for detailed discussion:

(1) Sodium sulfite method
(2) Alkaline peroxide method
(3) Iodimetric method
(4) Ammonium chloride method
(5) Mercurimetric method
(6) Potassium cyanide method
(7) Hydroxylamine hydrochloride method
(8) Methone or dimedon method

While many excellent papers criticizing and discussing formaldehyde analysis have been published from time to time, the exhaustive and critical study by Büchi[10] and the more recent assessment of currently available methods by Reynolds and Irwin[43] deserve particular attention. However, the latter is chiefly concerned with the analysis of impure formaldehyde solutions obtained in connection with studies of the mechanism of the combustion of hydrocarbons. Büchi's work deals with the analysis of commercial solutions and is probably the most important contribution to this subject. Following a detailed review of all the known methods of formaldehyde determination, including physical methods, Büchi concluded that, when judged on the basis of accuracy, simplicity, reliability, and economy with respect to time, apparatus, etc., only the first five of the chemical methods listed above were worthy of serious consideration. After an exhaustive laboratory study of these methods, he concluded that the sodium sulfite method was the most satisfactory for the analysis of commercial formaldehyde solutions.

The specific utility of the various methods which we have listed will be further discussed in the separate sections of this chapter. Those not chosen by Büchi have been included because of their utility in the presence of impurities which do not normally occur in commercial formaldehyde, but which are sometimes encountered in connection with its utilization. In general, these methods are also applicable to paraformaldehyde and alpha-polyoxymethylene, making it possible to determine the formaldehyde content of these polymers. This is definitely the case for methods (1), (2) and (3).

It should be noted that with the exception of method (8), none of these procedures are specific to formaldehyde. Methods (1), (4), (6), and (7) depend on characteristic aldehyde and ketone reactions. However, they are most sensitive to formaldehyde because of its high reactivity and least sensitive to ketones and the higher aldehydes. Methods (2), (3), and (5) are oxidation procedures and measure any chemical which can be oxidized under the conditions of the test. As to the effect of specific chemical impurities, the following table may serve as a rough index of the procedures to be considered when impurities of the type indicated are present.

Type of Impurity	Preferred Methods
Inorganic acids	(1) (2) (Acid must first be neutralized)
Inorganic bases (excluding ammonia)	(1) (2) (Base must first be neutralized)
Neutral salts	(1) (2)
Alcohols	(1) (2) (5)
Other aldehydes	(6-Schulek modification) (8)
Ketones	(1) (2) (5) (6) (8)
Simple aliphatic and aromatic acids	(1) (2) (Acid must first be neutralized)
Amines	(1) (5)
Amides	(1) (6)
Phenols	(2) (7)
Hydrocarbons	(1) (4)

In selecting a method, the more detailed information on the method indicated should be studied, since the relative amount of impurity that can be tolerated varies considerably for the different procedures. If the impurity has reacted irreversibly with the formaldehyde, only the free formaldehyde

(Courtesy, Heyden Chemical Corporation, Garfield, New Jersey)

Figure 31. Finished formaldehyde solution of standard concentration is weighed in storage tanks on scales prior to loading into tank cars or trucks.

can be determined. If a reversible reaction has taken place, part or all of the combined formaldehyde may be measured.

(1) **Sodium Sulfite Method.** Because of its accuracy, simplicity and rapidity, the sodium sulfite method, first developed by Lemme[30] and later improved by Seyewetz and Gibello[50] and Sadtler[48], is probably the best procedure for the determination of formaldehyde. The method is based on the quantitative liberation of sodium hydroxide when formaldehyde reacts

with sodium sulfite to form the formaldehyde-bisulfite addition product:

$$CH_2O \text{ (aq)} + Na_2SO_3 + H_2O \rightarrow NaOH + CH_2(NaSO_3)OH$$

Procedure: Fifty ml of a molar solution of pure sodium sulfite (prepared by dissolving 252 g of hydrated sodium sulfite or 126 g of the anhydrous salt in sufficient distilled water to make one liter of solution) and three drops of thymolphthalein indicator solution (0.1 per cent in alcohol) are placed in a 500-ml Erlenmeyer flask and carefully neutralized by titration with normal sulfuric or hydrochloric acid until the blue color of the indicator has disappeared (one or two drops is sufficient). An accurately measured and substantially neutral formaldehyde sample is then added to the sodium sulfite and the resulting mixture titrated slowly with the standard acid to complete decoloration. One ml of normal acid is equivalent to 0.03003 g formaldehyde and the per cent formaldehyde in the sample is determined by the following equation:

$$\% \text{ Formaldehyde} = \frac{\text{Acid titer} \times \text{Normality of acid} \times 3.003}{\text{Weight of sample}}$$

When analyzing 37 per cent commercial formaldehyde a sample of approximately 3 g may be conveniently weighed in a thimble beaker and dropped into the sodium sulfite. Büchi[10] recommends that 6 g of commercial formaldehyde be accurately weighed and diluted to exactly 100 ml in a volumetric flask, from which a 25-ml aliquot is then added to the sodium sulfite solution. The quantity of formaldehyde in an unknown solution may be readily determined by titration of an appropriate aliquot. The amount of sample may be varied depending on the formaldehyde content of the solution to be analyzed, but should be so adjusted that the solution titrated always contains excess sodium sulfite. A 50-ml volume of $2N$ sodium sulfite reacts quantitatively with approximately 1.5 g formaldehyde. Formaldehyde solutions containing substantial amounts of acid or alkali must always be neutralized before analysis.

Originally phenolphthalein was the recommended indicator for the alkaline titration; then Doby[14] advised the use of aurine (rosolic acid). Subsequently, Taüfel and Wagner[55] proposed thymolphthalein, which was also preferred by Kolthoff[27]. Both aurine and thymolphthalein give good results in practice and occasionally the use of one or the other may be indicated. Aurine solution is prepared by dissolving 1 g aurine in 400 ml 50 per cent methanol. The end point of this indicator is marked by a transition from pink to colorless.

The Holzverkohlungs Industrie, A.-G.[21] declares that it is not necessary to use an analytical grade of sodium sulfite for this determination, since reasonable quantities of neutral saline impurities such as sodium sulfate do not affect the titration. Sodium carbonate, however, must not be present. Since sodium sulfite solutions gradually absorb carbon dioxide on exposure to air, solutions which have stood for a long time should not be employed.

Analytical figures obtained for commercial formaldehyde solutions by the sodium sulfite method are accurate to within approximately ±0.02 per cent. Impurities commonly present in solutions containing formaldehyde have comparatively little effect on the analytical results. Methyl alcohol, methylal, and neutral formates do not interfere[6]. Büchi[10] agrees with Mach and Herrmann[33], and Borgstrom[6] that pure ethyl alcohol is also without influence. Acetaldehyde causes high results[25], particularly when in concentrations exceeding 0.5 per cent[10]. According to Büchi[10], acetone does not interfere in the analysis of 30 to 37 per cent formaldehyde until its concentration exceeds 7 per cent, whereas Mach and Herrmann declare it to have a greater influence. Reynolds and Irwin[43] point out that bisulfite compound equilibria are such that acetaldehyde estimation by this procedure is about 85 per cent, whereas acetone is only 60 to 65 per cent at room temperature.

The formation of the formaldehyde-bisulfite compound also serves as a basis for a procedure developed by Ripper[44] in which formaldehyde is determined by reaction with a sodium bisulfite solution of known iodine value followed by titration of the uncombined bisulfite with decinormal iodine[15, 45].

(2) **Alkaline Peroxide Method.** This method, which was developed by Blank and Finkenbeiner[3], is prescribed by The National Formulary[35] as formerly by the U. S. Pharmacopeia[42]. It is based on the oxidation of formaldehyde by hydrogen peroxide in the presence of a measured excess of alkali. The aldehyde is converted to formic acid which is immediately neutralized by the sodium hydroxide present so that the entire reaction proceeds according to the equation:

$$2CH_2O \text{ (aq)} + 2NaOH + H_2O_2 \rightarrow 2HCOONa + 2H_2O + H_2$$

The amount of sodium hydroxide consumed is determined by the titration of the unreacted alkali and the formaldehyde content is calculated from this figure.

Procedure: Fifty ml of normal sodium hydroxide solution and approximately 25 ml of neutral 6 to 7 per cent hydrogen peroxide (prepared by adding 5 ml of pure commercial 100-volume peroxide to 20 ml water) are placed in a 500-ml Erlenmeyer flask. To this mixture is added a carefully measured sample of the formaldehyde to be analyzed. The flask is then covered with a porcelain crucible cover, agitated thoroughly, and after one minute placed in a water bath kept at approximately 60°C for five minutes. After standing without heating for another five minutes, the mixture is cooled to room temperature and titrated with normal sulfuric or hydrochloric acid using 6 drops bromothymol blue solution* as indicator. The end point is taken when the color of the solution changes from blue to green. A blank titration is made with 50 ml of the normal alkali and 25 ml of the dilute peroxide. The formaldehyde content

---

* This indicator solution is prepared by dissolving 0.5 g dibromothymol sulfophthalein in 150 ml methanol, diluting with water to a volume of 300 ml and neutralizing to a distinct green.

of the sample is calculated by means of the following equation:

$$\% \text{ Formaldehyde} = \frac{(\text{Blank titer} - \text{Sample titer}) \times \text{Normality of acid} \times 3.003}{\text{Weight of Sample}}$$

The above procedure has been found satisfactory for industrial use. However, Büchi[10], who employs a 50-ml volume of 3 per cent peroxide for the oxidation, prefers a 15-minute heating period at approximately 60°C, since Mach and Herrmann[33] have found that impurities sometimes slow up the oxidation reaction. At room temperature a minimum reaction time of $1\frac{1}{2}$ hours is required. Sample requirements are similar to those for the sodium sulfite method (1). The quantity of CH_2O involved should be well under 1.5 g.

Phenolphthalein, thymol blue, and bromothymol blue are satisfactory indicators. Litmus and methyl red are less satisfactory because of their slow and uncertain color transformations. Methyl red is reported to give low results.

With reference to the effect of impurities on the alkaline peroxide method, Büchi[10] found that neither methyl nor ethyl alcohol interferes with the analysis. Homer[22], who prefers 2,4-dinitrophenol as indicator, states that commercial formaldehyde of abnormally high methanol content requires the application of a correction factor, *viz.*, 0.09 per cent for every 10 per cent methanol present. Acetone, when present at concentrations of 7 per cent or below, has a negligible effect[10]. Acetaldehyde gives high results when present in concentrations greater than 0.5 per cent. According to Kühl[28], hydrolysis products of proteins cause incorrect results, since hydrogen peroxide reacts with amino acids.

The hydrogen evolved in the alkaline peroxide oxidation of formaldehyde has been used as a basis for a gas volumetric method of formaldehyde determination by Frankforter and West[17] and Jaillet and Ouellet[23a].

(3) **Iodimetric Method.** Particularly suitable for the determination of small quantities of pure formaldehyde is the iodimetric procedure devised by Romijn[46]. This procedure is dependent on the oxidation of formaldehyde by hypoiodite formed when potassium hydroxide is added to a solution of formaldehyde to which a known amount of a standard iodine solution has been added. The iodine consumed is measured by back-titration of the iodine liberated when the reaction mixture is acidified. The reactions involve:

(1) Formation of hypoiodite,

$$6KOH + 3I_2 \rightarrow 3KI + 3KIO + 3H_2O$$

(2) Oxidation of formaldehyde to formate in the alkaline solution,

$$CH_2O + KIO + KOH \rightarrow HCOOK + KI + H_2O$$

(3) Formation of iodate from excess hypoiodite,

$$3KIO \rightarrow 2KI + KIO_3$$

(4) Liberation of iodine on acidification of the iodate-iodide mixture,

$$KIO_3 + 5KI + 6HCl \rightarrow 3I_2 + 6KCl + 3H_2O$$

(5) Titration of iodine with thiosulfate,

$$2Na_2S_2O_3 + I_2 \rightarrow Na_2S_4O_6 + 2NaI$$

The following procedure is recommended:

Procedure: A sample containing approximately 1 g dissolved formaldehyde, accurately weighed in a glass-stoppered weighing bottle, is diluted to the mark in a 500-ml volumetric flask. A 25-ml sample of this solution is then added by pipette to a 500-ml Erlenmeyer flask containing exactly 50 ml of decinormal iodine solution. An 8-ml volume of 2N sodium hydroxide is added immediately thereafter and the mixture is then allowed to stand for 10 minutes at room temperature. When this time has elapsed, 9 ml of 2N hydrochloric acid (sufficient to neutralize the 2N alkali) is added and the liberated iodine titrated with decinormal sodium thiosulfate. The iodine consumed by reaction with formaldehyde is determined by subtracting the final titer from the titer obtained from a blank analysis or by titrating 50 ml of the decinormal iodine plus 1 ml 2N hydrochloric acid with the same thiosulfate solution. One ml of N/10 iodine is equivalent to 0.0015015 g of formaldehyde.

If more concentrated solutions than those specified are employed for this method of analysis, an aldehyde-iodine compound is formed which introduces an error in the results[51].

This procedure gives accurate results only when applied to formaldehyde solutions substantially free from oxidizable organic impurities. Büchi[10] verified earlier findings that small quantities of ethyl alcohol, acetaldehyde, and acetone give high results. According to Borgstrom[6], the methanol and formic acid present in commercial formaldehyde have no effect on the iodimetric determination.

(4) **Ammonium Chloride Method.** Legler[29] and Smith[52] are probably the originators of this method which is based on the reaction:

$$6CH_2O \text{ (aq)} + 4NH_3 \rightarrow (CH_2)_6N_4 + 6H_2O$$

The following procedure for carrying out the analysis is a modification devised by Büchi[10].

Procedure: A 6-g sample of approximately 37 per cent formaldehyde (or its equivalent) is weighed accurately and diluted to exactly 100 ml in a 100-ml volumetric flask. A 25 ml volume of this solution is added by pipette to a glass-stoppered Erlenmeyer flask, together with three drops of 0.1 per cent bromothymol blue, and neutralized with normal caustic until the color changes to blue. Then 1.5 g of neutral ammonium chloride are added to the solution and 25 ml of normal caustic are added rapidly by means of a pipette. The flask is then tightly closed, using a little vaseline to insure a perfect seal, and allowed to stand for 1½ hours. The excess ammonia is then titrated with normal acid until the color changes to green. A blank may be run in which formaldehyde is not present. The consumption of ammonia is measured by the difference between the blank titer and the sample titer. The formaldehyde content of the solution is calculated by means of the following equation:

$$\% \ CH_2O = \frac{(Blank \ titer - Sample \ titer) \times Acid \ normality \times 18.02}{Weight \ of \ sample}$$

Each ml of normal acid is equivalent to 0.04503 g CH_2O.

Büchi reports that results obtained by this procedure tend, in general, to run 0.1 per cent low.

Since hexamethylenetetramine is not neutral to some indicators, *e.g.*, methyl orange, satisfactory results cannot be obtained when these indicators are employed. Kolthoff[27] designated phenol red, neutral red, rosolic acid (aurine) and bromothymol blue as the only satisfactory indicators. Norris and Ampt[39] recommend titrating to a full yellow with cresol red.

The time factor is important since the reaction of ammonia and formaldehyde is not instantaneous and a considerable period is required for the reaction to become quantitative at ordinary temperatures.

The presence of ethyl alcohol and acetone have comparatively little effect on the analytical results, whereas acetaldehyde in concentrations over 0.5 per cent leads to high values[10].

(5) **Mercurimetric Method.** This procedure is based on the determination of mercury quantitatively precipitated by the action of alkaline potassium mercury iodide (Nessler's reagent) on formaldehyde. The mercury is normally measured by iodimetry so that the net effect is similar to the iodimetric procedure previously described. The principal chemical reaction is as follows:

$$K_2HgI_4 + CH_2O \ (aq) + 3NaOH \rightarrow Hg + HCOONa + 2KI + 2NaI + 2H_2O$$

Romijn[47] was apparently the first to employ a method of this type, which he recommended for the determination of formaldehyde in air. Modifications of more general utility for formaldehyde analysis were later developed by Stüve[54] and Gros[19, cf. 7]. Stüve claims that highly accurate results may be obtained by this method for determining small amounts of formaldehyde with the use of $N/100$ solutions.

The following modification of this procedure is recommended by Büchi[10]:

Procedure: A 1.75-g sample of approximately 37 per cent formaldehyde solution or its equivalent is accurately weighed and diluted to the mark in a 500-ml volumetric flask. A 25-ml volume of this solution is then added by pipette to a 200-ml glass-stoppered Erlenmeyer flask containing a solution of 1 g mercuric chloride and 2.5 ml potassium iodide in 35 ml water, plus 20 ml $2N$ sodium hydroxide. The resulting mixture is shaken vigorously for 5 minutes. After this, it is acidified with 25 ml $2N$ acetic acid and 25 ml $0.1N$ iodine are immediately added by pipette. The solution is agitated until the precipitate is completely dissolved and the excess iodine back-titrated, using starch solution as indicator. The formaldehyde is calculated from the iodine consumption.

$$1 \text{ ml } N/10 \text{ iodine } = 0.0015015 \text{ g } CH_2O$$

Methanol and ethyl alcohol have little or no effect on the results obtained by this procedure. Acetone, even in large quantities, has no appreciable effect and the method in this respect is greatly superior to the iodimetric method. Acetaldehyde exerts a considerable effect even at concentrations as low as 0.5 per cent[10]. Organic hydroperoxides also interfere[43].

Büchi[11] recommends a modification of the mercurimetric method for determining formaldehyde in the presence of hexamethylenetetramine (page 431).

(6) **Potassium Cyanide Method.** Romijn[46] also devised this method of formaldehyde determination which is considered by Mutschin[34] to be as accurate as the sodium sulfite, alkaline peroxide, and iodimetric methods for pure solutions of formaldehyde, and superior to them in the presence of acetone, acetaldehyde, benzaldehyde, higher aldehydes, and ketones. The procedure is based on the quantitative formation of cyanohydrin when formaldehyde is treated with a solution containing a known excess of potassium cyanide.

$$CH_2O \text{ (aq)} + KCN + H_2O \rightarrow CNCH_2 \cdot OH + KOH$$

When the mixture obtained from this reaction is treated with a measured volume of standard silver nitrate in the presence of nitric acid, the unreacted cyanide is converted to silver cyanide and the excess silver nitrate measured by titration with thiocyanate.

A well known modification of this method devised by Elvove[16] is carried out as follows:

Procedure: A sample of formaldehyde solution (accurately measured) containing approximately 0.15 to 0.25 g dissolved formaldehyde is added to a glass-stoppered Erlenmeyer flask containing 100 ml of $N/10$ potassium cyanide solution. After shaking this mixture for a short time, it is carefully washed into a 200-ml volumetric flask containing exactly 40 ml $N/10$ silver nitrate and about 10 ml 10 per cent nitric acid. The volumetric flask is filled to the mark with distilled water, shaken thoroughly and filtered. The unreacted silver in 100 ml of the filtrate is then determined by titration with $N/10$ potassium thiocyanate, using ferric alum as an indicator. The amount of formaldehyde in the sample can then be calculated by the following equation in which n equals the ml of thiocyanate used in the titration:

$$(100 - [40 - 2n]) \ 0.003 = \text{g } CH_2O$$

A variant of this method developed by Schulek[49] is claimed to be superior to the above procedure in that it is even less sensitive to the presence of other aldehydes or ketones and is reported to give accurate results in the presence of 5 per cent acetaldehyde and 50 per cent acetone. This method involves determining unreacted cyanide by iodimetry and is carried out as follows:

Procedure: An accurately measured 5- to 10-ml sample of the formaldehyde solution to be analyzed containing 0.1 to 40 mg dissolved formaldehyde is added with constant agitation to exactly 25 ml $N/10$ potassium cyanide in a glass-stoppered flask and acidified after two to three minutes with concentrated hydrochloric acid. Bromine is then added until the solution develops a permanent yellow color. Excess bromine is removed by addition of 1 to 2 ml 5 per cent phenol solution, and approximately 0.3 g potassium iodide is added. Free iodine is titrated after $\frac{1}{2}$ hour with $N/10$ or $N/100$ thiosulfate, using starch as an indicator. The amount of potassium cyanide solution equivalent to formaldehyde equals the ml of $N/10$ cyanide originally added, minus one-half the titer of $N/10$ thiosulfate. One ml $N/10$ KCN is equivalent to 0.003002 g formaldehyde. The cyanide solution must be standardized iodimetrically against thiosulfate.

Pfeihl and Schroth[41] recommend determination of the excess cyanide by mercurimetry in another modification of this procedure. Diphenylcarbazone, which goes from colorless to violet, is employed as an indicator. It is claimed that sugar and other formaldehyde condensation products do not interfere when this procedure is used.

(7) **Hydroxylamine Hydrochloride Method.** Occasionally useful for the estimation of formaldehyde in the presence of alkali-sensitive products and impurities, such as cresol, phenol, and resinous materials, is the method of Brochet and Cambier[9, cf. 31, 32] based on the liberation of hydrochloric acid when hydroxylamine hydrochloride reacts with formaldehyde to form formaldoxime:

$$CH_2O \text{ (aq)} + NH_2OH \cdot HCl \rightarrow CH_2{:}NOH + H_2O + HCl$$

Procedure: Two 10-ml portions of a 10 per cent solution of hydroxylamine hydrochloride are added by pipette to two 125-ml Erlenmeyer flasks. To one of these flasks is then added a thimble beaker containing an accurately weighed sample of the solution to be analyzed. This sample should contain approximately 1 g formaldehyde. If resinous material is present and precipitates, a small quantity of methanol, *e.g.*, 10 ml, may be added to dissolve the precipitate. A similar amount of methanol should also be added to the solution in the other flask, which serves as a blank or control. After approximately 15 to 20 minutes, the contents of both flasks should be titrated with normal potassium hydroxide, using bromophenol blue as an indicator. The end-point is marked by a color change from yellow to light purple. The blank should require only a few cc of standard alkali and should be checked daily. The per cent formaldehyde is calculated by the following formula:

$$\% \text{ CH}_2\text{O} = \frac{(\text{Sample titer} - \text{Blank titer}) \times N \text{ of KOH} \times 3.003}{\text{Weight of sample}}$$

In a modified hydroxylamine method, Bryant and Smith[9a] use pyridine to make the oxime method go to completion and render the initial reaction mixture neutral to bromophenol blue. This is reported to eliminate the need of a blank titration.

(8) **Methone or Dimedon Method.** Methone (5,5-dimethyldihydroresorcinol, 5,5-dimethylcyclohexanedione-3,5, dimetol) is not only a valuable reagent for the detection of formaldehyde (page 372) but can also be employed for its quantitative determination both by gravimetric[23, 57, 60, 62] and volumetric technique. It is particularly valuable for the determination of formaldehyde in the presence of ketones, with which it does not react under ordinary conditions. A special procedure developed by Vorländer[57] makes it possible to employ methone for separating and analyzing both formaldehyde and acetaldehyde in mixtures. By this procedure it should also be possible to determine formaldehyde in the presence of other aliphatic aldehydes. Sugars such as glucose, lactose, and arabinose do not react with methone and accordingly do not interfere with its use in formaldehyde analysis.

Methone is a tautomeric compound as shown by the keto and enol forms indicated in the structural formulas shown below:

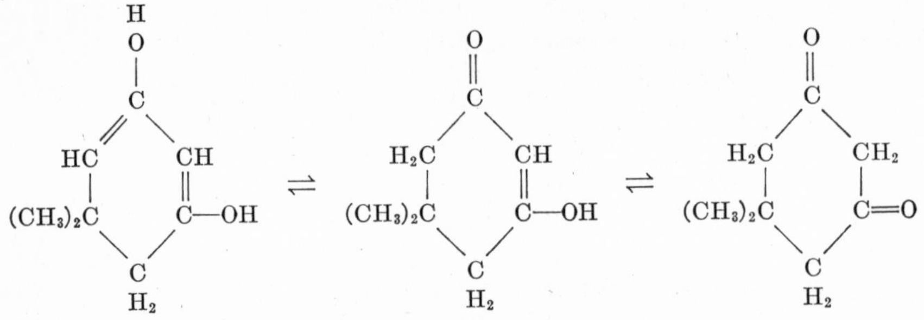

5,5-Dimethyldihydro-   *5,5-Dimethylcyclohex-*
resorcinol   *anedione-3,5*

It is a colorless to slightly yellow crystalline compound melting at 148 to 150°C, somewhat sparingly soluble in water but readily soluble in alcohol and acetone. Its solubility in water at various temperatures[58] is shown below:

Temperature (C°)	Solubility G Methone per 100 ml Solution
19	0.401
25	0.416
50	1.185
80	3.020
90	3.837

Methone reacts with alkali as a monobasic acid and may be titrated with standard alkali, using phenolphthalein as an indicator.

Formaldehyde and methone react quantitatively in neutral, alkaline, or mildly acidic aqueous or alcoholic solutions to form methylene bis-methone, as indicated in the following equation:

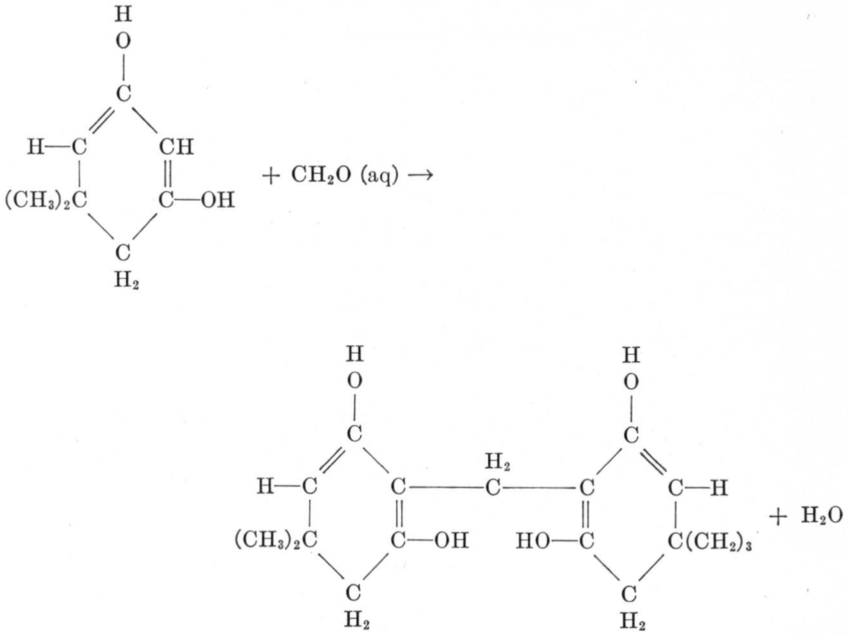

This product is almost completely insoluble in neutral or mildly acidic aqueous solutions; 100 ml water dissolve only 0.5 to 1.0 mg at 15 to 20°C. It is a crystalline material melting at 189°C. It is soluble in alkali, behaving as a monobasic acid, and can be titrated with standard alkali in alcohol solution.

Other aliphatic aldehydes such as acetaldehyde also form condensation products of low solubility by reaction with methone. These products differ from the formaldehyde derivative in that they are readily converted to cyclic hydroxanthene derivatives by treatment with glacial acetic or dilute sulfuric acid. These xanthene derivatives are not soluble in alkali and can thus be separated from the formaldehyde product. Acetaldehyde reacts with methone to give ethylidene dimethone (m.p. 139°C) which is 0.0079 per cent soluble in water at 19°C. The conversion of this product to the alkali-insoluble xanthene derivative is indicated below:

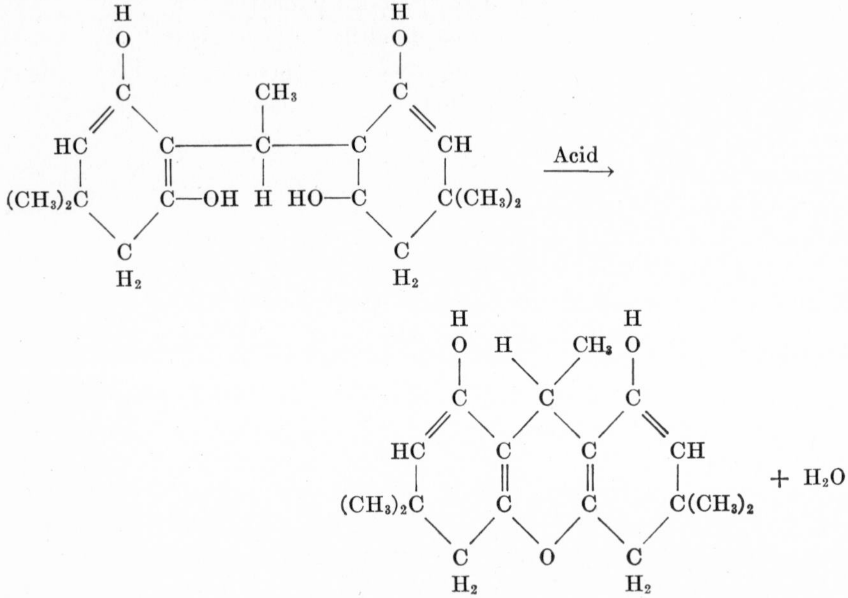

$$
\begin{array}{c}
\overset{\displaystyle H}{\underset{\displaystyle |}{O}} \qquad \overset{\displaystyle CH_3}{|} \qquad \overset{\displaystyle H}{\underset{\displaystyle |}{O}} \\
C \qquad\qquad C \\
\end{array}
$$

Acid →

+ H_2O

The acetaldehyde derivative, ethylidene dimethone, differs from the form-aldehyde product in that it behaves as a dibasic acid when titrated with alkali at 70 to 75°C.

In determining formaldehyde with methone the reagent is best added as a saturated aqueous solution or as a 5 to 10 per cent solution in alcohol. In the latter case, care must be taken to avoid addition of large quantities of reagent because of the limited solubility of methone in water and the fact that high concentrations of alcohol will interfere with the precipitation of the formaldehyde derivative. The formaldehyde solution to be analyzed should be neutral or mildly acid. Weinberger[60] claims that addition of salt increases the sensitivity of the test and states that agitation speeds up the precipitation of methylene derivative. Vörlander[57] allows a reaction period of 12 to 16 hours at room temperature for complete precipitation. By Weinberger's accelerated method[60] it is claimed that formaldehyde bis-methone is precipitated in 15 minutes when present at a concentration of 4 ppm. In the absence of other aldehydes the precipitate may be filtered off, washed with cold water, and dried to constant weight at 90 to 95°C for gravimetric measurement. Each gram of precipitate is equivalent to 0.1027 g formaldehyde. Yoe and Reid[62] report improved accuracy if the precipitation is carried at pH 4.6 in a sodium acetate-hydrogen chloride solution.

When acetaldehyde or other aliphatic aldehydes are present, their methone derivatives must be separated from the precipitate first obtained. This may be accomplished by shaking solution and precipitate with 1/15 volume of cold 50 per cent sulfuric acid for 16 to 18 hours; or the filtered precipitate (moist or vacuum-dried) can be heated with four to five times its volume of glacial acetic acid on a boiling water bath for 6 to 7 hours, after which it is treated with an excess of ice water to precipitate the products. The formaldehyde bis-methone can then be removed from the acid-treated precipitates with dilute alkali and reprecipitated by acidification[57]. When acetaldehyde is the only other aldehyde present, it may be determined by weighing the dried alkali-insoluble precipitate, one gram of which is equivalent to approximately 0.1180 g acetaldehyde. Since ethylidene bis-methone is more soluble than the methylene derivative, the method is not as accurate for acetaldehyde as formaldehyde.

The volumetric titration technique developed by Vörlander[57, 58] is readily applied to the measurement of moist precipitated methylene bis-methone by dissolving in alcohol and titrating with caustic at room temperature. Temperatures of 70°C do not affect the results of this titration. One ml of normal alkali is equivalent to 0.030 g formaldehyde. Another variant involves titration of a given volume of methone solution followed by titration of an equal volume of the same solution which has been reacted with the formaldehyde sample. The difference in these titers is equivalent to the formaldehyde titer since two mols of methone are equivalent to two liters of normal caustic, whereas one mol of methylene bis-methone (equivalent to two mols of methone) is equivalent to only one liter of normal caustic. If acetaldehyde is present, the final titration must be carried out at 70°C, since one mol of ethylidene bis-methone titrates as a dibasic acid at this temperature and thus does not interfere with the analysis, since it behaves the same as two mols of unreacted methone.

ASSAY OF COMMERCIAL FORMALDEHYDE

The assay of commercial formaldehyde normally involves the determination of formaldehyde, methanol, and formic acid. To these determinations are sometimes added the measurements of metallic impurities, particularly iron, aluminum and copper.

The concentration limits of formaldehyde, methanol and formic acid in commercial solutions measured in per cent by weight usually lie within the limits indicated below for low methanol and N.F. solutions.

	Low Methanol Solutions (%)	N.F. Solutions (%)
Formaldehyde	30–50	37.0–37.5
Methanol	0.2–1.0	6–15
Formic acid	0.01–0.05	0.01–0.04

It should be mentioned that formaldehyde produced by hydrocarbon oxidation sometimes possesses a foreign odor and may give a yellow to brown color when five ml of the commercial solution is mixed with an equal volume of concentrated sulfuric acid. Formaldehyde made from methanol gives little or no color when subjected to this test.

Formaldehyde Specifications

Specifications for 37 per cent formaldehyde are published in The National Formulary[35], the American Chemical Society's "Reagent Chemicals"[1a], Rosin's, "Reagent Chemicals and Standards"[47a], and the American Standards Association's specifications for photographic grade formaldehyde solution[1b]. In general, these specifications call for a formaldehyde content of 36 to 38 per cent as in the case of the A.C.S., or a 37.0 per cent minimum as in The National Formulary. The methanol content is usually listed as variable or from 10 to 15 per cent. Acidity, calculated as formic acid, ranges from 0.03 per cent (A.C.S.) to a 0.2 per cent maximum (N.F.). Maximum tolerances for total solids, when given, range up to 200 ppm (A.C.S.). The American Chemical Society specifies a maximum iron content of 10 ppm and a heavy metals content of 5 ppm.

All of the above specifications call for the hydrogen peroxide method of formaldehyde analysis. Acidity determinations involve titration of 10- or 20-ml samples diluted with one to two volumes of water using bromothymol blue or phenolphthalein as the indicator.

Determination of Formic Acid

Formic acid is determined by titration, following the procedure outlined by The National Formulary[35, cf. 42]:

Twenty ml of commercial formaldehyde is measured into an equal volume of distilled water, to which is added two drops of bromothymol blue solution, prepared by dissolving 0.10 g bromothymol blue in 100 ml 50 per cent alcohol. The formaldehyde solution is then titrated with normal or $0.1N$ sodium hydroxide (free of carbonate). One ml of normal alkali is equivalent to 0.04602 g formic acid. The acid content is calculated as per cent formic acid by weight, the weight of sample being determined from the approximate density of the formaldehyde analyzed.

Since formic acid is normally the only acid present in substantial amount, the assumption that all acidity is caused by formic acid is usually warranted.

Determination of Formaldehyde

The sodium sulfite (page 382) or the alkaline peroxide methods (page 384) are employed for this purpose. Samples need not be neutralized before analysis, if a suitable correction is made in the titration figure.

Determination of Methanol

Methanol content may be determined either by physical or chemical methods or by a combination of the two.

The chemical procedure of Blank and Finkenbeiner[4] has been found to possess a good degree of accuracy. In this procedure, the formaldehyde solution is completely oxidized by a measured excess of chromic acid and the consumption of oxidizing agent determined by titrating the unreacted chromic acid with a standard iodine solution. The oxidation equivalent of the formaldehyde in the sample analyzed is then calculated from its previously measured formaldehyde content and the methanol content calculated from the chromic acid consumed in excess of this figure. Naturally, the method is only applicable to solutions which are free of oxidizable matter other than methanol and formaldehyde. The traces of formic acid and methylal normally present in commercial formaldehyde are negligible. Under the test conditions, the oxidation reactions take place as indicated in the following equations:

$$CH_3OH + 3O^{--} \rightarrow CO_2 + 2H_2O$$

$$CH_2O + 2O^{--} \rightarrow CO_2 + H_2O$$

The following modification of Blank and Finkenbeiner's method is recommended:

A 50 ml volume of an approximately $2N$ test solution of chromic acid is added by means of an ordinary pipette to a 250-ml glass-stoppered Erlenmeyer flask to which is then added 40 ml of distilled water. The same pipette should be employed for chromic acid in all analyses. (The chromic acid test solution is prepared by dissolving 83.3 g pure chromium trioxide (CrO_3) in 688 ml distilled water, filtering through asbestos and adding gradually and with stirring 412 ml concentrated sulfuric acid.) Then a 1-gram sample of commercial N.F. formaldehyde is accurately weighed in a thimble beaker and dropped into the measured chromic acid solution which is lightly stoppered and gently agitated for one minute. The flask is allowed to stand overnight at room temperature, or let stand for one hour, heated at 60°C for one hour, and cooled. After this the contents of the flask are diluted with distilled water in a 1000-ml volumetric flask, made up to the mark, and shaken by reversing the flask 15 to 20 times. A 25-ml aliquot of this diluted solution is removed from the volumetric flask by pipette, treated with 6 ml 40 per cent potassium iodide, and titrated with $N/10$ sodium thiosulfate, using a 1 per cent solution of soluble starch as a titration indicator. The determination should be made in duplicate, blank determinations being run to determine the quantity of chromic acid employed. The per cent methanol is calculated by means of the following formula:

% Methanol by wt = (Blank titer − Sample titer) ×

Normality × 21.33 − (0.7114 × % CH_2O) − (0.232 × % HCOOH)

The simplest and most rapid method for methanol determination is by specific gravity measurement. The density of a formaldehyde solution is influenced both by its formaldehyde and methanol content. Accordingly, when the formaldehyde content has been accurately established, the methanol content can be determined from specific gravity data. Tables for this purpose have been prepared by Natta and Baccaredda,[36] Table 52, and Homer[22], Table 53. The former determined specific gravities at 18°/4°C. the latter at 15°/15°C. These data are in good agreement, as was demonstrated by Homer, who corrected Natta and Baccaredda's figures to 15°/15°C. for purposes of comparison. Average agreement is within ±0.02 per cent. For analysis of approximately 37 per cent commercial formaldehyde, the spe-

TABLE 52. DENSITY OF FORMALDEHYDE CONTAINING METHANOL AT 18°/40°*

% CH_2O	% Methanol				
	5	10	20	30	40
2.5	0.9975	0.9888	0.9731	0.9573	0.9405
5.0	1.0015	0.9965	0.9793	0.9634	0.9454
10.0	1.0203	1.0119	0.9932	0.9748	0.9556
15.0	1.0354	1.0261	1.0069	0.9865	0.9657
20.0	1.0507	1.0401	1.0207	0.9983	0.9756
25.0	1.0657	1.0551	1.0329	1.0100	0.9853
30.0	1.0803	1.0691	1.0460	1.0216	0.9959
35.0	1.0947	1.0830	1.0592	1.0325	—
40.0	1.1102	1.0981	1.0709	—	—
45.0	1.1254	1.1108	—	—	—

* Natta and Baccaredda[36].

TABLE 53. DENSITY OF FORMALDEHYDE CONTAINING METHANOL AT 15°/15°*

% CH_2O	% Methanol				
	5	10	20	30	40
10	1.0215	1.0124	0.9950	0.9769	0.9580
20	1.0523	1.0419	1.0228	1.0008	0.9785
30	1.0822	1.0713	1.0484	—	—

* H. W. Homer[22].

cific gravity tables on pages 68–72 (Table 15) are especially useful. Natta and Baccaredda advocate the determination of both formaldehyde and methanol by physical methods involving the measurement of density and refractive index[36]. Results are interpreted by means of their ternary diagram (Figure 30).

Methods employing both physical and chemical technique involve isolation of the methanol from the formaldehyde. The following procedure is based on a method of this type recommended by Homer[22].

Procedure: A 50-g sample of commercial formaldehyde solution is cooled to approximately 15°C in an ice-and-water bath. To this solution is then added an excess of aqua ammonia. The ammonia must be added gradually and with agitation, so that the temperature does not exceed approximately 20 to 25°C. The mixture is then allowed to stand in a closed flask for about 6 hours to insure complete reaction.

The treated solution, which should have a strongly ammoniacal odor, is distilled almost to dryness. The distillate containing alcohol, water, and excess ammonia is then acidified and distilled. The first 50 ml of distillate contains all the alcohol which is estimated by measurement of the specific gravity. Approximately 30 ml of 28 per cent aqua ammonia should be sufficient for use in this procedure.

Other variants of this type of procedure employ sodium bisulfite[2] and sulfanilic acid[18] in place of ammonia to facilitate formaldehyde removal.

Metallic Impurities

Aluminum, iron, and copper are occasionally determined in analyzing commercial formaldehyde. Normally not more than 3 ppm of aluminum are present, whereas copper and iron do not as a rule exceed about 1 ppm. Determinations are made by the usual methods of inorganic analysis following removal of formaldehyde. This may be accomplished in a satisfactory manner by boiling down a liter sample of formaldehyde solution in glass until solidification due to polymer formation takes place. The solid is then heated with a small flame until practically all has vaporized leaving a charred residue. This residue is cautiously dissolved by gradual addition of 50 ml of c.p. nitric acid, heating after addition of each increment. Approximately 10 ml of pure concentrated sulfuric acid is then added and heating is continued until a clear solution is obtained. Clarification and solution may be accelerated by gradually adding about 0.5 g of potassium chlorate. Iron and aluminum may both be determined in a solution prepared in this way. An additional liter of formaldehyde should be evaporated and oxidized for copper determination. Evaporation should be carried out under a hood and oxidations should be made with extreme caution since any undecomposed formaldehyde will react violently.

Assay of Commercial Paraformaldehyde

At present paraformaldehyde is the major formaldehyde polymer produced commercially. As has been previously pointed out, the term "trioxymethylene" is sometimes erroneously applied to this product. The assay of this product consists principally in the determination of its formaldehyde content. An ash determination is often included and occasionally metallic impurities, such as iron and aluminum, are determined. Screen analysis for particle size, melting-point determination, and solubility measurement give special information which is often of value, since these data differentiate various commercial grades of this polymer.

Formaldehyde Content. Commercial paraformaldehyde, $HO(CH_2O)_n \cdot H$, should contain 91 to 95 per cent or more formaldehyde by weight, the balance being principally combined water.

The American Standards Association's specification[1c] for photographic grade paraformaldehyde is a minimum of 95 per cent HCHO. The determination of formaldehyde content is usually made by means of the alkaline peroxide or sodium sulfite method. The procedures followed are identical with those already described for formaldehyde solution, except that the polymer sample (approximately 1 g) is accurately weighed out as a solid and added to the analytical reagents in which it quickly dissolves, reacting as though it were monomeric formaldehyde.

Ash. A 10-g sample is carefully ignited in a porcelain crucible which is heated to redness until all the carbon has burned off. The residue is carefully weighed and calculated as per cent ash. This figure normally does not exceed 0.1 per cent.

Metallic Impurities. These are determined from a nitric acid solution of the ash obtained from a suitable weight of sample by the standard methods employed for the determination of small quantities of metals. Iron and aluminum concentrations usually do not exceed 100 ppm.

Screen Analysis. This determination is carried out by the standard routine method. Regularly, paraformaldehyde powder passes a 30- to 40-mesh screen. The granular product falls between 4 and 40, whereas especially finely ground material passes a 100-mesh screen.

Melting Range. The melting range is taken in a sealed melting-point tube totally immersed in the heating bath employed. The melting range gives a rough index of the degree of polymerization of the polymer and, therefore, is also an index of reactivity. Ordinary paraformaldehyde usually melts in the range 120 to 160 C. Special, highly polymerized grades of low reactivity melt at 160 to 170°C.

Solubility. A rough measure of the rate of solubility in water also gives information concerning molecular weight and reactivity, although this figure is also influenced by pH (pages 124–126). A useful procedure[59] is to agitate 5 g of polymer with 25 ml distilled water at room temperature for one hour and filter. The undissolved polymer on the filter is washed with 10 ml water and the washings added to the original filtrate which is then analyzed by the sodium sulfite method. Results are calculated as grams of dissolved formaldehyde per 100 grams paraformaldehyde. That this is not a solubility measurement will be clear from the discussion in Chapter 7. The figures obtained are useful for comparative purposes since they furnish a rough index of product characteristics.

Determination of Formaldehyde in Special Compositions and in Products Treated with Formaldehyde

The isolation of formaldehyde from mixtures for analytical purposes is a knotty problem. In general, formaldehyde may be separated from water-insoluble materials by water extraction. Also material soluble in water-immiscible solvents can be removed from formaldehyde solutions with these solvents. However, it should be remembered that polar compounds such as alcohols, glycols, amines, and amides hold onto formaldehyde since they tend to form hemiacetals and their ammono analogs.

The distillation procedure described in connection with the isolation of formaldehyde for purposes of detection may also be employed for the quantitative isolation of formaldehyde for analysis. Formaldehyde can be quantitatively removed from aqueous solutions containing less than 10 per cent CH_2O by distillation or preferably by steam distillation. This treatment also hydrolyzes loose compounds of formaldehyde in many instances so that the combined formaldehyde can be isolated quantitatively.

The accuracy of these methods of separation is questionable in many instances and must be tested in specific cases. Loose compounds are often converted to irreversible reaction products under the influence of heat and long standing, or heating in the presence of alkalies results in the conversion of formaldehyde to formic acid and methanol by the Cannizzaro reaction. Methods employed for the analysis of formaldehyde in the presence of other compounds must be carefully chosen with reference to their utility in the presence of the impurities in question.

A complete review of the voluminous literature dealing with the determination of formaldehyde in miscellaneous compositions is impractical, since most mixtures will require special study on the part of the individual investigator. However, a few examples will be pointed out as illustrative of the methods employed for specific problems.

Mixtures Containing Phenol and Phenol-formaldehyde Condensates. As previously pointed out, the hydroxylamine hydrochloride method is of value for determining formaldehyde in the presence of phenols. Ormandy and Craven[40] also investigated the various analytical methods adaptable to the determination of formaldehyde in the manufacture of phenol-formaldehyde resins and found that both the alkaline peroxide and a special iodimetric method developed by Kleber[26] could be used. In the latter, formaldehyde is oxidized to formic acid by sodium hypochlorite in alkaline iodide solution, and the iodine back-titrated with thiosulfate. The peroxide method may be used for condensates containing little phenol, since up to 0.5 g phenol is reported to have no effect on the formaldehyde determination. In the case of high phenol concentrations, the condensate may be extracted with benzene or the phenol salted out.

Medicinal Soaps. Medicinal soaps containing formaldehyde may be analyzed by dissolving in four to five times their weight of water and precipitating the soap with barium chloride or sulfuric acid. After filtration, the filtrate is adjusted to a definite volume and the formaldehyde determined by the iodimetric method[1].

Dusting Powders. Weinberger[61] recommends determining small quantities of paraformaldehyde in dusting powders and other mixtures by treating the sample with sulfuric acid, distilling, and applying Romijn's potassium cyanide method to the distillate.

Fungicides and Seed-conserving Agents. Fungicides and insecticides containing copper sulfate should be treated with potassium ferrocyanide to remove the copper before the formaldehyde content is determined by the alkaline peroxide method[24]. In the case of seed-conserving agents containing mercury chloride, potassium chloride should be added prior to the distillation in order to prevent the mercury from distilling over with the aldehyde and causing erroneous results in alkaline peroxide or sodium sulfite analysis[5].

Determination of Combined Formaldehyde in Formals

A good procedure for the approximate determination of combined formaldehyde in formals was developed by Clowes[12]. It is carried out by heating a weighed sample of the formal plus 5 ml water with a mixture of 15 ml concentrated hydrochloric acid, 15 ml water and a slight excess of phloroglucinol for 2 hours at 70 to 80°C, allowing the mixture to stand overnight, then filtering off the reddish brown phloroglucinol formaldehyde resin in a Gooch crucible, washing it with 60 ml of water, and drying in an oven at 100°C for 4 hours. After cooling in a desiccator, the crucible containing the precipitate is weighed. The weight of the precipitated resin, whose empirical formula is approximately $C_7H_6O_3$, is divided by 4.6 to obtain the weight of combined formaldehyde. This procedure can naturally be employed only with formals whose hydrolysis products are soluble in water. Formaldehyde and phloroglucinol react in approximately equimolar proportions.

Soloveichik and Novikova[53] determine formaldehyde in polyvinyl formals by boiling an approximately 0.5-g sample with 100-ml 20 per cent sulfuric acid for 1 hour. This is followed by a steam distillation and analysis of the formaldehyde-containing distillate by the hydroxylamine hydrochloride method.

Determination of Combined Formaldehyde in Formaldehyde-treated Products

In general, heating with dilute acids followed by distillation makes it possible to isolate combined formaldehyde quantitatively or almost quantitatively from paper, textiles, leather and formaldehyde-protein compositions. This procedure is widely used in connection with the determination of combined formaldehyde in these materials. The following instances involving protein-formaldehyde compositions are typical.

In the determination of formaldehyde in fur, hair, and wool, the hairs are treated with dilute acid and the formaldehyde is distilled off in the customary manner. If the acid distillate gives a positive test for formaldehyde with Schryver's phenylhydrazine-potassium ferricyanide test (page 370), Romijn's iodine method of quantitative analysis may be used[56].

In the analysis of leather samples, it is recommended that the formaldehyde liberated by the acid treatment be distilled into an excess of sodium bisulfite. This excess is titrated with iodine, after which alkali is added to destroy the formaldehyde-bisulfite compound. Titration of the liberated bisulfite completes the procedure. It should be noted, however, that vegetable tannins interfere with the determination[27]. Theis recommends the procedure of Highberger and Retzsch[20] for this determination.

Similar methods are used by Nitschmann and co-workers[37, 38] for the determination of formaldehyde in formaldehyde-hardened casein. Here, the formaldehyde is distilled into 1 per cent sodium bisulfite solution and, after standing 1 hour, the excess bisulfite is titrated with decinormal iodine against starch. Treatment of the titrated liquor with 95 per cent ethyl alcohol, then with 5 per cent sodium carbonate, and retitration complete the analysis. It must be remembered in this connection that formaldehyde can react irreversibly with proteins under some circumstances and will not be detected by hydrolytic splitting when it has reacted in this way (see pages 316–317).

According to Bougault and Leboucq[8], combined formaldehyde in methylolamides can sometimes be determined by the mercurimetric method. Coppa-Zuccari[13] determines combined formaldehyde in urea resins by oxidation with alkaline hydrogen peroxide followed by acidification, steam distillation and titration of the formic acid liberated.

References

1. Allemann, O., *Z. Anal. Chem.*, **49,** 265 (1910); *Seifensieder Ztg.*, **40,** 49 (1913).
1a. American Chemical Society, "Reagent Chemicals," p. 158, Washington, D. C., American Chemical Society (1950).
1b. American Standards Association, "American Standard Specification for Photographic Grade Formaldehyde," A.S.A. pamphlet Z38.8.152, New York, American Standards Association (1949).
1c. American Standards Association, "American Standard Specification for Photographic Grade Paraformaldehyde," A.S.A. pamphlet Z38.8.153–1949, New York, American Standards (1949).
2. Bamberger, H., *Z. Angew, Chem.*, **17,** 1246–8 (1904).
3. Blank, O., and Finkenbeiner, H., *Ber.*, **31,** 2979–81 (1898); **32,** 2141 (1899).
4. *Ibid.*, **39,** 1326–7 (1906).
5. Bodnar, J., and Gervay, W., *Z. anal. Chem.*, **80,** 127–34 (1930); *C. A.*, **24,** 2825 (1930).
6. Borgstrom, P., *J. Am. Chem. Soc.*, **45,** 2150–5 (1923); *C. A.*, **17,** 3657 (1923).
7. Bougault, J., and Gros, R., *J. Pharm. Chim.*, **26,** 5–11 (1922); *C. A.*, **16,** 3281 (1922).

8. Bougault, J., and Leboucq, J., *Bull. Acad. Med.*, **108**, 1301–3 (1932); *C. A.*, **27**, 5480–1 (1932).

9. Brochet, A., and Cambier, R., *Compt. rend.*, **120**, 449 (1895); *Z. anal. chem.*, **34**, 623 (1895).

9a. Bryant, W. M. D., and Smith, D. M., *J. Am. Chem. Soc.*, **57**, 57–61 (1935).

10. Büchi, J., *Pharm. Acta Helv.*, **6**, 1–54 (1931).

11. *Ibid.*, **13**, 132–7 (1938); *C. A.*, **33**, 9550, (1939).

12. Clowes, G. H. A., *Ber.*, **32**, 2841 (1899).

13. Coppa-Zuccari, G., *Inds. plastiques*, **4**, 183 (1948); *C. A.*, **43**, 66 (1949).

14. Doby, G., *Z. angew. Chem.*, **20**, 354 (1907).

15. Donnally, L. H., *Ind. Eng. Chem., Anal. Ed.*, **5**, 91–2 (1933).

16. Elvove, E., *Am. J. Pharm.*, **83**, 455 (1910); *Chem. Zentr.*, **1911**, II, 1658.

17. Frankforter, G. B., and West, R., *J. Am. Chem. Soc*, **27**, 714–9 (1905).

18. Gnehm, R., and Kaufler, F., *Z. Angew. Chem.*, **17**, 673 (1904); **18**, 93 (1905).

19. Gros, R., *J. Pharm. Chim.*, **26**, 415–25 (1922); *C. A.*, **17**, 1402.

20. Highberger, J. H., and Retzsch, C. E., *J. Am. Leather Chemists' Assoc.*, **33**, 341 (1938); **34**, 131 (1939).

21. Holzverkohlungs Industrie, A. G., *Chem. Ztg.*, **54**, 582 (1930); *C. A.*, **25**, 1756.

22. Homer, H. W., *J. Soc. Chem. Ind.*, **60**, 213–8T (1941).

23. Ionescu, M. V., and Bodea, C., *Bull. soc. chim.*, **47**, 1408 (1930).

23a. Jaillet, J. B., and Ouellet, C., *Can. J. Chem.*, **29**, 1046–58 (1952).

24. Jakeš, M., *Chem. Ztg.*, **47**, 386 (1923); *C. A.*, **17**, 2252 (1923).

25. Kerp, W., *Arb. kaiserl. Gesundh.*, **21**, Pt. 2, 40 (1904); *Z. Unters. Nahr. u. Genussm.*, **8**, 53; *Z. Anal. Chem.*, **44**, 56–7 (1905).

26. Kleber, L. F., *J. Am. Chem. Soc.*, **19**, 316–20 (1897).

27. Kolthoff, I. M., "Die Massanalyze," Vol. 2, p. 186, Berlin, Julius Springer (1928).

28. Kühl, F., *Collegium*, **1922**, 133–42; *C. A.*, **16**, 3833 (1923).

29. Legler, L., *Ber.*, **16**, 1333–7 (1883).

30. Lemme, G., *Chem. Ztg.*, **27**, 896 (1903); *Chem. Zentr.*, **1903**, II, 911.

31. Lockemann, G., and Croner, F., *Desinfektion*, **2**, 595–616, 670–3 (1909); *Chem. Zentr.*, **1910**, I, 59, 203.

32. Lockemann, G., and Croner, F., *Z. anal. Chem.*, **54**, 22 (1915).

33. Mach, F., and Hermann, R., *Z. anal. Chem.*, **62**, 104 (1923).

34. Mutschin, A., *Z. anal. Chem.*, **99**, 346 (1934).

35. National Formulary, Ninth Edition, pp. 221–2, Town State Printer (1950).

36. Natta, G., and Baccaredda, M., *Giorn. Chim. Ind. Appl.*, **15**, 273–81 (1933).

37. Nitschmann, H., and Hadorn, H., *Helv. Chim. Acta*, **24**, 237–42 (1941).

38. Nitschmann, H., and Lauener, H., *Helv. Chim. Acta.* **29**, 174–9 (1946).

39. Norris, J. H., and Ampt, G., *Soc. Chem. Ind. Victoria, Proc.*, **33**, 801–10 (1933); *C. A.*, **29**, 3940.

40. Ormandy, W. R., and Craven, E. C., *J. Soc. Chem. Ind.*, **42**, 18–20T (1923).

41. Pfeihl, E., and Schroth, G., *Z. anal. Chem.*, **134**, 333–4 (1952).

42. "Pharmacopoeia of the United States of America," U. S. P. XII, pp. 267–8, Easton, Pa., Mack Printing Co. (1942).

43. Reynolds, J. G., and Irwin, M., Chemistry & Industry, **1948**, 419–424.

44. Ripper, M., *Monatsh.*, **21**, 1079–84 (1900).

45. Romeo, G., *Ann. chim. applicata*, **15**, 300–4 (1925); *C. A.*, **19**, 3446 (1925).

46. Romijn, G., *Z. anal. Chem.*, **36**, 19, 21 (1877).

47. Romijn, G., *Pharm. Weekblad*, **40**, 149 (1903).

47a. Rosin, J., "Reagent Chemicals and Standards," p. 188, New York, D. Van Nostrand & Co. (1946).

48. Sadtler, S. S., *Am. J. Pharm.*, **76,** 84–7 (1904); *Chem. Zentr.*, **1904,** I, 1176.
49. Schulek, E., *Ber.*, **58,** 732–6 (1925); *C. A.*, **20,** 727 (1926).
50. Seyewetz, A., and Gibello, *Bull. soc. chim.* (*3*), **31,** 691–4 (1904).
51. Signer, R., *Helv. Chim. Acta*, **13,** 44 (1930); *C. A.*, **24,** 1821 (1930).
52. Smith, B. H., *J. Am. Chem. Soc.*, **25,** 1028–35 (1903).
53. Soloveichik, L. S., and Novikova, E. M., *Zavodskaya Lab.*, **15,** 418–9 (1949); *C. A.*, **43,** 6946 (1949).
54. Stüve, W., *Arch. Pharm.*, **244,** 540 (1906).
55. Taüfel, K., and Wagner, C., *Z. anal. Chem.*, **68,** 25–33 (1926); *C. A.*, **20,** 1774 (1926).
56. Thuau, U. J., and Lisser, D., *Cuir tech.*, **28,** 212–3 (1939); *C. A.*, **34,** 1208 (1940).
57. Vorländer, D., *Z. anal. chem.*, **77,** 32–7 (1929).
58. *Ibid.*, **77,** 241–268 (1929).
59. Walker, J. F., *J. Am. Chem. Soc.*, **55,** 2825 (1933).
60. Weinberger, W., *Ind. Eng. Chem., Anal. Ed.*, **3,** 365–6 (1931).
61. *Ibid.*, **3,** 357–8 (1931).
62. Yoe, J. H., and Reid, L. C., *Ind. Eng. Chem., Anal. Ed.*, **13,** 238–40 (1941).

CHAPTER 19

HEXAMETHYLENETETRAMINE

Hexamethylenetetramine, $(CH_2)_6N_4$, known also as methenamine, hexamine, hexamethyleneamine, formin, aminoform, and urotropine, is an important formaldehyde product. Commercially, it is employed principally as a special form of anhydrous formaldehyde and is used for this purpose in the manufacture of synthetic resins, the hardening of proteins, etc. A small quantity is used for medicinal purposes as a urinary antiseptic and it is listed as methenamine in the U. S. Pharmacopeia. Its uses are discussed further in the section dealing with uses of formaldehyde.

Hexamethylenetetramine is the ammono-analog of trioxane, $(CH_2O)_3$. One of its chief virtues lies in the fact that, since it contains no oxygen, water is not liberated when it is employed as a methylenating agent. This is of special importance in the final hardening of phenol-formaldehyde resins where liberation of water leads to bubble formation. In addition, hexamethylenetetramine has neither the odor nor the chemical reactivity of formaldehyde and reacts as formaldehyde only in the presence of catalysts, under the influence of heat, or when brought in contact with an active formaldehyde acceptor. As a result, its reactions are in many cases more readily controlled than those of formaldehyde.

In addition to being an ammono-formaldehyde, hexamethylenetetramine is also a tertiary amine and shows the characteristic properties of such amines, forming innumerable salts, addition compounds, and complexes. In this it resembles pyridine, triethanolamine, etc., but differs in possessing a lower degree of basicity.

Approximately twenty-six million pounds of hexamethylenetetramine were manufactured in the United States in 1951. Available U. S. Tariff Commission figures for its production are given in Table 54.

Chemical Structure. Various chemical structures have been proposed for hexamethylenetetramine by different investigators. Of these, the formula of Duden and Scharff[47] seems to be in best agreement with the facts and is generally accepted as the probable structure:

404

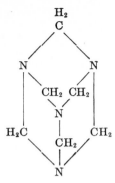

This structure is supported by the results of studies made with x-ray[44,] [102, 122, 125] and Raman[10] spectra. Measurements indicate that the carbon and nitrogen atoms in the molecule are all equivalent. In space, the nitrogen atoms occupy the summits of a tetrahedron, whereas the carbon atoms occupy the summits of an octahedron. The high symmetry possessed by the structural formula is demonstrated by Figure 32.

TABLE 54. U. S. PRODUCTION OF HEXAMETHYLENETETRAMINE*

Year	Production in lbs
1922	2,015,161
1928	1,661,645
1942	15,333,000
1943	24,733,000
1944	18,309,000
1945	11,430,000
1946	7,801,000
1947	8,317,000
1948	12,870,000
1949	8,810,000
1950	16,327,000
1951	25,939,000

* Data from U. S. Tariff Commission.

Measurements for bond angles and distances in the hexamethylene-tetramine molecule based on x-ray diffraction studies by Schomaker and Shaffer[122, 125] are:

	Gaseous $C_6H_{12}N_4$	Crystalline $C_6H_{12}N_4$
C–N distance	1.48 ± 0.01Å	1.45 ± 0.01Å
<C–N–C	$109.5 \pm 1°$	$107°$
<N–C–N	$109.5 \pm 1°$	$113° 30'$

Of the other structures advanced for hexamethylenetetramine that of

Lösekann[100] is of definite interest and is still accepted by some investigators:

$$CH_2-N=CH_2$$
$$N-CH_2-N=CH_2$$
$$CH_2-N=CH_2$$

This structure explains the behavior of hexamethylenetetramine as a mono-basic amine in that only one of the nitrogen atoms would be expected to

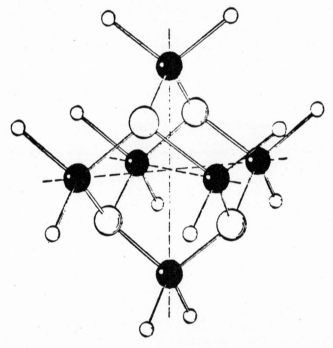

Figure 32. Structural diagram of hexamethylenetetramine. (From Dickinson R. G., and Raymond, A. L., *J. Am. Chem. Soc.*, **45**, 28 (1943).)

show monobasic characteristics. Spatial models showing this structure indicate that although it is less symmetrical than the Duden and Scharff structure, the general atomic configuration may be similar. The chief difference between the two structures lies in the fact that the atomic forces and the distances between atoms are not uniform in the Lösekann structure. Since the symmetry of the Duden and Scharff structure is probably lost when one of the nitrogen atoms becomes pentavalent, it is possible that the Lösekann structure gives us a picture of this change.

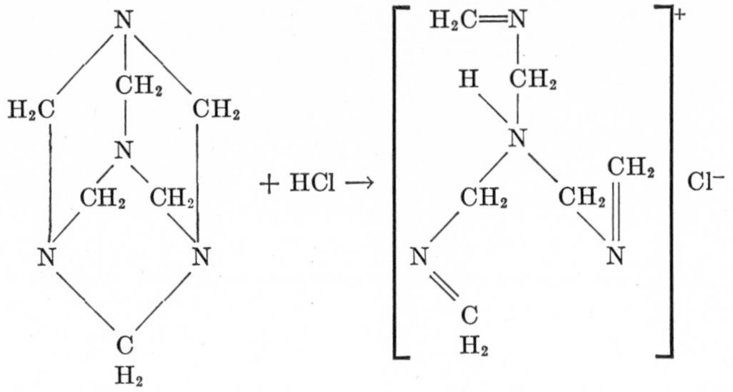

The structure first assigned to hexamethylenetetramine by Butlerov[22], together with structures postulated by Guareschi[66], Van't Hoff[137] and Dominikiewicz[45] are shown below:

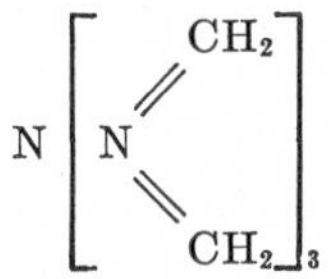

Butlerov

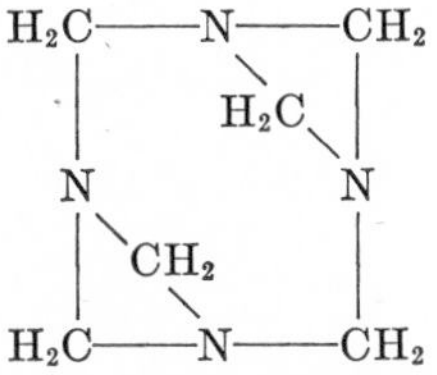

Van't Hoff

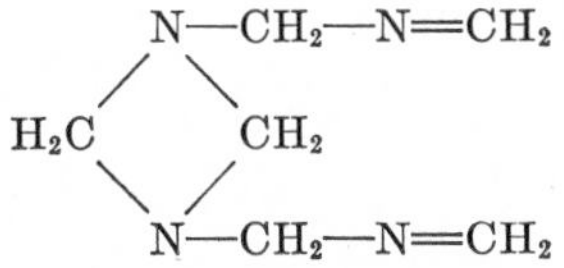

Guareschi

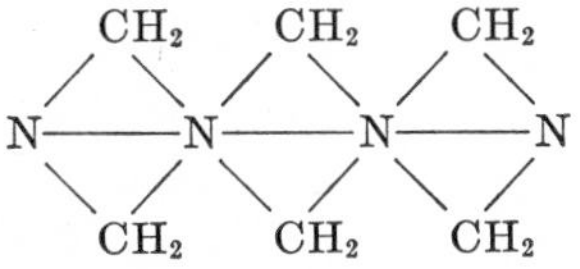

Dominikiewicz

Mechanism of Hexamethylenetetramine Formation. Hexamethylenetetramine was probably first prepared in 1859 by Butlerov[21] by reaction of gaseous ammonia and paraformaldehyde (Butlerov's "dioxymethylene"). As a result of further study, Butlerov[22] later identified it as possessing the empirical formula, $C_6H_{12}N_4$, and suggested the structural formula shown above. Preparation of hexamethylenetetramine by reaction of formaldehyde and ammonia in aqueous solution was reported by Hofmann[78] in 1869.

The formation of hexamethylenetetramine from aqueous formaldehyde and ammonia gas is indicated by the following equation, for which the heat of reaction has been calculated from the accepted thermochemical values:

$$6CH_2O \text{ (aq)} + 4NH_3 \text{ (g)} \rightarrow C_6H_{12}N_4 \text{ (aq)} + 6H_2O \text{ (l)} + 81 \text{ kcal}$$

At ordinary temperatures this reaction is rapid and proceeds quantitatively to completion.

Studies of the reaction mechanism by Duden and Scharff[47, 48] in 1895, coupled with the more recent studies of Richmond, Myers and Wright[116], indicate the formation of cyclo-trimethylenetriamine as an intermediate. However, the various steps in the reaction have not been completely demonstrated and alternate mechanisms, as well as competing reactions, are possible under various conditions of pH, temperature, reactant ratio and concentration. The formation of hexamethylenetetramine and methylamines by the action of formaldehyde on ammonium salts has already been discussed (pages 181–183).

Under alkaline conditions, it appears probable that the following mechanism predominates.

$$3NH_3 + 3CH_2O \text{ (aq)} \rightleftharpoons 3HO—CH_2—NH_2 \rightleftharpoons 3CH_2{:}NH + 3H_2O \qquad \text{I}$$

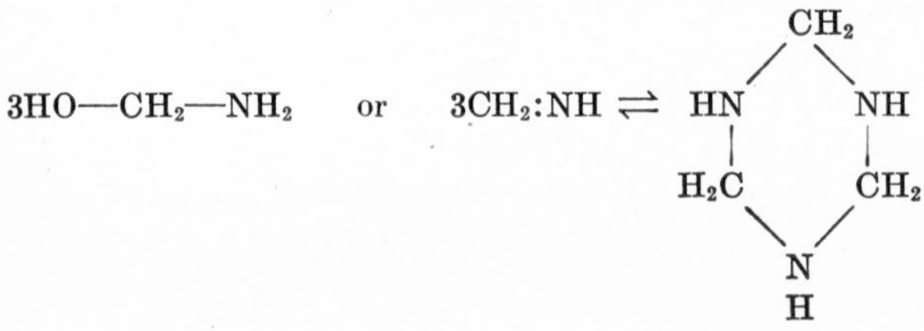

Methylolamine   *Methyleneimine*

$$3HO—CH_2—NH_2 \quad \text{or} \quad 3CH_2{:}NH \rightleftharpoons \qquad \qquad \text{II}$$

*Cyclotrimethylene-
triamine*

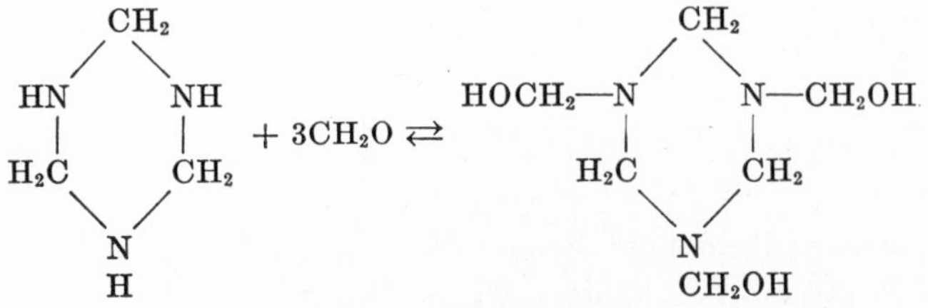

Trimethyloltrimethylenetriamine

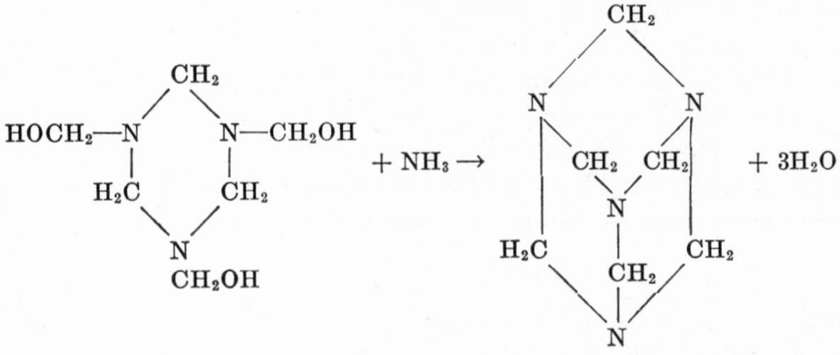

Hexamethylenetetramine

The final equation (IV) has been shown by Richmond, Myers and Wright[116] to be nonreversible in alkaline solution.

The formation of trimethylenetriamine is demonstrated by the fact that *freshly prepared* solutions containing equimolar proportions of formaldehyde and ammonia do not behave like aqueous hexamethylenetetramine and yield derivatives of trimethylenetriamine in chemical reactions. The hexamethylenetetramine salt of styphnic acid cannot be obtained when the solution is kept alkaline. Furthermore, 1,5-endomethylene-3,7-di-[*m*-nitrobenzenediazo]-1,3,5,7-tetrazacyclo-octane is not formed on reaction with diazotized *m*-nitroaniline as it is with hexamethylenetetramine. On reaction with benzoyl chloride and alkali, tribenzoyltrimethylenetriamine (1,3,5-tribenzoyl-1,3,5-triazacyclohexane) and some methylenedibenzamide is obtained with only a trace of 1,3,5-tribenzoyltriazapentane which is produced when hexamethylenetetramine is treated with these reagents. With an excess of nitrous acid, trinitrosotrimethylenetriamine (1,3,5-trinitroso-1,3,5-triazacyclohexane) is formed. This latter compound is also obtained from hexamethylenetetramine but in much lower yield. When less than an excess of nitrous acid is employed, both the formaldehyde-ammonia solution and hexamethylenetetramine yield dinitrosopentamethylenetetramine (1,5-endomethylene-3,7-dinitroso-1,3,5,7-tetrazacyclo-octane). Although the formation of these nitroso derivatives was apparently first reported by Griess and Harrow[64] and Mayer[104] in 1888, our knowledge of the above reactions is based largely on the work of Duden and Scharff[47, 48] as confirmed and elaborated by Richmond, Myers and Wright[116]. The major chemical relationships are indicated by the structural formulas on page 410.

The fresh ammonia-formaldehyde solution employed in the majority of the trimethylenetriamine reaction studies was prepared by neutralizing an equimolar solution of ammonium chloride in commercial formaldehyde with caustic soda. This is apparently a relatively pure solution of trimethylenetriamine. A more concentrated solution having similar properties was obtained by Henry[73] by adding an excess of anhydrous potassium carbonate to an equimolar solution of aqua-ammonia in commercial formaldehyde. Henry believed this was trimethylolamine, $N(CH_2OH)_3$, but Richmond, Myers and Wright conclude it to be essentially a water-methanol solution of trimethylenetriamine by its reactions as previously described. On reacting this solution with 60 per cent sulfuric acid, they obtained the sulfate of methylenediamine, (see page 292) and hexamethylenetetramine. On the basis of these findings, they believe that trimethylenetriamine tends to decompose into methylenediamine and the elements of hexamethylenetetramine. Strong acids partially stabilize these fragments as salts.

Studies of the reaction rate of formaldehyde and ammonia to form hexamethylenetetramine have been reported by Baur and Ruetschi[12], Boyd

$$
\begin{array}{ccc}
\text{NO} & \text{H} & \text{COC}_6\text{H}_5 \\
| & | & | \\
\text{N} & \text{N} & \text{N} \\
\diagup\,\diagdown & \diagup\,\diagdown & \diagup\,\diagdown \\
\text{H}_2\text{C}\quad\text{CH}_2 & \text{H}_2\text{C}\quad\text{CH}_2 & \text{H}_2\text{C}\quad\text{CH}_2 \\
|\qquad\quad| & |\qquad\quad| & |\qquad\quad| \\
\text{ON}-\text{N}\quad\text{N}-\text{NO} & \text{HN}\qquad\text{NH} & \text{C}_6\text{H}_5\text{CO}-\text{N}\quad\text{N}-\text{COC}_6\text{H}_5 \\
\diagdown\,\diagup & \diagdown\,\diagup & \diagdown\,\diagup \\
\text{C} & \text{C} & \text{C} \\
\text{H}_2 & \text{H}_2 & \text{H}_2
\end{array}
$$

$\xleftarrow{\text{HNO}_2}$ $\xrightarrow[\text{C}_6\text{H}_5\text{CO}]{\text{C}_6\text{H}_5\text{COCl} \;\; \text{NaOH}}$

Trinitrosotrimethylene-   *Trimethylene-*   *Tribenzoyltrimethylene-*
triamine   *triamine*   *triamine*

$\text{HNO}_2 \nwarrow$ $3\text{CH}_2\text{O} + \text{NH}_3$

$$
\begin{array}{c}
\text{C}_6\text{H}_4\text{NO}_2 \\
| \\
\text{N}_2 \\
| \\
\text{H}_2\text{C}-\text{N}----\text{CH}_2 \\
|\qquad\qquad| \\
\text{N}-\text{CH}_2-\text{N} \\
|\qquad\qquad| \\
\text{H}_2\text{C}-\text{N}----\text{CH}_2 \\
| \\
\text{C}_6\text{H}_4\text{NO}_2
\end{array}
$$

$\xleftarrow{\text{ClN}_2\text{C}_6\text{H}_4\text{NO}_2}$

$$
\begin{array}{c}
\text{N} \\
\diagup\;|\;\diagdown \\
\text{H}_2\text{C}\quad\text{CH}_2\quad\text{CH}_2 \\
|\qquad\text{N}\qquad| \\
\text{CH}_2\diagup\diagdown\text{CH}_2 \\
\text{CH}_2 \\
|\qquad\qquad| \\
\text{N}\qquad\qquad\text{N} \\
\diagdown\qquad\qquad\diagup \\
\text{C} \\
\text{H}_2
\end{array}
$$

$\xrightarrow[\text{NaOH}]{\text{C}_6\text{H}_5\text{COCl}}$

$$
\text{C}_6\text{H}_5\text{CON}\diagup^{\text{CH}_2\text{NHCOC}_6\text{H}_5}_{\diagdown\text{CH}_2\text{NHCOC}_6\text{H}_5}
$$

1,5-Endomethylene-   *Hexamethylene-*   *1,3,5-Tribenzoyltriazapentane*
3,7-di-[m-nitro benzenediazo]   *tetramine*
1,3,5,7-tetrazacyclo-octane

and Winkler[18], and Polley, Winkler, and Nicholls[113]. The former indicate that the rate is probably of the third order with an initial consumption of ammonia and formaldehyde in the stoichiometrical proportion of one to two. This is not supported by the work of other investigators. Richmond, Myers and Wright[116] note that an initial rapid reaction, probably the formation of trimethylenetriamine, is followed by a much slower phase which gives hexamethylenetetramine, possibly by formation of trimethyloltrimethylenetriamine and reaction of the latter with ammonia. Boyd and Winkler[18] observe that the reaction is characterized by a high initial rate but conclude that it is too complex to be classified as to order of reaction on the basis of existing data. Polley, Winkler, and Nicholls studied the reaction with ammonium salts in the absence and presence of buffering agents (see pages 182–183). Hypothetical mechanisms involving methylol amines and a hemi-hexamine are proposed by these latter investigators[18, 113].

As previously pointed out (page 182), ammonia is apparently the active reagent with which formaldehyde reacts to produce hexamethylenetetramine and the effect of hydrogen ions in reducing the reaction rate in salt solutions is determined by the equilibrium involving ammonia and the hydrogen ion. The equilibrium for hexamethylenetetramine formation was calculated for acid buffered solutions by Baur and Ruetschi[12] on the basis of the reaction,

$$6CH_2O + 4NH_4^+ \rightleftharpoons (CH_2)_6N_4 + 4H^+ + 6H_2O.$$

Values reported for the equilibrium constant $\dfrac{[CH_2O]^6 \cdot [NH_4^+]^4}{[(CH_2)_6N_4][H^+]^4}$, at various temperatures are as follows: 1.58×10^{10} at 25°C, 6.48×10^{10} at 38°C, 2.18×10^{11} at 50°C and 4.07×10^{11} at 64°C.

Manufacture of Hexamethylenetetramine. Hexamethylenetetramine is usually manufactured by the reaction of ammonia with aqueous formaldehyde. In a German process described by Chemnitius in 1928[31], 30 per cent aqueous formaldehyde containing 1 to 2 per cent methanol was placed in water-cooled clay vessels and ammonia gas passed in gradually, care being taken to keep the temperature of the mixture at or below 20°C to avoid loss of vapors. Addition of ammonia was continued until a slight excess had been added, after which a small amount of animal charcoal was added and the mixture filtered. The hexamethylenetetramine solution thus obtained was then concentrated in a steam-heated enamelled vacuum evaporator until most of the water had been evaporated and crystallization of the product had taken place. The damp crystals were separated from the mother liquor, washed with 25 per cent ammonia, centrifuged, washed with a little distilled water, and dried on cloth hurdles at 50°C. Mother liquor and washings were reworked to obtain additional product. Yields of 96 per cent theoretical based on formaldehyde were reported. The clay reactors employed in this procedure had a capacity of 250 liters and were charged with 180 kg of 30 per cent formaldehyde to which approximately 22 kg of ammonia was added. The complete process cycle was 12 hours and with ten reactors and two vacuum evaporators, approximately 11,000 pounds of hexamethylenetetramine were produced per month. In a recent semicontinuous German process[30], 30 per cent formaldehyde and 27 per cent aqua ammonia were fed continuously into a reactor in a weight ratio of approximately 2 to 1. The liquor from the reactor then passed into a three-compartment evaporator where it was concentrated to 40 per cent solids. This solution was fed in turn to a series of four batch evaporators in which distillation was carried out at 150 mm pressure and a temperature of 60 to 70°C. A suspension of hexamethylenetetramine crystals was pumped from the evaporators to two continuous centrifuges equipped with steam heated dryers through which the product was screw propelled at a temperature of about 105°C. The hexamethylenetetramine from these dryers was ground, screened and bagged. Mother liquor from the centrifuges was diluted with water, decolorized with active charcoal and returned to the concentrator. This plant produced 500 tons of 99.9 per cent hexamethylenetetramine per month.

For manufacture of hexamethylenetetramine, the technical literature on the materials of construction indicates that aluminum equipment[11] may be employed, and that several grades of chrome- and chrome-manganese steel, as well as cast iron containing 14.5 per cent silicon, are probably satisfactory[88].

Special methods for the isolation of hexamethylenetetramine from the aqueous solution in which it is normally obtained are described in patents. It is claimed, for example, that an extremely pure product may be precipitated by saturating the solution with ammonia[29]. Another patented process[108] describes addition of hexamethylenetetramine-water compositions to a hot, agitated liquid which is substantially nonvolatile and in which neither hexamethylenetetramine nor water is soluble. The liquid, preferably a white mineral oil, is continuously agitated at a temperature of about 275°F (135°C). By this procedure it is claimed that the volatile constituents of the hexamethylenetetramine solution are removed in the gaseous state and the solid hexamethylenetetramine is left in the oil, from which it is then separated. Schideler and Davis[120] add formaldehyde and ammonia to a mother liquor containing hexamethylenetetramine, cool to remove product crystals and recycle the reaction medium.

According to Novotny and Vogelsang[108], the common practice of producing hexamethylenetetramine by passing gaseous ammonia into concentrated formaldehyde may be modified by reacting anhydrous ammonia with vaporized aqueous formaldehyde. Kolosov[91] claims that hexamethylenetetramine of 98 per cent purity can be obtained by the reaction of gas containing 0.25 kg formaldehyde per cubic meter with anhydrous ammonia. Veller and Grigoryan[138] prepared hexamethylenetetramine by simultaneously admitting gaseous ammonia and formaldehyde into a vessel partly filled with water held at 20 to 25°C. A process variation described by Landt and Adams[94] involves the reaction of formaldehyde and ammonia in alcohol below 75°C. Precipitation or crystallization of hexamethylenetetramine from a nonaqueous medium is also employed by Eickmeyer[52] in a continuous process which involves pressure distillation of approximately 37 per cent formaldehyde to yield a formaldehyde rich vapor which is fed into a column with an entraining liquid, such as methanol or acetone. Water escapes from the bottom of the column and a mixture of formaldehyde and the entrainer is removed and reacted with ammonia.

In addition to the ammonia processes already discussed, a number of other procedures for the manufacture of hexamethylenetetramine are also described in the patent literature[2]. In a method employing ammonium salts[121], formaldehyde solution is added gradually to a solution of ammonium chloride which is supersaturated with sodium bicarbonate or to the ammonium bicarbonate-salt solutions obtained in the Solvay soda process. It is stated that the saline hexamethylenetetramine produced by this method can be used directly for resin manufacture or purified by extraction with alcohol. Another procedure of this type[42] describes addition of formaldehyde to an aqueous solution of ammonium sulfate and the oxide, hydroxide or carbonate of an alkaline-earth metal.

Processes for the preparation of hexamethylenetetramine by the reaction of ammonia and methylene chloride have been patented by Carter[27, 28].

Methods for obtaining hexamethylenetetramine from methane are also encountered in the patent literature. Plauson[112] oxidizes methane in admixture with ammonia gas by the action of oxygen in the presence of copper, silver and other catalysts, obtaining hexamethylenetetramine as a reaction product. Nashan[107] passes a mixture of methane and air through a visible, non-dampened discharge of high-tension, high-frequency alternating current to produce hexamethylenetetramine and formaldehyde. This procedure involves synthesis and reaction of both ammonia and formaldehyde in a one-step process.

Properties of Hexamethylenetetramine. Pure hexamethylenetetramine is a colorless, odorless crystalline solid with a sweet taste. In general, it appears commercially in two grades of purity: U.S.P., which is essentially a chemically pure product; and technical which although 99 per cent pure or better contains a small amount of moisture (usually not more than 0.3 per cent) and on combustion leaves a slight ash (usually not more than 0.3 per cent). Hexamethylenetetramine crystallizes regularly in rhombic dodecahedrons[58] which are stated to show piezoelectric properties[53]. Although it does not normally occur in hydrated form, a hexahydrate, $C_6H_{12}N_4 \cdot 6H_2O$, is reported[24, 36]. This hydrate is obtained in the form of crystalline prisms when a saturated aqueous solution is cooled to a temperature slightly above 0°C. The hydrate crystals decompose with efflorescence at 13.5°C.

Anhydrous hexamethylenetetramine sublimes on heating in the air with slight decomposition but does not melt. In a vacuum, it sublimes readily at 230 to 270°C[132] and vaporizes without appreciable decomposition when heated in a stream of hydrogen[62]. When heated in a sealed tube hexamethylenetetramine decomposes with charring at temperatures above 280°C.

On ignition, it burns slowly with a pale blue flame. Its heat of combustion is approximately 7.17 kcal/g. Values reported for the molal heat of combustion at constant pressure are 1006.7 (I.C.T.)[81], 1004.7 (Landolt-Bornstein)[93] and 1003.6 (Delépine and Badoche)[41]. The molal heats of formation calculated from these figures are −32.7 kcal, −30.7 kcal, and −29.6 kcal, respectively.

Hexamethylenetetramine dissolves readily in water with the evolution of heat. According to Delépine[37], the molal heat of solution of hexamethylenetetramine is 4.8 kcals per mol, or 20.1 kilojoules per mol at infinite dilution. Hexamethylenetetramine is unusual in that its water solubility decreases with increasing temperature, a fact which was apparently first reported by Grützner in 1898[65]. This property, which is shared by some of the other tertiary amines, is probably due to increasing association with water at low temperatures. The following solubility figures for U.S.P. hexamethylenetetramine at various temperatures were determined in the writer's laboratory:

Temp. (°C)	Wt. % Hexamethylene-tetramine in Solution
0	47.3
25	46.5
50	45.0

Recent measurements[85a] indicate that the solubility tends to increase gradually with further rise in temperature reaching a value of approximately 46.3 per cent at 100°C. The solubility of hexamethylenetetramine in water is lowered by dissolved ammonia. The magnitude of this effect is indicated by the following reported data[29] on the solubility of hexamethylenetetramine at room temperature in water and aqueous solutions containing ammonia:

Dissolved Ammonia (g per 100 cc)	Dissolved Hexamethylenetetramine (Saturation Value) (g per 100 cc)
None	52.0
18.4	22.2
35.7	6.4

Liquid ammonia is reported to dissolve 1.3 g hexamethylenetetramine per 100 cc[29].

Hexamethylenetetramine is somewhat soluble in alcohols and slightly soluble in ether and aromatic hydrocarbons. With the exception of chloroform, in which it is fairly soluble, it is only slightly soluble in chlorinated aliphatics. Solubility figures for representative solvents as determined by Ütz[135] are shown in Table 55. The solubility of hexamethylenetetramine in glycerol is reported as 26.5 per cent for 86.5 per cent glycerol and 20.9 per cent for 98.5 per cent glycerol[55].

Impure hexamethylenetetramine may be purified by recrystallization from solvents such as alcohol and chloroform, or by precipitation from aqueous solution by saturation with ammonia gas. According to Ohara[109], material of the highest purity is obtained by the latter procedure.

Aqueous solutions of hexamethylenetetramine are mildly basic, having pH values in the neighborhood of 8 to 8.5 for concentrations ranging from approximately 5 to 40 per cent. Russo[113] reports its dissociation constant as 1.4×10^{-9}.

Pure water solutions of hexamethylenetetramine are comparatively stable at ordinary temperatures, showing only an extremely slight degree of hydrolysis to formaldehyde and ammonia. According to Fincke[57], the formaldehyde content of a pure solution, after standing 15 hours at room temperature, is approximately 5 ppm. Hydrolysis may be appreciably reduced by

TABLE 55. SOLUBILITY OF HEXAMETHYLENETETRAMINE* IN VARIOUS SOLVENTS[124]

| Solvent | Solubility in g Hexamethylene-tetramine per 100 cc Solvent | |
	Room Temp.	Elevated Temp.
Petroleum ether...............	Insoluble	Insoluble
Ethyl ether..................	0.06	0.38
Trichlorethylene.............	0.11	—
Xylene.......................	0.14	—
Carbon bisulfide.............	0.17	—
Benzene.....................	0.23	—
Tetrachlorethane.............	0.50	—
Acetone.....................	0.65	—
Carbon tetrachloride.........	0.85	—
Amyl alcohol.................	1.84	—
Absolute ethanol.............	2.89	—
Methanol....................	7.25	11.93
Chloroform..................	13.40	14.84

* Data of Ütz[135].

the addition of small amounts (0.1 to 1 per cent) of sodium carbonate[118] and is accelerated by the addition of acid. According to Philippi and Löbering[99], the hydrolysis constant is a function of pH, decreasing from about 2×10^{-3} at pH values in the neighborhood of 2 to 1×10^{-4} at pH 6. Hydrolysis is also accelerated by heating. In a study of the effect of steam sterilization on ampoules of 25 per cent hexamethylenetetramine solution prepared for medicinal use, Toni[133] reports that a freshly prepared, formaldehyde-free solution whose pH was 8.32 showed a formaldehyde content of 0.07 per cent and a pH of 9.50 after 3 months' storage, whereas a solution which had been steam-treated showed a formaldehyde content of 0.12 per cent and a pH value of 10.01. The rise in pH is an index of ammonia liberation. Vickers[139] observed that ammonia and formaldehyde tend to recombine on standing at room temperature after heating in sealed containers.

Physiological Properties of Hexamethylenetetramine. Pure hexamethylenetetramine may be taken internally in small amounts, and is used in medicine as a urinary antiseptic (page 492). However, it must be pointed out that toxic action has been reported in some instances both in the medical and industrial use of this product. Some persons suffer a skin rash if they come in contact with hexamethylenetetramine or fumes from the heated material. It is not entirely clear whether this apparent toxicity is caused by the hexamethylenetetramine itself. It is possible that impurities in the technical material, reaction products formed in industrial use, decomposition products or hydrolysis products may be responsible. On hydrolysis, hexamethylenetetramine yields formaldehyde, which would naturally affect individuals who are sensitive to this aldehyde. However, this is not the whole story, since in some instances persons who are not sensitive to formaldehyde itself develop a rash on exposure to hexamethylenetetramine. In all cases involving toxic symptoms, a physician or dermatologist should be consulted. In general, methods of treatment involve removal from the zone of exposure and application of an anti-pruritic.

Chemical Reactions of Hexamethylenetetramine

In general, the reactions of hexamethylenetetramine fall into two classifications: Those in which it behaves as a tertiary amine and those in which it reacts as formaldehyde. Many reactions depend primarily on hydrolysis to formaldehyde which then behaves in a normal fashion.

Reactions with Inorganic Compounds. *Hydrolysis and Reduction.* When hexamethylenetetramine is heated with strong acids in aqueous solutions hydrolysis takes place, with liberation of formaldehyde and formation of ammonium salts. Reaction is practically quantitative and may be used for hexamethylenetetramine analysis (page 430):

$$(CH_2)_6N_4 + 4HCl + 6H_2O \rightarrow 6CH_2O + 4NH_4Cl$$

According to Graymore[63], a small amount of methylamine hydrochloride is also obtained when hexamethylenetetramine is refluxed for a short time with excess hydrochloric acid, and larger amounts are obtained with prolonged refluxing. This is apparently the result of secondary reactions of the type known to take place when formaldehyde is heated with ammonium chloride (page 182).

On reduction with zinc dust and hydrochloric acid, hexamethylenetetramine gives ammonium chloride and trimethylamine hydrochloride as the principal products[35, 63] Varying amounts of methylamine hydrochloride and in some cases a little dimethylamine hydrochloride were also obtained.

According to Graymore[63], the reaction mechanism probably involves formation of trimethylamine and trimethylenetriamine, which then breaks up to give ammonia and formaldehyde in equilibrium with methyleneimine and water in the acid media, as indicated in the following equations:

$$
\begin{array}{c}
\text{N} \\
\diagup \ \big| \ \diagdown \\
\text{H}_2\text{C} \quad \text{CH}_2 \quad \text{CH}_2 \\
\big| \quad \text{N} \quad \big| \\
\diagup \diagdown \\
\text{CH}_2 \ \text{CH}_2 \\
\diagup \diagdown \\
\text{N} \qquad \text{N} \\
\diagdown \ \ \diagup \\
\text{C} \\
\text{H}_2
\end{array}
\ + 3\text{H}_2 \rightarrow \text{N}(\text{CH}_3)_3 +
\begin{array}{c}
\text{H}_2 \\
\text{C} \\
\diagup \ \diagdown \\
\text{HN} \qquad \text{NH} \\
\big| \qquad \big| \\
\text{H}_2\text{C} \qquad \text{CH}_2 \\
\diagdown \ \diagup \\
\text{N} \\
\text{H}
\end{array}
$$

$$
\begin{array}{c}
\text{H} \\
\text{N} \\
\diagup \ \diagdown \\
\text{H}_2\text{C} \qquad \text{CH}_2 \\
\big| \qquad \big| \\
\text{HN} \qquad \text{NH} \\
\diagdown \ \diagup \\
\text{C} \\
\text{H}_2
\end{array}
\ + 3\text{H}_2\text{O} \rightarrow 3\text{CH}_2\text{O} + 3\text{NH}_3
$$

$$3\text{CH}_2\text{O} + 3\text{NH}_3 \rightleftharpoons 3\text{CH}_2{:}\text{NH} + 3\text{H}_2\text{O}$$

Reduction of methyleneimine would naturally give methylamine. Dimethylamine would result from reduction of formaldehyde derivatives of methylamine.

Alkaline reduction of hexamethylenetetramine with potassium hydroxide and zinc dust apparently proceeds in the same manner as the acid reduction, since Grassi[62] reports that trimethylamine and monomethylamine are produced.

Electrolytic reduction of acidic hexamethylenetetramine solutions results in the formation of varying amounts of the three methyl amines. Knudsen[89] reports that slow reduction with low current densities gives approximately equivalent quantities of the three amines, whereas high current densities result in a rapid reduction with an increased yield of methylamine.

Salt Formation with Mineral Acids. Salts of hexamethylenetetramine, which are the primary products formed when it is reacted with acids, may be isolated in many instances. In general, hexamethylenetetramine behaves as a monobasic compound and may in fact be titrated as such with mineral acids when methyl orange is used as indicator. Salts of hexamethylenetetramine and mineral acids may best be isolated when formed in nonaqueous solvents or, in some cases, in cold aqueous solutions. The hydrochloride of hexamethylenetetramine, $C_6H_{12}N_4 \cdot HCl$, may be obtained by addition of aqueous hydrochloric acid to an alcoholic solution of the base or by the action of hydrogen chloride on a hot solution in absolute alcohol[23]. According to Butlerov, it melts at 188 to 189°C. By the action of excess hydrogen chloride, a compound having the composition $C_6H_{12}N_4 \cdot 2HCl$ is formed[4]. This product is believed to be a molecular compound of hydrogen chloride and the neutral salt. When a solution of hexamethylenetetramine in cold hydrochloric acid is saturated with hydrogen chloride, dichloromethyl ether is obtained[131].

The sulfate of hexamethylenetetramine, $(C_6H_{12}N_4)_2 \cdot H_2SO_4$, is precipitated by the action of sulfuric acid on hexamethylenetetramine in cold alcoholic solution[24, 36]. It is acid to phenolphthalein but not to methyl orange.

Salts of hexamethylenetetramine and many other inorganic acids including the hydrobromide, $C_6H_{12}N_4 \cdot HBr$,[4, 65], the hydriodide, $C_6H_{12}N_4 \cdot HI$[98], the phosphates, $C_6H_{12}N_4 \cdot H_3PO_4$[105] and $5C_6H_{12}N_4 \cdot 6H_3PO_4 \cdot 10H_2O$[65, 66], the perchlorate, $C_6H_{12}N_4 \cdot HClO_4$[83], and the explosive chromates, $2C_6H_{12}N_4 \cdot H_2Cr_2O_7$ and $2C_6H_{12}N_4 \cdot H_2Cr_4O_{13}$[24, 36], have also been reported.

According to Delépine[36] and other investigators[24], a mononitrate of hexamethylenetetramine may be obtained by the action of nitric acid on an aqueous solution of hexamethylenetetramine at 0°C. With more concentrated acid a dinitrate, $C_6H_{12}N_4 \cdot 2HNO_3$ (m.p. 165°C), is produced[67, 105]. Hale[67] prepared this product in 82 per cent yield by gradual addition of C.P. nitric acid (density = 1.42) to a 25 per cent solution of hexamethylenetetramine at 0°C. The precipitated salt may be separated from acid by filtration through glass wool and dried after washing with cold 50 per cent ethyl alcohol and ether. It is readily soluble in water at room temperature but gradually decomposes when the solution is allowed to stand. The solution is acid and the total combined acid may be determined by titration with alkali. It is insoluble in alcohol, ether, chloroform, and carbon tetrachloride. With dilute nitric acid, Hale states that the hydrolysis reaction predominates and little salt formation takes place. Bachmann and Sheehan[9] obtained a 95 per cent yield of this product by a process essentially similar to that described by Hale.

Nitration. Nitration of hexamethylenetetramine gives the high explosive cyclo-trimethylenetrinitramine[43, 67] also called cyclonite, hexogen, and more recently designated as RDX (cf. Chapter 21, pages 473–474). This nitroamine was first prepared by Henning[72] in 1899 by the action of nitric acid on hexamethylenetetramine dinitrate. Its chemical structure was clarified by Herz[74] in 1920, who also recognized its nature as an explosive. In 1925, Hale[67] reported an improved method of preparation. Research carried out in connection with the development of a practical method for manufacturing this explosive for use in World War II resulted in outstanding technical improvements and a detailed knowledge of the chemistry involved. A popular account of this wartime development is described by Baxter[13] in his book, "Scientists Against Time."

Hale's studies[67] demonstrated that a 74 per cent yield of cyclonite could be obtained by gradual addition of hexamethylenetetramine to an excess of 99.8 per cent nitric acid at about −20°C, followed by drowning the mixture in ice water, washing the crystalline product precipitate with ice water and drying at 100°C. Highly concentrated acid was essential since the yield dropped to 15 per cent with 92 per cent acid. Wright and co-workers[144] have since demonstrated that an average yield of 83 per cent can be obtained by careful control of the rate of hexamethylenetetramine addition so that local decomposition is avoided. In this procedure, 1 mol of hexamethylenetetramine is added to 21 mols of 99.5 to 99.9 per cent nitric acid at 20 to 25°C over a half-hour addition period prior to drowning by addition to cracked ice. These investigators conclude that the reaction is a nitrolysis as indicated below:

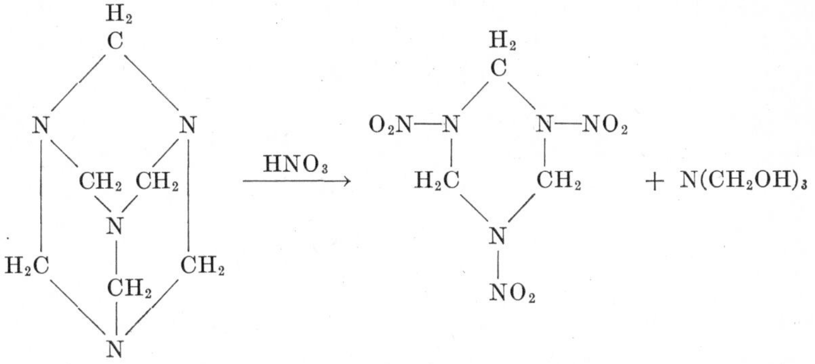

Cyclo-trimethylene-   Trimethylol
trinitramine   amine

According to their findings the aqueous diluate obtained when the nitration mixture is added to ice contains dimethylolnitramide, $NO_2 \cdot N(CH_2OH)_2$, its dinitrate or both.

Bachmann and Sheehan[9] developed an improved process which makes it possible to obtain two mols of cyclonite or RDX from one mol of hexamethylenetetramine by a nitration reaction in which ammonium nitrate and acetic anhydride are employed as indicated in the equation:

$$C_6H_{12}N_4 + 4HNO_3 + 2NH_4NO_3 + 6(CH_3CO)_2O \rightarrow 2C_3H_6N_3(NO_2)_3 + 12CH_3COOH$$
$$RDX$$

This reaction is reported to give a 70 per cent yield. It is carried out at 75°C by portion-wise addition of 98 per cent nitric acid and acetic anhydride, and a mixture of hexamethylenetetramine dinitrate and ammonium nitrate simultaneously and equivalently to a flask at such a rate that the temperature is kept at 75°C. The reagents are employed in approximately the same amounts indicated in the equation except for the nitric acid which must be employed in slight excess. This excess is kept to a minimum by adding hexamethylenetetramine as the dinitrate. The product crystals are filtered from the anhydrous reaction mixture.

Pure cyclo-trimethylenetrinitramine is a colorless, crystalline compound which melts at 205°C[143]. Its molecular heat of combustion at constant pressure is reported by Delépine and Badoche[41] as 503.9 kcal. Their calculated value for its molal heat of formation is −15.7 kcal. It is insoluble in cold water and alcohol but may be crystallized from 70 per cent nitric acid, acetic acid or acetone. Because of its nature as a high explosive, it must be handled with extreme care and with full knowledge of the hazards involved.

A by-product of hexamethylenetetramine nitrolysis studied by Wright and co-workers as well as Bachmann and Sheehan[9] is 1,3,5,7-tetranitro-1,3,5,7-tetrazacyclo-octane, known as HMX ("high melting explosive") and octogen.

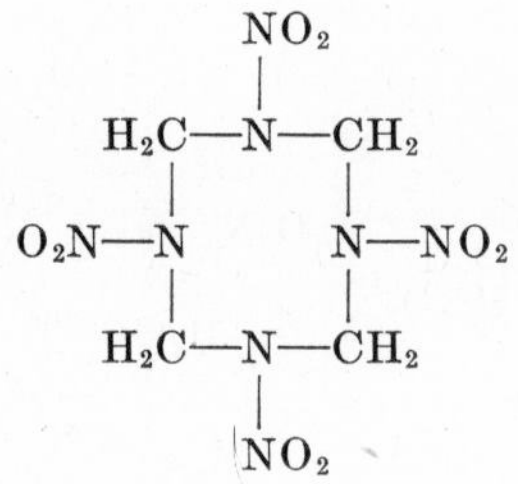

$$
\begin{array}{c}
NO_2 \\
| \\
H_2C\!-\!N\!-\!CH_2 \\
| \qquad\qquad | \\
O_2N\!-\!N \qquad N\!-\!NO_2 \\
| \qquad\qquad | \\
H_2C\!-\!N\!-\!CH_2 \\
| \\
NO_2
\end{array}
$$

The presence of this compound (m.p. 269°C) in crude cyclotrimethylenetrinitramine accounts for its low melting point of 202 to 203.5°C[143]. It can be prepared by the nitrolysis of dinitrosopentamethylenetetramine which Wright and co-workers[144] synthesized by the reaction of methylenediamine with dimethylol nitramide as indicated.

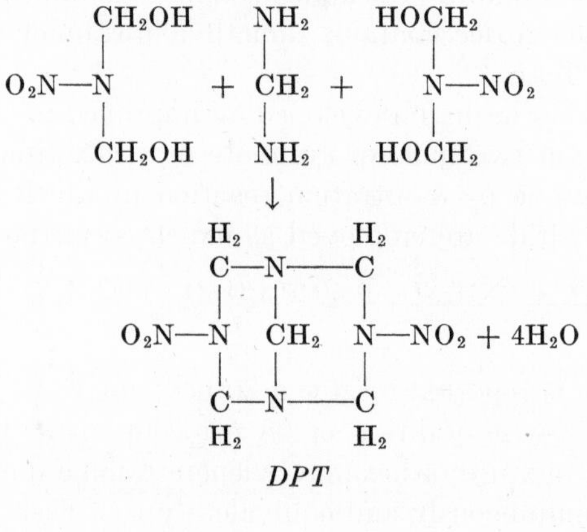

$$
\begin{array}{ccc}
CH_2OH & NH_2 & HOCH_2 \\
| & | & | \\
O_2N\!-\!N & +\ CH_2\ + & N\!-\!NO_2 \\
| & | & | \\
CH_2OH & NH_2 & HOCH_2
\end{array}
$$

$$\downarrow$$

$$
\begin{array}{ccc}
H_2 & & H_2 \\
C\!-\!N\!-\!\!-\!C & & \\
| \quad | \quad | & & \\
O_2N\!-\!N \quad CH_2 \quad N\!-\!NO_2\ +\ 4H_2O \\
| \quad | \quad | & & \\
C\!-\!N\!-\!\!-\!C & & \\
H_2 & & H_2
\end{array}
$$

DPT

The presence of dimethylol nitramide in the aqueous diluate obtained from the Hale process for cyclotrimethylenetrinitramine is indicated by the precipitation of DPT on neutralization with ammonia.

For details on the complex chemistry of hexamethylenetetramine nitrolysis, the reader is referred to the papers on this subject by Wright and co-workers[142, 143, 144, 145, 146, 147] and Gillies, Williams and Winkler[60].

Reactions with Nitrous Acid. The formation of dinitrosopentamethylenetetramine and trinitrosotrimethylenetriamine by reaction of hexamethylenetetramine with nitrous acid have already been described. Wright and co-workers obtained pure cyclo-trimethylenetrinitramine by oxidation of the trinitroso derivative with hydrogen peroxide in concentrated nitric acid.

Reactions with Hydrogen Peroxide. The primary reaction product of hexamethylenetetramine and hydrogen peroxide is the addition compound or hydroperoxide, $C_6H_{12}N_4 \cdot H_2O_2$. Von Girsewald[61] obtained this material in almost quantitative yield by vacuum evaporating at 40 to 50°C the solution obtained by dissolving hexamethylenetetramine in a slight excess of 30 per cent hydrogen peroxide. The stability of the somewhat unstable product is increased if a little acid (less than 3 per cent) is added to the reaction mixture. However, decomposition takes place if the evaporation temperature is allowed to exceed 70°C. The hydroperoxide is a colorless crystalline compound readily soluble in water or alcohol. It explodes when treated with concentrated sulfuric acid and oxidizes hydrochloric acid with liberation of chlorine.

The reaction of hexamethylenetetramine and hydrogen peroxide in the presence of substantial quantities of acid results in the formation of hexamethylenetriperoxidediamine, HMTD, a primary explosive which according to Bebie[14] may be employed in detonator compositions (page 474). Marotta and Alessandrini[103], who accept Lösekann's structural formula for hexamethylenetetramine, suggest that the compound is formed by the reaction:

$$
\begin{array}{ccc}
\text{CH}_2 \cdot \text{N} : \text{CH}_2 & & \text{CH}_2 \cdot \text{O} \cdot \text{O} \cdot \text{CH}_2 \\
\diagup & & \diagup \qquad\qquad \diagdown \\
2\text{N}\!\!-\!\!\text{CH}_2 \cdot \text{N} : \text{CH}_2 \;+\; 3\text{H}_2\text{O}_2 \xrightarrow{\text{Acid}} \text{N}\!\!-\!\!\text{CH}_2 \cdot \text{O} \cdot \text{O} \cdot \text{CH}_2\!\!-\!\!\text{N} \;+\; 6\text{CH}_2 : \text{NH} \\
\diagdown & & \diagdown \qquad\qquad \diagup \\
\text{CH}_2 \cdot \text{N} : \text{CH}_2 & & \text{CH}_2 \cdot \text{O} \cdot \text{O} \cdot \text{CH}_2
\end{array}
$$

An alternate mechanism involving the Duden and Scharff formula with liberation of trimethylenetriamine, $(\text{CH}_2:\text{NH})_3$, as a reaction intermediate could also be postulated. Von Girsewald[61] obtained hexamethylene-triperoxide-diamine in 66 per cent yield by addition of 28 g hexamethylenetetramine and 42 g citric acid to 140 g 30 per cent hydrogen peroxide. The reaction mixture warms somewhat and the peroxide separates in small rhombic crystals (27.5 g)*. Because of its extremely explosive nature, these crystals should be kept wet. The dry crystals explode violently when subjected to mechanical shock. This compound may also be produced by reactions involving formaldehyde (page 189).

Reactions with Hydrogen Cyanide. Hydrogen cyanide reacts gradually with hexamethylenetetramine in concentrated aqueous solution giving imidoacetonitrile, $HN(CH_2CN)_2$, (m.p. 75°C). Ammonia is liberated from the reaction mixture and dark-colored by-products are produced. Best results are obtained in the presence of a catalytic quantity of sulfuric or hydrochloric acid. If more than 2 per cent acid is employed, a small yield of nitriloacetonitrile $N(CH_2CN)_3$, is also produced. The latter product may be obtained in good yield by addition of 450 g of concentrated hydrochloric acid to a solution of 100 g hexamethylenetetramine in 500 cc water, plus 120 g hydrogen cyanide in the form of 30 to 40 per cent aqueous solution. The product crystallizes from the reaction mixture and may be purified by fractional crystallization from alcohol[55]. Nitriloacetonitrile is also obtained by addition of hydrochloric acid to a water solution of hexamethylenetetramine and potassium cyanide[46].

Reactions with Sulfur and Sulfur Derivatives. When hexamethylenetetramine is heated with sulfur at approximately 165°C, hydrogen sulfide is generated and a product is obtained which is partially soluble in water. Solutions of this material give highly colored precipitates with metallic salts. An orange-red precipitate obtained by addition of lead acetate possesses the empirical formula $Pb_2C_2N_2S_3$[87].

According to Wohl[141], hydrogen sulfide reacts with hexamethylenetetramine in hot aqueous or alcoholic solutions causing precipitation of an amorphous product (page 192).

Sulfur dioxide is reported to give the addition product, $C_6H_{12}N_4 \cdot SO_2$, when passed into a hot solution of hexamethylenetetramine in benzene. Crystals of this material, which are readily soluble in water, lose sulfur dioxide on heating to 60°C[69]. When the reaction is carried out in water or alcohols complex products are obtained. Hartung[69] reports that a compound with the empirical formula $C_6H_{22}N_4S_2O_{10}$ is obtained in hot methanol, whereas $C_5H_{11}N_3SO_3$ is produced in isopropanol or isobutanol.

Reactions with Halogens and Inorganic Halides. Chlorine decomposes hexamethylenetetramine in aqueous solution with formation of the explosive nitrogen trichloride[77]. With sodium hypochlorite, chloro derivatives are obtained. These products are apparently unstable and may explode on storage. According to Delépine[40], N-dichloropentamethylenetetramine, $C_5H_{10}N_4Cl_2$, may be obtained in the form of thin leaflets by addition of a

---

* Davis[32] recommends keeping the reaction mixture at 0°C or below for approximately 5 hours and employs only 90 g 30 per cent peroxide.

dilute sodium hypochlorite solution to a solution of hexamethylenetetramine. On heating to 78 to 82°C, the product explodes. Leulier[97] reports a dichlorohexamethylenetetramine (m.p. 77°C) produced by addition of 8.4 g potassium bicarbonate and 45 cc sodium hypochlorite (72 g active chlorine per liter) to 5 g hexamethylenetetramine in 40 cc water, claiming a yield of 77 to 80 per cent. A moist sample of this product prepared in the writer's laboratory exploded violently on standing. A tetrachloro- derivative, $C_6H_8N_4 \cdot Cl_4$, is claimed by Buratti[20].

Bromine reacts with hexamethylenetetramine in chloroform solution to produce a crystalline orange-red tetrabromide, $C_6H_{12}N_4Br_4$[33, 96]. On standing in the air, this product loses bromine and is converted to the yellow dibromide, $C_6H_{12}N_4Br_2$, which is also formed by the action of bromine water on aqueous hexamethylenetetramine solutions[79].

Hexamethylenetetramine reacts with iodine to give di- and tetraiodo derivatives. The diiodide, $C_6H_{12}N_4 \cdot I_2$, is described by Hoehnel[77] as red-brown needles or a yellow-green crystalline powder, soluble in hot alcohol but practically insoluble in cold alcohol, water, and ether. Tetraiodide occurs as dark-brown crystals or an orange-red powder, insoluble in water or ether, somewhat soluble in alcohol, and readily soluble in chloroform, acetone, and carbon bisulfide. The former is obtained by the action of an alcoholic iodine solution on a water solution of hexamethylenetetramine, whereas the latter is produced when an excess of alcoholic iodine is added to hexamethylenetetramine in alcohol. Hexamethylenetetramine tetraiodide is also formed by the addition of excess iodine-potassium iodide to aqueous hexamethylenetetramine solution. Scheuble[119] reports that a hexamethylenehexamine hexaiodide ($C_6H_{12}N_6I_6$) may be prepared by addition of nitrogen iodide to hexamethylenetetramine or by the joint reaction of ammonia, iodine, and hexamethylenetetramine:

$$C_6H_{12}N_4 + 2NI_3 \rightarrow C_6H_{12}N_6I_6$$

$$C_6H_{12}N_4 + 8NH_3 + 6I_2 \rightarrow C_6H_{12}N_6I_6 + 6NH_4I$$

This product is a violet-red powder which explodes on heating or sudden shock, but this can be avoided by mixing it with an inert material such as talc.

Mixed halides such as $C_6H_{12}N_4ICl$[77] are produced by the action of halogen compounds on hexamethylenetetramine.

Inorganic halides such as phosgene form addition compounds with hexamethylenetetramine. The phosgene compound has the composition $2C_6H_{12}N_4 \cdot COCl_2$, and melts at 187 to 190°C[115]. A mixture of caustic soda, phenol, glycerin, and hexamethylenetetramine in water solution was used in World War I for the neutralization of phosgene (page 477). With sulfur chloride, hexamethylenetetramine gives the addition compound $2C_6H_{12}$-$N_4 \cdot S_2Cl_2$[136].

Addition Compounds with Inorganic Salts. Hexamethylenetetramine forms addition compounds with a wide variety of inorganic salts including salts of alkali metals, alkaline earths, rare earths and metals. Although many of these compounds tend to conform to the type formula, $MX_n \cdot nC_6H_{12}N_4$, in which M stands for a metal ion of valence n, the number of molecules of combined hexamethylenetetramine per mol of salt is often lower than this value and several different complexes are often formed with the same salt.

Lead is reported to be almost unique in not forming compounds of this type, basic lead nitrate and chloride being the principal products formed when aqueous hexamethylenetetramine solutions are reacted with lead nitrate and lead chloride, respectively[49, 105]. Complexes are reported with salts of lithium, sodium, potassium, copper, silver, gold, magnesium, calcium, strontium, barium, zinc, cadmium, mercury, aluminum, titanium, lanthanum, cerium, neodymium, yttrium, erbium, thorium, tin, antimony, bismuth, chromium, molybdenum, tungsten, uranium, manganese, iron, cobalt, nickel, platinum, and palladium[1, 15].

Illustrative of the metallic salt complexes are the compounds formed with silver nitrate, mercuric chloride, and the chlorides of tin. When a moderately concentrated solution of hexamethylenetetramine in water is added to aqueous silver nitrate a white crystalline compound having the composition $AgNO_3 \cdot C_6H_{12}N_4$ is precipitated[34, 65]. The compound splits in boiling water and is readily soluble in cold ammonia or nitric acid. It gradually decomposes on exposure to light. Another compound, $2C_6H_{12}N_4 \cdot 3AgNO_3$, is reported as the reaction product when silver nitrate solution is added to excess hexamethylenetetramine. The mercuric chloride compound $C_6H_{12}N_4 \cdot 2HgCl_2$, is precipitated when hexamethylenetetramine is added to mercuric chloride in aqueous solution; when alcoholic solutions are employed, $C_6H_{12}N_4 \cdot HgCl_2$ is produced[65]. The products $2C_6H_{12}N_4 \cdot SnCl_2$ and $4C_6H_{12}N_4 \cdot SnCl_4$ are obtained from stannous and stannic chlorides respectively in cold absolute alcohol[49].

According to a recent patent, the colored addition compounds of hexamethylenetetramine and salts of cobalt and nickel are useful temperature indicators, *e.g.*: $CoCl_2 \cdot 2C_6H_{12}N_4 \cdot 10H_2O$ changes from rose to blue at 35°C; $CoBr_2 \cdot 2C_6H_{12}N_4 \cdot 10H_2O$, changes from rose to blue at 40°C; $CoI_2 \cdot 2C_6H_{12}N_4 \cdot 10H_2O$, changes from rose to green at 50°C; $Co(CNS)_2 \cdot 2C_6H_{12}N_4 \cdot 10H_2O$, from rose to blue at 60°C; $Co(NO_3)_2 \cdot 2C_6H_{12}N_4 \cdot 10H_2O$, from rose to purple-red at 75°C; $NiCl_2 \cdot C_6H_{12}N_4 \cdot 10H_2O$ from green to yellow at 60°C and from yellow to violet at 100°C[80].

Reactions with Organic Compounds. *Reactions with Alcohols (Alcoholysis).* Hexamethylenetetramine does not react with alcohols under neuttral or alkaline conditions. However, in the presence of acids it reacts as an ammono-formaldehyde to give formals and ammonium salts as indicated in the following equation, in which ROH represents an aliphatic alcohol[16]:

$$C_6H_{12}N_4 + 12ROH + 4HCl \rightarrow 6CH_2(OR)_2 + 4NH_4Cl$$

Tertiary alkoxymethylamines are produced by the joint reaction of aliphatic alcohols with hexamethylenetetramine and paraformaldehyde[140]. Tributoxymethylamine, $(C_4H_9OCH_2)_3N$, (b.p. 157°C @ 7 mm), for example, is obtained by refluxing butanol with $\frac{1}{6}$ mol hexamethylenetetramine and 1 mol formaldehyde as paraformaldehyde, removing the water formed by the reaction by distillation with butanol, and isolating the product by fractionation of the residue at reduced pressure.

Salt Formation and Other Reactions with Organic Acids. The primary reaction of hexamethylenetetramine with organic acids is salt formation. In general, these salts may be isolated in a pure state by combining base and acid in the theoretical proportions in concentrated aqueous solution and subjecting the product to vacuum evaporation. Heating should be avoided in the case of the stronger acids, since hydrolytic reactions may take place. Hexamethylenetetramine formate may be obtained by action of the pure acid on hexamethylenetetramine or by employing a slight excess of aqueous (for example 50 per cent) acid and removing water and excess acid under reduced pressure. The salt is reported to melt at 60°C. It is soluble in alcohol and water but insoluble in ether[5]. On heating with aqueous formic acid, hexamethylenetetramine is reduced with evolution of carbon dioxide and formation of methylamines. Salts of the higher fatty acids such as stearic or palmitic acids are obtained by heating with hexamethylenetetramine and a small quantity of water until the mixture sets to a crystalline mass which is then dried on a porcelain plate and recrystallized from alcohol. In preparing the stearate, 20 parts of stearic acid are heated with 9.8 parts hexamethylenetetramine and 10 parts water. Both salts give neutral aqueous solutions which evolve formaldehyde when heated[6].

Hexamethylenetetramine anhydromethylene citrate, which is also known as "Helmitol," is prepared by dissolving 50 g anhydromethylene citric acid in 170 g 95 per cent alcohol at 80°C and adding to this mixture a solution of 33 g of hexamethylenetetramine in 280 g 95 per cent alcohol prepared at 80°C. If both solutions are thoroughly mixed in 3 to 4 seconds, the salt separates in quantitative yield and is isolated by filtration and drying at 30 to 35°C[123]. If the solutions are mixed slowly a poor yield of an acid complex is obtained.

On heating hexamethylenetetramine with succinic acid or phthalic acid the methylene-bis derivatives of the imides are produced. Methylene bis-succinimide melts at 290 to 295°C; methylene bis-phthalimide at 226°C[110].

Reactions with Acetoacetic Ester and Acetyl Acetone. On refluxing hexamethylenetetramine with 5 parts of acetoacetic ester in absolute alcohol, reaction takes place with formation of the diethyl ester of dihydrolutidine dicarboxylic acid[82]. This is the same product which is formed when ammonia is added to methylene bis-(acetoacetic ester) (page 279). With acetyl acetone in boiling absolute alcohol, hexamethylenetetramine gives 2-imino-1,5-diaceto-4-methyl-1,2-dihydrobenzene[82].

Reactions with Halogenated Organic Compounds. With equimolar quantities of many organic halogen compounds, hexamethylenetetramine forms addition products in the manner characteristic of tertiary amines. Methyl iodide, for example, adds to hexamethylenetetramine in absolute alcohol to give lustrous needle-like crystals of the product, $C_6H_{12}N_4 \cdot CH_3I$[140]. The reaction takes place slowly at room temperature, more rapidly on heating. A yield of 90 per cent theoretical is said to be obtained in 5 to 10 minutes at 100°C. At higher temperatures (170 to 200°C) decomposition takes place, with formation of tetramethyl ammonium iodide. The methyl iodide addition product is readily soluble in water, slightly soluble in cold alcohol, but fairly soluble in hot alcohol and insoluble in cold ether and chloroform. On heating it melts at 190°C with decomposition. That the methyl iodide addition product is a quaternary nitrogen compound is demonstrated by the fact that it may be converted to the highly alkaline base, $C_6H_{12}N_4 \cdot CH_3OH$, by treatment with alkali hydroxides or silver oxide[76, 140]. Methylene iodide adds two mols of hexamethylenetetramine to give $2C_6H_{12}N_4 \cdot CH_2I_2$, (m.p. 165°C) on heating for 2 hours at 100°C in the presence of chloroform[140]. Benzyl chloride reacts readily, giving hexamethylentetramine benzyl chloride, m.p. 192°C[38]. Related addition products are produced by the action of hexamethylenetetramine on halogen derivatives of alcohols, aldehydes, ketones, esters, organic acids, amides, etc. Production of a betaine by the action of silver oxide and pyridine on the addition product of hexamethylenetetramine and chloroacetic acid is described in patents[84, 85]. Acid chlorides such as acetyl and benzoyl chloride give unstable addition products[68, 70].

On reduction, the hexamethylenetetramine addition products of alkyl and aralkyl halides give substituted dimethylamines. Dimethyl benzylamine is obtained in 60 to 70 per cent yield when hexamethylenetetramine benzyl chloride is heated with 25 to 100 per cent formic acid[130]. When subjected to the action of acids, the addition compounds give simple monoamines. This is an outstanding method for the preparation of primary amines. A 96 per cent yield of benzylamine is obtained by the action of alcoholic hydrogen chloride on hexamethylenetetramine benzyl chloride[39]. In a similar manner, amino ketones may be synthesized by treating the hexamethylenetetramine addition products of alpha halogen substituted ketones with alcoholic hydrogen chloride[101]. Formals are the by-products of these reactions. The corresponding amino acids can be synthesized by reacting α-halogenated fatty acids with hexamethylenetetramine in an inert solvent followed by saponification with an alcoholic solution of hydrogen chloride[75].

On heating in aqueous or dilute alcoholic solutions, the hexamethylene-tetramine compounds of many aryl methyl halides and their structural analogs undergo further reaction with the formation of carbonyl compounds. This is the basis of the Sommelet synthesis[56, 128] for aldehydes. The method is of fairly general utility for the production of aromatic and some heterocyclic aldehydes, e.g., 3-thenaldehyde can be prepared from 3-thenyl bromide and hexamethylenetetramine[25]. It is also claimed that it can be used for the conversion of allyl iodide to acrolein. The synthesis of acetophenone from alpha-chlorethylbenzene is reported. However, recent studies[8] indicate that contrary to early claims[56], saturated aliphatic aldehydes cannot be obtained by this process. Angyal and co-workers[7] have also demonstrated that the reaction is hindered in the case of benzyl halides containing substituents in two ortho positions.

Angyal and Rassack[8] have recently demonstrated the mechanism of this synthesis. Although this mechanism differs radically from the one originally suggested by Sommelet, their search of the early literature showed that Sommelet[129] later came to similar conclusions although these were not supported by experimental data.

In view of Angyal and Rassack's findings[8], it is apparent that the Sommelet synthesis involves the following reactions:

$$RCH_2X + C_6H_{12}N_4 \rightarrow [RCH_2 \cdot C_6H_{12}N_4]^+Cl^-$$

$$[RCH_2 \cdot C_6H_{12}N_4]^+Cl^- \rightarrow RCH_2NH_2$$

$$RCH_2NH_2 \xrightarrow{\text{CH}_2:\text{NH}} RCH:NH + CH_3NH_2$$

$$RCH:NH \xrightarrow{\text{H}_2\text{O}} RCHO + NH_3$$

In these reactions, methyleneamine stemming from hexamethylenetetramine acts as a hydrogen acceptor in the dehydrogenation of the amine, $R \cdot CH_2NH_2$ (e.g., benzylamine). Angyal and Rassack point out that methyleneimine is only one of the possible hexamethylenetetramine fragments that can be envisaged as the hydrogen acceptor. This acceptor could also be methylolamine or preferably, in their opinion, the mesomeric ion, $NH_2^+ \cdot CH_2 \leftrightarrow NH_2 \cdot CH_2^+$. Methylation of the benzylamine to methyl benzylamines is repressed in this synthesis by the methylation of ammonia which is more readily methylated. The mechanism indicates that the ratio of hexamethylenetetramine to halogen compound should be kept high for optimum results and this was confirmed by experiments. Under identical conditions 1, 2 and 3 mols of hexamethylenetetramine give 70, 77, and 80 per cent yields respectively with one mol of benzyl chloride. Best results were obtained by gradual addition of 50 g benzyl chloride to a boiling solution of 60 g hexamethylenetetramine in 100 cc water over a period of 1 hour. Refluxing was continued for 20 minutes followed by addition of 100 cc concentrated hydrochloric acid and another 15 minutes boiling. The benzaldehyde layer was separated after cooling and the remainder was isolated by steam distillation. The yield of distilled product was 82.3 per cent.

Reactions with Phenols. The primary products formed by the action of hexamethylenetetramine on phenols are molecular compounds. A large number of these have been characterized in which 1, 2, and 3 mols of hexamethylenetetramine are combined with 1 mol of a phenol[3, 15]. In the case of phenol itself, hexamethylenetetramine triphenol, $C_6H_{12}N_4 \cdot 3C_6H_5OH$, is obtained as a crystalline precipitate when concentrated aqueous solutions of hexamethylenetetramine and phenol are mixed at room temperature. After crystallization from water the product melts with decomposition at 115 to 124°C[106]. Its preparation, properties, and heat of formation were studied by Harvey and Baekeland in 1921[71]. A product having the composition $C_6H_{12}N_4 \cdot C_6H_5OH$ (m.p. 176°C) was also reported by Smith and Welch[127]. On heating, hexamethylenetetramine triphenol evolves ammonia with formation of an insoluble, infusible resinous mass similar to a "C" type phenol-formaldehyde resin[95]. Reactions of hexamethylenetetramine with phenols and phenol-formaldehyde condensates play an important part in the synthetic resin industry, and are discussed in detail by Ellis[54]. In general, hexamethylenetetramine reacts with phenols like an ammono-formaldehyde, differing from formaldehyde principally in that ammonia is evolved instead of water when methylene linkages are formed between phenol molecules.

Reactions of naphthols and thionaphthols with hexamethylenetetramine in ethanol and glacial acetic acid are reported by Galimberti[59]. Principal products are bis-(naphthoxymethyl) amines. In the case of alpha-thionaphthol, the product is bis-(alpha-naphthylthiomethyl) amine.

Duff[50] has recently developed a synthesis for hydroxyaldehydes based on the reaction of phenols with hexamethylenetetramine in glyceroboric acid. The glyceroboric acid is prepared by heating glycerol (150 g) with boric acid (35 g) at 170°C until free of water. In general this may be accomplished in approximately 30 minutes. In this liquid approximately 25 g of the phenol is heated with approximately 25 g hexamethylenetetramine at 150 to 160°C for 15 minutes with continuous agitation. The reaction mixture is cooled to 110°C and then added to a solution of 30 cc concentrated sulfuric acid in

100 cc water and steam-distilled. The following mechanism is suggested:

$$3C_6H_5OH + C_6H_{12}N_4 \rightarrow NH_3 + 3C_6H_4(OH) \cdot CH_2 \cdot N : CH_2$$
Phenol

$$C_6H_4(OH) \cdot CH_2 \cdot N : CH_2 \rightarrow C_6H_4(OH) \cdot CH : N \cdot CH_3$$

$$C_6H_4(OH) \cdot CH : N \cdot CH_3 + H_2O \rightarrow C_6H_4(OH)CHO + CH_3NH_2$$
Salicylaldehyde   Methylamine

The presence of methylamine may be demonstrated by its evolution on heating the residue from the steam distillation with alkali. The double bond shift indicated in the above equations appears unlikely. A similar shift was previously considered in connection with the Sommelet synthesis (page 427). It is possible that the mechanism is similar to that of the latter synthesis involving formation of a phenylmethylamine which is then dehydrogenated.

A related aldehyde synthesis also described by Duff[51] is the preparation of *p*-dialkylaminobenzaldehydes. A satisfactory yield of *p*-dimethylaminobenzaldehyde is obtained by heating dimethylaniline with hexamethylenetetramine, acetic and formic acid. The product is readily separated when the reaction mixture is added to dilute hydrochloric acid.

Reactions with Proteins, Cellulose, and Miscellaneous Natural Products. In general hexamethylenetetramine reacts with proteins, cellulose, lignin, etc., as a special form of formaldehyde and the results conform to those obtained with formaldehyde. Reactions are catalyzed by conditions which accelerate hydrolysis: *viz.*, heat and/or the presence of acids or acidic materials. These reactions, which have received little specific research study, are employed commercially for purposes similar to those for which formaldehyde itself is also used. Since the reactions are essentially formaldehyde reactions, they will not be treated specifically in this place. The fact that hexamethylenetetramine does not react rapidly with these materials under ordinary conditions is often a distinct technical advantage.

Analysis of Hexamethylenetetramine

Detection and Identification. In the absence of other formaldehyde derivatives, hexamethylenetetramine may be readily detected by heating with dilute sulfuric acid. Formaldehyde liberated by hydrolysis may then be detected by odor or by use of Schiff's reagent, etc. (Detection, Chapter 17). Hexamethylenetetramine may also be detected by precipitation of its mercuric chloride compound when an excess of mercuric chloride is added to a neutral solution or to a solution containing a little hydrochloric acid. This test is reported to have a sensitivity of approximately 10 ppm[114,][134]. Another widely used test involves the precipitation of the tetraiodide

by addition of a solution of iodine and potassium iodide in distilled water
$(1:1:100)^{148}$. Salts of hexamethylenetetramine may be detected in this way
after evaporation of their faintly acidic solutions over lime. Ammonium
salts are reported not to interfere with this procedure[149]. Another characteristic hexamethylenetetramine reaction is the precipitation of the orange
tetrabromide with bromine. When allowed to stand over caustic potash,
this product loses two bromine atoms to give the canary-yellow dibromide
(m.p. 198 to 200°C). The dibromide is also precipitated by addition of
sodium hypobromite to a solution of hexamethylenetetramine in dilute
hydrochloric acid[26].

Quantitative Determination of Hexamethylenetetramine. Unfortunately,
there is no method for the quantitative determination of hexamethylenetetramine which is both specific and accurate. Procedures which should
be relatively specific, such as those involving the formation of the tetraiodo
derivative, $C_6H_{12}N_4 \cdot I_4$,[92] the silver nitrate addition compound, $C_6H_{12}N_4 \cdot$
$2AgNO_3$[86], the picrate, $C_6H_{12}N_4 \cdot C_6H_2(OH)(NO_2)_2$[90], etc., have been demonstrated to be both inaccurate and undependable[17, 126]. Titration with
methyl orange, which is reasonably characteristic if strongly basic impurities such as ammonia are first determined by titration with phenolphthalein,
does not have a high degree of accuracy since the end point is not particularly distinct. In general, the most satisfactory analytical procedure depends
upon the consumption of acid by the ammonia liberated on acid hydrolysis,
but since basic impurities, including ammonia and ammonium salts of
volatile acids, as well as acidic impurities will affect the results, this method
of analysis is completely dependable only when applied to samples which
are substantially free of these impurities.

The hydrolytic method for hexamethylenetetramine determination is the
procedure prescribed by the United States Pharmacopeia[111]. It is based on
the fact that four equivalents of acid are consumed per mol of hexamethylenetetramine on hydrolysis with excess acid.

$$C_6H_{12}N_4 + 2H_2SO_4 + 6H_2O \rightarrow 2(NH_4)_2SO_4 + 6CH_2O$$

Procedure of U. S. Pharmacopeia XIV. Approximately 1 g of hexamethylenetetramine, accurately weighed and dried over sulfuric acid for 4 hours, is placed in a
beaker to which exactly 40 cc of normal sulfuric acid is added. The mixture is then
boiled gently until the odor of formaldehyde is no longer perceptible, distilled water
being added from time to time as necessary. After this, the solution is cooled, 20
cc of distilled water is added and the excess acid titrated with normal alkali using
methyl red as an indicator. Each cc of normal acid is equivalent to 35.05 mg of $C_6H_{12}N_4$.

Experience in the writer's laboratory indicates that the accuracy of this
procedure is improved if the acid hexamethylenetetramine solution is not
allowed to boil. This may be carried out in a satisfactory manner by plac-

ing the beaker on a steam-heated sand bath (135 to 140°C) under a fume hood. The time required for analysis may be reduced if water is not added to the evaporating solution. With an approximately 0.5 g sample of hexamethylenetetramine, the liberated formaldehyde is completely vaporized in approximately 2 to 3 hours. Methyl orange is recommended as a titration indicator.

According to Slowick and Kelley[126], the U. S. Pharmacopeia procedure, when carried out as directed, is slow, requiring about 9 hours, and their findings show that it tends to give somewhat high results with variations of approximately 0.2 per cent. Their studies also indicate that more accurate results may be secured by employing $N/5$ to $N/10$ acid and varying the weight of the sample proportionally.

An alternate method of analysis developed by Slowick and Kelley[126], which is claimed to be more accurate and less time-consuming, involves determining the ammonia formed on rapid hydrolysis by oxidation with calcium hypochlorite in the presence of sodium bicarbonate and bromide.

Determination of Impurities in Hexamethylenetetramine. *1. Ammonia and Ammonium Salts.* Ammonia in hexamethyltetramine or hexamethylenetetramine solutions may be roughly estimated by titration with phenolphthalein as an indicator if other strong bases are absent.

Ammonium salts can be determined with a fair degree of accuracy (± 2 per cent of the value obtained) by adding an excess of formaldehyde to an aqueous solution containing a weighed sample of hexamethylenetetramine and titrating with normal alkali, using phenolphthalien as an indicator. A blank analysis is then carried out by titrating a similar sample of hexamethylenetetramine in the absence of formaldehyde. The difference between the two titers is equivalent to the ammonia present in salt form. In the analysis of hexamethylenetetramine samples containing not more than 10 per cent by weight of ammonium salts, approximately 1 ml of 37 per cent formaldehyde should be added for each gram of hexamethylenetetramine involved. This formaldehyde must be neutralized to phenolphthalein before use. The procedure is based on the fact that the combined acid of an ammonium salt becomes titratable after it is reacted with formaldehyde (page 181). The accuracy of the method is impaired if substantial amounts of free ammonia are present.

2. Formaldehyde. Small quantities of formaldehyde in hexamethylenetetramine may be accurately estimated by use of Nessler's reagent following a method developed by Büchi[19]. To a hexamethylenetetramine solution containing less than 0.025 g CH_2O, 30 ml of Nessler's $HgCl_2$–KI reagent (pages 387–388) and 10 ml $2N$ sodium hydroxide are added. The mixture is thoroughly agitated and the precipitated mercury is then allowed to settle. The supernatant liquid is decanted and filtered, and the precipitate on the filter paper is washed twice with 30 to 40 ml of distilled water. The filter paper is then placed in the flask with the rest of the precipitated mercury and 5 ml of 2N acetic acid and 25 ml $N/10$ iodine solution are added. After agitating the contents of the flask for approximately 2 minutes, the excess iodine is titrated with $N/10$ thiosulfate and the result subtracted from the blank titer for 25 ml of the iodine solution to which 5 ml of 2N acetic acid have been added. Each ml of $N/10$ iodine consumed is equivalent to 0.001501 g formaldehyde.

3. Other Impurities. Sulfates, chlorides and heavy metals may be determined in hexamethylenetetramine by the usual analytical procedures. In general, no precipitation of silver chloride should result when 5 ml of a 33⅓ per cent hexamethylenetetramine solution is acidified with a few drops of nitric acid and treated with a drop of $N/10$ silver nitrate. Ten per cent barium chloride should not precipitate barium sulfate from a similar solution. The substantial absence of heavy metals is demonstrated if no coloration results when a few drops of 10 per cent sodium sulfide is added to a solution of hexamethylenetetramine in distilled water which has been made distinctly alkaline with ammonia.

Ash may be determined by igniting 5 g of hexamethylenetetramine in a weighed crucible. Technical hexamethylenetetramine should show not more than 0.3 per cent ash; U.S.P. hexamethylenetetramine not more than 0.05 per cent.

References

1. Altpeter, J., "Das Hexamethylenetetramine und Seine Verwendung," pp. 31–65, Halle, Verlag von Wilhelm Knapp (1931).
2. *Ibid.*, pp. 5–15.
3. *Ibid.*, pp. 80–83.
4. *Ibid.*, p. 26.
5. *Ibid.*, p. 88.
6. *Ibid.*, p. 89.
7. Angyal, S. J., Morris, P. J., Rassack, R. C., and Waterer, J. A., *J. Chem. Soc.*, **1949**, 2704–6.
8. Angyal, S. J., and Rassack, R. C., *J. Chem. Soc.*, **1949**, 2700–4.
9. Bachmann, W. E., and Sheehan, J. C., *J. Am. Chem. Soc.*, **71**, 1842–5 (1949).
10. Bai, K. S., *Proc. Ind. Acad. Sci.* **20A**, 71–6 (1944).
11. Bally, J., *Rev. Aluminum*, **15**, 1155–66 (1928); *C. A.*, **32**, 8839 (1938).
12. Baur, E., and Ruetschi, W., *Helv. Chim. Acta*, **24**, 754–67 (1941).
13. Baxter, J. P., 3rd., "Scientists Against Time," pp. 256–259, Boston, Little, Brown and Co. (1946).
14. Bebie, J. H., "Manual of Explosives, Military Pyrotechnics, and Chemical Warfare Agents," p. 81, New York, The Macmillan Co. (1943).
15. Beilstein-Prager-Jacobson, "Organische Chemie," Vol. I, pp. 586–7; I (suppl.), 308–11, Berlin, Julius Springer, 1918 and 1928 (suppl.).
16. Birchall, T., and Coffey, S., (to Imperial Chemical Industries Ltd.), U. S. Patent 2,021,680 (1930).

17. Bordeianu, C. V., *Ann. Sci. Univ. Jassy.*, **15**, 380–3 (1929); *C. A.*, **23**, 3189 (1929).
18. Boyd, M. L., and Winkler, C. A., *Can. J. Research*, **25B**, 387–96 (1947).
19. Büchi, J., *Pharm. Acta Helv.*, **13**, 132–7 (1938).
20. Buratti, R., Swiss Patent 90,703 (1921); *Chem. Zentr.*, **1922**, IV, 891.
21. Butlerov, (Butlerow) A., *Ann.*, **111**, 250 (1859).
22. *Ibid.*, **115**, 322 (1860).
23. *Ibid.*, **115**, 323 (1860).
24. Cambier, R., and Brochet, A., *Bull. soc. chim. France* (*3*), **13**, 394.
25. Campaigne, E. E., and LeSuer, Wm. M., (to Indiana University Foundation), U. S. Patent 2,471,092 (1949).
26. Carles, P. J., *Pharm. Chim.* (*7*), **13**, 279; *Repert. pharm.*, **28**, 129 (1916); *C. A.*, **10**, 1896 (1916).
27. Carter, C. B., and Coxe, A. E., (to S. Karpen & Bros.), U. S. Patent 1,499,001 (1924).
28. Carter, C. B., (to S. Karpen & Bros.), U. S. Patents 1,499,002 (1924); 1,566,822 (1925); 1,630,782 (1927); 1,635,707 (1927).
29. Carter, C. B., (to S. Karpen and Bros.), U. S. Patent 1,566,820 (1935).
30. *Chem. & Met. Eng.*, **53**, 218 (1946).
31. Chemnitius, F., *Chem. Ztg.*, **52**, 735 (1928).
32. Davis, T. L., "The Chemistry of Powder and Explosives," Vol. II, p. 451, New York, John Wiley & Sons (1943).
33. Delépine, M., *Bull. soc. chim. France* (*3*), **11**, 552 (1894).
34. *Ibid.*, (*3*) **13**, 74 (1895).
35. *Ibid.*, (*3*) **13**, 136 (1895).
36. *Ibid.*, (*3*) **13**, 353 (1895); (*3*) **17**, 110 (1897).
37. Delépine, M., *Compt. rend.*, **123**, 888 (1896).
38. Delépine, M., *Bull. soc. chim. France* (*3*), **17**, 292 (1897).
39. *Ibid.*, (*3*) **17**, 290 (1897).
40. *Ibid.*, (*4*) **9**, 1025 (1911).
41. Delépine, M., and Badoche, M., *Compt. rend.*, **214**, 777–81 (1942).
42. Dering, H., and Kelly, M., (to Superfine Chemicals, Ltd.), British Patent 396,467 (1933).
43. Desrergnes, L., *Chim. et Ind.*, **28**, 1038 (1932).
44. Dickinson, R. G., and Raymond, A. L., *J. Am. Chem. Soc.*, **45**, 22 (1923).
45. Dominikiewicz, M., *Arch. Chem. Farm.*, **4**, 1–7 (1939); *Chem. Zentr.*, **1940**, II, 1583.
46. Dubsky, J. V., and Wensing, W. D., *Ber.*, **49**, 1041 (1916).
47. Duden, P., and Scharff, M., *Ann.*, **288**, 218–220 (1895).
48. Duden, P., and Scharff, M., *Ber.*, **28**, 938 (1895).
49. Duff, J. C., and Bills, E. J., *J. Chem. Soc.*, **1929**, 411.
50. Duff, J. C., *J. Chem. Soc.*, **1941**, 547–50.
51. *Ibid.*, **1945**, 276–7.
52. Eickmeyer, A. G., U. S. Patent 2,542,315 (1951).
53. Elings, S. B., and Terpstra, P., *Z. Kryst. Mineral.*, **67**, 279 (1928); *C. A.*, **23**, 2863.
54. Ellis, C., "The Chemistry of Synthetic Resins," pp. 307–310, New York, Reinhold Publishing Corp. (1935).
55. Eschweiler, W., *Ann.*, **278**, 230 (1894).
56. Fabriques de Produits Chim. Org. de Laire, German Patent 268,786 (1913).
57. Fincke, H., *Z. Untersuch. Nahr.-u Genussm.*, **27**, König Festschrift, 246–53 (1914); *Chem. Zentr.*, **1914**, I, 1526.

58. Fittig, R., and Schwärtzlin, A., *Ann.*, **331**, 105 (1904).
59. Galimberti, P., *Gazz. chim. ital.*, **77**, 375–81 (1947); *C. A.*, **42**, 4558 (1948).
60. Gillies, A., Williams, H. L., and Winkler, C. A., *Can. J. Chem.*, **29**, 377–81 (1951).
61. Girsewald, C. von, *Ber.*, **45**, 2571 (1912).
62. Grassi, G., *Gazz. chim. ital.*, **36**, II, 505 (1906).
63. Graymore, J., *J. Chem. Soc.*, **1931**, 1490–4.
64. Griess, P., and Harrow, G., *Ber.*, **21**, 2737 (1888).
65. Grützner, B., *Arch. Pharm.*, **236**, 370–1 (1898); *Chem. Zentr.*, **1898**, II, 663.
66. Guareschi, J., "Einführung in das Studium der Alkaloide," p. 620, Berlin, H. Kunz-Krause (1897).
67. Hale, G. C., *J. Am. Chem. Soc.*, **47**, 2754 (1925).
68. Hartung, L., *J. Prakt. Chem. (2)*, **46**, 1 (1892).
69. *Ibid.*, **46**, 10 (1892).
70. Hartung, L., *Ber.*, **52**, 1467 (1919).
71. Harvey, M. T., and Baekeland, L. H., *Ind. Eng. Chem.*, **13**, 135 (1921).
72. Henning, G. F., German Patent 104,280 (1899).
73. Henry, L., *Bull. acad. roy. méd. Belg.*, 721 (1902).
74. Herz, Edmund von, Swiss Patent 88,759 (1920).
75. Hillmann, G., and Hillmann A., *Z. physiol. Chem.*, **283**, 71–3 (1948).
76. Hock, K., German Patent 139,394 (1903).
77. Hoehnel, M., *Arch. Pharm.*, **237**, 693 (1899).
78. Hofmann, A. W., *Ber.*, **2**, 153 (1869).
79. Horton, H. E. L., *Ber.*, **21**, 2000 (1888).
80. I. G. Farbenindustrie A.-G., Italian Patent 370,634 (1939); *Chem. Zentr.*, **1940**, II, 1514.
81. "International Critical Tables," Vol. V, p. 167, New York, McGraw-Hill Book Co. (1925).
82. Ionescu, M. V., and Georgescu, V. N., *Bull. soc. chim. France (4)*, **41**, 692 (1927); *C. A.*, **21**, 2872.
83. J. D. Riedel, A.-G., German Patent 292,284 (1916).
84. *Ibid.*, 336,154 (1921).
85. *Ibid.*, 337,939 (1921).
85a. Johnson, O., (E. I. du Pont de Nemours & Co.), unpublished communication.
86. Kaganova, F. I., and Shul'mann, A. A., *Farm. Zhur.*, **12**, No. 3, 15–16 (1939).
87. Kirchhof, F., *Gummi-Ztg.*, **39**, 892 (1925); *C. A.*, **19**, 2424 (1925).
88. Klinov, I. Ya., *Org. Chem. Ind. (U. S. S. R.)*, **7**, 45–8 (1940).
89. Knudsen, P., *Ber.*, **42**, 3994 (1909).
90. Kollo, C., and Angelescu, B., *Bull. soc. chim. (Rumania)*, **8**, 17–20 (1926); *C. A.*, **22**, 1302 (1928).
91. Kolosov, S., *Novosti Tekhniki*, **1936**, Nos. 40, 41, and 42; *C. A.*, **31**, 3002 (1937).
92. Korostishev'ska, L., *Farm. Zhur.*, **13**, No. 2, 23–27 (1940); *C. A.*, **35**, 848 (1941).
93. Landölt-Börnstein, "Physikalish-Chemischen Tabellen," 3rd. Suppl., 5th Ed., p. 2911, Berlin, Julius Springer (1936).
94. Landt, G., and Adams, W., (to Continental Diamond Fiber Co.), U. S. Patent 1,774,929 (1930).
95. Lebach, H., *Z. angew. Chem.*, **22**, 1600 (1909); *C. A.*, **4**, 391 (1900).
96. Legler, L., *Ber.*, **18**, 3350 (1885).
97. Leulier, A., *et al.*, *J. pharm. chim.*, **29**, 245–51 (1939); *C. A.*, **34**, 78 (1940).
98. Ley, H., *Ann.*, **278**, 59 (1894).
99. Löbering, J., and Philippi, E., *Biochem. Z.*, **277**, 365–75 (1935); *C. A.*, **29**, 4655 (1935).

100. Lösekann, G., *Chem. Ztg.*, **14**, 1409 (1890); *Chem. Zentr.*, **1890**, II, 814.
101. Mannich, C., and Hahn, F. L., *Ber.*, **44**, 1542 (1911).
102. Mark, H., *Ber.*, **57**, 1820 (1924); *C. A.*, **19**, 754 (1925).
103. Marotta, D., and Allessandrini, M. E., *Gazz. chim. ital.*, **59**, 942 (1929).
104. Mayer, F., *Ber.*, **21**, 2883 (1888).
105. Moschatos, H., and Tollens, B., *Ann.*, **272**, 273 (1893).
106. *Ibid.*, **272**, 280 (1893).
107. Nashan, P., (to Gutehoffnunghütte A.-G.), U. S. Patent, 1,930,210 (1933).
108. Novotny, E. E., and Vogelsang, G. K., (to Durite Plastics, Inc.), U. S. Patent 2,293,619 (1942).
109. Ohara, T., *J. Soc. Rubber Ind. Japan*, **10**, 438–41 (1937).
110. Passerini, M., *Gazz. chim. ital.*, **53**, 333 (1923).
111. "The Pharmacopeia of the United States," 14th Rev. pp. 360–1, Easton, Pa., Mack Printing Co. (1950).
112. Plauson, H., U. S. Patent 1,408,286 (1922).
113. Polley, J. R., Winkler, C. A., and Nicholls, R. V. V., *Can. J. Research*, **25B**, 525–34 (1947).
114. Puckner, W. A., and Hidpert, W. S., *J. Am. Chem. Soc.*, **30**, 1471 (1908).
115. Pushin, N. A., and Zivadinovic, R. D., *Bull. soc. chim. roy. Yougoslav.*, **6**, 165–8 (1936); *C. A.*, **30**, 4422 (1936).
116. Richmond, H. H., Myers, G. S., and Wright, G. F., *J. Am. Chem. Soc.*, **70**, 3659–64 (1948).
117. Russo, C., *Gazz. chim. ital.*, **44**, I, 18 (1914).
118. Salkowski, E., *Biochem. Z.*, **87**, 157 (1918); *C. A.*, **12**, 2379.
119. Scheuble, R., German Patent 583,478 (1933).
120. Schideler, P. and Davis, C. P., (to Chemical Construction Corporation), U. S. Patent 2,449,040 (1948).
121. Schieferwerke Ausdauer A.-G., British Patent 286,730 (1927).
122. Schomaker, V., and Shaffer, P. A., Jr., *J. Am. Chem. Soc.*, **69**, 1555–7 (1947).
123. Schwyzer, J., *Pharm. Ztg.*, **1930**, 46.
124. Seidell, A., "Solubilities of Inorganic and Organic Compounds," p. 1228, New York, D. Van Nostrand Co., Inc. (1928).
125. Shaffer, P. A., Jr., *J. Am. Chem. Soc.*, **69**, 1557–61 (1947).
126. Slowick, E. F., and Kelley, R. S., *J. Am. Pharm. Assoc.*, **31**, 15–19 (1942); *C. A.*, **36**, 2082 (1942).
127. Smith, L. H., and Welch, K. N., *J. Chem. Soc.*, **1934**, 729.
128. Sommelet, M., *Compt. rend.*, **157**, 854 (1913); *C. A.*, **8**, 660 (1914).
129. Sommelet, M., *Bull. Soc. Chim.*, (*4*), **17**, 82 (1915); (*4*), **23**, 96 (1917).
130. Sommelet, M., and Guioth, J., *Compt. rend.*, **174**, 687 (1922); *C. A.*, **16**, 1930 (1922).
131. Stephen, H., Short, W. F., and Gladding, G., *J. Chem. Soc.*, **117**, 510 (1920).
132. Tollens, B., *Ber.*, **17**, 654 (1884).
133. Toni, G., *Boll. chim. farm.*, **76**, 61–4 (1937); *C. A.*, **31**, 3633 (1937).
134. Twiss, D. F., and Martin, G., *Rubber Age*, **9**, 379–80 (1921); *India Rubber J.*, **61**, 1283–4 (1921); *C. A.*, **15**, 3567–8 (1921).
135. Ütz, F., *Süddeut. Apoth. Ztg.*, **59**, 832 (1919); *C. A.*, **14**, 3345.
136. Vanino, L., and Seitter, E., *Pharm. Zentr.*, **42**, 120 (1901).
137. Van't Hoff, J. H., "Ansichte über organische Chemie," I, p. 121, Braunschweig (1881).
138. Veller, S. M., and Grigoryan, M. M., *Ukrain. Khem. Zhur.*, **12**, 439–41 (1937); *C. A.*, **32**, 1864 (1938).

139. Vickers, M. H., *Pharm. J.*, **132**, 414 (1934); *C. A.*, **28**, 7427 (1934).
140. Wohl, A., *Ber.*, **19**, 1840 (1886).
141. *Ibid.*, **19**, 2345 (1886).
142. Wright, G. F., Aristoff, E., Graham, J. A., Meen, R. H., and Myers, G. S., *Can. J. Research*, **27B**, 520–44 (1949).
143. Wright, G. F., Brockman, F. J., and Downing, D. C., *Can. J. Research*, **27B**, 469–474 (1949).
144. Wright, G. F., Chute, W. J., Downing, D. C., McKay, A. F., and Myers, G. S., *Can. J. Research*, **28B**, 218–237 (1949).
145. Wright, G. F., Chute, W. J., McKay, A. F., Meen, R. H., and Myers, G. S., *Can. J. Research*, **27B**, 530–19 (1949).
146. Wright, G. F., McKay, A. F., and Richmond, H. H., *Can. J. Research*, **27B**, 462–8 (1949).
147. Wright, G. F., and Myers, G. S., *Can. J. Research*, **27B**, 489–502 (1949).
148. Zijp, C. van, *Pharm. Weekblad*, **55**, 45–7 (1918); *C. A.*, **12**, 889–90 (1918).
149. Zijp, C. van, *Pharm. Weekblad*, **57**, 1345–8 (1920); *C. A.*, **15**, 108 (1921).

CHAPTER 20

USES OF FORMALDEHYDE, FORMALDEHYDE POLYMERS, AND HEXAMETHYLENETETRAMINE

Part I

Introduction

Formaldehyde is a material of many uses and of increasing industrial importance. The fact that formaldehyde production has grown almost continuously since the first decade of the twentieth century, reaching a figure of almost one billion pounds per year in 1951 (page 2), is ample evidence of the magnitude of its use. This growth, which is the direct result of the unique properties of formaldehyde, has been made possible by the low cost and availability of the raw materials from which it is manufactured.

The properties which make formaldehyde of value are due principally to its high order of chemical reactivity, its colorless nature, its stability and the purity of its commercial forms. From a use standpoint, they are made manifest by its utility as a resinifying agent, synthetic agent, hardening agent, stiffening agent, tanning agent, disinfectant, bactericide, and preservative. When employed in organic syntheses, formaldehyde acts as a sort of chemical "button" with which similar or dissimilar molecules and radicals may be fastened together by means of methylene linkages.

Commercial formaldehyde is most commonly employed in the form of its aqueous solution. However, it is often used in the form of the solid polymer, paraformaldehyde, and may also be employed as the trimer, trioxane. It may thus be applied in the liquid phase as solution, in the solid phase as polymer, and in the gas phase produced by vaporization of either of the above-mentioned forms. The compound hexamethylenetetramine may also be regarded as a special form of formaldehyde since it functions as formaldehyde in the majority of its commercial applications. In this connection it should be noted that one pound of 100 per cent formaldehyde (CH_2O) is equivalent to 2.7 lbs of commercial 37 per cent solution, 1.05 lbs of 95 per cent paraformaldehyde, 1.00 lb of trioxane, and 0.78 lb of hexamethylenetetramine.

It is impossible to give exact figures covering the volume distribution of formaldehyde in all its applications. Completely accurate data are not available, and even if they could be obtained they would have little permanent utility. However, Table 56 may be taken as giving a rough indication of the approximate percentage distribution of formaldehyde solution employed in the principal fields of utilization in 1939–1942 and 1951. These figures deal solely with the formaldehyde sold as aqueous solution. Figures for 1951 based on total formaldehyde production, in which manufacture of paraformaldehyde and hexamethylenetetramine are included under chemical syntheses, indicate 65 per cent utilization in synthetic resins and 27 per cent in chemical syntheses. Due to wartime requirements for hexamethylenetetramine and pentaerythritol in explosives, about 35 per cent of the formaldehyde produced in the United States in 1943–44 was used for making these chemicals and resin production accounted for only 50 per cent of the total manufactured. The principal peacetime applications for formaldehyde solution, paraformaldehyde and hexamethylenetetramine are roughly indicated in Figure 33.

TABLE 56. APPROXIMATE VOLUME DISTRIBUTION OF FORMALDEHYDE SOLUTION IN MAJOR FIELDS OF APPLICATION

Field of Application	Average Percentage of Formaldehyde Consumed	
	1939–42	1951
Synthetic resins	78	72
Chemical syntheses	7	18
Textile and paper industries	3	2
Disinfection, embalming and agriculture	3	2
Tanning and hardening	1	1
Miscellaneous uses	8	5
Total	100	100

The major portion of formaldehyde is consumed by the resin industry. Uses in this field include not only the preparation of cast and molded products but also production of adhesives for plywood and other laminates, manufacture of oil-soluble lacquers, coating compounds, etc. The growing applications in chemical syntheses involve the manufacture of pentaerythritol, dyes, sulfoxylate reducing agents, drugs, textile modifiers, explosives, rubber accelerators, sequestering agents, surface-active agents, tanning chemicals, dienes for use in the production of synthetic rubber and many other useful products. A small fraction is also used as a bactericide for prevention of both plant and animal diseases and as a preservative in embalming. Still smaller amounts are employed for the tanning and hardening of protein materials. The remainder is used for a wide variety of miscellaneous uses. In considering the percentage volumes employed in the smaller applications it should be remembered that even a 1 per cent application involves about ten million pounds of formaldehyde on the basis of the 969 million pound figure reported for 1951.

The various uses of formaldehyde and the manner in which it is employed will be discussed in the following pages. Because of its overwhelming importance, the uses of formaldehyde in the resin industry will be discussed first. Other important fields of utilization will be treated in Chapter 21.

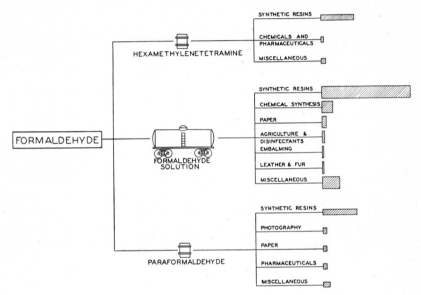

Figure 33. Uses of formaldehyde, paraformaldehyde, and hexamethylenetetramine.

Use of Formaldehyde in the Resin Industry

It is not the purpose of this book to cover in detail the use of formaldehyde in the resin field since it has already been ably handled in numerous books devoted to the industrial and technical aspects of the synthetic resin industry. The fundamental chemical reactions upon which the manufacture of formaldehyde resins are based have already been treated in the chapters dealing with formaldehyde reactions. Our chief object in this section is to review the principal resin types in the production of which formaldehyde plays an important part, with special emphasis on the newer developments in this field.*

Formaldehyde resins may be classified in two broad groupings: (1) Synthetic resins, or plastics synthesized from formaldehyde and simple com-

---

* Applications of resins or resin precursors in modifying the characteristics of textiles, paper and other materials will be reviewed in Chapter 21. The present discussion is limited principally to applications of resins in moldings, castings, extrusions and laminates.

pounds of low molecular weight such as phenol, urea, melamine, etc.; and (2) Resins modified by formaldehyde treatment. The latter are produced by the action of formaldehyde on high molecular weight resins or resin-like materials of natural or synthetic origin, such as casein, lignin, and polyvinyl alcohol.

Synthetic Resins

Thermosetting Resins. Resins synthesized with formaldehyde may be classified as either thermosetting or thermoplastic. The former, which in their primary forms may be molded, cast, etc., to give insoluble, infusible products, are from a volume standpoint the most important formaldehyde resins. However, the latter, which are softened or fused on heating, also include many commercially valuable materials.

Thermosetting formaldehyde resins are generally prepared by the action of formaldehyde on simple molecules which possess more than two active hydrogen atoms capable of reaction with formaldehyde. It is generally believed that they are made up of cross-linked macromolecules in which the simple parent molecules are bound together by methylene bonds. Infusibility and insolubility are probably inherent characteristics of compounds having a cross-linked structure. The formation of these resins is illustrated by the following equation in which $(A)H_3$ indicates a simple organic compound containing three reactive hydrogen atoms and x equals a number greater than 1 (usually about 1.5 to 2.0):

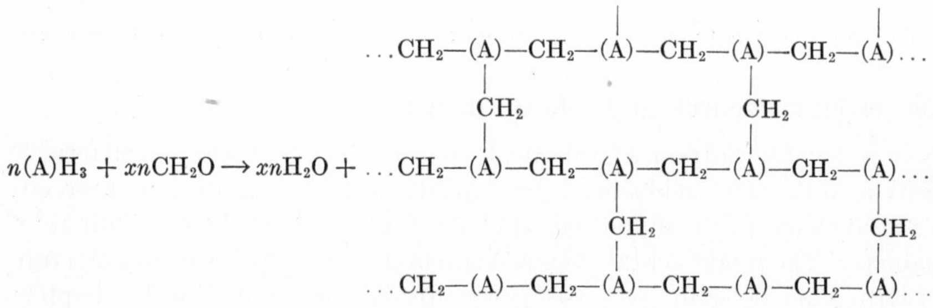

$$n(A)H_3 + xnCH_2O \rightarrow xnH_2O + \begin{array}{c} \ldots CH_2-(A)-CH_2-(A)-CH_2-(A)-CH_2-(A)\ldots \\ | \qquad\qquad\qquad\qquad | \\ CH_2 \qquad\qquad\qquad CH_2 \\ | \qquad\qquad\qquad\qquad | \\ \ldots CH_2-(A)-CH_2-(A)-CH_2-(A)-CH_2-(A)\ldots \\ | \qquad\qquad\qquad\qquad | \\ CH_2 \qquad\qquad\qquad CH_2 \\ | \qquad\qquad\qquad\qquad | \\ \ldots CH_2-(A)-CH_2-(A)-CH_2-(A)-CH_2-(A)\ldots \end{array}$$

Usually the reactions are so controlled that cross-linking is held in check and is not completed until the primary products are subjected to a final heating or curing operation. These primary products are the essential ingredients of molding powders and liquid casting resins. In many cases the thermosetting process involves reaction with additional formaldehyde or hexamethylenetetramine. (Cf. page 262.)

Although plastics of this type can be synthesized from a wide variety of diversified chemical compounds, those derived from phenols, urea and, more recently, melamine are produced in quantities far in excess of any other formaldehyde resins. Production figures in the U. S. for the principal varieties of synthetic resins are compared with those for phenol-formaldehyde and urea- plus melamine-formaldehyde resins in Table 57 which covers 1940, 1945, and 1950. Since melamine resins are of rather recent development, they are not included in the 1940 figure. These production data are based on values reported by the U. S. Tariff Commission.

Phenolic Resins. Although phenolic resins include those prepared from a wide variety of phenolic compounds, the total production consists largely of resins prepared from unsubstituted phenol. Phenol is particularly satisfactory because it can be readily obtained in a high state of purity and is accordingly uniform and dependable in behavior. A large fraction of the miscellaneous phenolics are derived from cresols, xylenols, tar acids and

TABLE 57. U. S. PRODUCTION OF PHENOL AND UREA-MELAMINE RESINS COMPARED WITH TOTAL PRODUCTION OF PRINCIPAL SYNTHETIC RESINS*

	1940		1945		1950	
	Million Pounds	Per cent of Total	Million Pounds	Per cent of Total	Million Pounds	Per cent of Total
Total resins........................	276.8	100.0	808.2	100.0	2,150.5	100.0
Phenolics...........................	93.4	33.7	186.5	23.1	451.1	20.9
Urea-melamine resins...............	21.5	7.8	71.4	8.8	219.2	10.2
Sum of above formaldehyde resins..	114.9	41.5	257.9	31.9	670.3	31.1

* Data from U. S. Tariff Commission. (1940 Figures do not include melamine resins.)

resorcinol. Cashew nut oil, which consists almost entirely of mono- and dihydric phenols containing a pentadecyldienyl substituent, is also important as a formaldehyde resin base which yields some cold-setting adhesives, coating compounds and resin modifiers.

Conventional applications of thermosetting phenolics in the production of cast and molded articles are discussed in detail in books dealing with plastics and synthetic resins. These include molded parts for radios, television cabinets, airplanes, telephones, electrical instruments, etc. Molded phenolics are strong, resilient, resistant to heat, water, oil and acids, and are good electrical insulators. Cast phenolics are specifically important because of their non-absorbent nature, transparency and decorative appearance. Other applications make use of their value as adhesives, binders and coating compounds, e.g., in the formation of hot press plywood adhesives, laminating varnishes, baking finishes, and grinding wheel binders. The following applications are reviewed in somewhat greater detail because of their bearing on special problems of phenol-formaldehyde chemistry.

Wood Modification. Phenol-formaldehyde condensates can be employed in the production of resin-containing wood having a high anti-shrink efficiency as demonstrated at the U. S. Forest Products Laboratory. Wood is thoroughly saturated with a solution of the primary water-soluble alkaline

(Courtesy, Admiral Corporation)

Figure 34. Molded phenol-formaldehyde resins. Production of a television console.

condensation products of phenol and formaldehyde and then cured under pressure so that the resin forms in the wood. Stamm and Seborg[53, 54] report that the water-resistance of wood treated in this way is superior to wood treated with preformed resins since the low molecular weight phenol alcohols can penetrate the wood substance more efficiently than the larger resin molecules. However, even so, the method is not practical for treating massive wood specimens and is best employed for the preparation of laminated panels from individually treated plies of veneer[34]. Resin-containing wood of this type may possess an anti-shrink efficiency of 70 per cent with only 30 to 50 per cent by weight of resin in the wood. Efficiencies as high as 50 per cent have been obtained with wood containing only 15 per cent resin. Treatment of veneer in this way makes it possible to use lower pressures in the production of plywood and allows production of material with compressed resin-treated faces and an uncompressed core in a single compression and assembly operation. Urea-formaldehyde condensates can

(Courtesy, Bakelite Company)

Figure 35. A laminated phenolic gear.

also be applied to wood by the same technique as that employed with phenolics. However, according to Millett and Stamm[34] the anti-shrink efficiency of the wood is reduced very little by this procedure.

Ion Exchange Resins. Thermosetting formaldehyde plastics prepared from a variety of phenols, aromatic amines, tannins, etc., so that they contain anionic and cationic groups are employed as synthetic zeolites for the purification of water and aqueous solutions of non-ionic materials. Ion exchange resins from tannins and other cationic compounds are of particular value for the removal of calcium, magnesium, and other cations, whereas aromatic amines give resins which absorb anions[2, 3]. Various plastics of this type are covered by a growing volume of patents[26, 27]. In discussing commercial developments in this field, R. J. Myers[38] points out that by treatment of water with resins of this type, it is possible to remove all the dissolved salts and obtain a product comparable to laboratory distilled water. Furthermore, the high original adsorptive power of the ion exchange resins is unimpaired after undergoing over 150 cycles of use. Typical applications for which they are now being evaluated include the "desalting" of aqueous solutions of carbohydrates, and gelatin, and removing acid from commercial formaldehyde[39] (page 28). Ion exchange resins have been used for the emergency de-salting of sea water and are also of apparent medical interest in connection with the treatment of ulcers. Pills containing an anion exchange plastic are reported to absorb acids in the stomach and return them to the body in the intestine[35].

Cold-Set Resorcinol Adhesives. Resorcinol-formaldehyde resins have become of outstanding industrial importance in recent years (since 1943) because of their unique value as cold-set adhesives. The extremely high reactivity of the polyhydric phenol makes it possible to produce an adhesive composition which can be cured at ordinary temperatures (70 to 90°F) and at a pH corresponding to approximate neutrality. The resins are of value as wood adhesives since they obviate the need of hot pressing, are not acidic, and have excellent resistance to out-door exposure. Phenol-modified or copolymer phenol-resorcinol resins also possess good low temperature reactivity at mildly alkaline pH conditions.

The discovery of practical methods for controlling the resorcinol-formaldehyde reaction is the basis of this development. Kinetic studies of the reaction by Raff and Silverman[43] indicate that the velocity of the uncatalyzed reaction is several order-of-magnitudes greater than that of phenol. They report that the reaction is of the first order, has an activation energy of approximately 19 kcals and a temperature coefficient of about 2.3 per 10°C. These investigators[44] have also made a quantitative study of the effect of alcohols in reducing the reaction rate. Their work indicates that this is apparently caused by the formation of hemiformals which control the rate at which the dissolved formaldehyde becomes available for reaction.

Numerous patents covering methods for the production of resorcinol and resorcinol-phenol adhesives have been granted to Norton[40], Rhodes[47], Spahr[51,52], Babcock[5] and others. In general these processes involve the preparation of a primary water or alcohol soluble resin produced by reacting resorcinol with less than one mol of formaldehyde. A catalyst is not essential for the reaction but alkaline or acidic catalysts are employed in some processes. Prior to use, the primary resin solution, which is adjusted to a pH of about seven, is treated with additional formaldehyde in the form of aqueous solution, paraformaldehyde or a formaldehyde donor, such as hexamethylenetetramine. In addition, a filler (e.g., walnut shell flour) may be added. Following this admixture with additional formaldehyde, the resin sets up in a few minutes to several hours as determined by the specific resin, the adhesive formulation employed and the curing temperature.

The resorcinol formaldehyde adhesives are also adaptable for applications involving rubber, plastics and metals. Hershberger[22] uses a resorcinol-formaldehyde binder to cause a polyvinyl coating to adhere to aluminum.

The chemistry, formulation and use of resorcinol-formaldehyde resins are discussed in review articles by Glavert[18], Hancock[20], and Rhodes[48].

Urea and Melamine Resins. Urea resins differ from phenolics in that they can be produced in light pastel colors and will even yield white moldings. They are hard, durable, light in weight, odorless, tasteless, and resistant to organic solvents. Urea-formaldehyde condensates also find particularly valuable applications in glues, plywood adhesives, varnishes, lacquers, and a wide variety of coating and sizing compositions.

Melamine resins possess an unusual degree of hardness, resistance to heat and water, and glass-like clarity[4, 49]. Their importance rests upon their useful properties and the development of improved processes for the industrial production of melamine. This chemical, which is a cyclic trimer of cyanamide, NH_2CN, is made by heating the dimeric dicyandiamide which is produced from calcium cyanamide. Melamine reacts with formaldehyde to form methylol derivatives (page 309) which may then be further condensed to melamine resins. In these resins the triazine nuclei of the melamine molecules are bound together by methylene radicals.

In recent years, the formaldehyde derivative, pentaerythritol (pages 221–223), has played an increasingly important role as a raw material for the production of alkyd resins, particularly the rosin-maleic alkyds.

Thermoplastic Synthetic Resins. Thermoplastic resins may be synthesized by reactions analogous to those which take place in the production of thermosetting resins. They are made by reaction of formaldehyde with simple molecules containing two or more hydrogen atoms capable of reacting in such a manner that linear or, in some cases, cyclic condensates are obtained. Compounds of this type are fusible and can be dissolved in some solvents. The formation of thermoplastic resins is indicated in the following equation in which $(A)H_2$ represents an organic raw material of the type required.

$$n(A)H_2 + nCH_2O \rightarrow \ldots (A)—CH_2—(A)—CH_2—(A)—CH_2\ldots + nH_2O$$

The hydrogen atoms indicated in $(A)H_2$ may be on hydroxyl groups, amino groups, amido groups, activated positions of an aromatic nucleus, etc.

When compounds having more than two active hydrogen atoms are employed, the reaction must be controlled to prevent cross-linking, which would result in the formation of an infusible or thermosetting resin. In some cases, this is not possible and raw materials employed for the production of thermosetting resins cannot always be utilized in the preparation of thermoplastics. True thermoplastic resins apparently cannot be prepared by reactions involving urea and formaldehyde.

(*Courtesy, American Cyanamid Co.*)

Figure 36. Radios with "Beetle" urea resin cabinets.

At present the most important resins of this group are the oil-soluble thermoplastics produced from formaldehyde and substituted phenols, such as those containing phenyl, tertiary butyl, and tertiary amyl groups in the para position. These resins find use in varnishes and other coating compounds, where they are characterized by their hardness, elasticity, and resistance to exterior exposure.

Thermoplastic resins formed by the acid condensation of formaldehyde and aniline are of importance because of their excellent electric resistance and low power factor. They are reddish-brown in color, have good structural strength, and are not appreciably affected by exposure to weather or ultraviolet light. They are used extensively for insulators and other electric equipment. Structurally they are stated to consist of the cyclic anhydroformaldehyde aniline, $(C_6H_5N:CH_2)_3$, and products produced by nuclear linking of two or more molecules of this compound by means of methylene groups.

Recent industrial interest in the thermoplastic resins produced from formaldehyde and aromatic hydrocarbons (pages 342–345) points to possible future development of these products.

As a result of the commercial development of tough, transparent, thermoplastic vinyl resins, the industrial importance of this type of resin has steadily increased. In this connection it should be pointed out that some vinyl monomers can be synthesized from formaldehyde and syntheses of this type may eventually prove of commercial interest. Polymers of methyl vinyl ketone, $CH_3COCH:CH_2$, and methyl isopropenyl ketone, CH_3COC-$(CH_3):CH_2$, obtained by reactions of formaldehyde with acetone and methyl ethyl ketone, respectively (see pages 226–229), have been studied by Morgan, Megson and Pepper[37]. Methylenemalonic ester, prepared from malonic ester and formaldehyde (pages 277–278) also polymerizes to a tough glass-like resin of high softening point. Copolymers of this ester with styrene and vinyl acetate have been described by Bachman and Tanner[6].

Resins Modified by Formaldehyde Treatment

The insolubilizing and hardening action of formaldehyde on proteins such as casein and gelatin has long been used for the conversion of these resinlike but water-sensitive materials to useful products. Formaldehyde has also been employed in recent years to produce modified resins from other cheap natural products and industrial wastes. New and useful materials have even been produced by the action of formaldehyde on some synthetic resins.

Protein Resins. Casein, zein, soybean protein, glue, and gelatin are perhaps the most important of the protein materials which have been employed industrially in the production of formaldehyde-modified products. Of these casein, zein and soybean protein are probably the most widely employed in the manufacture of resinous products,* whereas glue and gelatin are more commonly used for producing water-resistant bonds, sizes, and coatings. Although of importance, the utility of these modified resins is limited by the fact that they are somewhat water-sensitive and are accordingly unsatisfactory for uses where complete insensitivity to water is a prime necessity.

---

* Formaldehyde-modified casein, zein and soybean protein are also used for the preparation of synthetic fibers (pages 527–529).

It is estimated that approximately 73 million pounds of casein were employed by American industry in 1937 and that about 8 million pounds were consumed in the production of casein-formaldehyde resin[55]. The recent consumption of casein for this purpose is probably about 10 million pounds. In general, casein resins are produced by mixing powdered casein (preferably rennet casein) with approximately 25 per cent water to which plas-

Figure 37. Molded, cast, and laminated phenolics, urea- and melamine-formaldehyde resins, glue-cork compositions, casein buttons, protein fibers, polyvinyl formal sponge, formaldehyde-tanned, white leather and formaldehyde-set, dyed sheep pelt.

ticizers such as glycerol, acetic acid, ammonia, and alcohols are added together with dyes, pigments, etc. Rods and sheets extruded by the action of heat and pressure on compositions of the above type are hardened by soaking in 5 to 6 per cent formaldehyde solution for 2 to 60 days, depending on the thickness of the material. The treated resin is then washed, dried, polished, and machined for the production of buttons and other articles. Various processes have been developed for direct molding of these resins in recent years and may eventually supersede the older technique. In one procedure of this type described in the patent literature, casein is mixed in the presence of water with a condensation product of glycerol, glycol, etc., plus boric acid, dried, and hot-molded with formaldehyde. Similar results are also claimed when the formaldehyde is added to the initial mix[26]. Brother and McKinney[10] obtain a thermoplastic casein resin by mixing 100 parts of lactic acid casein with 250 parts of 30 per cent formaldehyde and 0.4 part of sodium hydroxide and allowing this mixture to stand for 20 hours at room temperature. After this, excess solution is removed and the moist product is air-dried at 70°C and ground in a ball mill. The powdered product obtained in this way is a thermoplastic molding powder.

The most promising developments in the field of direct molding casein plastics involve the preparation of modified resins of this type. Hipp and co-workers[23] have found that carbamide casein, prepared by reacting casein with potassium cyanate, gives excellent molding powders when pre-hardened with formaldehyde. These molding powders can be converted to simple objects having good strength and dimensional stability under ordinary circumstances. At least two thirds of the amino groups are converted to amido groups, —NHCONH$_2$, in carbamide casein. Calvert[12] obtains a resin stated to be suitable for continuous extrusion or injection molding by blending casein with a phenol-formaldehyde condensate.

Expansion of the use of soybean resins depends upon the development of better methods for the isolation of a high quality protein and the discovery of new methods for the production of resins with better physical properties. Soybean meal is used as an extender and modifier for phenol-formaldehyde resins[13, 57]. A product of this type described by Sweeney and Arnold[56] is made by mixing the following composition in the dry state and molding at 135 to 145°C:

Soybean meal	100 parts
Sodium hydroxide	2 parts
Phenol (crystals)	75 parts
Paraformaldehyde	25 parts

Resins are also produced by molding compositions containing soybean protein and formaldehyde. Procedures similar to those previously mentioned for the production of caseins may apparently be employed[10, 25]. For example, it is claimed that a thermoplastic resin is produced from a mixture of 100 parts soybean protein, 300 parts 37 per cent formaldehyde and 0.4 part sodium hydroxide; the mixture is allowed to stand for 20 hours at room temperature, freed of excess solution, dried at 70°C and ground to produce molding powder[8]. It is also reported that water solutions of soybean protein and formaldehyde may be prepared which on drying leave a thermoplastic resin[21]. According to Brother[8], fibrous materials, such as unsized kraft paper, which have been impregnated with this solution can be hot-pressed to give laminated sheets of good flexing and impact strength.

If single sheets of phenol-impregnated material are placed on the top and bottom of a stack of soybean-impregnated sheets, the laminated product will have the same water-resistance as phenolic plastic except on the edges, which may be protected. Both soybean protein and casein are also used in the production of adhesives, sizes, and miscellaneous coating compounds in some of which formaldehyde serves as a hardening agent.

The use of zein in the production of formaldehyde modified protein resins is steadily growing in importance. One application in the production of a molding resin is illustrated by Bers[7] preparation involving a mixture of 2 parts zein and 1 part of a pine wood resin with 5 per cent paraformaldehyde.

Aqueous dispersions of proteins such as casein, soybean protein and zein can be stabilized by controlled reaction with formaldehyde[9, 32, 30]. Brother and Smith[9] have obtained dispersions containing over 9 per cent formaldehyde modified protein which give tanned protein coatings on drying.

Browne and Hrubesky[11] employ paraformaldehyde in the preparation of waterproof glue. In order to obtain a glue composition which will not set up during the working period, they make use of paraformaldehyde whose reactivity has been reduced by heating at 100°C, which increases its degree of polymerization. This paraformaldehyde depolymerizes and reacts at a slower rate than the untreated polymer. They also employ oxalic acid as a retarding agent and maintain the glue bath at a temperature of 40 to 45°C. A glue of the type recommended by these investigators is prepared by soaking 100 parts of glue in 225 to 250 parts of water, melting the swollen glue at 60°C, adjusting to the desired bath temperature and adding 10 g paraformaldehyde and 5.5 g of oxalic acid, which must be stirred in thoroughly. Although acids in general act as retardants for the reaction of formaldehyde and proteins, oxalic acid is of particular value because it does not tend to cause any appreciable impairment of the dry glue joint. Plywood which has been bonded with glue containing paraformaldehyde is said to require a seasoning period of approximately one week after removing from the press to develop its maximum resistance to water.

Cork compositions in which formaldehyde-insolubilized glue or gelatin is employed as a binder are produced in considerable quantity in sheets, blocks and rods for use as insulating materials, gaskets, liner disks for beverage bottle caps, etc. A composition of this type developed by Cooke and Wilbur[15] contains 50 parts of granulated cork, 15 parts of glue gel, ⅛ part paraformaldehyde plus 2 parts of amine-treated tung oil, ½ part maleic anhydride, and 1 part glycerin.

Other protein materials utilized in the production of formaldehyde-modified resins and plastics include dried blood, yeast, leather waste, and vegetable proteins such as gluten, zein, ground-nut protein, cottonseed meal, and keratin from hoofs, horn, hair, and feathers.

Cheap resinous products may also be produced from a wide variety of crude protein materials. It is reported that fish meat can be converted to a thermoplastic resin by kneading with formaldehyde in the presence of water and under pressure at 80 to 150°C[1]. A molding composition prepared from formaldehyde and powdered animal tendons which have been washed, pulverized, degreased, and bleached[42] is also described.

Resins from Synthetic Materials. Synthetic polyamide plastics also give useful modified products by reactions involving formaldehyde (see pages 305, 528). N-alkoxymethyl polyamides can be employed as plywood adhesives[24] and film-forming coating compositions for flexible fabrics[19].

Resins from Wood Products. Lignin isolated from pulp-wood as a by-product of the paper industry is now utilized as a raw material for resins. Resins of this type are sometimes modified with formaldehyde, more commonly with phenol-formaldehyde condensates. In a process described by Scott[50], lignin precipitated by acidulation of the black liquor obtained in the alkaline soda or sulfate extraction of pulp is claimed to give a useful molding composition when dissolved in ammonia water and reacted with formaldehyde. According to Lundbäck[29], wood fiber containing lignin can be converted to a hard, board-like product by treatment with formaldehyde and subjection to heat and pressure. Approximately 0.1 to 10 per cent formaldehyde (based on dry wood) is employed. In another patented process[33], it is stated that useful molding powders can be produced by dissolving lignin in phenol and heating with formaldehyde or hexamethylenetetramine in the presence of an alkaline condensing agent. Collins, Freeman and Upright[14] report the preparation of acid-resistant resins from lignin-cellulose by reaction with formaldehyde and phenol. Reineck and Dunlap[45] prepare a thermosetting resin from furfuryl alcohol, redwood pulp and formaldehyde. Wood flour plays an important part in the resin industry as a filler for formaldehyde resins.

Natural tannin-containing materials such as quebracho wood or mimosa bark, when finely subdivided, can be compounded with formaldehyde or paraformaldehyde in the presence of a water-soluble plasticizer such as glycerin to give molding powders[41]. The plasticizer also serves as a solvent for the tannin, whose content in the mixture may be adjusted by adding tannin extract or wood flour.

Pinewood pitch is reported to condense with formaldehyde solution or paraformaldehyde to give useful resinous products[31].

Resins from Polyhydroxy Compounds. Sugar, starch, cellulose, and other high molecular weight polyhydroxy compounds may also be employed in the preparation of numerous resinous compositions in which formaldehyde is involved. Although they serve in many cases as fillers for other formaldehyde resins, it is probable that in some instances they are chemically involved in the resins produced. In some cases the resin-like characteristics of these materials are modified or enhanced by reaction with formaldehyde so that they also serve *per se* as modified resins.

Polyvinyl alcohol reacts with formaldehyde to give an almost colorless thermoplastic formal which is insoluble in water and alcohols but dissolves in chlorinated hydrocarbons, dioxan and other organic solvents. Its prin-

Figure 38. Testing thermosetting fir plywood adhesives.

cipal application is in the preparation of coating compositions but it is also used for the production of compression moldings and castings. It has a high molding temperature in the neighborhood of 330°F.

Excellent fine-grained polyvinyl formal sponges (Figure 39) are manufactured from polyvinyl alcohol solution. These sponges are hard when dry but soften rapidly in water. They can be sterilized and withstand boiling water, hot alkalies, dilute acids, soaps, and most detergents. Without added color, the polyvinyl sponge is snow white but can be produced in any color or texture desired.

Carbohydrates may also be converted to resins by processes involving formaldehyde. For example, mono- and disaccharides are reported to be converted to clear, glass-like resins by reaction with formaldehyde or paraformaldehyde in the presence of hexamethylenetetramine followed by condensation with acidic materials such as succinic, tartaric, or malic acid[17], and phthalic anhydride[16].

Starch is modified with formaldehyde to give pastes or powders which

(Courtesy, Ivano Inc.)

Figure 39. Polyvinyl formal sponge.

can be solidified to produce water-resistant products. Such materials may be employed as sizing agents, coating compounds, and adhesives. Procedures involve heating with formaldehyde under various conditions alone and in the presence of acid catalysts[28, 36]. Production of a rubber-like product from starch is also claimed by one investigator (page 210).

Miscellaneous

Formaldehyde is sometimes employed for vulcanizing vinyl resins containing reactive substituents. This procedure is of particular utility with some elastomers. An illustrative example is Reuter's[46] use of hexamethylenetetramine for vulcanizing a plasticizer polymer having a polyvinyl chloride base. This involves incorporating 0.5 to 10 per cent by weight of hexamethylenetetramine in the polymer composition and heating at 300 to 350°F.

References

1. Aason, A., British Patent 524,421 (1939); *J. Soc. Chem. Ind.*, **59**, 749.
2. Adams, B. A., and Holmes, E. L., *J. Soc. Chem. Ind.*, **54**, 1T-6T (1935).
3. Adams, B. A., and Holmes, E. L., U. S. Patent 2,151,883 (1939).
4. Anon., *British Plastics and Molded Products Trader*, 27–28 (June, 1939).
5. Babcock, G. E., (to U.S.A. as represented by the Secretary of Agriculture), U. S. Patent 2,494,537 (1950).
6. Bachman, G. B., and Tanner, H. A., (to Eastman Kodak Co.), U. S. Patent 2,212,506 (1940).
7. Bers, J. J., (to Armstrong Cork Co.), U. S. Patent 2,385,679 (1945).
8. Brother, G. H., and Smith, A. K., (to U. S. Secretary of Agriculture), U. S. Patent 2,210,481 (1940).
9. Brother, G. H., and Smith, A. K., U. S. Patent 2,380,020 (1945).
10. Brother, G. H., and McKinney, L. L., (to U. S. Secretary of Agriculture), U. S. Patent 2,238,307 (1941).
11. Browne, F. L., and Hrubesky, C. E., *Ind. Eng. Chem.*, **19**, 215 (1927).
12. Calvert, F. E., (to The Drackett Co.), U. S. Patent 2,424,383 (1947).
13. Chase, H., *British Plastics*, **7**, 516, 519-21 (1936).
14. Collins, W. R., Freeman, R. D., and Upright, R. M., (to Dow Chemical Co.). U. S. Patents 2,221,778-9 (1940).
15. Cooke, G. B., and Wilbur, S. I., (to Crown Cork & Seal Co.), U. S. Patent 2,104,692 (1938).
16. Ford, A. S., (to Imperial Chemicals Industries, Ltd.), U. S. Patent 1,949,832 (1934).
17. *Ibid.*, U. S. Patent 1,974,064 (1934).
18. Glavert, R. A., *British Plastics*, **19**, 333–9 (1947).
19. Graham, B., and Turner, H. S., (to E. I. du Pont de Nemours & Co., Inc.), U. S. Patent 2,443,450 (1948).
20. Hancock, E. G., *J. Soc. Chem. Ind.*, **66**, 337–40 (1947).
21. Heberer, A. J., and Phillips, C. H., (to The Glidden Co.), U. S. Patent 2,274,695 (1942).
22. Hershberger, A., (to E. I. du Pont de Nemours & Co.), U. S. Patent 2,403,077 (1946).
23. Hipp, N. J., Groves, M. L., Swain, A. P., and Jackson, R. W., *Modern Plastics*, **26**, 205 (1948).
24. Hoover, F. W., (to E. I. du Pont de Nemours & Co.), U. S. Patent 2,430,933 (1947).
25. Imperial Chemical Industries, Ltd., British Patent 523,759 (1939); *J. Soc. Chem. Ind.*, **59**, 749.
26. Kirkpatrick, W. H., (to National Aluminate Corp.), U. S. Patent 2,094,359 (1937).
27. *Ibid.*, U. S. Patent 2,106,486 (1938).
28. Leuck, G. J., (to Corn Products Refining Co.), U. S. Patents 2,222,872-3 (1940).
29. Lundbäck, T. A. I., (to Aktiebolaget Mo Och Domsjo Wallboard Co.), U. S. Patent 2,037,522 (1936).
30. Manley, R. H., and Evans, C. D., (to C. R. Wickard as Sec. of Agriculture), U. S. Patent 2,354,393 (1944).
31. Maters, C., (to Hercules Powder Co.), U. S. Patent 2,115,496 (1938).

32. McKinney, L. L., (to the United States of America), U. S. Patent 2,461,070 (1940).
33. Mead Corp., British Patent 484,248 (1937); *J. Soc. Chem. Ind.*, **57**, 1330B.
34. Millett, M. A., and Stamm, A. J., *Modern Plastics*, **25**, 125–7 (1947).
35. *Modern Plastics*, **25**, 182 (1947).
36. Moller, F. A., (to Naamlooze Vennvotschap: W. A. Scholten's Chemische Fabricken), U. S. Patent 2,246,635 (1941).
37. Morgan, G., Megson, N. J. L., and Pepper, K. W., *Chemistry & Industry*, **57**, 889 (1938).
38. Myers, R. J., *Chem. Met. Eng.*, **50**, 148–50 (1943).
39. Myers, F. J., *Ind. Eng. Chem.*, **35**, 861 (1943).
40. Norton, A. J., (to Penn. Coal Products Co.), U. S. Patent 2,385,370 (1945).
41. Phillips, R. O., and Werner Rottsieper, E. H., (to Forestal Land, Timber and Railways Co., Ltd.), U. S. Patent 2,286,643 (1942).
42. Potasch, F. M., Russian Patent 57,344 (1938); *Chem. Zentr.*, **1941**, I, 1889.
43. Raff, R. A. V., and Silverman, B. H., *Ind. Eng. Chem.*, **43**, 1423–7 (1951).
44. Raff, R. A. V., and Silverman, B. H., *Can. J. Research*, **29**, 857–62 (1951).
45. Reineck, E., and Dunlap, I. R., (to The Institute of Paper Chemistry), U. S. Patent 2,429,329 (1947).
46. Reuter, L. F., (to B. F. Goodrich Co.), U. S. Patent 2,427,070 (1947).
47. Rhodes, P. H., (to Penn. Coal Products Co.), U. S. Patents 2,385,371–4 (1945), 2,414,414 (1947).
48. Rhodes, P. H., *Modern Plastics*, **24**, 145–6, 238, 240 (1947).
49. Sanderson, J. McE., *Paint, Oil, Chem. Rev.*, **102**, No. 8, 7–9 (1940).
50. Scott, C. W., U. S. Patent 2,201,797 (1940).
51. Spahr, R. J., Moffitt, W. R., and Pryde, E. H., (to The Borden Co.), U. S. Patent 2,489,336 (1949).
52. Spahr, R. J., and Lieb, D. J., (to The Borden Co.), U. S. Patent 2,490,927 (1949).
53. Stamm, A. J., and Seborg, R. M., *Ind. Eng. Chem.*, **28**, 1164–9 (1936).
54. Stamm, A. J., and Seborg, R. M., *Trans. Am. Inst. Chem. Engrs.*, **37**, 385–98 (1941); *C. A.*, **35**, 5668 (1941).
55. Sutermeister, E., and Browne, F. L., "Casein and Its Industrial Applications," p. 402, New York, Reinhold Publishing Corp., (1939).
56. Sweeney, O. R., and Arnold, L. K., *Iowa Engr. Sta. Bull.* 154 (1942).
57. Taylor, R. L., *Chem. Met. Eng.*, **43**, 172 (1936).

General Reference

"Modern Plastics Encyclopedia," Plastics Catalogue Corp., New York (1949).

CHAPTER 21

USES OF FORMALDEHYDE, FORMALDEHYDE POLYMERS AND HEXAMETHYLENETETRAMINE

Part II

Miscellaneous Uses of Formaldehyde

The outstanding importance of formaldehyde solution, paraformaldehyde, and hexamethylenetetramine as raw materials for the manufacturer of synthetic resins has so overshadowed the other uses of these products that they have not received the publicity they deserve. Taken together, these miscellaneous applications of the various commercial forms of formaldehyde consume more than 25 per cent of the total quantity produced. Furthermore, although some of these uses are of a decidedly minor nature, many are not only of considerably immediate importance but represent rapidly expanding fields of industry. Many new uses for formaldehyde are also being discovered. The utility of formaldehyde as a tool for the research chemist in the construction of new chemical products may be compared to the utility of rivets as a tool for the structural engineer.

In the remaining pages, we shall review some of the miscellaneous uses of formaldehyde. This review is not complete, since to make it so would extend our discussion far beyond the intended scope of this book. In general, we have tried to limit the extent of our discussion for any one use or field of application with respect to its relative importance as a part of the general formaldehyde picture. However, a few extremely minor but important uses have been given a relatively extensive treatment because they illustrate how widespread are the applications of this versatile chemical. The following fields of application will be treated in alphabetical order:

Agriculture	Hydrocarbon products
Analysis	Insecticides
Concrete, plaster and related products	Leather
Cosmetics	Medicine
Deodorization	Metal Industries
Disinfection and fumigation	Metal sequestering
Dyes and dyehouse chemicals	Paper
Embalming and preserving	Photography
Explosives	Rubber
Fertilizers	Starch
Fireproofing	Solvents and plasticizers
Fuel specialties	Surface-active agents
Gas absorbents	Textiles
	Wood

Agriculture

The chief value of formaldehyde in agriculture lies in its ability to destroy microorganisms responsible for plant diseases. For this purpose it is employed as a disinfectant for seed, bulbs, roots, soil, hot bed frames, containers, bins and storage houses. In some cases, it is also employed to protect an agricultural product from decay on shipping or storage.

Seed, Bulb and Root Treatment. Seed disinfection is usually accomplished by dipping, sprinkling or spraying with dilute formaldehyde solution. It is of particular value for the prevention of the smut diseases of grain. It is estimated that diseases of this type are responsible for an annual loss of more than 25 billion bushels each of wheat and oats. Treatment with formaldehyde has proved effective in controlling the loose and covered smut of oats, the covered smut of barley, and the bunt of wheat[21]. A procedure recommended for disinfecting oat seeds is to spray lightly with commercial 37 per cent formaldehyde solution, which has been diluted with an equal volume of water, using approximately one quart of the dilute solution for 50 bushels of seed. The seed should then be covered tightly for four or five hours, aired thoroughly, and planted at once. Wheat and barley are usually disinfected by steeping in a solution of 1 pint of formaldehyde in 40 gallons of water for five minutes, followed by covering for two hours. As in the previous case, the seed should be planted immediately after treatment. If immediate planting is impossible, the seed should be washed with water and dried before storage[11]. If formaldehyde treatment is not carried out properly, the seed may be severely injured, since undue exposure to the disinfectant will result in decreased vitality or even complete loss of germinating ability[7, 11, 15]. In general, it is recommended that seed wet with dilute formaldehyde should never be allowed to dry before planting, and that planting in dry soil should be avoided. If it is necessary to plant in dry soil, the seed should first be washed. Seeds other than wheat, oats, and barley are also treated with formaldehyde for some purposes. For example, it is reported that in the case of corn, water-culture seedlings may be protected from fungi by soaking the seed for two hours in a solution of 5 cc of formaldehyde in one liter of water and then storing for 12 to 24 hours in a tightly closed container.

Leaf spot in beets may be prevented by dipping the seed for seven minutes in a solution of 1 pint 37 per cent formaldehyde in 8 gallons of water followed by rinsing. Celery or celeriac seeds require a 15 to 30 minute soak with a more dilute solution (1 pint 37 per cent formaldehyde to 32 gallons) to combat bacterial blight. In this case, also, the seed must be thoroughly rinsed after the treatment. Germination of celery and celeriac seed is usually delayed slightly by the treating process.

Seed potatoes are treated with formaldehyde for the control of scab and rhizoctonia. The latter disease, which is also known as black-scurf, stemrot, and rosette, is one of the most common potato diseases in the United States. A procedure reported to be successful for disinfecting seed potatoes against these diseases involves a 2- to 4-minute immersion in a solution of 1 pint of formaldehyde (37 per cent) in 15 gallons of water at temperatures ranging from 118 to 126°F. Authorities differ slightly with regard to temperature and time of treatment, some preferring the lower limits of the above given ranges. Some investigators recommend that the damp potatoes be covered for one hour after treatment.

Hawker[10] reports that addition of 0.5 per cent of 37 per cent formaldehyde (1 pint per 25 gallons) to the water used in the standard hot water treatment of narcissus bulbs reduces basal root rot effectively. However, if the bulbs are badly infected, a second treatment should be given in the following season. Dormant bulbs may be treated for four hours at 110 to 111.5°F using solution of the strength indicated above[17]. Foot and crown rot of rhubarb is controlled by soaking the roots for 30 minutes in a solution of 1 pint 37 per cent formaldehyde in 30 gallons of water.

Soil Disinfection. Bewley[2] states that formaldehyde is the best chemical disinfectant for soil. It destroys fungi and bacteria, and restores fertility. However, it is not as effective as disinfectants of the cresylic acid type in preventing damage from insects and soil animals. Tam and Clark[22] report increased root growth following disinfection of soil with steam or formaldehyde.

Soil disinfection has proved of particular value for controlling the damping-off diseases which are so destructive to seedlings. It is also generally recommended for the prevention of truck-crop diseases[6], such as scurf, black-rot, and stem-rot of sweet potatoes[7]. Disinfection may be accomplished by drenching the soil with a solution of 1 pint of 37 per cent formaldehyde in 12.5 gallons of water and then covering with burlap or paper for 12 to 24 hours[7]. Some authorities recommend that the soil be dried thoroughly before planting; others suggest that it merely be allowed to stand uncovered for several days. Treating solutions containing 1 pint of formaldehyde to 15 gallons of water are also commonly recommended. Damping-off is caused by soil-borne fungi of which pythium debaryanum is the most important member. Ogilvie and co-workers[16] report that damping-off of peas, tomatoes, cucumbers, and sweet peas may be controlled successfully by watering seed boxes with formaldehyde in dilutions of 1 to 200–600 after seeding. Doran and Guba[8] state that damping-off is also effectively controlled by treating with one quart of water containing 2 cc commercial 37 per cent formaldehyde for each square foot of soil and report no injury when using a solution of double the specified concentration. Results are the same whether the soil is seeded before or after treatment.

Another procedure recommended for the control of damping-off involves the use of formaldehyde dust produced by wetting a carrier substance, such as infusorial earth or ground charcoal, with formaldehyde solution and then drying[19, 20]. Anderson[1] states that pythium debaryanum may be controlled in tobacco seed beds by mixing each square foot of top soil with 1.5 ounces of a dust preparation made from 15 parts formaldehyde (37 per cent) and 18 parts ground charcoal just before planting.

Numerous plant diseases in addition to those already mentioned have also been reported to respond favorably to methods of control involving the use of formaldehyde. Reyneke[18] reports that formaldehyde vapor or aqueous solution prevents botrytis rot but does not control dry stalk rot in grapes. Other reported applications include the use of formaldehyde disinfection for preventing onion smut, mushroom diseases[14], and rust in-

Figure 40. Prevention of damping-off of seedlings by formaldehyde treatment of soil. Boxes were seeded simultaneously. Cards mark time of soil treatment. Check was untreated.

fection in olives[23]. Also of importance to agriculturists is the use of formaldehyde in bee culture for disinfecting equipment polluted with American foul-brood[9].

Formaldehyde is sometimes employed in conjunction with other agents which are stated to increase or modify its action as a seed or soil disinfectant. Braun[3] recommends a 1:320 formaldehyde solution containing copper sulfate in 1:80 dilution for treating wheat, barley, oats, and corn. Patented disinfectant compositions in which formaldehyde or its polymers are blended with other disinfectants such as phenol and mercuric chloride are also described[4, 5, 13]. One seed-disinfecting compound contains acidic, hygroscopic, or syrupy materials which are claimed to hinder the deposition of formaldehyde polymer on the seeds and thus prevent damage and increase the disinfecting action[5].

Prevention of Storage Rots and Infections. Disease and decay of root crops on storage may be prevented or substantially reduced by thorough disinfection and fumigation of bins, barrels, crates, etc., with formaldehyde.

Treatment of citrus fruit with sodium *o*-phenyl phenoxide and formaldehyde delays decay effectively[12]. When used alone, the phenolic compound controls decay but injures the fruit rinds, a difficulty that is apparently eliminated by using one part of 37 per cent formaldehyde to 4.5 parts sodium *o*-phenyl phenoxide. The fruit is treated by dipping in an aqueous solution of the disinfectants which may also contain a wax emulsion if desired.

Detailed information on seed and soil disinfection can be obtained from the U. S. Department of Agriculture as well as many State Agricultural Stations.

References

1. Anderson, P. J., *Conn. Agr. Exp. Sta. Bull.*, **359**, 336–54 (1934); *C. A.*, **28**, 6920 (1934).
2. Bewley, W. F., *Chemistry & Industry*, **63**, 237 (1944).
3. Braun, H., *J. Agr. Research*, **19**, 363–92 (1920); *Chem. Zentr.*, **1921**, II, 167.
4. Chemische Fabrik Ludwig Meyer, French Patent 565,198 (1924).
5. Chemische Fabrik Ludwig Meyer, German Patent 423,466 (1926); *Chem. Zentr.*, **1926**, II, 489.
6. Cook, H. T., and Nugent, T. J., *Virginia Truck Expt. Sta. Bull.*, **104**, 1663–1717 (1940).
7. Daines, R. H., *N. J. Agr. Expt. Sta.*, *Circular No. 437*, 8 pp. (1942).
8. Doran, W. L., and Guba, E. F., *Mass. Agr. Expt. Sta. Ann. Rept.*, **1939**, 23–4 (1940).
9. Garman, P., *Agr. Expt. Sta.*, *Conn. Rept.*, **48**, 305–7; *Chem. Zentr.*, **1926**, II, 1175.
10. Hawker, L. E., *Ann. Appl. Biol.*, **31**, 31–3 (1944).
11. Hurd, A. M., *J. Agr. Research*, **20**, 209–41 (1920); *C. A.*, **15**, 699.
12. Miller, E. V., Winston, J. R., and Meckstroth, G. A., *Citrus Ind.*, **25**, No. 10, 3, 15, 18 (1944).
13. Molz, E., (to Chemische Fabrik Ludwig Meyer), U. S. Patent 1,530,280 (1925).
14. Morris, R. H., U. S. Patent 1,694,482 (1928).
15. Neill, J. C., *New Zealand J. Agr.*, **31**, 24–5 (1925); *C. A.*, **20**, 472 (1926).
16. Ogilvie, L., Hickman, C. J., and Croxall, H. E., *Ann. Rept. Agr. Hort. Research Sta., Long Ashton, Bristol*, **1938**, 98–114; *C. A.*, **34**, 1121.
17. *Proc. Helminthol. Soc., Wash., D. C.*, **8**, 44–50 (1941).
18. Reyneke, J., *Farming in S. Africa*, **14**, 64–6 (1939); *J. Soc. Chem. Ind.*, **58**, 1171.
19. Sayre, J. D., and Thomas, R. C., *Science*, **66**, 398 (1927); *C. A.*, **22**, 299.
20. Sayre, J. D., and Thomas, R. C., *Ohio Agr. Expt. Sta., Bimonthly Bull.*, **13**, 19–21 (1928); *C. A.*, **22**, 1011.
21. Stewart, R., and Stephens, J., *Utah Expt. Sta. Bull.*, **108**, (1910); *C. A.*, **4**, 2972 (1910).
22. Tam, R. K., and Clark, H. E., *Soil Sci.*, **56**, 245–61 (1943).
23. Traverso, G. B., *Staz. sper. agrar. ital.*, **52**, 463–84 (1919); *Chem. Zentr.*, **1921**, I, 928.

Formaldehyde as a Reagent for Chemical Analyses

Small quantities of formaldehyde are employed for a variety of purposes in quantitative and qualitative analyses. Although for some of these purposes it has been replaced by other reagents, it remains a key chemical in several well known analytical procedures commonly employed in modern technical laboratories.

The Sörenson procedure for the titration of amino acids is perhaps the best known of the analytical procedures in which formaldehyde is employed. Reaction with formaldehyde makes it possible to carry out direct titration of these acids, since the basic properties of the amino radicals are nullified by formaldehyde condensation. The reactions upon which this procedure is based have been already discussed (page 311).

The liberation of anions, which takes place when formaldehyde reacts with ammonium salts, is employed for determining ammonium nitrate by means of acidometry as previously discussed (cf. page 181). The same principle is also recommended by Marcali and co-workers[6] for eliminating the distillation in the Kjeldahl nitrogen determination. After oxidation of the analysis sample with hot concentrated sulfuric acid in the presence of potassium sulfate and a mercury catalyst, the excess acid is neutralized to methyl red with $N/10$ alkali in the presence of sodium bromide, which prevents precipitation of mercury. The ammonium salt content of the mixture is then titrated with standard $N/10$ alkali in the presence of formaldehyde.

Formaldehyde is also commonly employed for the quantitative determination of halogen acids or inorganic halides in the presence of hydrocyanic acid or alkali cyanides. The solution to be analyzed is made alkaline with caustic and treated with excess formaldehyde. By this means cyanides are converted to formaldehyde cyanohydrin which will not interfere in the volumetric determination of halogen by the Volhard procedure, which can be applied after the solution has been acidified with nitric acid. This procedure was described by Polstorf and Meyer[11] in 1912. A recent paper by Mutschin[9] gives detailed directions for determining chlorides, bromides and iodides in the presence of CN^- with Mohr's method of titration following the addition of formaldehyde.

Vanino and Schinner[15] make use of formaldehyde in the gas volumetric determination of nitrous acid by means of the following reaction:

$$4HNO_2 + 3CH_2O \rightarrow 3CO_2 + 5H_2O + 2N_2$$

The reaction is carried out in the presence of hydrochloric acid. The carbon dioxide is absorbed by alkali and the nitrogen measured in a nitrometer.

Solutions of sodium sulfite and bisulfite can be analyzed by use of the reaction employed in the determination of formaldehyde with sodium sulfite[5]:

$$Na_2SO_3 + CH_2O + H_2O \rightarrow NaOH + NaSO_3 \cdot CH_2OH$$

The sulfite solution is titrated to neutrality with alkali, using phenolphthalein as an indicator; excess formaldehyde is then added and the alkali liberated is determined by another titration. If the titer obtained in the first titration is the same as that obtained in the second, the solution contains pure alkali bisulfite equivalent to the alkali used in either titration. If the titer is greater in the second titration, Na_2SO_3 is present and may be calculated. Pure sodium sulfite is neutral and the first titer is zero. If the first titration is the greater, the sulfite contains a foreign acidic impurity. Addition reactions of formaldehyde can also be employed in the quantitative separation of sulfides, sulfates, thiosulfates, and sulfites[4]. Carbonates can be detected in the presence of sulfites by using formaldehyde to prevent the evolution of sulfur dioxide on acidification of the formaldehyde-treated solution[13].

Precipitation of metals by the addition of formaldehyde and alkali to solutions containing salts or oxides of gold[13], silver[14], bismuth[3, 14] and copper[7] has been utilized in some instances for the determination of these metals. Gold may be determined colorimetrically by a method of this type based on the formation of colored colloidal gold. By this procedure, it is possible to detect one part of gold in 40,000 parts of solution[8].

Manganese is determined colorimetrically by the use of formaldoxime reagent and sodium hydroxide[16]. Formaldoxime reagent is prepared by dissolving 3 parts of paraformaldehyde and 7 parts of hydroxylamine hydrochloride in 5 parts of water.

Hexamethylenetetramine is of value in the detection and determination of metals because of its pronounced tendency to form complex products with metallic salts. Traub[12] makes use of hexamethylenetetramine in the separation of titanium and columbium. In this connection he points out the use of this reagent to separate divalent metal ions from numerous trivalent and tetravalent ions. Molybdenum is one of the few elements with a valence higher than two which is not precipitated by hexamethylenetetramine. Crowell and König[1] use hexamethylenetetramine for the microchemical detection of gold and platinum metals.

Nastyukov's "Formolite" reaction[10] is sometimes employed for the determination of unsaturated cyclic or aromatic hydrocarbons in oils. The sample is mixed with an equal volume of concentrated sulfuric acid, and a 0.5 to 1 volume of commercial formaldehyde is gradually added. The mixture is then shaken or otherwise agitated until reaction is complete (approximately 30 minutes to one hour) and diluted with petroleum ether free of hydrocarbons capable of giving a positive reaction. The diluted mixture is poured into ice-water, treated with excess ammonia, filtered and washed. Following a final wash with petroleum ether and water the solid "formolite" is dried at 105°C and weighed. The weight of unsaturated or aromatic hydrocarbon is equivalent to approximately 80 per cent of the weight of the "formolite" precipitate. (The weight of precipitate per 100 cc of oil is Nastyukov's "formolite number".) The procedure is reported to be satisfactory for the determination of terpenes and other hydroaromatics as well as aromatic hydrocarbons.

Tannins and related phenolic compounds may be determined by precipitation of insoluble products with formaldehyde in the presence of concentrated hydrochloric acid by a technique similar to that employed in the "formolite" reaction[2].

References

1. Crowell, W. R., and König, O., Mikrochem. ver. Mikrochim. Acta, 33, 303–9 (1948).
2. Glückmann, C., Pharm. Post., 37, 429–307, 533–4 (1904); Chem. Zentr., 1904, II, 740, 1261.
3. Hartwagner, F., Z. anal. Chem., 52, 17–20 (1918).
4. Hemmeler, A., Ann. Chim. applicata, 28, 419–27 (1938); C. A., 33, 4547 (1939).
5. Kühlsches Laboratorium, Collegium, 1922, 97; Chem. Zentr., 1922, IV, 475.
6. Marcali, K., Rieman, W., III, Ind. Eng. Chem., Anal. Ed., 18, 709–10 (1946).
7. Menzel, A., "Der Formaldehyde," p. 69, Vienna and Leipzig, Hartleben's Verlag (1927).
8. Muller, J. A., and Foix, A., Bull. soc. chim., 33, 717–720 (1922).
9. Mutschin, A., Z. anal. Chem., 99, 335–48 (1934); C. A., 29, 1356–7.
10. Nastyukov (Nastjukoff), A., J. Russ. Chem. Phys. Soc., 36, 881 (1904); 42, 1596 (1910).
11. Polstorff, K., and Meyer, H., Z. anal. Chem., 51, 601 (1912).
12. Traub, K. W., Ind. Eng. Chem., 18, 122 (1946).
13. Vanino, L., Ber., 31, 1763 (1898).
14. Vanino, L., and Treubert, F., Ber., 31, 1303 (1898).
15. Vanino, L., and Schinner, A., Z. anal. Chem., 52, 11 (1913).
16. Wagenaar, G. H., Pharm. Weekblad, 75, 641–8 (1938); C. A., 32, 6573.

Concrete, Plaster, and Related Products

According to the claims of some patents, formaldehyde and its derivatives have value in the preparation of concrete, plaster, etc. Griffiths[5] describes the use of aqueous formaldehyde as an addition agent in a concrete mix. A cellular cement, concrete, or plaster is produced by adding to the mix sodium carbonate, saponin, formaldehyde, and water[4]. Compositions for addition to plaster work and the like to make it impermeable to liquids and grease are described by Bauer[1] as containing water, formaldehyde, potassium dichromate, hydrochloric acid, sodium chloride and ferric chloride. F. J. Maas[7] obtains a plastic coating composition by adding 1 to 5 per cent of formaldehyde to a cement mix produced from magnesium oxide 20 to 50 parts, silica 10 to 50 parts, calcium carbonate 5 to 25 parts, fluorspar 1 to 10 parts and zinc oxide 1 to 15 parts, to which is then added a solution of magnesium chloride and water having a specific gravity of 20 to 40° Bé. A tough, flexible, self-hardening cement for use as a floor or wall covering is obtained by adding glue, water, and formaldehyde to a hydraulic (Portland) cement[6].

A small percentage of a water-soluble reaction product of formaldehyde and urea, when added to a water-cement mix (*e.g.*, plaster of Paris, Portland cement, etc.) improves the surface hardness of the set product and gives it increased strength after rapid drying[2]; cellular cements may be produced by inclusion of a gas-producing substance in the mix[3].

References

1. Bauer, M., British Patent 511,686 (1939); *C. A.*, **34,** 7084.
2. E. I. du Pont de Nemours & Co., Inc., British Patent 519,078 (1940).
3. E. I. du Pont de Nemours & Co., Inc., British Patent 523,450 (1940).
4. Etablissement Chlollet-Lefevre, French Patent 799,997 (1936); *C. A.*, **30,** 8561 (1936).
5. Griffiths, W. H., British Patent 488,550 (1937).
6. Griffiths, L. H., and Saul, W. (to Semtex Ltd.) British Patent 559,541 (1944).
7. Maas, F. J., U. S. Patent 2,228,061 (1941).

Cosmetics

Formaldehyde occasionally finds application in cosmetics for the prevention of excessive perspiration. Janistyn[2], for example, describes antiperspiration compositions for the feet containing 2 to 10 per cent of 30 to 35 per cent formaldehyde, 30 per cent cologne or lavender water, and 60 to 68 per cent distilled water. For the same purpose, powder-type products may employ small concentrations of paraformaldehyde as an active agent.

Small amounts of formaldehyde are sometimes used in antiseptic dentifrices and mouth washes[1, 4]. Germicidal soaps may also contain formaldehyde.

A German patent[3] covers a method of straightening curly hair whereby the hair is first treated with a keratin-softening substance and then with a keratin-hardening substance such as formaldehyde.

References

1. Augustin, J., *Deut. Parfüm. Ztg.*, **14,** 542 (1928); *Chem. Zentr.*, **1929,** II, 1323.
2. Janistyn, H., *Seifensieder-Ztg.*, **67,** 509–10 (1941).
3. Keil, F., German Patent 697,634 (1940).
4. Sandberg, M. H., U. S. Patent 2,527,686 (1950).

Deodorization

The deodorizing properties of formaldehyde probably depend in part upon its ability to react with ammonia, amines, hydrogen sulfide, mercaptans, etc., with the formation of less volatile products. Its germicidal action also arrests decay and decomposition. Formaldehyde is sometimes used as an air deodorant in public places where it is employed in the form of a dilute solution containing small percentages of essential oil compositions. Its use has been advocated in processes for deodorizing blubber[2], hydrolysis products of glue, leather, skins, hoofs, etc[6], cork[5], tar and mineral oils[7], deodorizing abattoirs[3], etc.

Special preparations involving formaldehyde are also employed as deodorants. A patented deodorant powder is prepared from paraformaldehyde, hexamethylenetetramine, a powdered emollient material, and ammonium chloride[8]. According to a British process, cakes or balls for disinfecting and deodorizing are formed from mixtures containing paraformaldehyde, rice flour, pine-needle oil, potassium permanganate, etc.[1] Kamlet[4] describes a solid deodorizing composition produced by blending paraformaldehyde and trioxane.

References

1. Brick, A., British Patent 258,110 (1925).
2. Holländer, M., German Patent 362,281 (1922).
3. Hopkinson, L. T., U. S. Patent 2,003,887 (1935).
4. Kamlet, J., (to Boyle-Midway, Inc.), U. S. Patent 2,464,043 (1949).
5. Milch, A., German Patent 264,305 (1911).
6. Plausons Forschungsinstitut G. m. b. H., German Patent 344,632 (1921); *Chem. Zentr.*, **1922**, II, 210.
7. Rutgerswerke-Aktiengesellachaft, German Patent 147,163 (1903).
8. Weber, F. C., U. S. Patent 1,813,004 (1931).

Disinfection and Fumigation

Formaldehyde destroys bacteria, fungi, molds and yeasts. Its application for the control of bacteria and fungi in soil and seeds has been treated in the section dealing with agriculture (pages 457–460). Its commercial importance as a fungicide is probably greater than as a bactericide. Such applications also include its use for disinfection of equipment in the fermentation industry, in the manufacture of antibiotics such as penicillin and streptomycin as well as its many services for the preservation of useful products (page 471).

Its use as a disinfectant dates from the last decade of the nineteenth century. Its powerful bactericidal properties were demonstrated by Loew and Fischer[13] in 1886, and practical methods for disinfecting sickrooms with formaldehyde solution were described by Aronson and Blum in 1892[16]. In recent years, its low cost and relatively innocuous nature as compared with mercury compounds and other powerful agents has enabled it to retain a recognized position as an important antiseptic and germicide.

Quantitative measurements of the disinfecting efficiency of formaldehyde show considerable variation, depending on the conditions of use and the microörganisms involved. Conservative figures which are probably generally acceptable indicate that a formaldehyde solution containing 50 parts CH_2O per million will act as an antiseptic to prevent the growth of bacteria and that a 4 per cent solution is sufficiently strong to destroy all vegetative and the majority of arthrogenous bacterial forms in less than 30 minutes[2]. Bactericidal efficiency increases with temperature and is augmented in some cases by the presence of soap, alcohol, or acid. Compounds which react with formaldehyde such as ammonia, sodium bisulfite, etc., reduce or destroy its bactericidal properties.

Comparison of the disinfecting characteristics of formaldehyde with those of phenol indicate that its bacteriostatic action is much more marked than that of the latter compound. According to Ganganella[6], the bactericidal action of formaldehyde is less affected by variations in its concentration: its dilution coefficient is about 1 at ordinary temperatures, whereas that of phenol under the same conditions is 6.5. The phenol coefficient of formaldehyde is reported by Tilley and Schaffer[26] to have a value of 1.05. Scott[23] reports that phenol in concentrations up to 5 per cent acts very slowly on anaerobic organisms, which are destroyed rapidly by 0.5 to 0.75 per cent formaldehyde. According to this investigator, a 0.5 per cent formaldehyde solution kills all aerobic bacteria, including sporophytes in 6 to 12 hours, saccharolytic anaerobes in 48 hours, and proteolytic anaerobes in 4 days.

Formaldehyde is employed as a disinfectant in both solution and gaseous forms for a wide variety of miscellaneous purposes. Common uses include: disinfection of sickrooms by sprinkling with 2 per cent formaldehyde[16] sterilization of surgical instruments by boiling 2–10 minutes with 2 to 4 per cent solution[7], destruction of anthrax bacteria in imported bristles by soaking for four hours in a 10 per cent solution maintained at 110°F.[15], etc. In the brewing industry a 0.5 per cent solution has been reported to be satisfactory for destruction of injurious microörganisms in the presence of enzymes and yeasts which must not be adversely affected[22]. A unique application described by Robinson[20] is the production of sterile maggots for surgical use from eggs which have been superficially sterilized with formaldehyde.

Numerous special disinfectant solutions containing formaldehyde are described in the trade and patent literature. A composition recommended for the treatment of barrels, butter tubs, etc., consists of a dilute solution of formaldehyde and water glass[24]. A disinfectant and deodorant for washrooms contains formaldehyde plus sufficient sodium metasilicate to give a definitely alkaline solution[28].

Disinfection with gaseous formaldehyde is most effective in the presence of moisture[17]. It is often accomplished merely by placing the articles to be disinfected in a closed cabinet containing formaldehyde solution at room temperature. Moldy leather goods or shoes which have been worn by persons with epidermophytosis (athlete's foot) may be sterilized in this way. According to O'Flaherty[18], effective sterilization can be secured by allowing the leather goods to stand for 12 to 18 hours in an air-tight box in which has been placed an open vessel containing a mixture of 1 part of commercial formaldehyde solution and 3 parts of water. After sterilization, the shoes or other articles should be aired thoroughly for several hours before use. Similar applications are found in cabinets for disinfecting gas masks[3] and barbers' cabinets for scissors, clippers, etc.

According to Dorset[4], outstanding advantages of formaldehyde as a fumigant include the following: (a) it is a strong germicide; (b) its action is not greatly hindered by albuminous or organic substances; (c) it does not injure fabrics, paint or metal; and (d) it can be used effectively and safely in households. The following disadvantages are also cited[4]: (a) fumigation is not reliable if the temperature of the room is below 65°F; (b) it has a penetrating odor and is irritating to the eyes and nose; (c) long exposure and careful sealing of inclosures are required. According to Horn and Osol[11], results of fumigation with formaldehyde are unsatisfactory if the relative humidity is below 60 per cent.

Many methods have been used for generating formaldehyde gas for purposes of fumigation. In some procedures, this is carried out simply by heating the aqueous solution or polymer. A typical apparatus[25] consisted of two sections, the inner containing formaldehyde solution or paraformaldehyde and the outer containing boiling water. These sections were heated by an alcohol lamp so that formaldehyde gas and water vapor were evolved. Paraformaldehyde should be employed in granular form, since this grade is more readily vaporized than the powdered product. Other methods for generating gaseous formaldehyde involve addition of the aqueous solution to a chemical agent with which it will react exothermically, evolving sufficient heat to vaporize the major portion of the solution involved. Agents used for this purpose include: potassium permanganate[14], sodium dichromate[12], bleaching powder[12], potassium or sodium chlorate[19], caustic soda, etc. Another procedure of this general type is carried out by adding water to a mixture of paraformaldehyde and a metallic peroxide[5]. Following a critical study of the utility of the above-mentioned chemical agents in fumigation with formaldehyde solution, Horn and Osol[11] concluded that the best results were obtained with bleaching powder.

A typical procedure for fumigation of 1000 to 1500 cu ft of space involves addition of 1 pound of U.S.P. formaldehyde to 4 to 8 ounces of solid potassium permanganate in a large porcelain dish. The room must be thoroughly sealed and the fumigator must make a rapid exit after adding the formaldehyde to the permanganate.

If fumigation is carried out by vaporizing commercial formaldehyde with applied heat, approximately 5 ounces is sufficient for fumigating 1000 cu ft. After fumigation the room should be thoroughly ventilated. Lingering odors of formaldehyde are readily dispelled by sprinkling with ammonia, which converts it to the odorless hexamethylenetetramine.

Volatile fungicides or fungistats are of value for the preservation of packaged materials when treatment with preservative solutions is impractical. Evaluation of various materials of this type for protecting military supplies and equipment as indicated by experiments on wood, leather, adhesive tape and maltagar showed that paraformaldehyde was effective[21]. However, combinations of paraformaldehyde with other volatile materials such as chlorinated trimethylacetonitrile[1], dihydronaphthalene[8] and camphor[9] are outstanding in their fungicidal activity. Tests show that whereas some species of fungi are not killed by either agent when used alone, the paraformaldehyde-containing mixtures kill essentially all varieties. This synergistic action of formaldehyde with other agents is also apparent in liquid disinfectants. Waugh[27] reports a germicidal formaldehyde solution in which thymol is added to increase the killing power of formaldehyde.

In addition to its direct use as a disinfectant, formaldehyde is also employed in the synthesis of other germicides and antiseptics. Methylene bis-trichlorophenol, prepared by the acid catalyzed reaction of formaldehyde with trichlorophenol, is a useful germicide which can be incorporated in antiseptic soaps[10, 28a] (see also page 259).

References

1. Carlisle, P. J., and Hinegardner, W. A., (to E. I. du Pont de Nemours & Co.), U. S. Patent 2,452,429 (1948).
2. Christian, M., "Disinfection and Disinfectants," pp. 74–104, London, Scott, Greenwood & Son (1913).
3. Dörge, H., *Gasmaske*, **12**, 82–5 (1940).
4. Dorset, M., *U. S. Dept. Agr., Farmers' Bull.*, **926**, (1918); *C. A.*, **12**, 1326.
5. Farbenfabriken vorm. Friedr. Bayer & Co., German Patent 117,053 (1906).
6. Ganganella, R., Chimie & Industrie, **43**, 147 (1939); *C. A.*, **34**, 4233, (1940).
7. Gelinsky, E., *Zentr. Bakt. Parasintenk., Abt. I, Orig.*, **146**, 27–48 (1940); *C. A.*, **35**, 3768 (1941).
8. Hinegardner, W. S., and Walker, J. F., (to E. I. du Pont de Nemours & Co.), U. S. Patent 2,425,678 (1947).
9. Hinegardner, W. S., (Electrochemicals Dept., E. I. du Pont de Nemours & Co.), Unpublished communication.
10. Gump, W. S. (to Burton T. Bush, Inc.), U. S. Patent 2,250,480 (1941).
11. Horn, D. W., and Osol, A., *Am. J. Pharm.*, **101**, 741–778 (1929).
12. Kline, E. K., *Am. J. Public Health*, **9**, 859–65 (1919); *C. A.*, **14**, 441.
13. Loew, O., and Fischer, E., *J. prakt. Chem.*, **33**, 321 (1886).
14. McClintic, T. B., *Hyg. Lab. Pub. Health and Marine Hospital Service Bull.*, **27**, (1907); *C. A.*, **1**, 1444, (1907).

15. McCullough, E. C., "Disinfection and Sterilization," p. 369–78, Philadelphia, Lea & Febiger (1936).
16. Menzel, "Der Formaldehyde von Vanino," p. 248, Vienna and Leipzig, Hartleben's Verlag (1927).
17. Nordgren, G., *Acta Path. Microbiol. Scand. Suppl.*, **40**, 1–165 (1939); *C. A.*, **33**, 6382, (1939).
18. O'Flaherty, F., *Shoe Leather Reptr.*, **226**, No. 3 (April 18, 1942).
19. Ressler, I., U. S. Patent 1,408,535 (1922).
20. Robinson, W. J., *Lab. Clin. Med.*, **20**, 77 (1934).
21. Scheffer, T. C., and Duncan, G., *Ind. Eng. Chem.*, **38**, 619–21 (1946).
22. Schnegg, H., *Z. gesa., Brauw.*, **28**, 807, 820 (1905).
23. Scott, J. P., *Infect. Dis.*, **43**, 90 (1928).
24. Szigethy, I., Hungarian Patent 126,983 (1941); *C. A.*, **35**, 7662 (1941).
25. Thöni, J., *Mitt. Lebensm. Hyg.*, **4**, 315–49; *C. A.*, **8**, 770 (1914).
26. Tilley, F. W., and Schaffer, J. M., *J. Bacteriol.*, **16**, 279 (1928).
27. Waugh, A. L., U. S. Patent 2,347,012 (1944).
28. White, A. R., (to Deodor-X Co.), U. S. Patent 2,077,060 (1937).
28a. Anon., *Chem. Ind.*, **66**, 504 (1950).

Dyes and Dyehouse Chemicals

Formaldehyde is employed in the synthesis of dyes, stripping agents, and various specialty chemicals used in the dye industry.

The use of formaldehyde as a synthetic agent for the production of coaltar colors is among the earliest of its commercial applications. Diaminodiphenylmethane was prepared around 1890 by the action of aniline hydrochloride on formaldehyde-aniline for use as an intermediate in the manufacture of rosaniline dyes[2]. Other rosaniline dye intermediates were also produced by similar processes involving formaldehyde and homologous arylamines. Related syntheses discovered at an early date include the production of aurine dyes from hydroxy aromatic acids (page 260), acridine dyes from meta-arylene diamines[6], and pyronines from meta-aminophenols[7]. An important application of formaldehyde is also found in the synthesis of indigo, where it may be employed for the preparation of the intermediate, phenylglycine[3, 8]. This acid is readily obtained by hydrolyzing the phenylglycine nitrile produced by reacting formaldehyde[3] or formaldehyde sodium bisulfite[8] with sodium cyanide and aniline, as indicated in the following equation:

$$C_6H_5NH_2 + CH_2O \text{ (or } CH_2OH \cdot NaSO_3) + NaCN \rightarrow$$

$$C_6H_5NHCH_2CN + NaOH \text{ (or } Na_2SO_3)$$

The above synthesis has replaced the older chloroacetic acid indigo process in some instances[13] and may also be employed in the preparation of other related indigoid dyes.

In addition to the syntheses described above, many other processes for the use of formaldehyde in the production of dyes and dye intermediates are described in the patent literature. Recent patents, for example, include the production of benzene-soluble azo dyes by coupling diazotized arylamines with aromatic hydroxy compounds plus formaldehyde[4], preparation of vat dyes by condensation of formaldehyde with N-dihydro-1,2,1',2'-anthraquinone azines[11], synthesis of nitroso dyes by reacting formaldehyde with p-N-alkylaminonitrosobenzene derivatives[5], formation of thiazole dyes from mercaptothiazoles, primary amines, and formaldehyde[10, 14], production of diphenylmethane intermediates[9], etc.

Arthur[1] describes a process for preventing the bleeding of azo pigments by reacting them with paraformaldehyde in concentrated sulfuric acid. A modified non-bleeding pigment is precipitated when the reaction mixture is drowned in water.

Formaldehyde sulfoxylates such as sodium and zinc formaldehyde sulfoxylate, whose preparation has been previously discussed (page 194), are manufactured in considerable amount for use as stripping agents and in the dyeing and printing of vat colors[15]. This agent is stated to be particularly effective for stripping textiles which are sun-faded or stained and leveling white fabrics prior to redyeing[12].

The application of formaldehyde for improving the color stability of dyed fabrics, etc., is discussed in connection with the formaldehyde treatment of textiles (pages 523-4).

References

1. Arthur, P., Jr., (to E. I. du Pont de Nemours & Co., Inc.), U. S. Patent 2,540,705 (1951).
2. Farbwerke vorm. Meister Lucius und Brüning, German Patent 53,937 (1889); *Chem. Zentr.*, **1891**, I, 480.
3. Farbwerke vorm. Meister Lucius und Brüning, German Patent 135,322 (1901).
4. Geller, L. W., (to National Aniline and Chemical Co.), U. S. Patent 1,870,806 (1932).
5. Kranz, F. H., (to National Aniline and Chemical Co.), U. S. Patent 2,185,854 (1940).
6. Leonhardt & Co., German Patent 52,324 (1890).
7. Leonhardt & Co., German Patents 58,955 (1891); 63,081 (1892).
8. Lepetit, R., *Chim. et Ind.*, **14**, 852 (1925); *C. A.*, **20**, 2585, (1926).
9. Mattison, F. L., (to E. I. du Pont de Nemours & Co., Inc.), U. S. Patent 1,954,484 (1934).
10. Messer, W. E., (to U. S. Rubber Co.), U. S. Patent 1,996,011 (1935).
11. Neresheimer, H., and Schneider, W., (to General Aniline Works, Inc.), U. S. Patent 1,830,852 (1931).
12. *Rayon Textile Monthly*, **25**, 102 (1944).
13. Ullmann, F., "Enzyklopädie der technischen Chemie," Vol. 6, p. 239, Berlin, Urban and Schwarzenberg (1930).
14. Williams, I., (to E. I. du Pont de Nemours & Co., Inc.), U. S. Patent 1,972,963 (1934).
15. Wood, H., *Chem. Age*, **38**, 85-6 (Jan. 29, 1938).

Embalming and Preserving

The use of formaldehyde in embalming is an important application of its preservative and hardening action on animal tissue. For this purpose formaldehyde is seldom used alone, but is employed in conjunction with various modifying agents in a wide variety of embalming fluids. The modifying agents in these compositions serve to facilitate penetration, increase preservative action, maintain moisture content and aid in the production of desired cosmetic effects. Most embalming fluids contain alcohol, glycerol and phenol as major ingredients in addition to aqueous formaldehyde. Special compositions described in the patent literature also contain water: calcium chloride, alcohol and a coloring agent[9]; magnesium citrate[10]; potassium nitrate and acetate[1]; ionizable aluminum salts[5]; soluble fluosilicates[6]; an emulsion with fatty substances containing lanolin[4]; bile salts[7]; etc. Compositions containing arsenic and mercury have been employed in some instances, but are now prohibited by law because of their masking influence in cases of suspected poisoning.

Formaldehyde and paraformaldehyde are used to preserve a large number of industrial products from the action of bacteria, molds and fungi. These include waxes, polishes, adhesives, fats, oils, starch, ferns, flowers, tankage, textiles, etc. Applications of formaldehyde for hides, rubber latex and oil well muds have been specifically described in the sections dealing with leather, rubber and hydrocarbon products. Miscellaneous applications include the preservation of anatomical specimens and bacterial cultures. Hollande[3] recommends a mixture of formaldehyde, copper acetate, picric acid and acetic acid as a special fixative for histological sections. The preservation of bacterial cultures is accomplished by treatment with formaldehyde vapor. This treatment kills the bacteria and hardens the gelatin media without changing the appearance of culture specimens, which may then be saved for future reference[2]. Inedible fish to be used in preparing fertilizer (page 476) may be preserved with formaldehyde.

The use of formaldehyde as a food preservative is prohibited by law, since formaldehyde-modified proteins are not readily assimilated and the aldehyde itself is a poisonous material. Dr. W. H. Park[8] points out in "Public Health and Hygiene" that although milk containing 1 part formaldehyde per 20,000 parts milk is apparently harmless, addition of small amounts of the preservative cannot be legally condoned since control of the concentrations used would be impractical.

References

1. Blum, F., *Pharm. Ztg.*, **41**, 468–9 (1896); *Chem. Zentr.*, **1896,** II, 356.
2. Hauser, G., *Munch. med. Wochnschr.*, **40**, 567–8 (1893); *Chem. Zentr.*, **1893,** II, 691.
3. Hollande, A.-Ch., *Ber. ges. Physiol.*, **4,** 162 (1920); *Chem. Zentr.*, **1921,** II, 209.

4. Jones, H. I., (to The Naselmo Corp.), U. S. Patent 2,048,008 (1936).
5. *Ibid.*, U. S. Patent 2,085,806 (1937).
6. Jones, H. I., (to National Selected Morticians), U. S. Patent 2,208,764 (1940).
7. *Ibid.*, U. S. Patent 2,219,927 (1940).
8. Park, W. H., "Public Health and Hygiene," p. 348, Lea and Febiger, Philadelphia and New York (1920).
9. Smith, G. Q., (to W. R. Clayton), U. S. Patent 1,215,210 (1917).
10. Speaber, T., U. S. Patent 1,390,392 (1921).

Explosives

A number of explosives have been prepared by syntheses involving formaldehyde as a key raw material. Two of these, according to Davis[8], stand out as being the most powerful and brisant of the solid high explosives which are suitable for military use. These are pentaerythritol tetranitrate and cyclo-trimethylenetrinitramine.

Although most of the important explosives derived from formaldehyde are prepared from formaldehyde derivatives, several explosive compounds can be prepared directly from formaldehyde solution and anhydrous formaldehyde. According to Travagli and Torboli (page 196), methylene nitrate, $CH_2(ONO_2)_2$, can be produced by the action of mixed acids on formaldehyde solution at 5°C. A close relative of methylene nitrate, nitromethoxymethyl nitrate, $NO_2 \cdot CH_2OCH_2NO_3$, was produced by Moreschi[18], who prepared it by the action of mixed acids on dichloromethyl ether, prepared from formaldehyde and hydrogen chloride (page 196). It is an oil which dissolves nitrocellulose and gives an explosive gelatin with 7 per cent of this material. A number of explosive peroxygen compounds such as hydroxymethyl hydroperoxide (methylol peroxide), dihydroxymethyl peroxide (dimethylol peroxide), etc., can be prepared from hydrogen peroxide and formaldehyde. These compounds have been discussed in detail in connection with the reactions of formaldehyde with inorganic agents.

An important group of explosive organic nitrates is produced by the action of nitric acid on the polyhydroxy compounds formed by the reaction of formaldehyde with other aldehydes, ketones, and nitroparaffins.

Pentaerythritol, which is prepared from formaldehyde and acetaldehyde (page 220), yields pentaerythritol tetranitrate, $C(CH_2ONO_2)_4$, which is also known as PETN, penthrite, penta, and niperyth. According to Davis[9], the technical product melts at approximately 138°C, whereas the pure material melts at 140.5 to 141°C. Other explosives derived from pentaerythritol include:

Pentaerythritol chlorohydrin nitrates (mixture of mono- and dichlorohydrin nitrates melting between 43–50°C)
Pentaerythritol diacetate dinitrate
Pentaerythritol diformate dinitrate

Pentaerythritol dimethyl ether dinitrate (m.p. 53–54°C)
Pentaerythritol trinitrate mononitrobenzoate (m.p. 50–55°C)
Pentaerythritol diglycollic ester tetranitrate (m.p. below room temperature)
Pentaerythritol monomethyl ether trinitrate (m.p. 79–80°C)
Pentaerythritol tetraacetate tetranitrate

The first seven are described in a German patent assigned to the West-falisch-Anhaltische Sprengstoff A.-G.[23]; the last is described by Wyler[30] in a U. S. patent. Dipentaerythritol, which is obtained as a by-product of the pentaerythritol process, yields dipentaerythritol hexanitrate, $(O_2NOCH_2)_3CCH_2OCH_2C(CH_2ONO_2)_3$, on nitration[12, 13].

Pentaglycerol, $CH_3C(CH_2OH)_3$, produced from propionaldehyde and formaldehyde (page 224), gives trimethylolmethylmethane trinitrate having the appearance of a yellow viscous oil[3,17]. Pentaglycol from formaldehyde and isobutyraldehyde (page 225) yields a liquid dinitrate, $(CH_3)_2C(CH_2ONO_2)_2$[25], and trimethylolethylethane, $C_2H_5C(CH_2OH)_3$, from n-butyraldehyde and formaldehyde, nitrates to give a waxy, solid dinitrate melting at 38 to 42°C[24].

Ketone derivatives of formaldehyde which yield explosives on nitration include anhydroennea-heptitol, which is prepared from formaldehyde and acetone (page 227). This polyhydroxy ether gives a pentanitrate, melting at 132°C, as well as lower nitrated products[16]. The tetramethylol ketones and alcohols obtained by reactions of formaldehyde with cyclopentanone and cyclohexanone respectively also give explosive nitrates, according to Friederich[11].

Nitrate explosives derived from methylol derivatives of nitroparaffins have attracted increasing attention with the development of practical processes for the manufacture of the parent nitrohydrocarbons. Trimethylolnitromethane trinitrate, prepared from trimethylolnitromethane (page 347) is a viscous, liquid explosive resembling nitroglycerol. This explosive is also known as "nitroisobutylglycerinetrinitrate" and "nib-glycerinetrinitrate". Nitroethane gives a dimethylol derivative with formaldehyde, dimethylolmethylnitromethane, $CH_3C(CH_2OH)_2NO_2$ (page 347). On nitration, this gives dimethylolmethylnitromethane dinitrate (2-methyl-2-nitro-propanediol 1,3-dinitrate) which melts at 139–140°C[2, 6, 10]. The formaldehyde derivative of 2-nitropropane, $(CH_3)_2C(CH_2OH)NO_2$, reacts with nitric acid to produce 2-nitroisobutanol nitrate[27].

Hexamethylenetetramine is the parent substance of a number of explosive compounds of which cyclotrimethylenetrinitramine (cyclonite, RDX, hexogen) is the most important member. The chemistry of this explosive has been previously reviewed (see pages 418–420). An account of Axis processes for its manufacture in World War II is given by Cooley[5]. A unique procedure not employing hexamethylenetetramine consisted in the reaction of potassium sulfamate with formaldehyde and nitric acid (W. Process[5]).

$$NH_2SO_3K + CH_2O \rightarrow CH_2:NSO_3K + H_2O$$

$$3CH_2:NSO_3K + 3HNO_3 \rightarrow (CH_2:NNO_2)_3 + 3KHSO_4$$

Numerous procedures involving hexamethylenetetramine have been patented by Schiessler and Ross[22], Wright[29], Roberts[21], Wyler[31], Caesar and Goldfrank[4], and others.

The peroxygen derivative, hexamethylenetriperoxidediamine (HMTD), whose preparation has been previously described (page 421) is a relatively stable organic peroxide having the properties of a primary explosive[19, 20, 23, 26].

Other explosives derived from hexamethylenetetramine include:

Hexamethylenetetramine perchlorates:
Mono, di, and tri perchlorate salts are reported by Hassel[15] as products of the reaction of hexamethylenetetramine and perchloric acid in aqueous solution.

Hexamethylenetetramine trinitro-*m*-cresolate:
This addition compound explodes at 325°C according to Datta[7].

Tetramethylenediperoxydicarbamide:
This white, finely crystalline explosive solid is obtained by the joint reaction of urea, formaldehyde, and hydrogen peroxide[14].

Cyclotrimethylenetrinitrosamine:
This explosive compound was described by Bellini[1] as a product of the action of nitrous acid on hexamethylenetetramine.

References

1. Bellini, L., *Ann. chim. applicata*, **31**, 125–9 (1941).
2. Bergeim, F. H., (to E. I. du Pont de Nemours & Co., Inc.), U. S. Patent 1,691,955 (1928).
3. Bombrini Parodi-Delfino, French Patent 771,599 (1934).
4. Caesar, G. V., and Goldfrank, M., (to Stein Hall & Co., Inc.), U. S. Patent 2,398,080 (1946).
5. Cooley, R. A., *Chemical Industries*, **59**, 645 (1946).
6. Crater, W., (to Hercules Powder Co.), U. S. Patent 2,112,749 (1938).
7. Datta, R. L., Misra, L., and Bardhan, J. C., *J. Am. Chem. Soc.*, **45**, 2430 (1923).
8. Davis, T. L., "The Chemistry of Powder and Explosives," Vol. II, p. 277, New York, John Wiley & Sons, Inc. (1943).
9. *Ibid.*, p. 279.
10. Ellis, C., (to Standard Oil Development Co.), U. S. Patent 2,274,629 (1942).
11. Friederich, W., U. S. Patent 1,962,065 (1934).
12. Friederich, W., and Brün, W., *Z. ges. Schiess.-u. Sprengstoffw.*, **27**, 73–6, 125–7 (1932); *C. A.*, **26**, 4176 (1932).

13. Friederich, W., and Brün, W., *Ber.*, **63**, 2681–90 (1930).
14. Girsewald, C. von, and Siegens, H., *Ber.*, **47**, 2466 (1914).
15. Hassel, O., Norwegian Patent 57,831 (1937).
16. Herz, E. R., German Patent 286,527 (1922).
17. Herz, E. R., German Patent 474,173 (1929).
18. Moreschi, A., *Atti R. Accad. Lincei Roma* (5), **28**, I, 277–80 (1919); *Chem. Zentr.*, **1922**, II, 291.
19. Nuevos Explosivos Indus. SA, French Patent 783,682 (1935).
20. Patry, M., *Z. ges. Schiess.-u. Sprengstoffw.*, **32**, 177 (1937).
21. Roberts, E., British Patent 609,761 (1948).
22. Schiessler, R. W., and Ross, J. H., British Patent 595,354 (1947).
23. Schmidt, A., *Z. ges. Schiess.-u. Sprengstoffw.*, **29**, 263 (1934).
24. Spaeth, C. P., (to E. I. du Pont de Nemours & Co., Inc.), U. S. Patent 1,883,044 (1932).
25. *Ibid.*, U. S. Patent 1,883,045 (1932).
26. Taylor, C. A., and Rinkenbach, W. H., *J. Franklin Inst.*, **204**, 374 (1927).
27. Vanderbilt, B. M., (to Standard Oil Development Co.), U. S. Patent 2,241,492 (1941).
28. Westfalisch-Anhaltische Sprengstoff A. G., German Patents 638,432–3 (1936).
29. Wright, G. G., (to Council for Scientific and Industrial Research, Ottawa, Canada), U. S. Patent 2,434,789 (1948).
30. Wyler, J. A., (to Trojan Powder Co.), U. S. Patent 2,086,146 (1937).
31. *Ibid.*, U. S. Patents 2,355,770 (1944) and 2,395,773 (1946).

Fertilizers

Formaldehyde has recently been employed for improving nitrogenous fertilizers containing urea so that the available nitrogen will not be readily soluble in water and will thus become available to plants gradually over a relatively extended period. Grass treated with a fertilizer of this type is reported to remain green throughout the growing season but requires less cutting because of the controlled growth. Clark, Yee and Love[1] point out that urea-formaldehyde condensates suitable for this use are known as 'urea-form" and are produced by reactions involving one to two mols of urea per mol formaldehyde. Reaction is carried out under mildly acidic conditions and the products are apparently methylene bis-ureas. Urea-formaldehyde resins, in whose production the reactant ratio is less than 1.0 mol urea per mol formaldehyde, are completely insoluble in water and hydrolyze too slowly to be of value as fertilizers. Keenen and Sachs[3, 4] prepare a urea-formaldehyde fertilizer by adding a nitrifying solution containing urea, ammonia and formaldehyde to an acidic fertilizer material such as super phosphate. Rohner and Wood[6] produce a nitrogenous fertilizer in a continuous manner by reacting an aqueous solution containing a urea-formaldehyde mol ratio in the range 0.8 to 1.3 at a pH of 3 to 5 while boiling at reduced pressure and removing the precipitated polymer by filtration. Kralovec and Huffman[5] manufacture a solid fertilizer by reacting a concentrated aqueous composition containing about 1.5 mols urea per mol formaldehyde at pH 2 to 6 on a heated drum; the solid product is scraped from the drum, neutralized and dried.

Hopkinson[2] uses formaldehyde in the preparation of fertilizer from inedible or "trash" fish. In this process formaldehyde serves both as a preservative and as a hardener of the solid fish substance. As a result, it is possible to obtain a dry, oil-free mash, which is not contaminated by products of putrefaction, on digesting and pressing the treated fish. This mash is a good fertilizer. By-product oil is relatively free of impurities such as proteins because the formaldehyde-treated fish substance does not tend to exude from the press. As a result, it is claimed that a high-grade oil which does not require special purification is obtained. One ton of fish requires 5 to 10 pounds 37 per cent formaldehyde for treatment.

References

1. Clark, K. G., Yee, J. Y., and Love, K. S., *Ind. Eng. Chem.*, **40**, 1178–1183 (1948).
2. Hopkinson, L. T., U. S. Patent 2,003,887 (1935).
3. Keenen, F. G., and Sachs, W. H., (to E. I. du Pont de Nemours & Co., Inc.), U. S. Patent 2,255,026 (1941).
4. *Ibid.*, U. S. Patent 2,255,027 (1941).
5. Kralovec, R. D., and Huffman, R. L., (to E. I. du Pont de Nemours & Co., Inc.), U. S. Patent 2,592,809 (1952).
6. Rohner, L. V., and Wood, A. P., (to the Solvay Process Co.), U. S. Patent 2,415,705 (1947).

Fireproofing

Formaldehyde may in some instances be employed in the preparation of fireproofing compositions. Stellmacher[3], for example, employs a waterglass solution in combination with hydrogen peroxide, sodium perborate or formaldehyde. A mixture of ammonium sulfate, ammonium chloride, magnesium carbonate, zinc sulfate and sodium phosphate containing formaldehyde is claimed to have value as a flameproofing agent[1].

A resinous substance produced by reactions involving formaldehyde, dicyandiamide and ammonia is also reported to function as a fireproofing agent[4]. Ethylidene-urea gives resins with formaldehyde which produce a voluminous char on exposure to heat and show promise in the production of fire-retardant coatings[2].

References

1. Almond, R., and Young, A. M., British Patent 487,969 (1938); *C. A.*, **33**, 412.
2. Kleeck, A. van., *Chem. Eng. News*, **19**, 626–8 (1941).
3. Stellmacher, H., German Patent 626,825 (1936).
4. Tramm, H., Clar, C., Kühnel, P., and Schuff, W., (to Ruhrchemie Akt.), U. S. Patent 2,106,938 (1938).

Fuels

Substantially anhydrous formaldehyde polymers (polyoxymethylenes), trioxane and hexamethylenetetramine burn readily without evolution of smoke and can accordingly be employed in the preparation of fuel tablets. Schilt[7] obtained polymers of the desired type by the action of phosphoric or sulfuric acid on formaldehyde solution or paraformaldehyde. A Swiss patent[4] describes the preparation of a stable solid fuel by mixing a formaldehyde polymer with nitrocellulose. The use of hexamethylenetetramine for the production of solid fuel tablets is described by Hanig[2] in a German patent. Modified fuels containing hexamethylenetetramine together with other agents have been patented by Schrimpe[8], Ringer[5, 6], Michels[3] and Speaker[9]. A trioxane fuel tablet has been developed by the U. S. Quartermaster Corps in collaboration with the Office of Scientific Research and Development. The tablet contains a binder to hold it in solid form while it burns and coloring matter is added to show that it is nonedible[1].

References

1. Anon., *Chem. & Met.*, **51,** 136 (1944).
2. Hanig, W., German Patent 325,711 (1920).
3. Michels, M., U. S. Patent 1,839,987 (1932).
4. Rau, M., Swiss Patent 192,104 (1937); *C. A.*, **32,** 4313 (1938).
5. Ringer, F., U. S. Patent 2,161,385 (1939); Swiss 203,653 (1939); *Chem. Zentr.*, **1940,** I, 2268.
6. Ringer, F., (to Burnol Products Inc.), U. S. Patent 2,289,040 (1942).
7. Schilt, W., British Patent 342,668 (1929).
8. Schrimpe, C. F., (to Perth Amboy Chemical Works), U. S. Patent 1,248,447 (1917).
9. Speaker, J. W., U. S. Patent 2,432,347 (1947).

Gas Absorbents

An absorbent carbon containing formaldehyde is claimed to have superior gas-absorbing properties and is recommended for use as a deodorizing composition[2].

In World War I, an aqueous solution containing hexamethylenetetramine, phenol, caustic soda, and glycerin was employed for absorbing and neutralizing phosgene. This composition was known as phenate hexamine and was employed in the so-called P. H. helmets for protection against phosgene[1]. More recently, Patrick[3] describes a poison gas absorbent suitable for use against phosgene which consists of a specially prepared alumina gel containing hexamethylenetetramine.

References

1. Bebie, J., "Explosives, Military Pyrotechnics and Chemical Warfare Agents," p. 120, New York, Macmillan Co. (1943).

2. Block, D. J., U. S. Patent 1,922,416 (1933).
3. Patrick, W. A., Jr., (to The Davison Chemical Co.), U. S. Patent 2,400,709 (1946).

Hydrocarbon Products

Formaldehyde serves the petroleum industry in many ways. It is of value in connection with the drilling and operation of oil wells, the purification and modification of petroleum fractions, and the synthesis of antioxidants, pour-point depressants, lubrication assistants, etc. It is also of interest for the treatment of coal tar hydrocarbons.

Oil Well Operations. Oil well drilling fluids or "muds" usually consist of a suspension of clay or starch or a combination of the two plus small quantities of special additives. Starch and related amylaceous colloids reduce loss of water but are degraded by the action of bacteria, yeasts and other microorganisms. Farrow[13] uses formaldehyde in a drilling fluid additive to prevent deterioration and states that less formaldehyde is required when it is employed in combination with borax. The drilling fluid additive consists of a mixture of paraformaldehyde and borax plus small amounts of starch, fuller's earth and magnesium carbonate. Approximately 0.2 to 0.8 pounds of formaldehyde are required per barrel of drilling fluid. Loss of drilling fluid can also be controlled by resin-forming combinations. Cardwell[9] makes use of the spontaneous curing of partially condensed reaction products of phenol and cresylic acid with formaldehyde at 100°F to form sealing deposits in wells. Post and Oldham[28] use water-soluble condensates of formaldehyde and tannins, such as quebracho. Ahe and Zweifel[1] state that leaks can be prevented by successive addition of casein dispersions and formaldehyde.

Loss of oil field operating equipment through the corrosive action of hydrogen sulfide in well brine can be effectively controlled by feeding a little formaldehyde between casing and tube with a drip-type lubricator. This will be treated further in connection with the use of formaldehyde as an acid inhibitor (pages 495–496). Formaldehyde is also used in a process deviesd by Hershma[18] for demulsifying and desalting crude petroleum-brine emulsions. This is accomplished by adding a composition prepared by mixing formaldehyde solution, glycerin, and naphthenic or sulfonaphthenic acids to the emulsion. Approximately 0.1 to 1 per cent of this mixture based on the weight of oil treated causes a rapid separation of oil and brine.

Refining of Hydrocarbons. The reactivity of acid formaldehyde with some unsaturated hydrocarbons, aromatics and organic sulfur derivatives has long been recognized as of potential value in refining. Such procedures are now receiving increasing attention. In addition, it is known that formaldehyde improves the stability of gasolines and fuel oils. Downs and Walsh[12] report that formaldehyde exerts an antiknock influence in iso-octane, which knocks by a low temperature combustion process, but causes knocking in benzene and methane.

An early process for the separation of thiophene from benzene rests on the fact that this compound condenses more rapidly with acid formaldehyde than does benzene. According to this procedure, 10,000 parts of benzene containing thiophene are agitated with 1500 parts of 73 per cent sulfuric acid and 45 parts of 30 per cent formaldehyde at ordinary temperature for several hours. When the benzene layer ceases to give a test for thiophene, it is removed from the acid liquor and steam-distilled to separate it from amorphous condensates[6].

Following his discovery of the "formolite" reaction (page 462), Nastyukoff[27] patented a procedure for removing unsaturated aromatics and hydroaromatics from petroleum fractions by treatment with formaldehyde in the presence of concentrated sulfuric acid and a current of hydrogen gas.

Later patents covering the use of formaldehyde for refining hydrocarbon oils include: the use of formaldehyde and ammonia in a mixture for treating mineral oil[8], removal of sulfur from gasoline vapor by passing it through a mass of copper turnings in the presence of hydrogen chloride and formaldehyde[24], and refining cracked hydrocarbon vapors by treatment with formaldehyde in combniation with an acid condensing agent such as hydrogen chloride and zinc or aluminum[20].

Recently Arundale and co-workers[2, 3, 4, 5, 33] have developed processes for removing sulfur compounds, olefins and other unsaturates from motor fuels by treatment with formaldehyde in the presence of an acid catalyst. Pilot plant evaluation of these techniques is reported to demonstrate that product losses are less than with unmodified acid treatments and that the fuels obtained are more stable, less corrosive and display improved lead susceptibility. Sixty per cent sulfuric acid is stated to be preferable as the acid catalyst and about 1.5 per cent formaldehyde based on the naphtha feed appears sufficient in most cases. Treating temperatures may be in the range 20 to 50°C. In the absence of acid, treating with 0.5 to 2.0 per cent of formaldehyde at about 250°C gave a product with improved resistance to oxidation and caused no appreciable loss of octane number. Hartough, and Badertscher[17] use formaldehyde and ammonium chloride for refining a highly aromatic petroleum fraction principally of polycyclic aromatics to obtain a light-colored product possessing improved color stability. The treatment is carried out at around 100°C and the product contains combined nitrogen.

The stabilizing action of formaldehyde on gum-formers in gasoline was described by Somerville[31] in 1923. Paraformaldehyde, "trioxymethylene," was also stated to have a protective action on oils in a German patent that issued in 1925. More recently, Carnell[10] has found that furnace and diesel fuels consisting of cracked distillates which boil in the range 380 to 640°F can be stabilized against color formation and sludging by the addition of small amounts of formaldehyde. The mechanism of this effect is unknown. The formaldehyde may be added to oil at 50 to 300°F as an aqueous solution, polymer or gas so that 0.001 to 0.2 per cent of formaldehyde is dissolved or dispersed. Since oil treated in this way is stated to be corrosive in the presence of oxygen, Stedman[32] adds an aliphatic amine, such as butylamine, with the formaldehyde to prevent corrosion. Cook[11] has found that a relatively noncorrosive diesel fuel can be obtained by adding 0.01 to 0.05 volume per cent formaldehyde to a sour crude oil fraction. Hartough, Caesar and Lukasiewiez[16] remove acidic constituents from mineral oil by treatment with an ion-exchange resin prepared by condensing thiophene with formaldehyde and a base such as ammonia, urea or ethylenediamine.

Low-viscosity oils are said to be converted to high-viscosity oils of better lubricating properties by treatment with a mixture of formaldehyde and acetic acid containing a catalyst such as sulfuric acid or ferric chloride[15]. According to a German patent[30], the carcinogenic materials in tar oil residues are rendered harmless by treating them with formaldehyde after separation of the phenol fraction.

Addition Agents. Particularly illustrative of hydrocarbon addition agents prepared by syntheses involving formaldehyde are certain pour-point depressors and anticorrosive agents for use with high pressure lubricants.

Methylene-linked aromatic derivatives containing aliphatic substituents have value as pour-point depressors for lubricating oils. Reiff[29] obtains a product of this type by condensing a chlorinated ester wax with an aromatic and reacting the resultant product with formaldehyde employing a strong acid catalyst. Mikeska[23] treats petroleum aromatics with paraformaldehyde in glacial acetic acid using zinc chloride as a catalyst and alkylates the product by reaction with stearyl chloride in carbon bisulfide in the presence of aluminum chloride. Leiber[21] introduces an alkyl group containing less than 7 carbon atoms in a naphthalene-formaldehyde resin.

Aminophenol derivatives act as antioxidants in lubricating oils and improve the life of bearings by preventing the formation of corrosive products and limiting their action. Fuller and Hamilton[14] obtain an oxidation inhibitor by reacting formaldehyde with an N-substituted arylamine. McCleary[22] describes the use of an alkylated aminomethyl phenol prepared by a Mannich reaction involving an alkyl or aralkyl secondary amine, formaldehyde and a phenol. Bartleson[7] protects copper-lead bearings by use of a high boiling acid condensation product of formaldehyde and N-ethylaniline. Illustrative of another product type are lubricant antioxidants produced by reacting dialkyl dithiophosphoric acids with formaldehyde plus an alcohol, glycol, mercaptan or phenol[25, 26]. The formation of a product of this type as described by Moss and co-workers is indicated below:

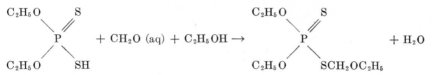

Hexamethylenetetramine has itself been reported as a stabilizer for lubricating and insulating oils[19].

References

1. Ahe, K. L. V., and Zweifel, H. C., (to Richfield Oil Corp.), U. S. Patent 2,338,217 (1944).
2. Arundale, E., (to Standard Oil Development Co.), U. S. Patent 2,569,216 (1951).
3. Arundale, E., and Haworth, J. P., (to Standard Oil Development Co.), U. S. Patent 2,567,174 (1951).
4. Arundale, E., and Juterbock, E. E., (to Standard Oil Development Co.), U. S. Patent 2,567,175 (1951).
5. Arundale, E., and Mikeska, L. A., (to Standard Oil Development Co.), U. S. Patent 2,567,173 (1951).
6. Badische Anilin-und Soda-fabrik, German Patent 211,239 (1909).
7. Bartleson, J. D., (to Standard Oil Co. of Ohio), U. S. Patent 2,432,713 (1947).
8. Behm, H., U. S. Patent 1,813,884 (1931).
9. Cardwell, P. H. (to The Dow Chemical Co.), U. S. Patent 2,485,527 (1949).
10. Carnell, P. H., (to Leonard Refineries, Inc.), U. S. Patent 2,560,632 (1951).
11. Cook, F. V., (to Stanolind Pipe Line Co.), U. S. Patent 2,496,444 (1950).
12. Downs, D., and Walsh, A. D., *Nature*, **163**, 370–1 (1949).
13. Farrow, J. R., (to National Lead Co.), U. S. Patent 2,525,783 (1950).
14. Fuller, E. W., and Hamilton, L. A., (to Socony-Vacuum Oil Co.), U. S. Patent 2,223,411 (1940).
15. Fulton, S. C., (to Standard Oil Development Co.), U. S. Patent 2,018,715 (1935).
16. Hartough, H. D., Caesar, P. D., and Lukasiewicz, S. J., (to Socony-Vacuum Oil Co.), U. S. Patent 2,585,652 (1952).
17. Hartough, H. D., and Badertscher, D. E., (to Socony-Vacuum Oil Co.), U. S. Patent 2,501,602 (1950).
18. Hershman, P. R., (to Petro Chemical Co.), 2,153,560 (1939).
19. I. G. Farbenindustrie A. G., French Patent 636,332 (1927).
20. Levine, I., (to Universal Oil Products Co.), U. S. Patent 1,974,311 (1934).
21. Lieber, E., (to Standard Oil Development Co.), U. S. Patent 2,412,589 (1946).
22. McCleary, R. F., (to The Texas Co.), U. S. Patent 2,363,134 (1944).
23. Mikeska, L. A., (to Standard Oil Development Co.), U. S. Patent 2,352,280 (1944).
24. Morrell, J. C., (to Universal Oil Products Co.), U. S. Patent 2,098,059 (1937).
25. Moss, P. H., and Cook, E. W., (to American Cyanamid Co.), U. S. Patent 2,589,675 (1952).
26. Moss, P. H., and Hook, E. O., (to American Cyanamid Co.), U. S. Patent 2,586,555 (1952).
27. Nastyukoff, (Nastudoff) A. M., German Patent 486,022 (1929).
28. Post, E. F., and Oldham, W. N., (to American Cyanamid Corp.), U. S. Patent 2.579,453 (1951).

29. Reiff, O. M., (to Socony Vacuum Oil Co.), U. S. Patent 2,147,546 (1939).
30. Robinson, H. W., German Patent 320,378 (1920); *Chem. Zentr.*, **1920,** IV, 56.
31. Somerville, A. A., British Patent 269,840 (1923).
32. Stedman, R. F., (to Leonard Refineries, Inc.), U. S. Patent 2,560,633 (1951).
33. Weber, G., *Oil Gas J.*, **50,** 38–9 (Feb. 4, 1952).

Insecticides

Although formaldehyde is not generally effective as an insecticide when used in the absence of other agents, it has some value as a stomach poison for flies. A 0.5 to 1 per cent solution of formaldehyde is reported to be $2\frac{1}{4}$ times as effective as standard arsenic solution when used for this purpose[8]. In general, the formaldehyde solution is mixed with water to which sugar or milk is added to attract the flies. The solution is then exposed in shallow dishes or on absorbent paper[4]. Morrill reports that beer, ethyl alcohol, or vinegar is more attractive as bait than sugar or milk[11].

Prendel[12] claims that a spray containing 1 per cent commercial formaldehyde and 0.5 per cent soft soap is effective for the destruction of mosquitoes in cow houses, barns, etc.

Formaldehyde is employed in a number of insecticidal compositions. An aqueous emulsion of carbon bisulfide containing dissolved formaldehyde and a sulfonated fatty alcohol is stated to be of value for combating both plant and animal pests[5]. Another composition of this type contains sulfur, carbon bisulfide, fatty acids, alcohols and protein-formaldehyde reaction product[9]. A formaldehyde solution mixed with alcohol and methyl chloride is also said to have insecticidal properties[13].

Fur can be made moth-resistant by immersing for about 5 hours in an approximately 1 per cent solution of formaldehyde containing sufficient salt (ca. 3 mols per liter) to prevent swelling at a pH of not over 2.5. Traill and McLean[14] claim that this process kills or prevents the growth of larva and substantially reduces damage on exposure to moths.

Formaldehyde finds indirect use for pest extermination in the preparation of various derivatives employed in making insecticides, e.g., References 1, 2, 3, 7, 10. Forge[6] has found that the formal of alpha-(3,4-dihydroxyphenyl)-tetrahydropyran augments the insecticidal action of pyrethrum.

References

1. Ainley, A. D., and Davies, W. H., (to Imperial Chemical Industries, Ltd.), U. S. Patent 2,395,440 (1946).
2. Casaburi, U., U. S. Patent 1,986,044 (1935).
3. Chemische Fabrik Ludwig Meyer, German Patent 579,858 (1933); *Chem. Zentr.*, **1933,** II, 1746.
4. Davidson, J., *Bull. Ent. Res.*, **8,** 297–309 (1918); *C. A.*, **13,** 3265.
5. Deutsche Guld-und Silber-Scheidenstalt vorm. Roessler, British Patent 510,519 (1939); *J. Soc. Chem. Ind.*, **58,** 1186.
6. Forge, F., (to United States of America), U. S. Patent 2,421,570 (1947).

7. Hook, E. O., and Moss, P. H., (to American Cyanamid Co.), U. S. Patent 2,586,555 (1952).
8. Lloyd, L., *J. Entmol. Research*, **11**, 47–63 (1920); *C. A.*, **15**, 1579.
9. McQuiston, R. C., British Patent 420,068 (1935).
10. Moore, W., (to Tobacco By-Products and Chemical Corp.), U. S. Patent 2,041,298 (1936).
11. Morrill, A. W., *J. Econ. Entomol.*, **7**, 268–74 (1914).
12. Prendel, A. R., *Med. Parisitol (U.S.S.R.)*, **9**, 637–8 (1940); *Rev. Applied Entomol.*, **31B**, 56–7 (1943); *C. A.*, **38**, 6028 (1944).
13. J. D. Riedel-E. de Haen A. G., German Patent 587,747 (1933); *Chem. Zentr.*, **1934**, I, 753.
14. Traill, D., and McLean, A., (to Imperial Chemical Industries, Ltd.), U. S. Patent 2,424,068 (1947).

Leather

The most important application of formaldehyde in the leather industry is in the tanning process. For this purpose formaldehyde is employed directly as formaldehyde solution and indirectly in the form of formaldehyde derivatives. Formaldehyde is also used as a disinfectant both for crude hides and finished leather. In addition, formaldehyde products are sometimes employed in dressing and finishing leather goods, fur, and hair.

Formaldehyde as a Tanning Agent. The direct use of formaldehyde as a tanning agent is principally in the production of white washable leather. Leather of this type is particularly useful for gloves, sportswear, and white military leathers. The outstanding virtues of formaldehyde-tanned leather are its color and its high degree of water-resistance. Its disadvantage is a tendency to become brittle and non-stretchy, but this can be prevented if the tanning is carried out correctly. Smith[38] states that when properly dressed, formaldehyde-tanned leather will keep indefinitely.

Formaldehyde is also used for tanning both hide and hair (*e.g.*, in the case of sheepskins with the wool on), for pretanning, and for blending with other tanning agents.

It is employed as a pretanning agent for heavy leathers that are to be subjected to vegetable tanning, because it accelerates penetration of the vegetable tannins[50]. Very rapid tannage is reported when skins pretanned with formaldehyde are treated with quebracho extract and borax. After penetration, the borax is neutralized with an organic acid[23]. In a process described by Turley and Somerville[47], white leather is produced by tanning hides lightly with formaldehyde and then completing the tannage with an acidified solution of water-glass.

Blends of formaldehyde with metal tanning agents such as aluminum or chromium compounds are advocated in some processes[21, 30, 36, 47, 48]. The addition of formaldehyde to pickle liquor is stated to reduce the chromium oxide (Cr_2O_3) fixed and raise the shrinkage temperature of the finished leather[17, 43]. Mason[25] also reports that formaldehyde serves well in combination with mineral or salt-acid tanning solutions where coloring is to be avoided. Because of its tendency to form insoluble compounds with vegetable tannins, blends with these agents are apparently not entirely satisfactory. Neutral synthetic tanning agents are usually compatible.

Although the exact mechanism of formaldehyde tanning is still not definitely known, it appears probable that it is the result of the reaction of leather-collagen with formaldehyde. The general nature of this reaction has been previously discussed in connection with formaldehyde-protein reactions (pp. 312–317). Chemical studies of aldehyde tanning indicate that it is characterized by two reaction types: a rapid reaction involving free amino groups and a slow reaction involving the amido groups of the peptide linkages[41]. Cross linkage of these amido groups by methylene radicals involving contiguous protein molecules is believed by Küntzel[20] and Theis[45] to be of paramount importance in the production of water resistance and other characteristic properties of the tanned leather. The amount of fixed formaldehyde in the tanned collagen is in the order of magnitude of 0.5 per cent. Balfe[2] points out that hydrogen bonding of methylolamino groups with an appropriate group in another polypeptide chain may also be involved in the tanning process.

As has been previously noted, formaldehyde tanning must be carried out properly if good results are to be obtained. This involves control of the tanning solution, washing of the tanned leather, and adequate fat liquoring. Best results are obtained when the formaldehyde tanning bath is neutral or mildly alkaline. This is illustrated by Table 58, based on the work of Theis and Esterly[44]. As indicated, the optimum pH values are in the range 6 to 8. The preferred formaldehyde concentration seems to be of the order of 0.1 to 1 per cent, and it is apparently advisable to add the formaldehyde gradually in several increments throughout the process. The presence of salt in the tanning solution is important, since it appears evident from the work of Theis and his co-workers that it is desirable to repress the swelling of the hide during the tanning process, as this swelling may result in grain surfaces which are readily cracked[51]. Other tanning addition agents mentioned in the literature include sodium thiosulfate and soap[35], borax plus substances which assist formaldehyde penetration[29], etc. Formamide is recommended as an agent for delaying the action of formaldehyde until the solution has penetrated the hide[8]. Rapid tanning is claimed for a process employing a solution containing a substantial quantity of alcohol[22].

The temperatures employed in formaldehyde tanning processes usually range from around 60 to 100°F. The time required depends both on the type of bath employed and the nature of the leather to be tanned. It may range from a few hours in the case of light leathers to one or two days for heavy leathers. Tanning is usually followed by washing. In this connection, Woodroffe[57] claims that brittleness in the leather can be prevented if unreacted formaldehyde is removed by washing the tanned skins with a solution of ammonium chloride or sulfate. Proper fat-liquoring is extremely important for a good quality of finished leather[3].

The Pullman-Payne process[32], patented in Great Britain in 1898, appears to be the first practical method of tanning with formaldehyde. It involves gradual addition of a solution containing formaldehyde and sodium carbonate to a mixture of hides and water. The temperature is gradually raised to 118°F as the process nears completion.

Illustrative of more recent processes is a method for tanning rabbit skins which was described by Gellée[12] in 1931. Skins are soaked for 2 days, degreased with 2 per cent sodium carbonate at 86°F for 10 minutes, and then added to an aqueous solution containing 0.5 per cent commercial formaldehyde at 54 to 61°F. After 24 hours, 20 to 30 g of sodium carbon-

TABLE 58. INFLUENCE OF pH FORMALDEHYDE TREATMENT OF HIDES*

pH of Treating Solution	Amount of CH₂O Fixed	Thermal Stability of Leather	Probable Nature Formaldehyde–Nitrogen Bonding in Product
1–3	Small	Fair	$-N \cdot CH_2 \cdot N-$
6–8	Greater	Excellent	$-N \cdot CH_2 \cdot N-$ and $-NH \cdot CH_2 \cdot NH-$
9–12	Much greater	Less	$-N(CH_2OH)_2$, $-NHCH_2OH$, and $-N:CH_2$

* Data of E. R. Theis and A. R. Esterly[44].

ate are added for each liter of solution and the treatment continued for another day.

Wilson[52] reports good results for calfskins, goatskins, and sheepskins when 1000 lbs of pickled skin are drummed with 100 gals of water containing 80 lbs of salt for five minutes at 70°F, after which 30 lbs of commercial formaldehyde and 10 lbs of wood alcohol are added slowly in the course of one hour. After another hour, a solution of 10 lbs of soda ash in 10 gals of water is gradually added in one hour, and after an additional hour's drumming the mixture is allowed to stand overnight. The next day, the pH value is raised gradually to 7.5 by adding more soda ash until it remains at this value for one hour.

A review of the patent literature indicates that tanning may also be accomplished by derivatives of formaldehyde in which its characteristic reactivity is modified by various organic groups. Tanning processes are described in which the skins are treated with solutions or emulsions of polymerizable methylol compounds[18]. A process claiming the use of a solution prepared from 3 parts urea, 1.5 parts formaldehyde, 2 parts sodium carbonate, 16 parts sodium chloride and 128 parts water is illustrative[31]. In this procedure the hides, skins or pelts are prepared by any of the common methods and are then immersed in the urea-formaldehyde solution at approximately 35°C and agitated for 5 hours. The solution is then warmed to 45°C, acidified with sulfuric acid to a pH of approximately 3, and agitated for one-half hour. After this, the temperature is raised to 55°C and cooled after working the skins for a quarter of an hour. Finally the skins are washed with cold water, fat-liquored, and dried.

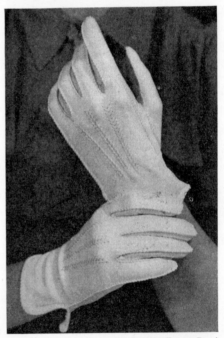

(Courtesy, Fownes Bros.)

Figure 41. White gloves prepared from leather tanned with formaldehyde.

Melamine-formaldehyde derivatives and related compounds have also proved of value in tanning. Windus[53] submerges the untanned hide for a minimum of three hours in a dilute aqueous solution containing formaldehyde and a suspension of melamine which gradualy dissolves and reacts as the tanning process progresses. Preformed methylolmelamines can also be employed[9]. Grim and Niedercorn[15] use formaldehyde and ammeline in a process for the treatment of chrome-tanned leather.

Syntans. In addition to its direct use in the tanning process, formaldehyde plays an important role in the production of synthetic tanning agents, or syntans. In general, these materials are water-soluble phenol formaldehyde condensation products containing sulfonic acid groups. At the present time they are used almost universally in the tanning industry. Early syntans were usually employed in combination with other tanning agents to modify the tanning process and save natural tannins. When used alone, they produced a poor yield of weak, low quality leather. Recent developments in this field have resulted in the production of synthetic agents that are stated to be capable of completely replacing natural tannins. A number of different varieties of syntans have been produced for use in the leather industry. These materials vary somewhat in their action and are often adapted for different specialty applications. They are employed for pretanning, modifying the action of established tan baths, and retanning imported leathers. Their use is stated to result in the production of stronger, softer and lighter-colored leathers with an improved degree of stability. They are also said to accelerate the penetration of the tanning agents with which they are used and to cause a more even tanning. Some syntans have a filling action and tighten the grain of loose skins.

Synthetic tanning agents were apparently first prepared by Edmund Stiasny[40] and their industrial use dates with his discovery which was somewhat prior to 1911. Stiasny used the term "syntan" as a designation for condensation products prepared either by heating phenols with acidified formaldehyde and solubilizing the resinous product by sulfonation or by reacting sulfonated phenols with formaldehyde. More recently the term has also been applied to other soluble polynuclear phenol-formaldehyde derivatives which are not sulfonated products, but nevertheless show tanning characteristics: for example, synthetic tanning agents which owe their solubility to a plurality of hydroxy groups may be obtained by the acid condensation of formaldehyde with resorcinol and pyrogallol[49]. The fact that a reaction product of formaldehyde with pyrogallol in the presence of an acid catalyst precipitates glue from solution and behaves somewhat like a tannin was reported by Baeyer[1] in 1872, but the product was apparently not evaluated as a tanning agent at that time.

Most of the syntans which have been prepared since Stiasny's discovery have been sulfonic-acid derivatives; and considerable study has been given to the effect of different methods of synthesis, variations in substituent groups and other modifications. Although generally used as free acids, neutral syntans are also employed. Whereas mineral acids may cause trouble in tanning, this is not true of the acidic syntans. As a result, it has been found desirable to remove free mineral acids from these products. The effect of variations in methods of synthesis has been studied in considerable detail by Wolesensky[56], who points out that products obtained from sulfonated phenols and formaldehyde have little or no filling action, whereas syntans with filling properties are obtained when the preformed phenol-formaldehyde condensates are sulfonated. He also demonstrates[55] that non-phenolic compounds such as the condensation product of toluene sulfonic acid and formaldehyde show little or no tanning action.

Syntans for the total replacement of natural tannins were developed in Germany prior to World War II[4]. An agent called "Tannigan Extra B" was produced by sulfonating a phenol-formaldehyde novolac using approximately 0.45 mol sulfuric acid per mol phenol. Other "Tannigan" type compounds were produced later based on cresols, resorcinol, catechol, dihydroxydiphenyl sulfone, "brown oil", and beta-naphthol. Phenol-formaldehyde condensates were also reacted with lignin sulfonic acids to obtain still other products. In this country, melamine-formaldehyde resins were produced for tanning white leather, particularly in combination with chromium salts[4].

It is impossible to give exact information concerning the chemical structure and synthesis of the syntans which are now in commercial use. However, examination of the patent literature indicates that in addition to traditional syntan types, considerable attention is being given to nitrogenous products prepared by reactions involving urea, thiourea, melamine, etc. The following examples are illustrative. Geigy[11] produces water-soluble products by the action of concentrated sulfuric acid on phenols in the presence of urea, condenses these with formaldehyde and aromatic hydroxy carboxylic acids in strong sulfuric acid, and neutralizes the resulting materials. Somerville and Raterink[39] react phenolsulfonic acid with formaldehyde in the presence of thiourea. Hassler[16] condenses a sulfonated phenol with formaldehyde in the presence of ammonia or an alkylamine. Swain and Adams[42] treat sulfonated phenols or sulfonated aromatics with formaldehyde and melamine. Robinson and Beach[34] produce a chromated product by reacting a water-soluble dichromate with a formaldehyde derivative of lignin sulfonic acid. Freund and Mahler[10] sulfonate a condensation product of distilled cashew shell liquid with formaldehyde and combine it with an unsaturated fatty acid. Riehs[33] refluxes two mols of phenol with one mol of formaldehyde in the presence of strong hydrochloric acid and sulfonates the resultant resin with concentrated sulfuric acid at 100°C.

A recent patent by Nagy[27] describes the use of small concentrations of vegetable tannins as a stabilizer to prevent low-cost phenolformaldehyde syntans from darkening on storage. This change is undesirable since the aged syntans darken the leather in tanning.

Miscellaneous Uses of Formaldehyde in the Leather and Hide Industry. The preservative and disinfecting action of formaldehyde is also employed in the leather industry. For example, it is stated that extensive deterioration of skins may be avoided by treating with formaldehyde prior to lime pit or tanner's pit processes[19]. Treatment of raw calf hide with solutions containing 1 to 2 per cent formaldehyde alone or with 8 per cent sodium chloride plus 0.1 per cent soda kills bacteria, but is reported to make soaking difficult after storage and to prevent easy unhairing. This undesirable effect is reduced when soda is omitted from the solution[37].

Formaldehyde Treatment of Fur and Hair. Formaldehyde finds use in giving fur or hair a permanent set. Conversion of wooled sheepskins to "Mouton," "Bonmouton," "Beaver Lamb," and "Nutria Lamb" which resemble natural furs is an application of this type. Since the treated woolen fibers will not return to their natural crimped condition on wetting, the procedure is often described as waterproofing. These fur-setting processes usually involve treating the combed fibers of a sheepksin with either formaldehyde or a formaldehyde donor, alone or in the presence of an acid catalyst, or with a resin-forming formaldehyde condensate and then subjecting to heat by the action of a hot iron[54]. Thuau and Lisser[46] state that paraformaldehyde ("trioxymethylene"), formaldehyde bisulfite and hexamethylenetetramine may be employed as formaldehyde donors. Patented processes describe treating with formaldehyde[28] or formaldehyde vapors[14] and subjecting to the combined action of mechanical pressure and heat. Damaging action to the skin is avoided by introducing water-repellents such as fat or paraffin. Another procedure[24] involves application of formaldehyde solution containing an organic or inorganic acid catalyst, a waiting period to allow for complete penetration of the fibers and hot ironing at 190°C. Muller[26] employs a treating solution containing an emulsified wax, formaldehyde and phthalic acid. The use of resin-forming solutions containing formaldehyde derivatives of urea or melamine is described by Gottfried[13] and Calva[5]. Specific formulations of treating solutions and process "know-how" are closely guarded commercial secrets. The treated fur is often dyed to resemble beaver but may also be produced in white, black, red and other colors. Dyeing is sometimes carried out before the setting process and sometimes after setting. Fur setting processes are not confined to use with sheepskin. However, the major commercial use is in this connection. Calva[6] uses a procedure of this general type for setting growing hair on living beings.

Formaldehyde treatment is also used for improving the felting and dyeing properties of hair. Casaburi[7] accomplishes this by first tanning the hair in a solution containing 0.2 to 0.4 per cent citric acid, 1.5 to 3 per cent caustic soda, and 10 to 14 per cent formaldehyde, then centrifuging and heating at 70°C.

References

1. Baeyer, A., *Ber.*, **5,** 1904 (1872).
2. Balfe, M. P., *Chemistry & Industry*, **1950,** 50.

3. Bowes, J. H., *J. Intern. Soc. Leather Trades' Chemists*, **20**, 50–60 (1936).
4. *British Plastics*, **1949**, 641–2.
5. Calva, J. B., U. S. Patents 2,211,645 (1940); 2,240,388 (1941).
6. Calva, J. B., U. S. Patent 2,390,073 (1945).
7. Casaburi, V., Italian Patent 368,274 (1938); *Chem. Zentr.*, **1940**, I, 2101.
8. Dangelmajer, C., and Perkins, E. C., (to E. I. Cu Pont de Nemours & Co., Inc.), U. S. Patent 2,061,063 (1936).
9. Dawson, W. O. (to American Cyanamid Co.), U. S. Patent 2,316,740 (1943).
10. Freund, E. H., and Mahler, P., (to General Foods Corp.), U. S. Patent 2,415,347 (1947).
11. Geigy, J. R., French Patent 39,845 (1931); addition to French Patent 660,008 (1928); *C. A.*, **23**, 5349 (1929).
12. Gellée, M. R., *Halle aux cuirs*, **1931**, 43–8.
13. Gottfried, S., U. S. Patent 2,225,267 (1940).
14. Gottfried, S., (to Pannonia Ltd.), U. S. Patent 2,323,751 (1943).
15. Grim, J. M., and Niedercorn, J. G., (to American Cyanamid Co.), U. S. Patent 2,353,556 (1944).
16. Hassler, F., (to I. G. Farbenindustrie A. G.), U. S. Patent 2,013,928 (1935).
17. Holland, H. C., *J. Intern. Soc. Leather Trades' Chemists*, **24**, 221–234 (1940).
18. I. G. Farbenindustrie, A. G., French Patent 838,188 (1939).
19. Kohl, F., (to I. G. Farbenindustrie, A. G.), German Patent 694,488 (1940).
20. Küntzel, A., *Angew. Chem.*, **50**, 307 (1937).
21. Lloyd, G. F., French Patent 811,306 (1937); *C. A.*, **31**, 8985, (1937).
22. Lloyd, G. F., British Patent 482,286 (1938); *C. A.*, **32**, 6904 (1938).
23. McCandlish, D., Atkin, W. R., and Paulter, R., *J. Intern. Soc. Leather Trades' Chemists*, **18**, 509–11 (1934).
24. Martin, C. W., Sons Ltd., British Patent 528,459 (1940).
25. Mason, C. F., *Chem. Ind.*, **47**, 260–3 (1940).
26. Muller, O. F., (to Dri-Wear, Inc.), U. S. Patent 2,140,759 (1938).
27. Nagy, D. E., (to American Cyanamid Corp.), U. S. Patent 2,592,587 (1952).
28. Pannonia Báránybörnemesitö és Kereskedelmi R. T., British Patent 505,487 (1939).
29. Pensel, G. R., (to The Ritter Chemical Co.), Canadian Patent 327,675 (1932).
30. Pensel, G. R., U. S. Patent 2,071,567 (1937).
31. Porter, R. E., (to National Oil Products Co.), U. S. Patent 1,975,616 (1934).
32. Pullman, J., Pullman, E. E., and Payne, E. E., British Patent 2872 (1898).
33. Riehs, C., (Vested in the Attorney General of the United States), U. S. Patent 2,469,787 (1949).
34. Robinson, E. A., and Beach, R. M., (to National Oil Products Co.), U. S. Patent 2,401,508 (1946).
35. Rogers, A., (to Roessler & Hasslacher Chemical Co.), U. S. Patent 1,845,341 (1932).
36. Röhm, O., and Schell, H., (to Rohm & Haas Co.), German Patent 686,655 (1940).
37. Simoncini, E., *Boll. uffic. staz. sper. ind. pelli e. mat. concianti, Suppl. tec.*, **7**, 81–7 (1932); *C. A.*, **27**, 624 (1933).
38. Smith, P. I., "Principles and Processes of Light Leather Manufacture," pp. 83–310, Chicago Hide and Leather Publishing Co. (1912).
39. Somerville, I. C., and Raterink, H. R., (to Rohm & Haas Co.), U. S. Patent 1,951,564 (1934).
40. Stiasny, E., U. S. Patent 1,237,405 (1914); German Patent 262,558 (1911); Austrian Patent 58,045 (1911).

41. Stiasny, E., *J. Intern. Soc. Leather Trades' Chemists*, **20**, 50–60 (1936); *C. A.*, **30**, 2793 (1936).
42. Swain, R. C., and Adams, P., (to American Cyanamid Co.), U. S. Patent 2,282,536 (1942).
43. Theis E. R., *J. Am. Leather Chemists' Assoc.*, **35**, 452–70 (1940).
44. Theis, E. R., and Esterley, A. R., *J. Am. Leather Chemists' Assoc.*, **35**, 563 (1940).
45. Theis, E. R., and Ottens, E. F., *J. Am. Leather Chemists' Assoc.*, **35**, 330–47 (1940).
46. Thuau, U. J., and Lisser, D., *Cuir tech.*, **28**, 212–3 (1939); *Chem. Zentr.*, **1939**, II, 3358.
47. Turley, H. G., and Somerville, I. C., (to Rohm & Haas Co.), U. S. Patent 2,129,748 (1938).
48. Vogel, F. A., and Ernest, M., U. S. Patent 1,982,586 (1934).
49. Wilson, J. A., "The Chemistry of Leather Manufacture," Vol. II, p. 765, New York, Chemical Catalog Co., (Reinhold Publishing Corp.), 1929.
50. Wilson, J. A., "Modern Practice in Leather Manufacturing," p. 420, New York, Reinhold Publishing Corp., 1941.
51. *Ibid.*, p. 421.
52. *Ibid.*, p. 422.
53. Windus, W., (to John R. Evans & Co.), U. S. Patent 2,470,450 (1949).
54. Winton, E. R., *J. Soc. Dyers Colourists*, **65**, 333–5 (1949).
55. Wolesensky, E., "Investigation of Synthetic Tanning Materials," Bur. Standards Tech. Paper No. 302, 45 pp. (1925).
56. Wolesensky, E., "Behavior of Synthetic Tanning Materials Toward Hide Substances," Bureau Standards Tech. Paper No. 309 (1926).
57. Woodroffe, E., *J. Intern. Soc. Leather Trades' Chemists*, **26**, 122–4 (1942).

Medicine

Formaldehyde is seldom used for medicinal purposes. However, it is employed as a dilute solution in water or alcohol to prevent excessive sweating[11]. The principal application is in the treatment of tinea pedis (athlete's foot, ringworm of the foot) to reduce sweating of the feet. For this purpose the feet may be soaked for five minutes in a solution made by diluting 6 oz. of commercial 37 per cent formaldehyde with water to a volume of 2 quarts. Small quantities of formaldehyde and paraformaldehyde are also encountered in some proprietary preparations recommended for the prevention of excessive sweating, for the treatment of athlete's foot, etc. Paraformaldehyde is sometimes added to medicated talcums.

Of special interest is the growing use of formaldehyde as an agent for modifying the properties of vaccines, antibiotics, vitamins and other medicinal preparations. In this connection, it has been found to have definite value for reducing toxicity in some instances. One such application is the conversion of toxins to toxoids which are nontoxic, although still capable of stimulating typical antigenic response[10]. Schultz and Gebhardt[16] believe that formaldehyde combines with the amino groups of the toxins and after injection into the body is gradually removed by oxidation, thereby restoring the toxin or virus to its original state. Research directed to inactivating an influenza virus without loss of immunizing potency has shown that procedures employing ultraviolet light and formaldehyde make it possible to obtain modified vaccines that are as powerful as equivalent quantities of the active virus[3]. Carter[4] uses formaldehyde in the preparation of a detoxified pollen extract. Boquet and Vendrely[2] report the conversion of cobra venom to anavenin by the action of formaldehyde at a pH of approximately 7. The detoxifying action of formaldehyde has also been studied in the treatment of antibiotics such as tyrothricin, gramicidin and tyrocidin whose use is hindered by their toxicity and lack of solubility. Fraenkel-Conrat and co-workers[6] have found that the reaction product of gramicidin with formaldehyde is less toxic, less hemolytic and somewhat more soluble in water. Moureu and co-workers[12] have also studied this problem and describe new derivatives produced by the action of succinic, maleic and other acid anhydrides on formaldehyde-treated tyrocidin and gramicidin.

Druey[5] reports that the condensation of sulfathiazole with formaldehyde and thioglycolic acid gives a carboxymethyl-mercaptomethyl-sulfathiazole whose neutral salts decompose rapidly to give sulfathiazole when given parenterally. Schoen and Gordon[14, 15] obtain riboflavin preparations of improved solubility in water by reaction with aqueous formaldehyde at pH values greater than 7. These products are believed to be hemiacetals.

Lang and Buck[7] prepare formaldehyde-amine derivatives of insulin which are stated to extend its physiological action by more than 24 hours.

Pacini[13] uses formaldehyde in a method for making non-bitter cascara sagrada extract[13]. This procedure is carried out by boiling with dilute aqueous formaldehyde. Unreacted formaldehyde, not removed by distillation during the treating process, is subsequently eliminated by reaction with ammonia.

Hexamethylenetetramine is a well known urinary antiseptic for the treatment of pyelitis, cystitis, and other diseases of the urinary tract. Its action apparently resides in the hydrolytic liberation of formaldehyde in acid urine, and for this reason acidity is often insured by giving the patient alternate doses of hexamethylenetetramine and sodium acid phosphate. When prescribed for medicinal use, hexamethylenetetramine is given the name of methenamine, formin, or urotropine.

A unique application of hexamethylenetetramine in improving the activity of vitamin E preparations is described by Baxter[1]. Alpha, beta, gamma and delta tocopherols exhibit vitamin E activity but the alpha compound is greatly superior to the others in this respect. Low activity appears to be associated with an unsubstituted hydrogen atom on the aromatic phenolic nucleus of beta and gamma tocopherol and two such atoms on the delta compound. Reaction with hexamethylenetetramine and an organic acid such as acetic, formic or benzoic acid replaces this hydrogen with a formyl group (see pages 428–9) and raises the vitamin E activity.

Formaldehyde, paraformaldehyde and hexamethylenetetramine have also found applications in the synthesis of drugs and other medicinal products among which are methylene ditannin (Tannoform), hexamethylenetetramine sodium benzoate (Cystazol), and hexamethylenetetramine anhydromethylene citrate (Helmitol). Illustrative recent syntheses include the use of formaldehyde and isobutyraldehyde in the preparation of pantothenic acid by Lawson and Parke[8] and the synthesis of chloromycetin by Long and Troutman[9] in which hexamethylenetetramine is employed to introduce an amino group in place of a bromine atom in an alpha-bromoacetophenone intermediate (see also page 426).

The application of anion exchange resins for the treatment of peptic and duodenal ulcers has already been discussed (page 444).

References

1. Baxter, J. G., (to Eastman Kodak Co.), U. S. Patent 2,592,531 (1952).
2. Boquet, P., and Vendrely, R., *Compt. rend. soc. biol.*, **137**, 179 (1943).
3. Can. Chem. Process Ind., **30**, 102, March 1946.
4. Carter, E. B., (to Abbott Laboratories), U. S. Patent 2,019,808 (1935).
5. Druey, J., *Helv. Chim. Acta*, **27**, 1776–82 (1944).
6. Fraenkel-Conrat, H. L., Humfield, H., Lewis, J. C., Dimick, K. P., and Olcott, H. S., (to U. S. Dept o.f Agriculture), U. S. Patent 2,438,209 (1948).
7. Lang, E. H., and Buck, J. S., (to Burroughs Wellcome Co.), U. S. Patent 2,354,211 (1944).
8. Lawson, E. J., and Parke, H, C., (to Parke, Davis & Co.), U. S. Patent 2,399,362 (1946).
9. Long, L. M., and Troutman, H. D., *J. Am. Chem. Soc.*, **71**, 2473 (1949).
10. McCulloch, E. C., "Disinfection and Sterilization," pp. 369–78, Philadelphia, Lea & Febiger (1936).
11. Merck & Co., Inc., "The Merck Manual," Eighth Ed., pp. 1267, 1342 (1950).
12. Moureu, H., Chovin, P., and Rivoal, G., *Bull. soc. chim. biol.*, **31**, 1062–9 (1949).
13. Pacini, A. J., U. S. Patent 1,917,598 (1933).
14. Schoen, K., and Gordon, S. M., *Arch. Biochem.*, **22**, 149 (1949).
15. Schoen, K., and Gordon, S. M., (to Endo Products Inc.) U. S. Patent 2,587,533 (1952).
16. Schultz, E. W., and Gebhardt, L. P., *Proc. Soc. Exper. Biol. Med.*, **32**, 1111 (1935).

Metal Industries

Applications of formaldehyde and its products in the metal industries include their use as acid inhibitors, reducing agents, and electroplating addition agents. Since 1945, formaldehyde has played an outstanding role in reducing the hydrogen sulfide corrosion of oil well equipment.

Use of Formaldehyde and its Derivatives as Pickling Addition Agents. Formaldehyde exerts an inhibiting effect on the action of acid on ferrous metals without preventing the quick solution of rust and scale. This property gives it value in the pickling and descaling of metal parts, although in recent years it has been largely replaced by more effective inhibitors, some of which are formaldehyde derivatives. This inhibiting action, at least under some conditions, is not observed in the case of aluminum and zinc. Table 59 based on the data of Mascré[18] shows the effect of formaldehyde on iron, aluminum, and zinc in various pickling and scaling solutions. According to Holmes[12], formaldehyde is most effective in relatively concentrated acids (*e.g.*, 1 vol. conc. H_2SO_4 to 5 vols. water, or 1 vol. conc. HCl to 1 vol. water) since in dilute acids the inhibiting action is practically nil at the high temperatures (50 to 80°C) used in mill practice to secure rapid cleaning. Mason[20] cleans metal parts under the influence of an alternating current in a solution containing 200 g of hydrogen chloride per liter and 0.2 per cent formaldehyde. Trioxane (alpha-trioxymethylene), the cyclic trimer of formaldehyde, is an effective pickling inhibitor for ferrous metals when employed with sulfuric, phosphoric or hydrochloric acid[35].

Formaldehyde is of special interest for use in pickling processes since it reduces hydrogen embrittlement. Rosenfeld's[31] studies indicate that it is superior to some commercial inhibitors in this respect when employed for pickling SAE 1020 steel with sulfuric and hydrochloric acid. Inhibition of embrittlement is enhanced when it is employed in combination with quinoline and *p*-toluidine.

Hexamethylenetetramine has an acid-inhibiting effect somewhat similar to that of formaldehyde. Chamberlain[3] states that two to five pounds of this agent act effectively in a ton of 60° sulfuric acid. It has been also reported[6] that hexamethylenetetramine inhibits the action of hydrochloric and phosphoric acids on aluminum. Trithioformaldehyde and the products obtained by passing hydrogen sulfide into formaldehyde solution are reported to be superior to formaldehyde as pickling addition agents in nonoxidizing acid baths in a process described by Sebrell[33]. According to Schmidt[32], an extremely effective inhibitor for use in pickling iron or steel with sulfuric acid is produced by adding 256 g of diorthotolyl thiourea to 150 g of commercial formaldehyde and 200 cc of water and refluxing for 16 to 18 hours. One part of this oily product is said to be effective in 22,000 parts of 5 per cent sulfuric acid. Other effective inhibitors include phenol-formaldehyde-ammonia resins[7], reaction products of formaldehyde with ammonium thiocyanate[13] aniline thiocyanate[11], and formaldehyde derivatives of nitrogen-sulfur compounds produced by treating diphenylguanidine with carbon bisulfide[17], and guanidine with hydrogen sulfide[10].

The use of formaldehyde in compositions for the prevention of rusting of iron or steel is illustrated by Eberhard's process[5], in which the metal is coated with a solution of a tungsten compound in a mixture of formaldehyde and uric acid. This treatment is stated to convert rust to magnetic oxide and leave a protective coating on evaporation. Burke[2] claims that a noncorrosive refrigerant may be obtained by combining a small amount of formaldehyde with sulfur dioxide.

Control of Hydrogen Sulfide Corrosion of Oil Well Equipment. The corrosive action of hydrogen sulfide containing brines in oil wells is extremely severe resulting in the loss of costly equipment and leading to

TABLE 59. INFLUENCE OF FORMALDEHYDE ON METALLIC CORROSION IN PICKLING AND SCALING SOLUTIONS*

Metal Treated	Solutions Used				Temp. (°C)	Relative Wt. Loss (g/sqm/hr)	Duration of Test (hrs)
	Type	Acid	Acid Concn. (g/100 cc)	CH₂O Concn. (%)			
Drawn iron tube	Pickling	HCl	23	None	20	389.0	50
" " "	"	"	23	1.5	20	6.2	50
" " "	"	H₂SO₄	30.6	None	20	262.0	50
" " "	"	"	30.6	0.3	20	5.4	50
" " "	"	"	21.9	None	60	1030.0	50
" " "	"	"	21.9	0.3	60	107.0	50
" " "	Scaling	HCl	3.5	None	—	78.2	24
" " "	"	Water plus 10% commercial scaling compound with a CH₂O basis			—	0.4	24
Sheet aluminum	"	HCl	3.5	None	—	4.0	24
" "	"	"	3.5	0.3	—	10.6	24
Zinc sheet	"	"	3.5	None	—	800.0	2
" "	"	"	3.5	0.3	—	900.0	2

* Data of J. Mascré[18].

complicated operating difficulties. Tubing must be pulled frequently to repair leaks and corrosion products sometimes lead to severe emulsification problems. These corrosion problems are further enhanced by electric currents and are accelerated by air although the latter is usually absent in wells which deliver sour crude. As a result of a research study of agents for controlling hydrogen sulfide corrosion, Menaul[1, 21, 22] discovered that small amounts of formaldehyde were extremely effective for this purpose. Laboratory tests carried out to simulate oxygen-free subsurface conditions demonstrated that 125 and 250 ppm formaldehyde reduced corrosion losses from 0.92 lb/sq ft per year to 0.113 and 0.105 respectively. Field tests proved even more successful in that corrosion was reduced by nearly 100 per cent in contrast to the 85 per cent reduction in the laboratory. In practice, the formaldehyde is added to the well in the annulus between casing and tubing using a lubricator or drip system. Results indicate that one pound of formaldehyde per 100 barrels of well water or brine gives satisfactory protection. Although the exact mechanism of the protective action has not been proved, Menaul and Dunn[22] conclude that a reaction product of formaldehyde with the sulfide brine acts as the inhibitor. This compound is not a mercaptan since mercaptans were found to be ineffective. It is observed that iron or steel assumes a bluish gun metal appearance on immersion in the formaldehyde treated brine and that this film is substantially unaffected after cleaning with 1 per cent hydrochloric acid.

Other patented processes for inhibition of the corrosive action of hydrogen sulfide brine include addition of formaldehyde and ammonia[25], formaldehyde and hexamine[26], and water soluble urea-formaldehyde condensation products[30]. Moyer and co-workers[27] prevent hydrogen sulfide brine corrosion with formaldehyde and amino compounds including urea and melamine. Murray[28] describes the mitigation of the corrosive action of ammonium sulfides by reaction with formaldehyde to form a neutral inhibitor.

The reducing action of formaldehyde has long been employed in connection with the preparation of silver mirrors. An early process[4] of this type is carried out by covering the articles to be silvered with a mixture prepared by adding a solution of 6 g of silver nitrate in 3 cc of water to a mixture of 6 cc of commercial formaldehyde and 7 cc of glycerin, and then exposing the treated surface to ammonia vapors. The silver mirror forms immediately when the film of treating solution becomes ammoniacal. Good wetting of the surface is extremely important and can be improved by adding a little acetone to the treating solution. The thickness of the coating can be controlled by variations in solution concentration. According to a recent publication by Misciattelli[24], copper mirrors can be produced by the action of cold formaldehyde on alkaline solutions of copper tartrate and sulfate containing glycerin in the presence of colloidal silver or traces of precipitated silver which act as a reduction catalyst. A practical procedure of this type[23] makes use of a solution prepared by dissolving 4 g of copper sulfate, 15 g of sodium potassium tartrate and 6 g of caustic soda in one liter of distilled water and adding 15 cc of a 1 per cent solution of gum arabic. This solution is mixed with 100 cc of commercial formaldehyde immediately before use. Best results are obtained by washing a sheet of glass with a stannous chloride solution and then treating with a silvering solution so that an extremely thin layer of silver is precipitated. The plate is then washed and flooded rapidly with the copper solution. Approximately 1.5 hours are necessary to produce a satisfactory coating.

Narcus[29] reports that the chemical reduction method is the most practical procedure for metallizing plastics prior to plating. The silvering of cellulose acetate parts is accomplished by wet tumbling with pumice to remove surface gloss, cleaning, treating with an acid solution of stannous chloride and then silvering with ammoniacal silver nitrate. The silvered parts may then be barrel plated with copper.

Processes based on the production of metals by reduction of salts with alkaline formaldehyde have also been employed for the coloring of plaster of Paris[34]. A recent procedure for the coloring of oxide-coated aluminum or aluminum-alloy surfaces may be carried out by impregnating the surface with a soluble salt of silver or gold and then treating with formaldehyde solution or formaldehyde gas[19].

Formaldehyde has sometimes been used as a reducing agent in working up residues containing noble metals, for example, in precipitating silver and gold (page 179–180).

In electroplating, formaldehyde is sometimes employed indirectly when formaldehyde derivatives are used as plating addition agents. That formaldehyde itself may act as an addition agent for some purposes is indicated by a process[16] claiming its use as an addition agent for the production of bright coatings of nickel and cobalt from solutions of their sulfates or chlorides. A formaldehyde concentration of one gram per liter is said to be effective. Illustrative of formaldehyde derivatives which are reported to be effective addition agents are: a cresol-formaldehyde condensate solubilized by reaction with sulfuric acid which is said to facilitate the rapid formation of thick tin coatings of fine structure from an acid-type bath[8] and a reaction product of ammonium thiocyanate and formaldehyde which enables one to obtain bright zinc coatings from a cyanide plating bath[9]. Also, Kardos[15] uses the formaldehyde condensation product of the dithiocarbamate of a straight chain alkylene diamine as a zinc brightener.

Formaldehyde is also claimed to be of value for the purification of zinc cyanide plating baths[14].

Formaldehyde has also been discovered to be of value in the anodic treatment of aluminum, where its presence in the electrolyte solution maintains the porosity of the oxide film which is produced. A concentration of 0.3 to 0.5 per cent formaldehyde in a bath containing chromic acid, chromates, and glycerin is stated to be sufficient to achieve the desired results[36].

References

1. Barnes, K. B., and Deegan, C. J., *Oil Gas J.*, **44**, 84–6 (Oct. 27, 1945).
2. Burke, F. D., U. S. Patent 2,019,559 (1935).
3. Chamberlain, G. D., (to R. T. Vanderbilt Co., Inc.), U. S. Patent 1,719,618 (1929).
4. Chemische Fabrik von Heyden, German Patent 199,503 (1908); *Chem. Zentr.*, **1908**, II, 554.

5. Eberhard, R., U. S. Patent 1,893,495 (1933).
6. Hamor, W. A., *Chem. Eng. News*, **18**, 53 (1940).
7. Harmon, J., and McQueen, D. M., (to E. I. du Pont de Nemours & Co., Inc.), U. S. Patent 2,165,852 (1939).
8. Harshaw Chemical Co., British Patent 404,533 (1934).
9. Henricks, J. A., (to Udylite Co.), U. S. Patent 2,101,580 (1937).
10. Hill, W. H., (to American Cyanamid Co.), Canadian Patent 408,212 (1942); U. S. Patent 2,384,467 (1945).
11. Hill, W. H., (to Koppers Co., Inc.), U. S. Patent 2,586,331 (1952).
12. Holmes, H. N., U. S. Patent 1,470,225 (1923).
13. Horst, P. T., (to Wingfoot Corp.), U. S. Patent 2,050,204 (1936).
14. Imperial Chemical Industries, Ltd., British Patent 474,449 (1937).
15. Kardos, O., (to Hanson-Van Winkle-Munning Co.), U. S. Patent 2,589,209 (1952).
16. Louis Weisberg, Inc., British Patent 464,814 (1937).
17. Magoun, G. L., (to Rubber Service Laboratories), U. S. Patent 1,868,214 (1932).
18. Mascré, J. E. J. G., (to Establissements Lambiotte Freres Premery), U. S. Patent 2,160,406 (1939).
19. Mason, R. B., (to Aluminum Co. of America), U. S. Patent 1,988,012 (1935).
20. Mason, S. R., (to Western Electric Co.), U. S. Patent 1,839,488, (1932).
21. Menaul, P. L., (to Stanolind Oil and Gas Co.), U. S. Patent 2,426,318 (1947).
22. Menaul, P. L., and Dunn, T. H., *Am. Inst. Mining Met. Engrs., Tech. Pub No. 1970* (1946).
23. Misciattelli, P., U. S. Patent 2,183,202 (1939).
24. Misciatelli, P., *Atti congr. intern. chim. Rome*, **4**, 689; *Chem. Zentr.*, **1941**, I, 345.
25. Moyer, M. I., and Hersh, J. M., (to Cities Service Oil Co.), U. S. Patent 2,496,594 (1950).
26. *Ibid.*, U. S. Patent 2,496,595 (1950).
27. *Ibid.*, U. S. Patent 2,496,596 (1950).
28. Murray, C. A., (to The Pure Oil Co.), U. S. Patent 2,589,114 (1952).
29. Narcus, H., *Proc. Am. Electroplaters' Soc.*, **31**, 76–92 (1944).
30. Nunn, L. G., Jr., (to Standard Oil Dev. Co.), U. S. Patent, 2,514,508 (1950).
31. Rosenfeld, M., *Iron Age*, **161**, 82 (June 17, 1948).
32. Schmidt, J. G., (to E. F. Houghton & Co.), U. S. Patent 1,807,711 (1931).
33. Sebrell, L. B., (to Goodyear Tire & Rubber Co.), U. S. Patent 1,805,052 (1931).
34. Vanino, L., German Patent 113,456 (1900).
35. Walker, J. F., (to E. I. du Pont de Nemours & Co., Inc.), U. S. Patent 2,355,599 (1944).
36. Windsor-Bowen, E., and Gower, C. H. R., British Patent 537,474 (1941).

Metal Sequestering Agents

Ethylenediamine tetraacetic acid and some of its simple derivatives are outstanding metal sequestering agents that are sold under a number of trade names. They are made by syntheses involving the reaction of formaldehyde and alkali cyanides or glycolonitrile with the parent amine (see pages 286–287).

Paper

Formaldehyde is employed by the paper industry for improving the wet-strength, water-resistance, shrink-resistance, grease-resistance, etc., of paper, coated papers, and paper products. For these purposes formaldehyde is used both directly and in the form of reactive formaldehyde derivatives. In addition, formaldehyde serves as a disinfectant and preservative in connection with some phases of paper manufacture. Less direct applications are found in the preparation of finishes, sizing agents, etc.

Direct Treatment of Paper and Paper Pulp with Formaldehyde. Direct treatment of paper with formaldehyde is, in general, employed for improving water-resistance and wet-strength, but also has value in connection with parchmentizing processes. Treating methods normally require acid catalysts and may accordingly involve the formation of methylene ethers of the paper cellulose (page 211). These treatments are similar in many respects to the related processes used for making textiles crease-proof and water-resistant.

In 1911, Ernst Fues[16] was granted a U.S. patent covering a process for making a parchment paper by treating paper with an acid solution and formaldehyde. This was followed in 1926 by a process for increasing the water resistance of paper[17]. The latter process is carried out by treating paper with an aqueous solution of formaldehyde containing an acid-forming salt and heating the dried paper to a temperature of 60 to 100°C. A related patent[21] covers a similar procedure for increasing the strength and water-resistance of paper by treatment with an acidic formaldehyde solution and drying at 110 to 120°C. Filter paper is impregnated with 2 cc of concentrated nitric acid in 100 cc of commercial formaldehyde, excess solution squeezed out, and the paper dried on heated rolls at 120°C. Paper subjected to this process is said to retain substantially all of its strength when wet. Paper which has been made water-resistant may lose its ability to absorb water. According to Kantorowicz[22], paper towels and handkerchiefs which have a high wet-strength and are also water-absorbent can be prepared from a mixture of untreated cellulose and cellulose which has been treated with formaldehyde and nitric acid as described above. Richter and Schur[31] claim the production of absorbent papers of high wet-strength by treating a paper web prepared from a mixture of gelatinized and ungelatinized pulp with dilute formaldehyde and then immersing in 72 per cent sulfuric acid for a short time at room temperature. This treatment is followed by washing and drying.

In some processes for wet strength paper, a modifying agent is added which may give a reactive methylol derivative. Schenck treats paper with a solution of formaldehyde containing an acid catalyst plus about 1 per cent by weight of a linear polyamide[33].

A parchment-like paper said to be water-proof, leathery, and tough is obtained by a process of Rockwood and Osmun[32] in which the fibers are treated with a solution of zinc chloride in an equal weight of commercial formaldehyde, dried at 100°C, calendered on hot rolls, and finally washed and dried.

According to DeCew[13], treating paper pulp with formaldehyde accomplishes two purposes: (a) it prevents fermentation of the mucilaginous materials in the pulp and thus prevents slime and gel formation; (b) it coagulates the mucilaginous material in the pulp and gives a harder, snappier paper which is more readily handled on the paper machines. One pound of formaldehyde is stated to be sufficient for 1000 pounds of paper stock in this application, and stock so treated improves with age instead of deteriorating.

Treatment of Paper with Formaldehyde and Proteins. The insolubilizing action of formaldehyde on gelatin, glue, casein, and other protein materials is also utilized for improving the water-resistance and wet strength of paper. A method of this type was patented in Germany as early as 1893[11]. Paper was impregnated with a gelatin or glue solution, after which it was subjected to the action of formaldehyde gas or aqueous formaldhyde solution. The product was recommended for antiseptic bandages, for which purpose the disinfecting action of formaldehyde was stated to make it specially suitable. In a more recent procedure, Harrigan and Krauss[18] obtain an absorbent paper of high wet-strength by treating an absorbent paper with a solution containing 6 ounces of animal glue and 3 ounces of 37 per cent formaldehyde per gallon of water. The effect of this treatment is immediate, but the tensile strength improves on ageing and may show a 200 per cent increase after several days. A modified procedure which is said to produce a high-strength, water-permeable, flexible paper involves impregnation with a glue solution containing glycerin followed by drying, after which the paper surface is given a light treatment with aqueous formaldehyde[2]. Morita[27] claims the preparation of an improved tracing paper by impregnating the fibers with a solution of sodium alginate and formaldehyde containing a mixture of calcium or aluminum acetate or aluminum sulfate and subjecting the treated material to a heating or drying process.

Indirect Methods of Formaldehyde Treatment of Paper. Indirect applications of formaldehyde for the production of paper having a high degree of water-resistance and a good wet-strength are illustrated by Schur's procedure.[34] This involves treating with a reactive urea-formaldehyde product and is reported to give a flexible, absorbent paper containing less than 2 per cent urea-formaldehyde resin. A somewhat similar product is obtained by making paper from pulp to which 5 per cent of a urea-formaldehyde compound has been added[6]. Melamine-formaldehyde condensates are also suitable for the production of water-resistant papers and, according to the patent literature, may also be applied to the finished paper[19] or added to the paper pulp in the hollander[35].

Melamine-formaldehyde resins are outstanding for wet strength paper in that they can be economically added to the pulp in the beater rather than as tub size to the finished paper or as a spray on the wet freshly formed sheet. Since melamine resins can be readily produced as a cationic colloidal dispersion, they are efficiently taken up by the paper pulp. Methods of preparing a cationic melamine resin of this type are described by Maxwell[25]. Landes[24] points out that cationic urea resins have also been produced for addition in the beater but little has been published concerning them. He states that it is believed that these resins are made by combining molecules containing additional basic groups with the urea-formaldehyde condensate. The use of guanidine is mentioned in this connection.

Bursztyn[7] has shown with dyed color photomicrographs that melamine-formaldehyde resin is more uniformly distributed on paper fibers than the ordinary urea resin. Also, it is effective with bleached pulp where urea-formaldehyde is very inefficient. A slightly acidic pH is essential for the development of wet strength and Bursztyn[8] states that it is not possible to make a paper with an acceptable wet strength with a pH higher than 5 when urea-formaldehyde is employed. However, this investigator[9] also points out that urea-formaldehyde has an advantage in that the paper can be repulped with hot water whereas melamine-formaldehyde gives a boilproof paper which makes it impossible to return broke to the production cycle.

Shrinkproof paper is produced, according to Newkirk[29], by wetting paper somewhat below saturation with a solution containing 50 parts cresol, 36 parts of commercial formaldehyde, an alkaline material, and 25 parts of water and heating the moist paper to a temperature above 200°F. This paper is said to show little shrinkage after repeated wetting and drying. Shrinkproof paper may also be obtained by a related process in which the paper is treated with a solution containing formaldehyde and urea[1]. Kvalnes[23] obtains cardboard for business machine cards that is dimensionally stable by dipping in an aqueous solution of methylolureas at a pH of 7 to 9 and curing under pressure at 280 to 320°F.

Coated Papers. Production of water-resistant coatings on paper is readily accomplished by the use of coating compositions containing water-soluble adhesives, such as glue and casein, which can be insolubilized by reaction with formaldehyde under proper conditions of temperature, concentration, catalysis, etc. Formaldehyde may in some cases be added to the coating compound; in others it may be applied separately to the coated paper. Soluble formaldehyde derivatives which can be converted to water-insoluble resins or which act as insolubilizers for other materials in the coating composition are also of considerable technical importance in the preparation of resistant coatings. Special applications of coatings involving formaldehyde are found in pigmented coated papers, washable wallpaper, grease-proof containers, etc.

An example of the use of formaldehyde in this field is found in paper coated with clay and crystals of "gamma gypsum" embedded in a formaldehyde hardened film of casein. Paper coated in this way is claimed by Offutt and Gill[30] to be water-resistant and to possess an outstanding degree of smoothness, whiteness, and flexibility. Such a paper is obtained when an alkaline casein solution and formaldehyde equivalent to 0.5 to 2 per cent of the weight of dissolved casein is added to the slurry of clay and hydrated gypsum used for coating. Fleck[15] produces a water-resistant, decorated paper by applying a coating compound containing clay, pigment, a little pine oil, and an alkaline solution of casein and treating the printed paper with a solution of 1.5 pounds of commercial formaldehyde and 3 pounds of alum in 10 gallons of water. This formaldehyde solution is applied by a roll, after which the paper is dried. A simplified process for producing washable wallpaper also employing a protein which is insolubilzed with formaldehyde in the course of the coating process is claimed by Bright[5]. Hexamethylenetetramine can be used as an insolubilizer for the coating binder and may be employed in much larger proportion than formaldehyde without causing the coating mixture to thicken[36]. Formaldehyde is released by the action of heat in the drying process provided the pH is not too high at this point. Bennett[4] controls the action of formaldehyde in a casein coating mixture by first adding acetaldehyde or some other aldehyde capable of reacting with formaldehyde to form an aldol.

Compositions containing urea-formaldehyde, melamine-formaldehyde, and polyvinyl alcohol are finding increasing use as binders in combination with starch in coating compositions. In one application of this type, Nelson[28] applies a pigmented composition containing a clay and starch, dries to produce a preformed set coating and then treats with a urea or melamine resin and heats to insolubilize the coating. Another method of application[20] is to use dimethylolurea in connection with polyvinyl alcohol to increase the water-resistance of polyvinyl alcohol films, coatings, and sizing compositions. The dimethylolurea and a catalyst, e.g., ammonium chloride or sulfates, are added to the polyvinyl alcohol solution either plain or pigmented, and the coated product is subsequently heated to bring about the insolubilizing action. A gelatinous dispersion prepared from dimethylolurea, an amylaceous substance, and a small amount of acid is also claimed to give a hard, tenacious, water-resistant coating.[3] The use of melamine-formaldehyde condensates in paper-coating compositions may be illustrated by a process patented by Widmer and Fisch[37].

Greaseproof paper and cardboard are obtained by making use of the action of formaldehyde on proteins. Calva[10] produces a greaseproof cardboard container by coating the interior surface with a water solution of glue, glycerin, and formaldehyde and heating above 130°F to complete insolubilization of the glue. Casein, soybean protein, or glue plus a plasticizer and formaldehyde may also be used in producing a primary coating on paper over which a cellulose-ester film is then deposited to give a flexible grease-proof coating[14]. McKee[26] employs glue and hexamethylenetetramine to obtain greaseproofness on the inner wall of a container.

Coggeshall[12] produces a paper liner, which can be used in contact with a tacky rubber composition and which will not stick or "pick" on separation, by deeply impregnating paper with a solution containing casein and formaldehyde, drying, coating with flexibilized starch, and friction-calendering to produce a glossy surface.

References

1. American Reinforced Paper Co., British Patent 501,514 (1939).
2. Anderson, C. A., U. S. Patent 2,146,281 (1939).
3. Bauer, J. V., and Hawley, D. M., (to Stein Hall Manufacturing Co.), U. S. Patent 2,212,314 (1940).
4. Bennett, E. G., (to The Champion Paper and Fibre Co.), U. S. Patent 2,369,427 (1945).
5. Bright, C. G., (to Paper Patents Co.), U. S. Patent 2,123,399 (1938).
6. Brown Co., Belgian Patent 436,053 (1939).
7. Bursztyn, I., *British Plastics*, **20**, 299–304 (1948).
8. Bursztyn, I., *Paper-Maker (London)*, **117**, 26, 28–32 (1949); *World's Paper Trade Rev.*, **130**, TS105–11 (1948).
9. Bursztyn, I., *Finnish Paper Timber J.*, **28**, 207–10 (1946).
10. Calva, J. B., (to General Products Corp.), U. S. Patent 2,122,907 (1938).
11. Chemische Fabrick auf Actien (vorm E. Schering), German Patent 88,114 (1893).
12. Coggeshall, G. W., (to S. D. Warren Co.), U. S. Patent 2,173,097 (1939).
13. DeCew, J. A., (to Process Engineers, Inc.), U. S. Patent 1,483,630 (1924).
14. Fischer, H. C., Thompson, J. F., and Sooy, W. E., (to The Gardner-Richardson Co.), U. S. Patent 2,205,557 (1940).
15. Fleck, L. C., (to Paper Patents Co.), U. S. Patent 1,995,626 (1934).
16. Fues, E., U. S. Patent 1,033,757 (1912).
17. Fues, E., U. S. Patent 1,593,296 (1926).
18. Harrigan, H. R. and Krauss, J. M., U. S. Patent 1,997,487 (1935).
19. Hofferbert, R. P., (to American Cyanamid Co.), Canadian Patent 409,429 (1942
20. Izard, E. F., (to E. I. du Pont de Nemours & Co., Inc.), U. S. Patent 2,169,250 (1939).
21. Kantorowicz, J., U. S. Patent 1,816,973 (1931).
22. Kantorowicz, J., U. S. Patent 2,010,635 (1935).
23. Kvalnes, H. M., (to E. I. du Pont de Nemours & Co., Inc.), U. S. Patent 2,422,423 (1947).
24. Landes, C. G., *TAPPI*, **33**, No. 9, pp. 465–70 (Sept., 1950).
25. Maxwell, C. S., (to American Cyanamid Co.), U. S. Patents 2,559,220–1 (1951).

26. McKee, R. H., U. S. Patent 2,544,509 (1951).
27. Morita, T., Japanese Patent 133,002 (1939); C. A., **35**, 3369 (1941).
28. Nelson, H. E., (to Stein, Hall & Co., Inc.), U. S. Patent 2,460,998 (1949).
29. Newkirk, F. F., (to American Reinforced Paper Co.), Canadian Patent 405,245 (1942).
30. Offutt, J. S., and Gill, J. W., (to U. S. Gypsum Co.), U. S. Patent 2,231,902 (1941).
31. Richter, G. A., and Schur, M. O., (to Brown Co.), U. S. Patent 2,096,976 (1937).
32. Rockwood, C. D., and Osmun, K. L., (to The Union Selling Co.), U. S. Patent 2,107,343 (1938).
33. Schenck, W. A., (to Munising Paper Co.), U. S. Patent 2,540,352 (1951).
34. Schur, Mo. O., (to Brown Co.), Canadian Patent 393,326 (1940) (to Reconstruction Finance Corp.), U. S. Patent 2,338,602 (1944).
35. Society of Chemical Industry in Basle, British Patent 480,316 (1938).
36. Sutermeister, E., and Browne, F. L., "Casein and Its Industrial Applications," A.C.S. Monograph No. 30, p. 309, New York, Reinhold Publishing Corp. (1939).
37. Widmer, G., and Fisch, W., (to Ciba Products Corp.), U. S. Patent 2,197,357 (1940).

Photography

The sensitized surfaces most generally employed on photographic films, papers, etc., consist of gelatin coatings which carry light-sensitive silver salts. Formaldehyde and compounds which liberate formaldehyde find wide and varied applications in photography due to (a) their hardening and insolubilizing action on gelatin, (b) their reducing action on silver salts, and (c) the fact that these actions can be controlled within a considerable range of intensity by variations in the degree of acidity or alkalinity attendant in their use. Methods of applying formaldehyde and its products in photography include their use as additions to photographic elements and to processing baths applied to photographic elements.

Ehrenfried[18] has recently called attention to the fact that formaldehyde can be employed as an independent photographic developer. In this connection, he has demonstrated that acceptable pictures can be produced with alkaline formaldehyde solutions containing some sodium chloride. However, when used alone formaldehyde is a low potential developer with low selectivity. In addition the stability of alkaline formaldehyde solutions is notoriously poor and their exhaustion rate as developers is rapid.

Although applications of formaldehyde and its products, such as paraformaldehyde, formaldehyde bisulfite, etc., have increased considerably in recent years, their use in this field is not new. In 1896, Lumiere and Seyewetz[23] reported that formaldehyde accelerated the action of hydroquinone-sulfite developers. A short time later these same investigators[29] recommended the use of combinations of paraformaldehyde and sodium sulfite in place of sodium hydroxide or sodium carbonate in developing solutions. This use is apparently based on the formation of alkali in the well-known reaction:

$$Na_2SO_3 + CH_2O + H_2O \rightarrow NaOH + NaSO_3 \cdot CH_2OH$$

However, except in developers containing hydroquinone as the sole developing agent, formaldehyde tended to produce a high fog, and preference was shown for acetone-sulfite combinations as an alkali generator. However, in recent years, Muehler[35] has found that non-fogging "tropical" developers, particularly suited for use in hot climates, can be obtained by using halogen substituted hydroquinones such as 2-chlorohydroquinone ("Adurol" of Hauff) and 2-bromohydroquinone ("Adurol" of Schering), dichlorohydroquinone, etc., with formaldehyde or "trioxymethylene." In addition, Crabtree and Ross[15] have patented non-fogging developers containing a small amount of formaldehyde, an additional alkali and "Metol," pyrocatechol, or para-aminophenol as the developing agent.

Developers relying on the formaldehyde-sulfite reaction for alkali were not widely used until the introduction of the extremely high contrast, silver chloride "litho" films for photolithography. Where maximum contrast and emulsion hardness are prime requisites, as in the photolithographic half-tone reproduction processes, the formaldehyde-hydroquinone-sulfite developers are in common use. Willcock[49] summarizes the case for developers containing formaldehyde and points out that they have characteristics which are considerably different from those energized by direct addition of sodium or potassium hydroxide. Dry powder mixtures which may be dissolved in water to give "litho" developers of this type require the use of paraformaldehyde. However, such compositions are not completely satisfactory because of the difficulties encountered in dissolving this polymer, as well as the troublesome polymerization of formaldehyde in the highly alkaline solutions. Muehler[36] claims that a mild acid, preferably boric acid, or a buffering salt will prevent the slow polymerization.

The use of formaldehyde-bisulfite combinations as both preservatives and accelerators for photographic developing solutions was described in 1889 by Schwartz and Mercklin[42]. The sulfite reaction is reversible and in the presence of a strong alkali, such as sodium hydroxide, proceeds as shown below until equilibrium is attained.

$$NaSO_3 \cdot CH_2OH + NaOH \rightleftharpoons Na_2SO_3 + CH_2O + H_2O$$

Recently Donovan and Wadman[16] patented dry powder developing compositions comprising hydroquinone, a buffer salt, and formaldehydebisulfite.

Special developers are sometimes produced by combining known developing agents with formaldehyde and its derivatives. Lüttke and Arndt[31] condensed *p*-aminophenol developing agents with formaldehyde and acetaldehyde in the presence of potassium bisulfite and reported that the new agents possess greater developing action than ordinary p-aminophenol developers. Paris[38] combined hexamethylenetetramine with pyrogallol, pyrogallol-monomethyl ether or methyl-pyrogallol-monomethyl ether, obtaining crystalline complexes easily soluble in alkaline solutions. The solutions were stated to be more resistant to aerial oxidation than the parent pyrogallols.

The hardening action of formaldehyde on the gelatin of photographic emulsions permits high temperature development to be carried out. To prevent excessive hardening which retards development by slowing penetration of the developer, Agnew and Renwick[2] first bathe the film or plate in a solution containing formaldehyde and a relatively high concentration of sodium sulfate or disodium orthophosphate. For high-temperature development, Perutz[39] adds 10 to 20 grams of hexamethylenetetramine per liter of developer. Gevaert[21] adds hexamethylenetetramine to developers to counteract the action of "blue-black" agents already present in the emulsion to be developed or also present in the developer itself.

High temperature permits short developing times where minimum processing delay is of paramount importance such as in photographing the finish of races, x-ray films and films for television transmission. For obtaining full development in about 10 seconds, Jaenicke[24] recommends an alkaline-formaldehyde bath between fixation and washing. He also claims that the addition of very small amounts of aldehydes to the alkali bath of his split developer reduces the required bathing time considerably. For ultra-rapid development and fixation requiring no more than 20 seconds for development, fixation and drying of photographic papers, Fischer[19] hardens the emulsion in 10 per cent formalin solution and develops in a warm developer containing 25 per cent potassium iodide. Capstaff[13] employs a high-contrast hydroquinone-bisulfite-sodium hydroxide-formaldehyde developer as the "first" developer of his reversal process. For fixing photographic films, plates, and papers at temperatures up to 90°F, Crabtree[14] recommends a sodium thiosulfate-sodium sulfite-formaldehyde fixer.

In 1889 Schwartz and Mercklin[42] added formaldehyde-bisulfite products to photographic emulsions for increasing the sensitivity of the emulsion. Subsequent disclosures[9, 10, 20, 43, 44, 45] recognize that it is common practice to harden gelatin photographic emulsions with formaldehyde, formaldehyde-formers and derivatives thereof. Merckens[33] points out that a minimum of gelatin is prerequisite to quick setting of silver-gelatino-halide emulsions upon coating, and describes how a 2 per cent gelatin solution treated at 50°C with formaldehyde becomes as viscous as an ordinary solution containing 8 to 10 per cent gelatin.

Processed photographic films, particularly cinematographic films which are subject to much wear, are treated with renovating and preserving solutions containing formaldehyde as an essential ingredient, according to Bodine[7], Mackler[32] and Stewart[48], who also[47] described treating the processed film with formaldehyde gas. Roth[40] describes a method for rendering nitrocellulose films fireproof in ordinary projecting machines by applying a layer of gelatin to the base side of the film and forming the layer into a heat-protecting coating by treating it with a mixture of equal parts of a 5 per cent formaldehyde solution and a 2 per cent solution of sodium bisulfite. Adamson[1] suggests a nonflammable film support comprising a nucleus or core of formaldehyde-hardened gelatin containing glycerin as the plasticizer and coated on both sides with first a rubber varnish and finally a celluloid varnish.

Formaldehyde is widely used in color photography for hardening the emulsion layer before treating it with silver, bleaching, mordanting, toning, etc.[8, 12, 26, 37, 46], and in place of alum[30] for toning developed silver images. Jennings[25] and Schneider[41] prepare aldehyde-condensed dimeric and polymeric dye-intermediates for use in film emulsions to be processed by color-forming developers. In methods of color photography involving the use of diffusely dyed silver halide emulsion layers, the dyes tend to diffuse from one stratum to another causing false, or at least contaminated, color reproduction. According to Dreyfuss[17], dyes, dye-formers and catalysts for local dye-bleaching processes can also be rendered non-diffusing by incorporating in the layer as a precipitant, a polymeric condensation product obtained from formaldehyde, or a formaldehyde compound, dicyandiamide and a nitrogenous base such as aniline, melamine, etc. For the same type of color process, Kodak[51] describe non-diffusing, bleach-out dyestuffs obtained by reacting formaldehyde, or a formaldehyde-forming material, with an azo dye containing a benzene ring carrying a hydroxyl or amino group. Gutekunst[23] prepares carbocyanine dyes for sensitizing photographic emulsions by condensing 6-acetamino-quinaldine alkyl halides with formaldehyde in alcoholic solution in the presence of a strong base.

For intensifying silver images, Van Monckhoven[34] in 1879 bleached the image with a solution of silver nitrate and potassium cyanide and then redeveloped. More recently, Blake-Smith and Garle[6] and Geiger[22] recommended formaldehyde or acetaldehyde with potassium hydroxide as a redeveloper for the Van Monckhoven intensification process.

In the manufacture of silk-screen stencils by the carbro or ozobrome transfer method, a photographic print is treated first with a potassium ferricyanide-bichromate-bromide bleach bath, transferred to an acid-formaldehyde bath, and then squeegeed into contact with the carbro sheet[5].

In recent years there has been a tendency to replace formaldehyde or paraformaldehyde in some photographic applications with a formaldehyde releasing agent. For example, Bryce[11] describes sizing paper for photographic purposes by treating with a mixture of gelatin and dimethylol urea and then heating to a temperature which will release the formaldehyde without decomposing the cellulose. Baldsiefen[3] describes obtaining blue-black image tones by the addition of methylol amides including dimethylol urea to gelatino silver halide emulsion. Baldsiefen[4] also attributes similar properties to methylol nitromethane. Lowe and Fowler[27] describe hardening hydrolyzed polyvinyl acetate silver halide emulsions with a water soluble melamine-formaldehyde hardener resulting from the reaction of one mol of melamine and 1 to 10 mols of formaldehyde. Wood[50] claims to have achieved improved stability for high contrast developers by substituting a mixture of trioxane and dimethylol or monomethylol urea for the previously used paraformaldehyde.

References

1. Adamson, A. G., (to W. M. Still & Sons, Ltd.), U. S. Patent 1,904,113 (1933).
2. Agnew, J., and Renwick, F. F., (to Ilford Ltd.), U. S. Patent 1,424,062 (1922).
3. Baldsiefen, W. D., (to E. I. du Pont de Nemours & Co., Inc.), U. S. Patent 2,363,493)1944).
4. *Ibid.*, U. S. Patent 2,364,017 (1944).
5. Biegeleisen, J. I., and Busenbark, E. J., "The Silk Screen Printing Processes," pp. 129–130, New York, McGraw-Hill, (1938).
6. Blake-Smith, R. E., and Garle, J. L., *Phot. Mitt.*, **1901**, 38, 360.
7. Bodine, H. O., (to Agfa Ansco Corp.), U. S. Patent 2,150,757 (1939).
8. Brewster, P. D., U. S. Patents 1,992,169 (1935) and 2,070,222 (1937).
9. Brunken, J., (to Agfa Ansco Corp.), U. S. Patents 1,870,354 (1932); to I. G. Farbenindustrie A. G., British Patent 373,829 (1932).
10. Brunken, J., (to Agfa Ansco Corp.), U. S. Patent 2,169,513 (1939).
11. Bryce, R. S., (to Eastman Kodak Co.), U. S. Patent 2,354,662 (1944).
12. Bunting, A. L., (to Union Research Corp.), U. S. Patents 2,312,874–5 (1943).
13. Capstaff, J. G., (to Eastman Kodak Co.), U. S. Patent 1,460,703 (1923).
14. Crabtree, J. I., *Photo Miniature*, **15**, No. 173,204 (1918).
15. Crabtree, J. I., and Ross, J. F., (to Eastman Kodak Co.) U. S. Patent 1,933,789 (1933).
16. Donovan, T. S., and Wadman, W. V., (to Eastman Kodak Co.), U. S. Patent 2,313,523 (1943).
17. Dreyfuss, P., (to Chromogen, Inc.), U. S. Patent 2,317,184 (1943).
18. Ehrenfried, G., *Phot. Sci. and Technique*, **18B**, 2–5 (1952).
19. Fischer, R., German Patent 681,737 (1939).
20. Fricke, H., and Brunken, J., (to Agfa Ansco Corp.), U. S. Patent 2,154,895 (1939).
21. Gevaert Photo-Producten N. V., British Patent 524,592 (1940).
22. Geiger, J., *Photo News*, **51**, 235 (1907).
23. Gutekunst, G. O., (to Eastman Kodak Co.), U. S. Patent 1,532,814 (1925).
24. Jaenicke, H., *Photo Ind.*, **35**, 514–6; 540–2 (1937).
25. Jennings, A. B., (to E. I. du Pont de Nemours & Co., Inc.), U. S. Patent 2,294,909 (1942).

26. Kelly, W. V. D., (to Prizma, Inc.), U. S. Patent 1,411,968 (1922).
27. Lowe, W. G., and Fowler, W. F., Jr., (to Eastman Kodak Co.), U. S. Patent 2,367,511 (1945).
28. Lumière, L., and Seyewetz, A., *Bull. Soc. Chim.* **15**, 1164 (1896).
29. Lumière, L., and Seyewetz, A., *Monit. Scient.*, **17**, 109 (1903).
30. A. Lumière et ses Fils, French Patent 324,921 (1902).
31. Lüttke, H., and Arndt, P., British Patent 23,729 (1904); French Patent 347,396 (1904).
32. Mackler, A. I., U. S. Patent 2,053,621 (1936).
33. Merckens, W., German Patent 301,291 (1919).
34. Monckhoven, D. van, *Bull. Belge.*, **6**, 178 (1879); *Phot. Korr.*, **16**, 208 (1879).
35. Muehler, L. E., (to Eastman Kodak Co.), U. S. Patent 1,857,515 (1932).
36. *Ibid.*, U. S. Patent 2,184,053 (1939).
37. Newens, F. R., "Colour Photography," p. 61, London, Blackie and Son, Ltd. (1931); Wall, E. J., "Photographic Facts and Formulas," p. 240–1, Boston, American Photographic Publishing Co. (1940).
38. Paris, L. F., French Patent 763,035 (1934).
39. Perutz, O., *Phot. Ind.*, **1931**, 1131.
40. Roth, C. A., (to Powers Photo Engraving Co.), U. S. Patent 1,496,325 (1924).
41. Schneider, W., (to Agfa Ansco Corp.), U. S. Patent 2,186,734 (1940).
42. Schwartz, A. F. Y., and Mercklin, F. H., British Patent 741 (1889).
43. Sheppard, S. E., and Houck, R. C., (to Eastman Kodak Co.), U. S. Patent 2,059,-817 (1936); (to Kodak Ltd.), British Patent 463,427 (1937).
44. Sheppard, S. E., and Houck, R. C., (Eastman Kodak Co.), U. S. Patent 2,139,774 (1938); (to Kodak Ltd.), British Patent 479,419 (1938).
45. Sheppard, S. E., and Houck, R. C., (to Kodak Ltd.), British Patent 504,378 (1939).
46. Snyder, F. H., and Rimbach, H. W., (to Technics, Inc.), U. S. Patents 2,171,609 (1939) and 2,231,201 (1941).
47. Stewart, V. A., U. S. Patent 1,569,151 (1926).
48. Stewart, V. A., (to Louis A. Solomon), U. S. Patent 1,997,269 (1935).
49. Willcock, R. B., *British J. Phot.*, **86**, 195 (1939).
50. Wood, W. H., (to Harris-Seybold-Potter Co.), U. S. Patent 2,477,323 (1949).
51. Young, R. V., (to Eastman Kodak Co.), U. S. Patent 2,331,755 (1943).

Rubber

Formaldehyde is used in connection with the handling of rubber latex, the production of crude rubber, the modification of rubber for special purposes and the synthesis of rubber accelerators and antioxidants. It may also find applications in the production of synthetic rubber and rubber substitutes.

Rubber Latex. Small quantities of formaldehyde (0.1 to 1 per cent) are sometimes added to rubber latex for the purpose of preventing putrefaction of its protein constituents[41]. Unfortunately, however, formaldehyde-preserved latex thickens after a short time and cannot be shipped. Ammonia, which is commonly used as a latex preservative, is an excellent anti-coagulant but is said to be inferior to formaldehyde in that the rubber obtained from ammonia-preserved latex shows a higher degree of water adsorption. According to McGavack[23], the advantageous properties of both agents can be obtained by adding ammonia to formaldehyde-preserved latex which has been allowed to stand for approximately 24 hours and consequently contains substantially little free formaldehyde. It is reported that latex preserved in this way will withstand shipment and on coagulation yields an improved rubber of low water absorption. The latter property is believed to be a result of the insolubilizing action of formaldehyde on the rubber protein. Rubber obtained from formaldehyde-preserved latex is also reported[24] to have a good resistance to oxidation since it retains a high proportion of the natural antioxidants present in the latex. In addition, it is stated to have a lower viscosity than the usual pale crepe or smoked sheet rubbers, and an improved resistance to abrasion.

In addition to the procedure described above a number of other techniques have been developed for stabilizing formaldehyde-treated rubber latex. Rhines[33] removes carbon dioxide from the treated latex by passing air through it, Linscott[21] adds the sodium salt of an arylsulfonic acid and Bevilacqua[4] describes the addition of methyl cellulose. Rumbold[34] reports that formaldehyde-preserved latex is not readily concentrated by creaming with conventional hydrophilic creaming agents and claims that this can be remedied by adding the formaldehyde condensation product of an aryl sulfonic acid in the form of its alkali metal salts or by use of certain nonionic surface-active agents.

Patents indicate that formaldehyde may also be used as an agent for gelling and coagulating dispersed rubber compositions. Hayes, Madge and Lane[15] obtain a dispersion capable of rapid gelling at ordinary temperatures by adding excess formaldehyde to an ammonia-preserved latex to which sulfur, zinc oxide, accelerators, etc., have been added. It is stated that articles can be prepared from this dispersion by spraying, impregnating or spreading on forms, etc. To obtain the desired results 0.3 to 1.0 per cent formaldehyde in excess of the amount which reacts with ammonia must be added. Gelling is also accomplished by adding phenol prior to the addition of formaldehyde[16]. According to Woodruff[43], increased yields of a rubber-like coagulum may be obtained from latex by addition of a water-soluble gum and dispersed mineral oil, which acts as extenders, and then coagulating with formaldehyde. Riatel[32] produces a self-coagulating vulcanizable bath of latex by addition of an ammoniacal solution of zinc acetate mixed with a water-dispersion of colloidal sulfur containing casein and formaldehyde. Neiley[30] produces rubber articles by dipping forms into an ammonia-stabilized latex composition containing sulfur, antioxidants, accelerators, etc., plus a potential coagulant such as zinc ammonium chloride, and then immersing the coated forms in a solution of formaldehyde, which reacts with the ammonia stabilizer and activates the coagulant. Threads are formed from aqueous rubber dispersions containing glue which have been subjected to a formaldehyde treatment[39]. Ebers produces useful combinations of rubber latex and resorcinol formaldehyde resins[12].

Rubber Derivatives. Formaldehyde probably reacts with rubber in much the same manner that it reacts with other unsaturated hydrocarbons (pages 326–335). In 1923, Kirchhof[19] obtained what he designated as rubber-formolite by reacting pale-crepe rubber dispersed in petroleum ether with sulfuric acid and then heating with formaldehyde. The product after treatment with hot water and ammonia was a yellow-brown powder which decomposed without melting on being heated, swelled in carbon bisulfide and pyridine but did not dissolve in water or related solvents. Approximately 2 g of rubber-formolite were obtained for each gram of crepe rubber. Related processes involving controlled reactions of formaldehyde and rubber under acid conditions yield various forms of modified rubber. According to McGavack[22], a product similar to hard rubber may be obtained by the action of formaldehyde on rubber in the presence of sulfuric acid. A patented adhesive for bonding rubber or rubber-like products to metal, glass or hard rubber may be prepared by the action of formaldehyde or paraformaldehyde on rubber dispersions in ketones such as cyclohexanone, camphor or benzophenone[40]. Products ranging from soft, tacky materials to hard resins are described in a German process[18] involving an acid-catalyzed reaction of formaldehyde with an aqueous dispersion of rubber containing a protective colloid. These products may be vulcanized to produce leather or linoleum-like materials. It is also claimed[11] that paint-retentive surfaces can be obtained by subjecting rubber articles to a superficial hardening treatment with paraformaldehyde and boron fluoride.

Combinations obtained by reacting or vulcanizing rubber with phenol-formaldehyde resins are proving of interest. A detailed study of this subject is reported by Van der Meer[42]. Pophan[31] describes a resin produced by the condensation of formaldehyde with phenol and oxidized rubber.

Accelerators. The synthesis of rubber accelerators by processes involving formaldehyde forms the basis of numerous industrial patents. Hexamethylenetetramine has long been recognized as a mild rubber accelerator most commonly employed in combination with other materials[3, 26, 36]. Its application in the vulcanization of both natural and synthetic rubber was described in 1915 by Hofmann and Gottlob[17] in a patent covering the use of ammonia derivatives showing an alkaline reaction at vulcanization temperatures.

The condensation products of formaldehyde and aromatic amines possess good accelerating powers in many instances. Formaldehyde-aniline and methylene dianiline are early examples of this type. Bastide[2] patented the use of methylene dianiline as a vulcanization agent in France in 1913, attributing its utility to its action as a sulfur solvent. According to Naylor[29], anhydroformaldehyde-*p*-toluidine, $(CH_3C_6H_4N:CH_2)_3$, produced by the action of formaldehyde on a dilute aqueous emulsion of *p*-toluidine, is superior to formaldehyde-aniline and has the advantage of being less poisonous and less likely to cause prevulcanization. Condensation products of formaldehyde and aliphatic amines have also proved useful, as indicated by Bradley and Cadwell[5] in a patent describing the use of trialkyl trimethylenetriamines derived from formaldehyde and primary aliphatic amines and tetralkylmethylenediamines derived from formaldehyde and secondary alkylamines. More recent accelerators of this type include materials, prepared by reacting formaldehyde with polyethylenepolyamines[7, 8] and the compound obtained by treating the methylene bis- derivative prepared from formaldehyde and piperidine with carbon bisulfide[27], and the reaction product of formaldehyde, carbon bisulfide, a primary or secondary aryl amine and certain secondary aliphatic amines[13].

Also of importance in connection with rubber accelerators is the use of formaldehyde as a hardening and modifying agent for other aldehyde-amine accelerators. According to Scott[35], hard brittle products having improved accelerating properties are obtained by the action of formaldehyde on the syrupy and sticky resinous accelerators produced by the condensation of aniline with acetaldehyde, propionaldehyde, etc. The modified accelerators are not only easier to handle but are reported to impart a better tensile strength and elongation at break to the finished rubber than the original aldehyde amines. Sebrell[37] has patented an accelerator produced by treating a condensation product of acetaldol and aniline with formaldehyde.

An entirely different type of accelerator, which has proved of considerable value is obtained by Zimmermann's process[44] involving refluxing commercial formaldehyde with mercaptobenzothiazole. The product is a white crystalline material which melts at 130°C. Although highly active, it is reported to possess the advantageous property of not accelerating vulcanization until the rubber mix is heated so that partial vulcanization on compounding and milling is avoided. Reaction products of secondary amines, such as morpholine and piperidine, with mercaptobenzothiazole and formaldehyde are also claimed to be of value as rubber accelerators[6]. Numerous other formaldehyde derivatives involving mercaptobenzothiazole have been patented. Recent patents by Dean[10] and Harman[14] are illustrative.

Antioxidants. Antioxidants are obtained by reactions involving formaldehyde and amines. According to Semon[38], tetraphenylmethylenediamine, produced by refluxing a mixture of 36 per cent formaldehyde with an excess of diphenylamine in benzene solution, acts as a good antioxidant when employed to the extent of 0.1 to 5 per cent in a rubber composition. An antioxidant is also obtained by the reaction of diphenylamine, formaldehyde, and cyclohexanol[25]:

$$(C_6H_5)_2NH + CH_2O + C_6H_{11}OH \rightarrow (C_6H_5)_2NCH_2OC_6H_{11} + H_2O$$

Other antioxidants are prepared by reacting formaldehyde with amino-acenaphthene[20], amino derivatives of diphenylene oxide[9], and a mixture of alpha-naphthylamine and the amine derived by the action of ammonia on beta-beta-dichloroethyl ether[28].

Synthetic Rubber. Potential applications of formaldehyde in the synthetic-rubber industry may be found in its use as a raw material for the synthesis of dienes (see page 330). Arundale[1] has found that creaming of synthetic latices is improved by adding formaldehyde in conjunction with an ammonium soap emulsifier.

As we have previously pointed out (pages 190–191), rubber-like materials can be obtained by the reaction of formaldehyde with alkali sulfides.

References

1. Arundale, E., (to Standard Oil Dev. Co.), U. S. Patent 2,462,591 (1949).
2. Bastide, J., French Patent 470,833 (1914).
3. Bedford, C. W., (to The Goodyear Tire and Rubber Co.), U. S. Patent 1,380,765 (1921).
4. Bevilacqua, E. M., (to U. S. Rubber Co.), U. S. Patent 2,587,281 (1952).
5. Bradley, C. E., and Cadwell, S. M., (to The Naugatuck Chemical Co.), U. S. Patent 1,444,865 (1923).
6. Bunbury, H. M., Davies, J. S. H., Naunton, W. J. S., and Robinson, R., (to Imperial Chemical Industries, Ltd.), U. S. Patent 1,972,918 (1934).
7. Cadwell, S. M., (to The Naugatuck Chemical Co.), U. S. Patent 1,840,392 (1932).
8. *Ibid.*, U. S. Patent 1,843,443 (1932).
9. Clifford, A. M., (to Wingfoot Corp.), U. S. Patent 2,026,517 (1936).
10. Dean, R. T., (to American Cyanamid Corp.), U. S. Patent 2,385,335 (1945).
11. Dunlop Rubber Co., Ltd., British Patent 523,734 (1939).
12. Ebers, E. S., (to U. S. Rubber Co.), Canadian Patent 435,754 (1946).
13. Hardman, A. F., (to Wingfoot Corp.), U. S. Patent 2,496,941 (1950).
14. Harman, M. W., (to Monsanto Chem. Co.), U. S. Patent 2,470,555 (1949).
15. Hayes, C., Madge, E. W., and Lane, F. H., (to Dunlop Rubber Co.), U. S. Patent 1,887,201 (1932).
16. Hayes, C., Madge, E. W., and Jennings, F. C. (to Dunlop Rubber Co), U. S. Patent 1,890,578 (1932).
17. Hofmann, F., and Gottlob, K., (to Synthetic Patents Co.), U. S. Patents 1,149,580 (1915).
18. I. G. Farbenindustrie A. G., British Patent 486,878 (1939).
19. Kirchhof, F., *Chem. Ztg.*, **47**, 513 (1923).
20. Lauter, W. M., (to The Goodyear Tire and Rubber Co.), U. S. Patent 1,838,058 (1931).
21. Linscott, C. E., (to U. S. Rubber Co.), U. S. Patent 2,534,359 (1950).
22. McGavack, J., U. S. Patent 1,640,363 (1927).

23. McGavack, J. ,(to The Naugatuck Chemical Co.), U. S. Patent 1,872,161 (1932).
24. McGavack, J., and Linscott, C. E., (to United States Rubber Co.), U. S. Patent 2,213,321 (1940).
25. Martin, G. D., (to Monsanto Chemical Co.), U. S. Patent 2,054,483 (1936).
26. Miller, T. W., U. S. Patent 1,551,042 (1925).
27. Moore, W. A., (to The Rubber Service Laboratories Co.), U. S. Patent 1,958,924 (1934).
28. Morton, H. A., U. S. Patent 1,847,974 (1932).
29. Naylor, R. B., (to Fisk Rubber Co.), U. S. Patent 1,418,824 (1922).
30. Neiley, S. B., (to Almy Chemical Co.), U. S. Patent 2,172,400 (1939).
31. Popham, F. J. W., (to The British Rubber Producers' Research Ass'n.), U. S. Patent 2,392,691 (1946).
32. Riatel, M. M., (to Cela Holding S. A.), U. S. Patent 2,120,572 (1938).
33. Rhines, C. E., (to U. S. Rubber Co.), U. S. Patent 2,371,544 (1945).
34. Rumbold, J. S., U. S. Patent 2,534,374-5 (1950).
35. Scott, W., (to E. I. du Pont de Nemours & Co., Inc.), U. S. Patent 1,638,220 (1927).
36. Scott, W., (to The Rubber Service Laboratories Co.), U. S. Patent 1,743,243 (1930).
37. Sebrell, L. B., (to Wingfoot Corp.), U. S. Patent 1,994,732 (1935).
38. Semon, W. L., (to B. F. Goodrich Co.), U. S. Patent 1,890,916 (1932).
39. Shepherd, T. L., U. S. Patent 2,203,701 (1940).
40. Twiss, D. F., and Jones, F. A., (to Dunlop Rubber Co.), British Patent 348,303 (1932).
41. Ultée, A. J., *Arch. Rubbercult.*, I, 405–12 (1917); *C. A.*, **12**, 1010 (1918).
42. Van der Meer, S., *Rec. trav. chim.* **63**, 147–56 (1944).
43. Woodruff, F. O., (to H. H. Beckwith), U. S. Patent 1,929,544 (1933).
44. Zimmerman, M. H., (to The Firestone Tire and Rubber Co.), U. S. Patent 1,960,-197 (1934).

Starch

The ways in which the properties of starch can be modified by treatment with formaldehyde for various applications have been discussed in Chapter 10 (pages 209–210). Reaction on the neutral or alkaline side gives cold-swelling soluble starches, whereas acid catalyzed reaction gives water-insoluble non-swelling products. The former are probably hemiacetals, the latter are methylene ethers. Various procedures can be used to obtain a wide range of properties for specific uses as indicated by the patents previously cited. Neumann[3] inhibits the gellation of starch by employing a hexamethylenetetramine-phenol complex as an additive. Starch cooked with 10 per cent of this composition becomes viscous but does not gel on cooling to room temperature. The use of starch in combination with urea- and melamine-formaldehyde resins in paper coatings has also been mentioned. Hewett[2] points out that starch probably forms a chemical union with these formaldehyde condensation products as well as formaldehyde resins derived from phenol and ketones in adhesive compositions.

Bauer[1] states that the protein components of wheat and rye flours prevent it from dispersing in water as starch does but that this can be corrected by reaction with formaldehyde. Flour treated in this way can be used for the same purposes as starch itself. The process of modification is carried out by blending the flour with somewhat less than one per cent paraformaldehyde and heating 1.0 to 1.5 hrs at 250°F. The same reaction requires 7 days at room temperature.

Starch and flour are also used extensively as extenders of urea-formaldehyde adhesives for hardwood plywood.

References

1. Bauer, J. V., (to Stein, Hall & Co.), U. S. Patent 2,443,290 (1948).
2. Hewett, P. S., Carter, R. E., Church, R. J., *Paper Trade J.*, **124**, (TAPPI Sect.), 45–50 (1947).
3. Neumann, H. T., U. S. Patent 2,417,515 (1947)

Solvents and Plasticizers

Formaldehyde serves as a raw material for the synthesis of polyhydroxy compounds, formals and other methylene derivatives which have value as solvents and plasticizers.

An important industrial use is found in the synthesis of ethylene glycol. This involves reaction at high pressures (over 300 atm.) with carbon monoxide and water in the presence of an acid catalyst to produce hydroxyacetic (glycolic) acid[5, 6]. This acid is then reacted with methanol to give the methyl ester which is converted to the glycol by catalytic hydrogenation[1, 7, 8].

Dioxolane or glycol formal, prepared by the action of formaldehyde on ethylene glycol under acidic conditions, is perhaps the best known of the formal solvents (page 205). Related solvents are also obtained by reacting formaldehyde with diethylene glycol[10], ethylene glycol monomethyl ether[12], 1,3-butylene glycol[4], and other polyhydric alcohols[2, 14]. As was previously pointed out (page 206), glycerol formal is a good solvent for cellulose esters and other plastics. It should be remembered that formals can also be prepared by the reaction of formaldehyde on olefins (page 330).

Stanley[13] employs hexamethylenetetramine as a catalyst for the production of mono-alkyl ethers of ethylene glycol or polyethylene glycol by the reaction of ethylene oxide and an aliphatic alcohol.

Glycol, glycerol, erythritol and higher polyhydroxy compounds can be synthesized by hydrogenating the products of formaldehyde aldolisation[3]. Wallerstein and Alba[15] purify fermentation glycerol by treatment with a small concentration of formaldehyde in the presence of alkali followed by distillation.

Plasticizers can be prepared by reacting the chloromethyl alkyl ethers, prepared by the action of formaldehyde and hydrogen chloride on alcohols (page 212), with the sodium salts of dibasic acids such as sodium phthalate[9]. The alkyl oxymethyl esters prepared in this way are similar to the related ethylene glycol derivatives in their plasticizing action, but are more readily hydrolyzed.

According to Seymour[11] the phthalate ester of diethylene glycol formal is useful as a plasticizer for cellulose derivatives. This formal is prepared by condensing two mols of diethylene glycol with one mol of paraformaldehyde in the presence of sulfuric acid.

References

1. Cockerill, R. F., (to E. I. du Pont de Nemours & Co., Inc.), U. S. Patent 2,316,564 (1943).
2. Dreyfuss, C., French Patent 745,525 (1933); *C. A.*, **27,** 4238 (1933).
3. Hanford, W. E., and Schreiber, R. S., (to E. I. du Pont de Nemours & Co., Inc.), U. S. Patent 2,224,910 (1940).
4. I. G. Farbenindustrie, A. G., French Patent 828,417 (1938); *C. A.*, **33,** 174 (1939).
5. Larson, A. T., (to E. I. du Pont de Nemours & Co., Inc.), U. S. Patent 2,153,064 (1939).
. Loder D. J., (to E. I. du Pont de Nemours & Co., Inc.), U. S. Patent 2,152,852 (1939).
7. *Ibid.*, U. S. Patent 2,285,448 (1942).
8. *Ibid.*, U. S. Patent 2,331,094 (1943).
9. Nicholl, L., (to Kay-Fries Chemicals, Inc.), U. S. Patents 1,984,982–3 (1934).
10. Seymour, G. W., (to Celanese Corp. of America), U. S. Patent 2,031,619 (1936).
11. Seymour, G. W., (to C. Dreyfuss), Canadian Patent 368,772 (1937).
12. Seymour, G. W., and Baggett, J. L., (to C. Dreyfuss), Canadian Patent 390,733 (1940).
13. Stanley, H. M., (to Philip Eaglesfield Distillers Co., Ltd.), French Patent 947,250 (1949).
14. Soc. Nobel Francoise, British Patent 481,951 (1938).
15. Wallerstein, J. S., and Alba, R. T., (to The Overly Biochemical Research Foundation, Inc.), U. S. Patent 2,366,990 (1945).

Surface-active Agents

Preparation of Surface-active Agents. In many instances, the utility of formaldehyde in the preparation of surface-active compounds is based on the fact that it can be used to introduce solubilizing groups into water-insoluble compounds containing long aliphatic hydrocarbon radicals. In one process, wetting agents are made by reacting formaldehyde with an olefin or an unsaturated halogen derivative of an olefin in the presence of sulfuric acid or an acid sulfate and solubilizing the primary reaction products by sulfonation when this is necessary[8]. Compounds that have wetting, cleansing, foaming, and emulsifying properties are produced by reacting aliphatic amides containing at least 8 carbon atoms with paraformaldehyde and sulfur dioxide in the presence of tertiary amines[1]. Useful surface-active agents of this type are also prepared by heating a fatty amide, such as oleic acid amide, with the sodium bisulfite compound of formaldehyde using a secondary aliphatic amine as a catalyst[12].

$$C_{17}H_{33}CONH_2 + HOCH_2SO_3Na \rightarrow C_{17}H_{33}CONHCH_2SO_3Na + H_2O$$

Other surface-active agents are synthesized by reaction of thiourea with chloromethyl amides produced by the joint action of formaldehyde and hydrogen chloride on amides such as stearamide[10].

Surface-active agents are also synthesized by using formaldehyde to produce products of increased molecular weight by linking molecules which already contain solubilizing polar groups. A wetting agent for use in separating minerals is obtained by condensing sulfonated naphthalene with formaldehyde[7]. The condensation product of beta-naphthalene sulfonic acid and formaldehyde is a useful non-foaming dispersant[11, 15]. Polymeric detergents are obtained by reacting a phenol-formaldehyde condensate with an alkylene oxide and esterifying the alcohol thus obtained with phosphoric acid[2].

Frothing agents are prepared by the reaction of 2,4-toluylenediamine and formaldehyde[13] or nonaromatic secondary amines, formaldehyde, and phenol[4, 5, 6]. Reaction products of naphthenic acid soaps and formaldehyde are claimed to be of special value for breaking petroleum emulsions[9, 14]. These emulsions can also be broken with the mineral acid salt of the complex amines produced by the joint reaction of a phenol, formaldehyde and a nonaromatic secondary amine[3].

References

1. Balle, G., Rosenbach, J., and Ditters, G., (to I. G. Farbenindustrie A. G.), U. S. Patent 2,210,442 (1940).
2. Bock, L., and Rainey, J. L., (to Röhm & Haas Co.), U. S. Patent 2,454,542 (1948).
3. Bond, D. C., and Savoy, M., (to Pure Oil Co.), U. S. Patent 2,457,634 (1948).
4. Bruson, H. A., (to The Resinous Products and Chemical Co.), U. S. Patent 1,952,008 (1934).
5. Bruson, H. A., (to Röhm & Haas Co.), U. S. Patent 2,033,092 (1936).
6. *Ibid.*, U. S. Patent 2,036,916 (1936).
7. E. I. du Pont de Nemours & Co., Inc., British Patents 501,655–6 (1939).
8. Haussman, H., and Dimroth, H., (to I. G. Farbenindustrie A. G.), German Patent 672,370 (1939); *C. A.*, **33**, 3928, (1939).
9. Hershman, P. R., (to Petro Chemical Co.), U. S. Patent 2,153,560 (1939).
10. I. G. Farbenindustrie A. G., British Patent 507,207 (1939); *C. A.*, **34**, 551 (1940).
11. Kalber W. A., (to Dewey and Almy Chem. Corp.), U. S. Patent 2,056,924 (1936).
12. Mack, L., (to General Aniline & Film. Corp.), U. S. Patent 2,366,415 (1945).
13. Pollak, E., British Patent 519,710 (1940); *J. Soc. Chem. Ind.* **59**, 515 (1940).
14. Suthard, J. G., U. S. Patent 2,206,062 (1940).
15. Tucker, G. R., (to Dewey and Almy Chem. Corp.), U. S. Patent 2,046,757 (1936).

Textiles

In the textile industry, formaldehyde is employed for the production of fabrics which are crease-resistant, shrink-proof, possess modified dyeing characteristics, etc. For these purposes, it is used alone, in the presence of other agents, and in the form of simple, reactive formaldehyde derivatives. In the case of "artificial wool" or synthetic protein fibers, formaldehyde treatment is an integral part of the process of fiber production. Formaldehyde is also used as a disinfectant for preventing the deleterious action of mold and mildew on textiles. Less direct applications are found in the use of resins and other products synthesized with formaldehyde which serve as sizing agents, textile assistants, and agents for the production of water-repellency.

Complete coverage of the immense volume of chemical literature patents and trade publications dealing with the utilization of formaldehyde and its derivatives is beyond the scope of this book. The objective of the following discussion is merely to review outstanding types of application and the references cited are illustrative rather than exhaustive.

Treatment of Cellulosic Fabrics. The chemical reactions of formaldehyde with cellulose under both alkaline and acidic conditions have been previously described (pages 210–212). In the presence of alkalies, hemiformals are probably produced but stable condensation products are not formed; in the presence of acids, water is split out with the production of methylene ethers. Modification of cellulosic fibers by formaldehyde treatment is apparently dependent on chemical reaction, and permanent alteration of fiber properties is chiefly affected under acidic conditions. The nature of the modification is determined by the degree of reaction and is controlled by the amount of moisture present during the treating process, the presence or absence of dehydrating agents, the nature and concentration of acidic catalysts, and the temperature employed. According to Wood[105] maximum combination of formaldehyde and cellulose is approximately 17.2 per cent but this degree of union is neither possible nor desirable in ordinary treating processes in which the extent of combination normally lies in the range of 0.2 to 3.0 per cent[42]. Although fiber treatment is usually carried out with aqueous formaldehyde solutions, Minaev and Frolov[71] report that modification may also be obtained with formaldehyde vapors or with a solution of formaldehyde in acetone.

Indirect processes in which textiles are treated with reactive formaldehyde derivatives are controlled by the conditions which cause these derivatives to take part in chemical reactions and naturally vary with the nature of derivative involved.

Commercial interest in the modification of cellulosic fabrics by treatment with formaldehyde was first aroused by the observation of Eschalier[29] in 1906 that the wet strength of regenerated cellulose or rayon fibers could be improved by an acid formaldehyde treatment. This was called "sthenosage" from the Greek word "sthenos" meaning strength. Interest at that time was based on the fact that early fabrics of this type were notoriously low in wet strength. As this property has since been improved by developments in rayon manufacture, the process is no longer of practical importance. Today the principal utility of formaldehyde treatment of cellulosic textiles is based on improvement in crease-resistance, shrink-resistance, elasticity, feel, and dyeing properties. In the case of artificial velvets, crush-resistance is of particular importance.

Crease- and Crushproofing. Perhaps one of the largest of the present uses for formaldehyde in the textile industry is in the production of cellulosic fabrics and fibers possessing various degrees of resistance to creasing and crushing. Specific procedures and processes for procuring these effects covering many methods of direct and indirect treatments have been patented[4, 21, 34, 44, 46, 57, 79, 80, 81, 90, 103, 17, 75]. When a direct treatment is employed, these procedures usually follow a general pattern of the sort indicated below:

(1) The fibers are impregnated with a formaldehyde solution containing an acidic catalyst and perhaps a modifying agent.
(2) Excess solution is squeezed out or removed by centrifuging.
(3) The moist fibers are dried at a moderate temperature (below 100°C).
(4) The dried fibers are given a curing or baking treatment at an elevated temperature not exceeding approximately 170°C.
(5) Catalyst and unreacted reagents are removed by washing.

Steps 3 and 4 are often combined so that the fabric is dried and then cured at the same treating temperature. The heating process may involve carrying the fabric over heated rolls, applying a heated iron, use of a heated oven, passing through a hot nonaqueous fluid, etc. With indirect treatments involving formaldehyde derivatives, the pattern of treatment is often identical to that shown above. However, in such treatments, alkaline catalysts may sometimes be employed and in some instances the use of a catalyst may not be required.

Unless creaseproofing processes involving direct treatment with free formaldehyde are carefully controlled, there is danger of excessive tendering, which will be manifested by loss of tensile strength and reduction of wearing properties. Whether this effect is brought about solely by the action of the acid catalyst or is partially caused by the action of the formaldehyde itself is uncertain. However, according to Hall[43], rayon and cotton fabrics are now being subjected commercially to acid-formaldehyde processes indicating that any appreciable degree of tendering may be avoided by proper control. In general, cotton is more susceptible to tendering than regenerated cellulose and is seldom subjected to direct methods of treatment. Indirect treatments involving urea- and melamine-formaldehyde compositions are reported to result in less tendering. Less acid is required in these processes and it is also believed that the resins formed or employed in the treating solutions protect the fiber from the action of the acid.

As previously stated, the first step in processes for the direct treatment of cellulosic fibers or fabrics consists in impregnating the fibers with a solution containing formaldehyde, an acidic catalyst, and in some cases an addition agent. Formaldehyde concentrations may vary from approximately 1 to 40 per cent. Catalyst concentrations range from a few hundredths of a per cent to several per cent, depending upon the catalyst, which may be a mineral acid, an oxidizing acid, an organic acid, or a salt having an acidic reaction. Hydrochloric and sulfuric acids are used in many instances. One English process[60] recommends small concentrations of nitric, perchloric, and sulfuric acid. It has also been claimed[7] that formic, acetic, citric, tartaric, adipic, phthalic, and aromatic sulfonic acids may be employed. Ammonium chloride[46] and aluminum thiocyanate[44] may be cited as acidic salts used as catalysts. Addition agents are used in some cases to modify, moderate, or enhance the effect of the formaldehyde on the fabric or to prevent tendering. Hydroxy compounds, alkanol-amines, amine salts, amides, inorganic neutral salts, etc., are employed. The impregnation period may vary from a few minutes to 24 hours and the temperature of the treating bath may be in the range of 10 to 100°C or even higher. In some cases addition agents may be applied in a separate bath before or after the formaldehyde treatment. Following the drying process, the curing or baking treatment may vary both in temperature and time. Pressure may also be used. Roberts and Watkins[82] cure the treated and dried fabric for 30 minutes at a pressure of 30 to 50 psi.

Indirect methods of treatment usually involve treating the fabric with a solution containing (a) formaldehyde and a compound capable of forming a resin with formaldehyde, (b) a simple methylol derivative of formaldehyde and a resin-forming material, or (c) a soluble formaldehyde resin. Processes of this type employing various combinations of formaldehyde with urea and with phenols were developed in England by Foulds, Marsh, Wood and others[37] about 1925. A typical procedure described in the patent literature consists in impregnating a cotton fabric with an aqueous solution containing formaldehyde, urea, and boric acid, squeezing out excess solution, and drying at 130°C. An intermediate condensation product of phenol and formaldehyde may be applied in an aqueous or alcoholic solution containing caustic soda, dried, and then heated to 180°C. In another procedure[3] crease-proofing is obtained by saturating a textile fabric with a solution of 28 g of monomethylol urea in 75 cc water to which one gram of tertiary ammonium phosphate has been added. The wet fabric is then dried at a low temperature and cured by heating for two minutes at 120°C.

Melamine has proved of value in creaseproofing processes of the indirect type because the primary formaldehyde condensation product is readily resinified without the use of an acid catalyst[41, 51, 89, 17].

Aqueous glyoxal may be used in conjunction with an amino-formaldehyde thermosetting resin and an acid catalyst to prevent wrinkling, crushing, creasing and laundry shrinkage[74]. Urea- or melamine-formaldehyde resins are employed in a process of this type and the impregnated cloth is cured at a temperature in the range 212 to 350°F.

Gruntfest and Gagliardi[40] point out that formaldehyde, glyoxal and selected resins which can penetrate the textile fibers augment their recovery from deformation and improve stiffness, whereas resins which are deposited on the surface of the fiber reduce its resistance to creasing and impair its multifilament characteristics.

According to a German patent, crease-proofing can also be effected with polyvinyl alcohol which forms a water-insoluble resin with acid formaldehyde. A solution containing 50 parts polyvinyl alcohol, 300 to 500 parts of 35 per cent formaldehyde solution, 20 parts of 10 per cent hydrochloric acid and 2000 parts water is stated to produce crease-proofing when the impregnated cloth is dried and baked at 110°C[50].

Reduction of Water Sensitivity: Shrinkproofing. Although, under some circumstances direct formaldehyde treatment can be employed to increase the affinity of cellulose for water, it is more commonly used to produce the opposite effect, which is a characteristic sequel of treatments completed in the absence of water. Useful properties resulting from reduction of water sensitivity are: shrink-resistance, reduction of swelling due to moisture and other swelling agents, freedom from water-spotting, and wash-resistance of mechanical effects such as embossing. In general, the methods of treatment used to achieve these effects are similar to those employed for creaseproofing but are less severe in nature. In a few cases, an alkaline catalyst is employed[56]. How alkalies function in such cases is not known. However, since alkali catalyzes the Cannizzaro reaction in which formaldehyde is converted to formic acid and methanol, it is possible that the formaldehyde may become acidic during treatment.

Hall[43] states that the usual preshrinking processes are less successful on certain types of viscose fabrics and that acid-formaldehyde treatments offer an escape from this difficulty. He also observes that the amount of formaldehyde entering into combination must be limited or the fabrics may become highly crease-resistant and resist pressing into pleats, etc.

The use of formaldehyde derivatives, resin-formers and water-soluble resins in shrinkproofing is illustrated by processes employing hexamethylenetetramine[101], alkylated methylolmelamines[64] and urea-formaldehyde resins plus glyoxal[74].

Water Repellency. Early processes for producing water repellency in cellulosic textiles involved impregnation with nonaqueous solutions of hydrophobic materials (e.g., metallic soaps). These treatments weight the fabric, change its appearance and feel, and are removed by dry-cleaning. More recent processes make use of formaldehyde derivatives which chemically or physically attach hydrophobic molecules to the cellulosic fibers. In some cases, the treating solution may contain free formaldehyde which reacts in processing. In general, processes for producing water repellency are similar to those for shrink-proofing or creaseproofing involving impregnation followed by drying and curing. These processes yield water-repellent textiles which show little if any differences in feel, appearance and weight from the original fabric and which are not readily altered by washing and dry-cleaning.

In general, the formaldehyde derivatives employed in the production of water-repellent textiles are methylol or methylene derivatives of polar compounds containing long chain aliphatic radicals. Illustrative of the various chemical types are: quaternary salts derived from tertiary amines and alpha-chloromethyl ethers of fatty alcohols[97], the hexamethylenetetramine derivative of stearoyl chloride[59], amine derivatives of reaction products of fatty acid amides, formaldehyde and hydrogen chloride[104], products obtained on reacting phosphorus trichloride with compounds prepared by treating fatty acid nitriles with anhydrous formaldehyde in the presence of strong acids[77], methylol derivatives of substituted phenols[100], and a tetradecylphenol-formaldehyde resin emulsion[25].

A so-called "splash proofing" of fabrics may be accomplished by treating textiles with gelatin or water-soluble resins and formaldehyde which insolubilizes the gelatin or resin on the fabric. A lubricating agent such as a sulfonated alcohol or oil is usually present in the treating solution. Only a fine film of resin is necessary so that dilute solutions are employed. As usual, the insolubilization takes place when the fabric is heated[78]. Glue and formaldehyde are also employed for this purpose[63].

Increase of Water Affinity. The water imbibition of cellulosic fabrics is reported to be greatly increased without increasing its resistance to creasing by an acid formaldehyde treatment in which the unreacted formaldehyde and acid catalyst are removed before drying the fabric[43]. Results of this type are claimed for a process[72] in which a cellulosic yarn is immersed in 30 times its volume of 40 per cent formaldehyde acidified with enough sulfuric acid to produce a normal acid solution and maintained at a temperature of 80°C for ten minutes. After this, the fabric is washed in cold and then hot water until all the formaldehyde and acid are removed. It is finally dried at 105°C for 30 minutes.

Effects of this type, probably result from a swelling action of formaldehyde solution which is made permanent by a minimal amount of methylenic cross-linking.

Improvements Relating to Dyeing and Dyed Fabrics. As previously pointed out, the affinity of cotton for direct dyes can be slightly increased by treating with acid formaldehyde in the presence of water[72]. Improved dye penetration and fastness are claimed to result when cellulose acetate fabrics are treated with strong aqueous formaldehyde containing 10 to 30 per cent by volume of a monatomic aliphatic alcohol at 50°C and then washed free of the treating solution[6]. It is stated that the effects produced by this treatment also result from swelling of the fiber which is retained after treatment. The reduced affinity of cellulose for direct dyes after dry-heating with formaldehyde in the presence of an acid is employed for the production of two-toned effects in dyed fabrics. According to Bowen and co-workers[10], this may be accomplished by printing the fabric with a solution containing 6 per cent formaldehyde and 0.6 per cent ammonium thiocyanate, drying at 70°C, baking for 5 minutes at 150°C, and then dyeing with a direct dye. The treated portions of the fabric are only slightly colored.

Goldthwait[38] obtains methylenated cotton, containing only 0.5 to 1.5 per cent combined formaldehyde, that is almost completely dye resistant, by reaction with formaldehyde in nonaqueous solution (e.g., acetone) in the presence of an acid catalyst. This process appears unique in that the treated fabric is not seriously embrittled or weakened in tensile strength and is resistant to biological rotting.

Processes involving formaldehyde and nitrogenous compounds are often of special value in improving the dyeing characteristics of cellulosic textiles. A patented process of this type involves treating a regenerated cellulose fabric with a solution containing formaldehyde, thiourea, and potassium tetroxalate, drying, baking at 140 to 150°C, washing to remove acid, and then boiling for 5 minutes with a 4 per cent solution of cyanamide[9]. Other processes for improving the dye affinity of cellulosic textiles include treating with a water solution of formaldehyde and guanidine salts and then heating the treated fabric[65]. Basic radicals can also be combined with cellulose by reaction with a formaldehyde derivative of a quaternary pyridine-type base[85].

Treatment with formaldehyde-urea condensates before or after dyeing is said to enhance the affinity of the fabric for dyes and improve fastness. According to Schneevoigt and Nowak[86], local pattern effects can be obtained by printing treatments involving a methylol urea. Cellulose fibers impregnated with a solution containing formaldehyde, a long-chain aliphatic amine, and a tetralkyl ammonium base are also reported to acquire a good affinity for wool dyes[58].

Formaldehyde likewise has applications in dyeing with naphthol colors. Schroy[89] claims that the addition of small amounts of formaldehyde as a stabilizing agent to a naphtholate bath greatly improves the wash fastness and insures a uniform affinity of the naphthol dye for cotton and vegetable fibers.

Improvements in the fastness of some direct dyes to washing and perspiration can be obtained by treatment of the dyed fabric with formaldehyde. Direct colors whose molecules contain two or more hydroxyl groups, one or more amine groups, and one or more azo (—N:N—) linkages are highly soluble in water and tend to bleed easily in water or alkaline solution. It is this type of dye which is claimed to be particularly susceptible to improvement by an after treatment of the fabric with formaldehyde[92]. It is probable that formaldehyde acts as a color stabilizer by reacting with the hydroxyl or amino groups of the dye molecules to form methylene derivatives of low solubility[78]. There is also the possibility that formaldehyde may actually couple the cellulose to the dye. According to Geigy[53] cotton dyed with an azo dye should be immersed after washing in a solution of 0.5 kg of 40 per cent formaldehyde in 200 liters water and allowed to stand for one-half hour at room temperature. The formaldehyde treatment may also be carried out at elevated temperatures in slightly acid media or in alkaline solution. Another method[32] employs a warm bath containing 3 to 35 per cent formaldehyde and 1 to 2 per cent bichromate in which the dyed fabric is soaked for 20 to 30 minutes, washed, and dried[63]. Quinn[78] states that the insolubilizing action of formaldehyde on the amino groups of special direct colors appears to give better wash fastness on filament and spun viscose rayon than on cotton and mercerized cotton. The fastness of dyed cellulose fibers to water, acid boil, and wet ironing is also reported to be improved by treatment with condensation products of a polyethylene polyamine, *e.g.*, diethylenetriamine, formaldehyde, and a ketone such as acetone[96]. The quaternary addition product of hexamethylenetetramine and ethyl chloride can be used to improve dye fastness by application in aqueous solution with or without formaldehyde followed by drying and curing at an elevated temperature[52].

Flameproofing. Flameproofing treatments generally involve using basic noncombustible amino-plast resins in combination with inorganic phosphorus and sulfur containing acids or other compounds known to have flame-proofing or fire-retardant characteristics as textile modifiers. The general method of application usually involves impregnation followed by curing at an elevated temperature.

Representative patents describe: (1) the application of treating solutions made up of ortho-phosphoric acid, urea, ammonia and formaldehyde[36]; (2) aqueous melamine-formaldehyde resin solutions applied to cloth previously impregnated with pyrophosphoric acid[76]; (3) application of protein dispersions containing hexamethylenetetramine, ammonia, sulfuric acid, and phosphoric acid[48]; and (4) combinations of fire retardants, such as chlorinated naphthalene, with amino-plast resins[35]. The amino-plast resins are said to act principally as glow-retardants.

Miscellaneous Effects. A large number of special effects other than those which have already been mentioned can be obtained by formaldehyde treatment of a wide variety of cellulosic fibers including esters, ethers, and other special derivatives. The following examples based on the claims of patent processes are illustrative:

Fabrics containing warp and highly twisted weft yarns of an organic cellulose derivative are treated with, *e.g.*, 1 per cent, formaldehyde to harden the warp yarns[31].

The elasticity of cellulose acetate fibers is increased by treatment with thiourea, guanidine or dicyandiamide, and formaldehyde or hexamethylenetetramine[27]. Cellulose acetate fabrics are treated with formaldehyde-urea resins to increase their resiliency[26].

Cellulose acetate fibers are curled by treatment with a hot aqueous solution containing hexamethylenetetramine and urea[28].

Formaldehyde derivatives are often employed to render various finishes more lasting. It is claimed, for example, that: (a) starch finishes may be improved by applying a 1 per cent methoxymethylpyridinium chloride solution to the finished fabric, drying and baking[30], (b) flame-proofing finishes may be rendered more permanent by applying in a solution containing gelatin and finally treating with formaldehyde[61].

A process for improving the strength of cotton and other vegetable fiber cords is carried out by treating with a hexamethylenetetramine solution containing urea, sulfonated esters of dicarboxylic acids and ammonium phosphate, $(NH_4)_2HPO_4$, drying under tension and then curing in a relaxed condition[47].

Wool-like rayon and cotton fibers are obtained by treating with an amino-plast solution in a twisted condition and then detwisting after curing[45].

A regenerated cellulosic fabric, which is resistant to mustard gas penetration, is produced by coating the fabric with polyvinyl alcohol and formalizing this film by treating with dilute formaldehyde and an inorganic acid in a non-solvent mixture of acetone, dioxane or dioxolane and water[83].

Treatment of Protein Fibers. Reaction with formaldehyde decreases the sensitivity of proteins to water and other solvents and increases their resistance to chemical reagents. The nature of the reactions involved has already been discussed (pages 312–317). These same reactions are responsible for the modifying action of formaldehyde and reactive formaldehyde derivatives on protein fibers and related fibers containing amido and amino radicals.

Wool. The improved resistance of formaldehyde treated wool to the action of boiling water, alkalies, alkali sulfides, etc., may be utilized to protect wool in processes such as bleaching, dyeing, washing, etc., which involve exposure to these agents. In a process patented by Kann[55] in 1905, it is stated that satisfactory resistance of woolen fibers can be developed by treatment with formaldehyde vapors or dilute neutral or weakly acid solutions containing as little as 0.03 per cent formaldehyde. Good results are reported when a hot 4 per cent formaldehyde solution is employed.

Since formaldehyde treatment also increases the resistance of wool to certain dyes, pattern effects may be secured by making use of this fact in the dyeing process. Both formaldehyde and condensation products of formaldehyde with cresol-sulfonic acid may be employed for this purpose according to a German patent[5]. Dyed woolen fabrics have been claimed to show greater fastness and stability after treatment with formaldehyde alone or in combination with tannic acid[16].

Formaldehyde and its derivatives are employed in several shrink-proofing processes for wool. A British process[15] claims that this effect can be secured by treating the wool with a solution containing formaldehyde and an acid at a pH of not more than 2, drying, baking at a temperature sufficient to complete reaction without tendering, washing, and drying. Johnstone and van Loo state that felting and shrinking can be effectively reduced by treatment with a substantially unpolymerized dispersion of an alkylated methylol melamine in water so that 2.5 to 15 per cent by weight is deposited on the wool which is then dried and heated to 200 to 300°F[54]. Another process[69] recommends treating wool with a water solution containing formaldehyde, boric acid, urea, and glycerol. It is also claimed that the shrinkage of wool can be reduced by treatment with special formaldehyde derivatives. Dichloromethyl ether which is produced from formaldehyde and hydrogen chloride is reported to effect a substantial reduction in wool shrinkage if applied to the fabric in an inert nonaqueous solvent such as Stoddard solvent or trichloroethylene[98].

Yakima and Shivrina[106] report that woolen textiles may be rendered resistant to bacterial degradation by soaking in aqueous formaldehyde, washing and drying. Wool treated in this way stands up well on use in paper machines. Resistance of treated samples was measured by treating with 0.4 per cent solution of trypsin in a solution buffered to pH 8.3–8.5. Cloth treated with 12 to 15 per cent formaldehyde for 24 hrs gave complete resistance for 10 days; cloth treated with 5 per cent potassium dichromate before and after formaldehyde treatment showed complete resistance for 18 days and was not completely destroyed in 39 days. Humfeld, Elmquist and Kettering[49] state that wool sterilized with formaldehyde retains a sufficient amount of this agent to render it resistant to the action of bacteria.

According to a patent of J. W. Brown[14], it is claimed that wool can be rendered moth-proof by treating with an aqueous solution containing urea, formaldehyde and glycerin. The wet fabric is oven-dried and finally subjected to a scouring treatment. The treating solution may be acidified with boric, acetic or formic acid or with sodium bisulfite.

A. E. Brown and M. Harris[13] report that wool can be made more resistant to chemical agents, such as alkalies, and more elastic by reduction of its cystine disulfide cross-linkages followed by simultaneous introduction of a more stable methylene linkage. This is done by subjection to a reducing agent, such as sodium hydrosulfite or sodium formaldehyde sulfoxylate, in the presence of formaldehyde.

Artificial Fibers from Natural Proteins. Practically all artificial protein fibers or "synthetic wools" are given a stabilizing treatment with formaldehyde in the course of manufacture[2]. This procedure confers water-resistance, strength and protection from biological attack. At present, the majority of these fibers is produced from casein and zein. Soybean and peanut proteins are also of commercial interest in this connection and considerable research has also been carried out with other proteins. In general, protein fibers have many of the characteristics of natural wool but are somewhat lacking in strength. As a result they are usually blended with wool, rayon or other fibers for commercial use. A detailed review of the reactions which take place during the critical stabilizing or hardening process has been published by Traill[94]. Although these reactions undoubtedly involve some form of cross-linking, this investigator concludes that we still have no definite proof of what actually takes place.

Although the production of "synthetic wool" from casein has been carried out on a commercial scale for only a few years, the preparation of these fibers dates back about half a century. In 1898 Millar[70] patented a process for spinning casein from an acetic acid solution, and insolubilizing the fiber in a bath containing formaldehyde and aluminum salts. The present practice of spinning casein from an alkaline solution appears to have been first proposed by Todtenhaupt in 1904[93]. Partial hydrolysis of the protein in the alkaline solution is believed to break cross-linkages in the protein molecule giving it a more linear structure, better suited to fiber production. In general, casein fiber processes involve forcing an alkaline dispersion of the protein through spinnerettes into an acid coagulating bath, hardening with formaldehyde and finally washing and drying the fibers thus produced[8, 68]. The acid coagulating bath may contain formaldehyde so that initial coagulation and formaldehyde hardening are accomplished in one step. However, the more recent practice appears to favor coagulating in an acid bath containing such inorganic salts as aluminum sulfate, zinc chloride, sodium sulfate, sodium chloride, etc., and hardening with acid formaldehyde in a later step. Recent procedures also tend to subject the fibers to a stretching process before final hardening. Laboratory studies of fiber production involving variations of this type are described by Peterson and co-workers[73]. Various procedures described in patents include numerous variations in processing technique[33, 62]. In some instances plasticizers have been added to the casein dispersion[102]. Dry spinning from an alkaline casein mass and hardening the fibers with gaseous formaldehyde in the alkaline state is described by Signer[91].

Partial acetylation of casein fibers is reported to improve resistance to boiling solutions, as under dye-bath conditions, and causes dyeing properties to approximate those of natural wool[12].

Processes for producing fibers from soybean protein[11], peanut protein[95] and zein[19, 20, 22, 23] are similar to those described for casein in the type of processing involved. In the case of zein, some formaldehyde may be added to the protein dispersion before spinning since it does not gel the alkaline dope. Croston[24] has recently shown that superior zein fibers can be obtained by curing the fiber under substantially anhydrous conditions. In this process, the fiber is passed through a curing mixture containing formaldehyde and a strong acid in an inert solvent, such as dioxane, toluene and tetrachlorethane at a temperature of around 100 to 110°C. Sulfuric and hydrochloric acid are satisfactory catalysts. This cure is reported to produce irreversible formaldehyde cross-links so that the fiber shows reduced shrinkage and unusual strength after boiling in acid dye baths.

It has been reported that other protein materials may also be used as the basis for fiber preparations. These include castor bean globulin[1] and keratin from hoofs, horn, hair, etc.

Nylon. Since nylon is a synthetic polyamide, nylon fibers may be classified as protein types. However, the homogeneity of their chemical structure and freedom from amino groups cause them to differ in many respects from natural or reconstituted protein fibers. As previously pointed out (pages 305–306), reaction with formaldehyde in the presence of an acid catalyst leads to the formation of methylol groups and methylene cross-linkages. Joint reactions with alcohols and formaldehyde result in the introduction of alkoxymethyl groups. These reactions can be employed to modify preformed fabrics to obtain useful variations in properties.

McCreath[67] obtains nylon filaments, which cannot be cold drawn to over 75 per cent of their original length, by treating an unoriented or undrawn fiber with aqueous formaldehyde in the presence of an acid catalyst and baking at 100 to 150°C. Walz[99]uses a similar treatment to raise the melting point and lower the stretch of nylon thread. Graham[39] produces nylon filaments with a wool-like crimp by heating relaxed filaments of formaldehyde-treated oriented filaments at 10 to 50°C below their melting point until shrinkage has taken place to the extent of at least 50 per cent. The liveliness, crease-resistance, solvent-resistance, dye-receptivity and melting point of a heat-set nylon are also increased by a formaldehyde treatment. Claymont and Schupp[18] carry out this process by heating an acid treated fabric in anhydrous formaldehyde gas while holding it at fixed dimensions to prevent shrinkage.

Introduction of methylol and alkoxymethyl groups by reacting nylon yarn with an alcohol and formaldehyde in the presence of an acid catalyst also produces useful modifications. Schneider[87] obtains a stronger, higher melting, more resilient fiber with improved dye receptivity by treating in this way so that not over 20 per cent of the —CONH— groups are substituted, cold drawing and cross-linking by an acid baking process. When 35 to 55 per cent of the —CONH— groups are substituted by methylol or alkoxymethyl groups, an elastic fiber is obtained[88]. Cold drawn fibers do not become elastic when subjected to the same process so that cloth with a one-way stretch may be obtained by treating a fabric with a filling of unoriented fibers and a warp of oriented fibers.

The affinity of nylon for direct dyeing can be increased by treatment with amino-plast resins[66]. This process can be carried out by impregnating the fabric with an aqueous solution containing formaldehyde and cyanamide or a cyanamide derivative and then baking.

References

1. Astbury, W. T., Bailey, K., and Chibnall, A. C., British Patent 467,704 (1937).
2. Atwood, F. C., *Ind. Eng. Chem.*, **32**, 1547–9 (1940).
3. Battye, A. E., Marsh, J. T., Tankard, J., Watson, W. H., and Wood, F. C., (to Tootal Broadhurst Lee Co. Ltd.), U. S. Patent 2,088,227 (1937).
4. Battye, A. E., Candlin, E. J., Tankard, J., Corteen, H., and Wood, F. C., (to Tootal Broadhurst Lee Co., Ltd.), British Patent 506,721 (1939).
5. Bayer & Co., German Patents 337,887 and 340,455 (1921).
6. Beck, K., U. S. Patent 1,947,928 (1934).
7. Böhme Fettchemie G. m. b. H., French Patent 805,504 (1936).
8. Borghetty, H. C., *Am. Dyestuff Reptr.*, **25**, 538–9 (1936).
9. Boulton, J., and Morton, T. H., (to Courtaulds Ltd.), U. S. Patent 2,234,889 (1941).
10. Bowen, H. H., Majerus, V. H., and Kellett, S., British Patent 452,149 (1936).
11. Boyer, R. A., *Ind. Eng. Chem.*, **32**, 1549–51 (1940).
12. Brown, A. E., Gordon, W. G., Gall, E. C., and Jackson, R. W., *Ind. Eng. Chem.*, **36**, 1171–5 (1944).
13. Brown, A. E., and Harris, M., *Ind. Eng. Chem.*, **40**, 317 (1948); (to Harris Research Laboratories), U. S. Patent 2,508,713 (1950).

14. Brown, J. W., British Patent 475,422 (1937); French Patent 807,021 (1937).
15. Calico Printers' Assoc. Ltd., British Patent 519,361; *J. Soc. Chem. Ind.*, **59**, 439 (1940).
16. Casella & Co., German Patent 303,223 (1919).
17. Caspé, J., (to American Cyanamid Co.), U. S. Patent 2,501,435 (1950).
18. Claymont, B. G., and Schupp, O. E., Jr., (to E. I. du Pont de Nemours & Co., Inc.), U. S. Patent 2,540,726 (1951).
19. Cline, E. T., (to E. I. du Pont de Nemours & Co., Inc.), U. S. Patent 2,475,879 (1949).
20. Cline, E. T., and Biehn, G. F., (to E. I. du Pont de Nemours & Co., Inc.), U. S. Patent 2,429,214 (1947).
21. Corteen, H., *et al.*, (to Tootal Broadhurst Lee Co., Ltd.), U. S. Patent 2,158,494 (1939).
22. Croston, C. B., and Evans, C. D., (to U. S. Of America as represented by the Sec. of Agriculture), U. S. Patent 2,478,248 (1949).
23. Croston, C. B., and Evans, C. D., *Textile Research J.*, **19**, 202–11 (1949).
24. Croston, C. B., *Ind. Eng. Chem.*, **42**, 482–4 (1950).
25. Dodd, C. G., (to the U. S. A. as represented by the Sec. of War), U. S. Patent 2,582,239 (1952).
26. Dreyfuss, H., *et al.*, (to Celanese Corp. of America), U. S. Patent 2,196,256 (1940).
27. Dreyfuss, H., *et al.*, British Patent 500,804 (1939); *Chem. Zentr.*, **1940**, I, 484.
28. Deile, H., (to W. H. Duisberg), U. S. Patent 2,260,513 (1941).
29. Eschalier, X., British Patent 25,647 (1906); French Patent 347 724 (1906); U. S. Patent 995,852 (1911).
30. Evans, J. G., and Salkeld, C. E., (to Imperial Chemical Industries Ltd.), British Patent 472,389 (1936).
31. Ewing, H., British Patent 519,986 (1940); *J. Soc. Chem. Ind.*, **59**, 440 (1940).
32. Farbewerke vorm. Meister Lucius and Brünning, German Patents 296,141 (1917).
33. Ferretti, A., (Vested in Alien Property Custodian), U. S. Patents 2,338,916; 2,338,917; 2,338,918; 2,338,919 and 2,338,920 (1944).
34. Finlayson, D., and Perry, R. G., (to British Celanese Ltd.), British Patent 462,599 (1937).
35. Fischer, E. K., (to Interchemical Corp.), U. S. Patent 2,461,538 (1949).
36. Ford, F. M., and Hall, W. P., (to Joseph Bancroft and Sons, Co.), U. S. Patent 2,482,756 (1949).
37. Foulds, R. P., Marsh, J. T., and Wood, F. C., (to Tootal Broadhurst Lee Co., Ltd.), British Patent 291,473–4 (1926).
38. Goldthwait, C. F., *Textile Research J.*, **21**, (1), 55–62 (1951).
39. Graham, B., (to E. I. du Pont de Nemours & Co., Inc.), U. S. Patent 2,516,562 (1950).
40. Gruntfest, I. J., and Gagliardi, D. P., *Ind. Eng. Chem.*, **41**, 760 (1949).
41. Hall, A. J., *Chem. Inds.*, **41**, 159 (1937).
42. Hall, A. J. *Can. Textile J.*, **58**, No. 6, 34–6, 50 (1941).
43. *Ibid.*, **58**, No. 8, 35–8, 50 (1941).
44. Heberlein, G., Jr., Weiss, E., and Hemmi, H. (to Heberlein Patent Corp.), U. S. Patent 2,205,120 (1940).
45. Heberlein, Georges, Heberlein, Georg, Weiss, E., Odinga, T., and Risch, K., (to Heberlein Patent Corp.), U. S. Patent 2,463,618 (1949).
46. Heckert, W. W. (to E. I. du Pont de Nemours & Co., Inc.), U. S. Patent 2,080,043 (1937).

47. Henning, L. G., and Sessions, F. L., (Henning to Sessions), U. S. Patent 2,466,808 (1949).
48. Hochstetter, F. W., (to Hochstetter Research Lab.), U. S. Patent 2,368,660 (1945).
49. Humfeld, H., Elmquist, R. E., and Kettering, J. H., *Tech. Bull.*, **588**, U. S. Dept. of Agriculture, Washington, D. C. (1937).
50. I. G. Farbenindustrie A. G., British Patent 431,704 (1935).
51. I. G. Farbeindustrie A. G., British Patent 458,877 (1936).
52. Imperial Chemical Industries, Ltd., French Patent 883,253 (1943).
53. J. R. Geigy & Co., German Patent 114,634 (1900).
54. Johnstone, E. P., Jr., and van Loo, W. J., Jr., (to American Cyanamid Co.), U. S. Patent 2,329,622 (1943); Re-issue U. S. Patent 22,566 (1944).
55. Kann, A., U. S. Patent 787,923 (1905).
56. Karplus, H., U. S. Patent 1,591,922 (1926).
57. Keyworth, C. M., (to Sir Thomas and Arthus Ward, Ltd.), British Patent 493,938 (1938).
58. Kösslinger, K., Klare, H., and Rein, H., (to I. G. Farbeindustrie A. G.), U. S. Patent 2,200,452 (1940).
59. Krefeld, E. W., (to Heberlein Patent Corp.), U. S. Patent 2,242,565 (1941).
60. Lantz, L. A., Whitfield, J. R., and Miller, W. S., (to Calico Printers' Assoc. Ltd.), British Patent 460,201 (1937).
61. Leroy, Y. A. R., (to Brick Trust Ltd.), U. S. Patent 2,052,886 (1936).
62. Lis, L., and Horton, R., (to Aralac, Inc.), U. S. Patent 2,432,776 (1947).
63. Loewenthal, R., "Handbuch der Farberei der Spinnfasern," 3rd Ed., Vol. 2, p. 932, Berlin, Verlag U. and S. Loewenthal (1920).
64. Lynn, J. E., (to American Cyanamid Co.), U. S. Patent 2,466,457 (1949).
65. MacGregor, J. H., (to Courtaulds Ltd), U. S. Patents 2,356,677 (1944) and 2,417,312 (1947).
66. *Ibid.*, U. S. Patent 2,458,397 (1949).
67. McCreath, D., (to Imperial Chemical Industries, Ltd.), U. S. Patent 2,425,334 (1947).
68. Malard, J., *Textile Colorist*, **61**, 195–8 (1939).
69. Meister, O., German Patent 223,883 (1910).
70. Millar, A., British Patent 6700 (1898).
71. Minaev, V. I., and Frolov, S. S., *Trans. Inst. Chem. Tech., Ivanovo U. S. S. R.*, **1**, 166–73 (1935); *C. A.*, **30**, 855.
72. Morton, T. H. (to Courtaulds Ltd.), U. S. Patent 2,049,217 (1936).
73. Peterson, R. F., Caldwell, T. P., Hipp, N. J., Hellback, R., and Jackson, R. W., *Ind Eng. Chem.*, **37**, 492–6 (1945).
74. Pfeffer, E. C., Jr., and Epelberg, J., (to Cluett, Peabody and Co.), U. S. Patent 2,412,832 (1946).
75. Pinkney, P. S., (to E. I. du Pont de Nemours & Co., Inc.), U. S. Patent 2,311,080 (1943).
76. Pollak, F. F., U. S. Patent 2,421,218 (1947).
77. Pollock, M., and Zerner, E., (to Sun Chemical Corp.), U. S. Patent 2,493,371 (1950).
78. Quinn, D. P., *Textile Bull*, No. 12, Aug. 15, 1941.
79. Rein, H., (to I. G. Farbenindustrie A. G.), U. S. Patent 2,166,325 (1939).
80. Ripper, K., British Patent 503,670 (1939).
81. Roberts, A. E., and Watkins, W., British Patent 452,766 (1936).
82. *Ibid.*, British Patent 500,184 (1939).

83. Rooney, J. H., Sharphouse, J. H., and Hawtin, P. R., (Celanese Corp. of America), U. S. Patent 2,452,152 (1948).
84. Shaeffer, A., *Melliand Textil Ber.*, **26**, 18–20 (No. 1, Jan.) 37–41, (No. 2, Feb.), (1945).
85. Schirm, E., (to I. G. Farbenindustrie A. G.), U. S. Patent 2,120,267 (1938).
86. Schneevoigt, A., and Nowak, A., (to I. G. Farbenindustrie A. G.), U. S. Patent 1,871,087 (1932).
87. Schneider, A. K., (to E. I. du Pont de Nemours & Co., Inc.), U. S. Patent 2,430,953 (1947).
88. *Ibid.*, U. S. Patent 2,441,085 (1948).
89. Schroy, P. C., (to American Cyanamid Co.), U. S. Patent 2,284,609 (1942).
90. Schubert, E., *et al.*, British Patent 519,734; *J. Soc. Chem. Ind.*, **59**, 440 (1940).
91. Signer, R., U. S. Patent 2,404,665 (1946).
92. Sparks, C. E., (to E. I. du Pont de Nemours & Co., Inc.), U. S. Patent 2,234,201 (1941).
93. Todtenhaupt, F., German Patent 170,051 (1904).
94. Traill, D., Chemistry & Industry, **1950**, 23–30.
95. Traill, D., *Ind. Chem.*, **21**, 71–4, 95 (1945).
96. Trebaux, J., (to J. R. Geigy, A. G.), U. S. Patent 2,272,783 (1942).
97. Wakelin, J., *Chem. Inds.*, **43**, 53 (1938).
98. Walker, J. F., (to E. I. du Pont de Nemours & Co., Inc.), U. S. Patent 2,253,102 (1941).
99. Waltz, J. E., (to E. I. du Pont de Nemours & Co., Inc.), U. S. Patent 2,477,156 (1949).
100. Weeldenburg, J. G., (to American Enka Corp.), U. S. Patent 2,468,530 (1949).
101. Weisberg, M., Stevenson, A. S., and Beer, L., (to Alrose Chemical Co.), U. S. Patent 2,441,859 (1948).
102. Whittier, E. O., and Gould, S. P., U. S. Patent 2,225,198 (1940).
103. Wolf, B., Kling, W., and Rau, M., (to Böhme Fettchemie G. m. b. H.), U. S. Patent 2,108,520 (1938).
104. Wolf, E., (to Heberlein Patent Corp.), U. S. Patent 2,296,412 (1942).
105. Wood, F. C., *Nature*, **124**, 762 (1929).
106. Yakima, P. A., and Shivrina, A. M., *J. Applied Chem.* (*U. S. S. R.*), **14**, 560–5 (1941); *C. A.*, **36**, 3364 (1942).

Wood

Formaldehyde Treatment of Wood. For the modification of wood and wood products, formaldehyde is generally employed in the form of resins or simple derivatives which are converted to resins by heat. Impregnation processes of this type have already been mentioned in connection with the use of formaldehyde in synthetic resins. Although these processes are of the greatest technical importance, other processes involving the treatment of wood with formaldehyde have also been the subject of industrial patents. A process for preserving wood is carried out by impregnation with an aqueous solution of preservative salt such as sodium fluoride in which a small quantity of a formaldehyde polymer has been dissolved[2]. Formaldehyde is employed with potassium cyanide in a composition for wood preservation[9]. Another process for preserving wood from the attack of insects or other animal or vegetable pests involves impregnating, painting, soaking or spraying with a solution of formaldehyde in kerosene or turpentine. It is claimed that an approximately 2 per cent formaldehyde solution may be obtained in these solvents by saturating with formaldehyde vapors obtained by methanol oxidation[1]. Formaldehyde is also employed in an impregnating composition containing neutral mineral oil, mineral spirits, manganese drying oils, turpentine, rosin, dichlorobenzene, and an alum[7]. The hardening action of formaldehyde on gelatin is employed in a procedure for preparing wooden veneers for wall paper in which the veneer is first steeped in a bath containing cellulose acetate, chrome alum solution, formaldehyde, and water, after which it is dried and then immersed in a water solution of glycerol and gelatin[3]. Processes for the artificial ageing of wood by heating with gaseous mixtures of formaldehyde, ammonia and air are also described[4, 5, 6].

Recently Stamm and Tarkow[8] have shown that the shrinkage and swelling of wood can be substantially reduced by direct kiln treatment with formaldehyde gas (paraformaldehyde vapors) in the dry state at 140 to 230°F in the presence of a gaseous mineral acid catalyst. Apparently, under these conditions it is possible to react the hydroxyl groups of cellulose and lignin so that methylene cross-linkages are formed and these prevent swelling and shrinkage. Anti-shrink efficiencies of 80 to 40 per cent are attainable by this process. The weight increase of the treated wood is only 3 to 6 per cent on the dry basis. All processes involving treatment of cellulose with mineral acids cause embrittlement of the wood and this process is no exception. Embrittlement is minimized by introducing the catalyst after the aldehyde has been taken up by the wood and by employing nitric acid rather than hydrogen chloride as the catalyst.

References

1. Berkeley, R. T. M., and Stenhouse, E., British Patent 221,599 (1924).
2. Grubenholzimprägnierung Gesellschaft, German Patent 73,218 (1918).
3. Hellmers, H. A., British Patent 388,593 (1933); *C. A.*, **27**, 5931, (1933).
4 Kleinstück, M., German Patent 323,973 (1920).
5. Kleinstück, M., German Patent 324,159 (1920).
6. Kleinstück, M., German Patent 325,657 (1920).
7. Nielson, H. C., (to National Wood Products Co.), U. S. Patent 1,886,716 (1932).
8. Stamm, A. J., and Tarkow, H., (to the U. S. A. as represented by the Sec. of Agriculture), U. S. Patent 2,572,070 (1951).
9. Tabary, A. R., French Patent 765,153 (1934).

AUTHOR INDEX

Note: Superscripts indicate reference numbers for author's work.

SUBJECT INDEX

Formaldehyde solutions
  in nonpolar solvents, 46
  in polar solvents, 46–47, 63, 65
    in alcohol, 61–63
    in water, See Aqueous formaldehyde
      solutions
Formaldehyde-urea resins, 401
Formaldoxime, 185, 389, 462
Formalhydrazine, 184
"Formalin," 65, 72, 131, 140
Formals, 282, 327, 335, 425, 427, 515
  combined formaldehyde in determina-
    tion of, 400
  formation of, 203–211
Formamide, 63, 130, 185, 314
  reaction with formaldehyde, 291
Formic acid, 4, 6, 8, 16, 23, 66, 105, 126,
    160, 161, 162, 163, 179, 182, 183, 185,
    188, 198, 285, 292, 305, 312, 359, 367,
    369, 375, 376, 386, 399, 401, 425, 426,
    429, 521, 527
  effect on polymerization of monomeric
    formaldehyde, 35
  in commercial formaldehyde solutions
    concentration limits of, 393
    determination of, 394
  in formaldehyde solutions
    determination of, 86
    formation of, 74, 78
  removal of, from formaldehyde, 28, 30
Formin, 404, 492
"Formol," 65
Formolite reaction, 345, 462–463, 479
Formose, 165, 179
Formthionals, 190
9–Formyl fluorene, reduction of, 167
N-Formyl-nitro-hydroxybenzylamine,
  293
Fractional condensation of formaldehyde
  and water vapor, 108–109, 127
Fractional crystallization, 118
Fractionation (adiabatic) of formalde-
  hyde, 109–111
Free energy
  of dissolved formaldehyde, 91
  of monomeric formaldehyde gas, 38, 91
Freezing point
  of aqueous formaldehyde solutions, 88
  of methylal, 203
Friedel-Crafts reaction, 214
Fructose, 165, 369

Fuel, formaldehyde as, 477
Fumigant, formaldehyde as a, 467–468
Fungicides
  formaldehyde in, 458, 468
  determination of, 400
Fur
  formaldehyde in, determination of, 401
  formaldehyde treatment of, 489
Furans
  reaction with formaldehyde, 353–354
Furfural, 313, 353, 354, 367, 369
Furfuraldehyde, 373
Furfuryl alcohol, 353, 354
2-Furoic acid, 354
Fusible resins, 263–264

Gallic acid, 244, 369, 376
Gelatin, 315, 316
Glucose, 369, 390
Glue, waterproof, use of formaldehyde
  in, 450
Glyceraldehyde, 165, 368
Glycerin, 450, 477, 478, 497, 527
Glyceroboric acid, 428
Glycerol, 100, 329, 369, 414, 428, 471, 515,
  526, 533
  reaction with formaldehyde, 206
Glycerol formal, 515
Glycinamide, 299
Glycine, 186, 286
  reaction with formaldehyde, 311
Glycine ester triformal, 312
Glycine ethyl ester hydrochloride, 312
Glycine methylamide, 299–300
Glycolaldehyde triacetate, 142
Glycol formal, 205, 515
Glycolic acid, 166, 186, 187, 195, 369, 515
Glycolic aldehyde, 165, 166
Glycolonitrile, 185, 186, 498
Glycols, reaction with formaldehyde,
  205–206
Glyoxal, 160, 313, 368, 521
Glyoxalic acid, 367
Gold, determination of, 462
Gold chloride, 180
Gold salts, reduction of, 167
Gramicidin, 316, 492
Grignard reagents, reaction with form-
  aldehyde, 345–346
Guaiacol, 261
Guanidine, 494, 501, 525